Thermochemical Processing of Biomass

Wiley Series in Renewable Resources

Series Editor:

Christian V. Stevens, Faculty of Bioscience Engineering, Ghent University, Belgium

Titles in the Series:

Wood Modification: Chemical, Thermal and Other Processes
Callum A. S. Hill

Renewables-Based Technology: Sustainability Assessment
Jo Dewulf and Herman Van Langenhove

Biofuels
Wim Soetaert and Erik Vandamme

Handbook of Natural Colorants
Thomas Bechtold and Rita Mussak

Surfactants from Renewable Resources
Mikael Kjellin and Ingegärd Johansson

Industrial Applications of Natural Fibres: Structure, Properties and Technical Applications
Jörg Müssig

Thermochemical Processing of Biomass: Conversion into Fuels, Chemicals and Power
Robert C. Brown

Biorefinery Co-Products: Phytochemicals, Primary Metabolites and Value-Added Biomass Processing
Chantal Bergeron, Danielle Julie Carrier and Shri Ramaswamy

Aqueous Pretreatment of Plant Biomass for Biological and Chemical Conversion to Fuels and Chemicals
Charles E. Wyman

Bio-Based Plastics: Materials and Applications
Stephan Kabasci

Introduction to Wood and Natural Fiber Composites
Douglas D. Stokke, Qinglin Wu and Guangping Han

Cellulosic Energy Cropping Systems
Douglas L. Karlen

Introduction to Chemicals from Biomass, 2nd Edition
James H. Clark and Fabien Deswarte

Lignin and Lignans as Renewable Raw Materials: Chemistry, Technology and Applications
Francisco G. Calvo-Flores, Jose A. Dobado, Joaquín Isac-García and Francisco J. Martín-Martínez

Sustainability Assessment of Renewables-Based Products: Methods and Case Studies
Jo Dewulf, Steven De Meester and Rodrigo A. F. Alvarenga

Cellulose Nanocrystals: Properties, Production and Applications
Wadood Hamad

Fuels, Chemicals and Materials from the Oceans and Aquatic Sources
Francesca M. Kerton and Ning Yan

Bio-Based Solvents
François Jérôme and Rafael Luque

Nanoporous Catalysts for Biomass Conversion
Feng-Shou Xiao and Liang Wang

Forthcoming Titles:

Chitin and Chitosan: Properties and Applications
Lambertus A.M. van den Broek and Carmen G. Boeriu

Biorefinery of Inorganics: Recovering Mineral Nutrients from Biomass and Organic Waste
Eric Meers and Gerard Velthof

The Chemical Biology of Plant Biostimulants
Danny Geelen and Lin Xu

Waste Valorization: Waste Streams in a Circular Economy
Sze Ki Lin

Thermochemical Processing of Biomass

Conversion into Fuels, Chemicals and Power

Second Edition

Edited by

ROBERT C. BROWN
Iowa State University
Ames, Iowa
USA

This edition first published 2019
© 2019 John Wiley & Sons Ltd

Edition History

Thermochemical Processing of Biomass: Conversion into Fuels, Chemicals and Power, First Edition, Wiley 2011.

All rights reserved. No part of this publication may be reproduced, stored in a retrieval system, or transmitted, in any form or by any means, electronic, mechanical, photocopying, recording or otherwise, except as permitted by law. Advice on how to obtain permission to reuse material from this title is available at http://www.wiley.com/go/permissions.

The right of Robert C. Brown to be identified as the author of the editorial material in this work has been asserted in accordance with law.

Registered Offices
John Wiley & Sons, Inc., 111 River Street, Hoboken, NJ 07030, USA
John Wiley & Sons Ltd, The Atrium, Southern Gate, Chichester, West Sussex, PO19 8SQ, UK

Editorial Office
The Atrium, Southern Gate, Chichester, West Sussex, PO19 8SQ, UK

For details of our global editorial offices, customer services, and more information about Wiley products visit us at www.wiley.com.

Wiley also publishes its books in a variety of electronic formats and by print-on-demand. Some content that appears in standard print versions of this book may not be available in other formats.

Limit of Liability/Disclaimer of Warranty
In view of ongoing research, equipment modifications, changes in governmental regulations, and the constant flow of information relating to the use of experimental reagents, equipment, and devices, the reader is urged to review and evaluate the information provided in the package insert or instructions for each chemical, piece of equipment, reagent, or device for, among other things, any changes in the instructions or indication of usage and for added warnings and precautions. While the publisher and authors have used their best efforts in preparing this work, they make no representations or warranties with respect to the accuracy or completeness of the contents of this work and specifically disclaim all warranties, including without limitation any implied warranties of merchantability or fitness for a particular purpose. No warranty may be created or extended by sales representatives, written sales materials, or promotional statements for this work. The fact that an organization, website, or product is referred to in this work as a citation and/or potential source of further information does not mean that the publisher and authors endorse the information or services the organization, website, or product may provide or recommendations it may make. This work is sold with the understanding that the publisher is not engaged in rendering professional services. The advice and strategies contained herein may not be suitable for your situation. You should consult with a specialist where appropriate. Further, readers should be aware that websites listed in this work may have changed or disappeared between when this work was written and when it is read. Neither the publisher nor authors shall be liable for any loss of profit or any other commercial damages, including but not limited to special, incidental, consequential, or other damages.

Library of Congress Cataloging-in-Publication Data applied for

ISBN: 9781119417576

Cover Design: Wiley
Cover Images: Courtesy of Peter Ciesielski; Education globe © Ingram Publishing/Alamy Stock Photo

Set in 10/12pt TimesLTStd by SPi Global, Chennai, India

Printed and bound by CPI Group (UK) Ltd, Croydon, CR0 4YY

10 9 8 7 6 5 4 3 2 1

TP
339
.T476
2019

This book is dedicated to former and current students and staff who helped build the thermochemical processing programs of the Bioeconomy Institute and its predecessor, the Center for Sustainable Environmental Technologies, at Iowa State University.

Contents

List of Contributors	xv
Series Preface	xvii
Preface	xix

1 Introduction to Thermochemical Processing of Biomass into Fuels, Chemicals, and Power — 1
Xiaolei Zhang and Robert C. Brown
- 1.1 Introduction — 1
- 1.2 Thermochemical Conversion Technologies — 5
  - 1.2.1 Direct Combustion — 5
  - 1.2.2 Gasification — 6
  - 1.2.3 Pyrolysis — 7
  - 1.2.4 Solvent Liquefaction — 8
- 1.3 Diversity of Products: Electric Power, Transportation Fuels, and Commodity Chemicals — 8
  - 1.3.1 Biopower — 9
  - 1.3.2 Biofuels — 10
  - 1.3.3 Bio-Based Chemicals — 11
- 1.4 Economic Considerations — 11
- 1.5 Environmental Considerations — 12
- 1.6 Organization of This Book — 13
- References — 14

2 Condensed Phase Reactions During Thermal Deconstruction — 17
Jake K. Lindstrom, Alexander Shaw, Xiaolei Zhang, and Robert C. Brown
- 2.1 Introduction to Condensed Phase Reactions During Thermal Deconstruction of Biomass — 17
- 2.2 Thermochemical Processes — 19
  - 2.2.1 Processes Yielding Chiefly Solids — 20
  - 2.2.2 Processes Yielding Chiefly Liquids — 22

		2.2.3	Processes Yielding Chiefly Gases	24
	2.3	Understanding Condensed Phase Reactions		28
		2.3.1	Challenges in Investigating Condensed Phase Reactions	28
		2.3.2	The Role of Cell Wall Structure in Thermal Deconstruction	30
		2.3.3	Use of Computational Chemistry to Understand Thermal Deconstruction	34
	2.4	Conclusions		41
		References		42
3	**Biomass Combustion**			**49**
	Bryan M. Jenkins, Larry L. Baxter, and Jaap Koppejan			
	3.1	Introduction		50
	3.2	Combustion Systems		51
		3.2.1	Fuels	51
		3.2.2	Types of Combustors	53
	3.3	Fundamentals of Biomass Combustion		59
		3.3.1	Combustion Properties of Biomass	59
		3.3.2	Combustion Stoichiometry	65
		3.3.3	Equilibrium	68
		3.3.4	Rates of Reaction	68
	3.4	Pollutant Emissions and Environmental Impacts		71
		3.4.1	Oxides of Nitrogen and Sulfur	72
		3.4.2	Products of Incomplete Combustion	74
		3.4.3	Particulate Matter	74
		3.4.4	Dioxin-Like Compounds	74
		3.4.5	Heavy Metals	76
		3.4.6	Radioactive Species	76
		3.4.7	Greenhouse Gas Emissions	77
		References		77
4	**Gasification**			**85**
	Karl M. Broer and Chad Peterson			
	4.1	Introduction		85
		4.1.1	History of Gasification	85
		4.1.2	Gasification Terminology	87
	4.2	Fundamentals of Gasification		88
		4.2.1	Heating and Drying	89
		4.2.2	Pyrolysis	89
		4.2.3	Gas–Solid Reactions	90
		4.2.4	Gas-Phase Reactions	91
	4.3	Feed Properties		91
	4.4	Classifying Gasifiers According to Method of Heating		95
		4.4.1	Air-Blown Gasifiers	95
		4.4.2	Oxygen/Steam-Blown Gasifiers	96
		4.4.3	Indirectly Heated Gasifiers	96
	4.5	Classifying Gasifiers According to Transport Processes		98
		4.5.1	Fixed Bed	99

| | | 4.5.2 | Bubbling Fluidized Bed (BFB) | 101 |
| | 4.6 | 4.5.3 | Circulating Fluidized Bed (CFB) | 103 |

		4.5.2	Bubbling Fluidized Bed (BFB)	101
		4.5.3	Circulating Fluidized Bed (CFB)	103
		4.5.4	Entrained Flow	104
	4.6	Pressurized Gasification	106
	4.7	Products of Gasification	106
		4.7.1	Gaseous Products	106
		4.7.2	Char and Tar	109
	4.8	System Applications	110
		4.8.1	Process Heat	110
		4.8.2	Combined Heat and Power (CHP)	117
		4.8.3	Fuel and Chemical Synthesis	117
		Acknowledgement	118
		References	118
5	**Syngas Cleanup, Conditioning, and Utilization**	**125**
	David C. Dayton, Brian Turk, and Raghubir Gupta	
	5.1	Introduction	125
	5.2	Syngas Cleanup and Conditioning	126
		5.2.1	Particulates	128
		5.2.2	Sulfur	130
		5.2.3	Ammonia Decomposition and HCN Removal	132
		5.2.4	Alkalis and Heavy Metals	132
		5.2.5	Chlorides	133
		5.2.6	Tars and Soot	134
	5.3	Syngas Utilization	137
		5.3.1	Syngas to Gaseous Fuels	138
		5.3.2	Syngas to Liquid Fuels	145
	5.4	Summary and Conclusions	159
		References	164
6	**Fast Pyrolysis**	**175**
	Robbie H. Venderbosch	
	6.1	Introduction	175
	6.2	Fundamentals of Pyrolysis	175
		6.2.1	Effects of the Chemical and Physical Structure of Biomass and Intermediate Products	176
		6.2.2	Effects of Ash	179
	6.3	Properties of Pyrolysis Liquids	184
	6.4	Fast Pyrolysis Process Technologies	186
		6.4.1	Ensyn (CFB)	186
		6.4.2	Valmet/UPM (CFB)	188
		6.4.3	BTG-BtL (Rotating-cone)	188
		6.4.4	Dynamotive Technologies Corp	190
	6.5	Applications of Pyrolysis Liquids	192
		6.5.1	Combustion	192
		6.5.2	Diesel Engines	193
		6.5.3	Co-refining Options	194

		6.5.4	Gasification	199
	6.6	Chemicals		200
	6.7	Catalytic Pyrolysis		202
	6.8	Concluding Remarks		202
		Acknowledgement		203
		References		203

7 Upgrading Fast Pyrolysis Liquids — 207
Karl O. Albrecht, Mariefel V. Olarte, and Huamin Wang

	7.1	Introduction		207
	7.2	Bio-oil Characteristics and Quality		208
		7.2.1	Feedstock Factors Affecting Bio-oil Characteristics	209
		7.2.2	Effect of Pyrolysis Operating Conditions on Bio-oil Composition	210
		7.2.3	Need for Upgrading Bio-oil	211
	7.3	Norms and Standards		212
	7.4	Physical Pre-treatment of Bio-oil		213
		7.4.1	Physical Filtration	213
		7.4.2	Solvent Addition	213
		7.4.3	Fractionation	214
	7.5	Catalytic Hydrotreating		214
		7.5.1	Stabilization Through Low Temperature Hydrotreating	214
		7.5.2	Deep Hydrotreating	217
	7.6	Vapor Phase Upgrading via Catalytic Fast Pyrolysis		218
		7.6.1	CFP Chemistry	221
		7.6.2	Key Factors Impacting Catalytic Fast Pyrolysis	221
		7.6.3	Practical Catalytic Fast Pyrolysis of Lignocellulosic Biomass	223
	7.7	Other Upgrading Strategies		223
		7.7.1	Liquid Bio-oil Zeolite Upgrading and Co-processing in FCC	223
		7.7.2	Reactions with Alcohols	227
	7.8	Products		228
		7.8.1	Liquid Transportation Fuel Properties	228
		7.8.2	Chemicals	232
		7.8.3	Hydrogen Production	235
	7.9	Summary		235
		References		238

8 Solvent Liquefaction — 257
Arpa Ghosh and Martin R. Haverly

	8.1	Introduction		257
		8.1.1	Definition of Solvent Liquefaction	257
		8.1.2	History of Solvent Liquefaction	257
	8.2	Feedstocks for Solvent Liquefaction		259
		8.2.1	Feedstock Types	259
		8.2.2	Benefits of Liquid Phase Processing	259
		8.2.3	Reaction Types	261
		8.2.4	Processing Conditions	262

8.3	Target Products		263
	8.3.1	Bio-oil	263
	8.3.2	Production of Fuels and Chemicals	264
	8.3.3	Co-products	265
8.4	Processing Solvents		265
	8.4.1	Inorganic Solvents	268
	8.4.2	Polar Protic Solvents	271
	8.4.3	Polar Aprotic Solvents	274
	8.4.4	Ionic Liquids	276
	8.4.5	Non-Polar Solvents	277
	8.4.6	Influence of Process Conditions	278
8.5	Solvent Effects		283
	8.5.1	Physical Effects	283
	8.5.2	Solubility Effects	284
	8.5.3	Structural Effects	287
	8.5.4	Chemical Effects	287
8.6	Engineering Challenges		292
	8.6.1	High Pressure Feed Systems	292
	8.6.2	Separation of Solid Residue	293
	8.6.3	Solvent Recovery and Recycle	293
8.7	Conclusions		294
	References		294

9 Hybrid Processing — 307
Zhiyou Wen and Laura R. Jarboe

9.1	Introduction		307
9.2	Thermochemical Conversion of Lignocellulosic Biomass for Fermentative Substrates		308
	9.2.1	Fast Pyrolysis for Production of Pyrolytic Substrates	308
	9.2.2	Gasification of Biomass for Syngas Production	309
9.3	Biological Conversion of Fermentative Substrates into Fuels and Chemicals		310
	9.3.1	Fermentation of Pyrolytic Substrates	310
	9.3.2	Fermentation of Syngas	313
9.4	Challenges of Hybrid Processing and Mitigation Strategies		318
	9.4.1	Pyrolysis–Fermentation Process	318
	9.4.2	Gasification–Syngas Fermentation Process	320
9.5	Efforts in Commercialization of Hybrid Processing		322
9.6	Conclusion and Perspectives		323
	References		323

10 Costs of Thermochemical Conversion of Biomass to Power and Liquid Fuels — 337
Mark M. Wright and Tristan Brown

10.1	Introduction		337
10.2	Electric Power Generation		338
	10.2.1	Direct Combustion to Power	338

		10.2.2	Gasification to Power	339
		10.2.3	Fast Pyrolysis to Power	339
	10.3	Liquid Fuels via Gasification		340
		10.3.1	Gasification to Fischer-Tropsch Liquids	340
		10.3.2	Gasification to Mixed Alcohols	341
		10.3.3	Gasification to Gasoline	342
		10.3.4	Gasification and Syngas Fermentation to Ethanol	343
		10.3.5	Gasification and Syngas Fermentation to PHA and Co-product Hydrogen	343
	10.4	Liquid Fuels via Fast Pyrolysis		344
		10.4.1	Fast Pyrolysis and Hydroprocessing	344
		10.4.2	Catalytic Fast Pyrolysis and Hydroprocessing	345
		10.4.3	Fast Pyrolysis and Gasification to Fuels	346
		10.4.4	Fast Pyrolysis and Bio-oil Fermentation to Ethanol	346
	10.5	Liquid Fuels via Direct Liquefaction		348
	10.6	Liquid Fuels via Esterification		349
	10.7	Summary and Conclusions		349
		References		350

11 Life Cycle Assessment of the Environmental Performance of Thermochemical Processing of Biomass   355
Eskinder Demisse Gemechu, Adetoyese Olajire Oyedun, Edson Norgueira Jr., and Amit Kumar

	11.1	Introduction		355
	11.2	Life Cycle Assessment		356
		11.2.1	Introduction to LCA and Life Cycle Thinking	356
		11.2.2	Goal and Scope Definition	357
		11.2.3	Life Cycle Inventory	357
		11.2.4	Life Cycle Impact Assessment	358
		11.2.5	Life Cycle Interpretation	359
		11.2.6	Sensitivity and Uncertainty Analyses	359
	11.3	LCA of Thermochemical Processing of Biomass		360
		11.3.1	Overview of the Thermochemical Processing of Biomass	360
		11.3.2	The Use of LCA to Promote Low Carbon Technologies	360
		11.3.3	Review of LCA Studies on Thermochemical Processing of Biomass	360
	11.4	Discussion on the Application of LCA for Thermochemical Processing of Biomass		369
		11.4.1	Establishing Goal and Scope	369
		11.4.2	Life Cycle Inventory Analysis	370
		11.4.3	Life Cycle Impact Assessment	371
	11.5	Conclusions		372
		Acknowledgements		373
		References		373

Index   **379**

List of Contributors

Karl O. Albrecht   Chemical and Biological Process Development Group, Pacific Northwest National Laboratory, Richland, WA, USA

Larry L. Baxter   Department of Chemical Engineering, Brigham Young University, Provo, UT, USA

Karl M. Broer   Gas Technology Institute, Des Plaines, IL, USA

Robert C. Brown   Department of Mechanical Engineering, Iowa State University, Ames, IA, USA

Tristan Brown   Department of Forest and Natural Resources Management, SUNY ESF, Syracuse, NY, USA

David C. Dayton   RTI International, Cornwallis Road, Research Triangle Park, NC, USA

Eskinder Demisse Gemechu   Department of Mechanical Engineering, Donadeo Innovation Centre for Engineering, University of Alberta, Edmonton, Alberta, Canada

Arpa Ghosh   Chemical and Biological Engineering Department, Iowa State University, Ames, IA, USA

Raghubir Gupta   RTI International, Cornwallis Road, Research Triangle Park, NC, USA

Martin R. Haverly   Technical Services, Renewable Energy Group, Ames, IA, USA

Laura R. Jarboe   Department of Chemical and Biological Engineering, Iowa State University, Ames, IA, USA

Bryan M. Jenkins   Department of Biological and Agricultural Engineering, University of California, Davis, CA, USA

Jaap Koppejan   Procede Biomass BV, Enschede, Netherlands

Amit Kumar   Department of Mechanical Engineering, Donadeo Innovation Centre for Engineering, University of Alberta, Edmonton, Alberta, Canada

Jake K. Lindstrom Department of Mechanical Engineering, Iowa State University, Ames, IA, USA

Edson Norgueira, Jr., Department of Mechanical Engineering, Donadeo Innovation Centre for Engineering, University of Alberta, Edmonton, Alberta, Canada

Mariefel V. Olarte Chemical and Biological Process Development Group, Pacific Northwest National Laboratory, Richland, WA, USA

Adetoyese Olajire Oyedun Department of Mechanical Engineering, Donadeo Innovation Centre for Engineering, University of Alberta, Edmonton, Alberta, Canada

Chad Peterson Department of Mechanical Engineering, Iowa State University, Ames, IA, USA

Alexander Shaw School of Mechanical and Aerospace Engineering, Queen's University Belfast, Belfast, UK

Brian Turk RTI International, Cornwallis Road, Research Triangle Park, NC, USA

Robbie H. Venderbosch Biomass Technology Group BV, AV Enschede, the Netherlands

Huamin Wang Chemical and Biological Process Development Group, Pacific Northwest National Laboratory, Richland, WA, USA

Zhiyou Wen Department of Food Science and Human Nutrition, Iowa State University, Ames, IA, USA

Mark M. Wright Department of Mechanical Engineering, Iowa State University, Ames, IA, USA

Xiaolei Zhang School of Mechanical and Aerospace Engineering, Queen's University Belfast, Belfast, UK

Series Preface

Renewable resources, their use and modification are involved in a multitude of important processes with a major influence on our everyday lives. Applications can be found in the energy sector, paints and coatings, and the chemical, pharmaceutical, and textile industry, to name but a few.

The area interconnects several scientific disciplines (agriculture, biochemistry, chemistry, technology, environmental sciences, forestry, etc.), which makes it very difficult to have an expert view on the complicated interaction. Therefore, the idea to create a series of scientific books that will focus on specific topics concerning renewable resources, has been very opportune and can help to clarify some of the underlying connections in this area.

In a very fast changing world, trends are not only characteristic for fashion and political standpoints; science is also not free from hypes and buzzwords. The use of renewable resources is again more important nowadays; however, it is not part of a hype or a fashion. As the lively discussions among scientists continue about how many years we will still be able to use fossil fuels – opinions ranging from 50 to 500 years – they do agree that the reserve is limited and that it is essential not only to search for new energy carriers but also for new material sources.

In this respect, renewable resources are a crucial area in the search for alternatives for fossil-based raw materials and energy. In the field of energy supply, biomass and renewables-based resources will be part of the solution alongside other alternatives such as solar energy, wind energy, hydraulic power, hydrogen technology, and nuclear energy. In the field of material sciences, the impact of renewable resources will probably be even bigger. Integral utilization of crops and the use of waste streams in certain industries will grow in importance, leading to a more sustainable way of producing materials. Although our society was much more (almost exclusively) based on renewable resources centuries ago, this disappeared in the Western world in the nineteenth century. Now it is time to focus again on this field of research. However, it should not mean a "retour à la nature," but it should be a multidisciplinary effort on a highly technological level to perform research towards new opportunities, to develop new crops and products from renewable resources. This will be essential to guarantee a level of comfort for a growing number of people living on our planet. It is "the" challenge for the coming generations of scientists to develop more sustainable ways to create prosperity and to fight poverty and hunger in the world. A global approach is certainly favoured.

This challenge can only be dealt with if scientists are attracted to this area and are recognized for their efforts in this interdisciplinary field. It is, therefore, also essential that consumers recognize the fate of renewable resources in a number of products.

Furthermore, scientists do need to communicate and discuss the relevance of their work. The use and modification of renewable resources may not follow the path of the genetic engineering concept in view of consumer acceptance in Europe. Related to this aspect, the series will certainly help to increase the visibility of the importance of renewable resources. Being convinced of the value of the renewables approach for the industrial world, as well as for developing countries, I was myself delighted to collaborate on this series of books focusing on different aspects of renewable resources. I hope that readers become aware of the complexity, the interaction and interconnections, and the challenges of this field and that they will help to communicate on the importance of renewable resources.

I certainly want to thank the people of Wiley's Chichester office, especially David Hughes, Jenny Cossham and Lyn Roberts, in seeing the need for such a series of books on renewable resources, for initiating and supporting it, and for helping to carry the project to the end.

Last, but not least, I want to thank my family, especially my wife Hilde and children Paulien and Pieter-Jan, for their patience and for giving me the time to work on the series when other activities seemed to be more inviting.

Christian V. Stevens,
Faculty of Bioscience Engineering
Ghent University, Belgium
Series Editor 'Renewable Resources'
Sep 2018

Preface

The genesis of this book was an invitation by Christian Stevens to describe the "thermochemical option" for biofuels production at the Third International Conference on Renewable Resources and Biorefineries at Ghent University in 2007. At that time, many people working in the biofuels community viewed thermochemical processing as little more than an anachronism in the age of biotechnology. I was very appreciative of Chris' interest in exploring alternative pathways. After the conference, he followed up with an invitation to submit a book proposal on thermochemical processing to the Wiley Series in Renewable Resources, for which he serves as Series Editor. I was happy to accept, although the press of other responsibilities slowed publication until 2011. In the intervening decade since the book was first proposed, the subject of thermochemical processing has moved from relative obscurity to prominence, offering several pathways to advanced biofuels, bio-based chemicals, and biopower.

The first edition having been well received, the publisher contacted me about preparing a second edition with updated material on thermochemical processing. Retirements and changes in interests among the original contributors to the first edition have resulted in several changes in lead authorship of chapters. Xiaolei Zhang has substantially rewritten the introductory chapter on thermochemical processing. Jake Lindstrom has prepared a perspective on condensed phase reactions during thermal deconstruction. Karl Broer, co-author of the original chapter on gasification, led revisions for the second edition. Karl Albrecht has updated the chapter on bio-oil upgrading. Arpa Ghosh offers a comprehensive review of solvent liquefaction, including the production of sugars as well as of bio-oil, the focus of the original chapter. Zhiyou Wen has rewritten the chapter on hybrid processing, which has advanced significantly in the last decade. Amit Kumar agreed to prepare a chapter on the sustainability of thermochemical processing, a new topic in the second edition. I was extremely pleased to have Bryan Jenkins, David Dayton, Robbie Venderbosch, and Mark Wright return as lead authors on the chapters dealing with combustion, syngas upgrading, pyrolysis, and techno-economic analysis, respectively. I am impressed with the team of co-authors that each of the lead authors assembled to help them prepare their chapters.

The project editors at Wiley were extremely patient and helpful as I worked through the second edition of *Thermochemical Processing of Biomass* – many thanks to Emma Strickland, Sarah Higginbotham, Rebecca Ralf, and Lesley Jebaraj. I am also indebted to several people who helped me with administrative and management responsibilities at the Bioeconomy Institute (BEI) at Iowa State University while the second edition was being prepared: Ryan Smith, Jill Euken, Mary Scott-Hall, and Scott Moseley. Finally, I wish to acknowledge my wife, Carolyn, who has been the most steadfast of all during the preparation of both editions.

Robert C. Brown
Iowa State University
Ames, Iowa, USA

1

Introduction to Thermochemical Processing of Biomass into Fuels, Chemicals, and Power

Xiaolei Zhang[1] and Robert C. Brown[2]

[1]*School of Mechanical and Aerospace Engineering, Queen's University Belfast, Belfast, BT9 5AH, UK*
[2]*Department of Mechanical Engineering, Iowa State University, Ames, IA, USA*

1.1 Introduction

Thermochemical processing of biomass uses heat and catalysts to transform plant polymers into fuels, chemicals, or electric power. This contrasts with biochemical processing of biomass, which uses enzymes and microorganisms for the same purpose. In fact, both thermochemical and biochemical methods have been employed by humankind for millennia. Fire for warmth, cooking, and production of charcoal were the first thermal transformations of biomass controlled by humans, while fermentation of fruits, honey, grains, and vegetables was practiced before recorded time. Despite their long records of development, neither has realized full industrialization in processing lignocellulosic biomass. While petroleum and petrochemical industries have transformed modern civilization through thermochemical processing of hydrocarbons, the more complicated chemistries of plant molecules have not been fully developed.

Ironically, the dominance of thermochemical processing of fossil resources into fuels, chemicals, and power for well over a century may explain why thermochemical processing of biomass is sometimes overlooked as a viable approach to bio-based products. Smokestacks belching pollutants from thermochemical processing of fossil fuels is an

Table 1.1 Comparison of biochemical and thermochemical processing.

	Biochemical processing	Thermochemical processing
Products	Primarily alcohols	Range of fuels and chemicals
Reaction conditions	Less than 70 °C, 1 atm	100–1200 °C, 1–250 atm
Residence time	2–5 d	0.2 s–1 h
Selectivity	Can be made very selective	Depends upon reaction
Catalyst/biocatalyst cost	$0.50/gal ethanol	$0.01/gal gasoline
Sterilization	Sterilize all feeds	No sterilization needed
Recyclability	Difficult	Possible with solid catalysts
Size of plant (biomass input)	2000–8000 tons/d	5–200 tons/d (fast pyrolysis)

Source: Adapted from Reference [1].

indelible icon from the twentieth century that no one wishes to replicate with biomass. However, as described in a report released by the US Department of Energy in 2008 [1], thermal and catalytic sciences also offer opportunities for dramatic advances in biomass processing. Actually, thermochemical processing has several advantages relative to biochemical processing, as detailed in Table 1.1. These include the ability to produce a diversity of oxygenated and hydrocarbon fuels, reaction times that are several orders of magnitude shorter than biological processing, lower cost of catalysts, the ability to recycle catalysts, and the fact that thermal systems do not require the sterilization procedures demanded for biological processing. The data in Table 1.1 also suggest that thermochemical processing can be done with much smaller plants than is possible for biological processing of cellulosic biomass. Although this may be true for some thermochemical options (such as fast pyrolysis), other thermochemical options (such as gasification-to-fuels) are likely to be built at larger scales than biologically based cellulosic ethanol plants when the plants are optimized for minimum fuel production cost [2].

The first-generation biofuels industry, launched in the late 1970s, was based on biochemically processing sugar or starch crops (mostly sugar cane and maize, respectively) into ethanol fuel and biochemically processing oil seed crops into biodiesel. These industries grew tremendously in the first 15 years of the twenty-first century, with worldwide annual production reaching almost 26 billion gallons of ethanol [3] and 5.3 billion gallons of biodiesel in 2016 [4]. The development of first-generation biofuels has not been achieved without controversy, including criticism of crop and biofuel subsidies, concerns about using food crops for fuel production, and debate over the environmental impact of biofuels agriculture, including uncertainties about the role of biofuels in reducing greenhouse gas (GHG) emissions [5]. Many of these concerns would be mitigated by developing second-generation biofuels that utilize high-yielding nonfood crops that can be grown on marginal or waste lands. These alternative crops are of two types: lipids from alternative crops and lignocellulosic biomass.

Lipids are a large group of hydrophobic, fat-soluble compounds produced by plants and animals. They are attractive as fuel for their high energy content. The most common of these are triglycerides, which are esters consisting of three fatty acids attached to a backbone of glycerol. Triglycerides can be converted into transportation fuels in one of two ways. Biodiesel is produced by transesterification of the triglycerides to methyl esters, which are blended with petroleum-derived diesel. Renewable diesel is produced

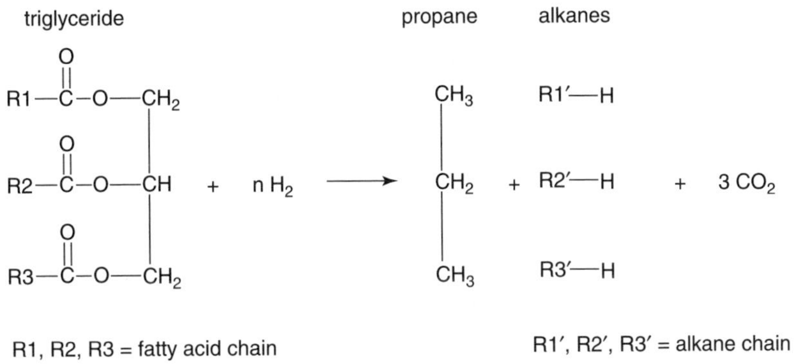

Figure 1.1 Simplified representation of hydrogenation of triglyceride during hydrotreating.

by hydrotreating triglycerides to yield liquid alkanes and co-product propane gas (see Figure 1.1). Biodiesel has dominated most lipid-based fuel production because of the relative simplicity of the process, which can be done at small scales. Biodiesel is not fully compatible with petroleum-derived diesel, an advantage of renewable diesel. However, hydrotreating requires higher capital investment, with economics favoring larger facilities that may be incompatible with the distributed nature of lipid feedstocks [6].

Soybeans were originally thought an attractive feedstock for biodiesel production, reducing GHG emissions by 41% compared to conventional diesel and producing 93% more energy output compared to corn ethanol [7, 8]. However, use of soybeans and other edible oils for fuel has been criticized as competing with their use as food [8, 9]. Soybeans are also an expensive energy source, representing 85% of the cost of producing biodiesel [8]. For this reason, most first-generation biodiesel and renewable diesel have been produced from low-cost waste fats and oils.

Wider use of biodiesel and renewable diesel will require alternatives to traditional seed crops, which only yield 50–130 gal/acre [10]. Suggestions have included jatropha (200–400 gal/acre) [11] and palm oil (up to 600 gal/acre) [12], but the most promising alternative is microalgae, which are highly productive in natural ecosystems with oil yields as high as 2000 gal/acre in field trials and 15 000 gal/acre in laboratory trials [13]. Lipids from algae also have the advantage of not competing with food supplies. However, the process is currently challenged by the high costs associated with harvesting and drying algae and the practical difficulties of cultivating algae with high lipid content [14]. Considerable engineering development is required to reduce capital costs, which are as high as $1 million/acre, and to reduce production costs, which exceed $10/gal. The challenge of lipid-based biofuels is producing large quantities of inexpensive lipids rather than upgrading them to fuels.

Lignocellulosic biomass is a biopolymer of cellulose, hemicellulose, and lignin (Figure 1.2) [16]. Lignocellulosic biomass dominates most terrestrial ecosystems and is widely managed for applications ranging from animal forage to lumber. Cellulose is a structural polysaccharide consisting of a long chain of glucose molecules linked by glycosidic bonds. Glycosidic bonds also play a vital role in linking pentose, hexose, and sugar acids contained in hemicellulose. Breaking these bonds releases monosaccharides,

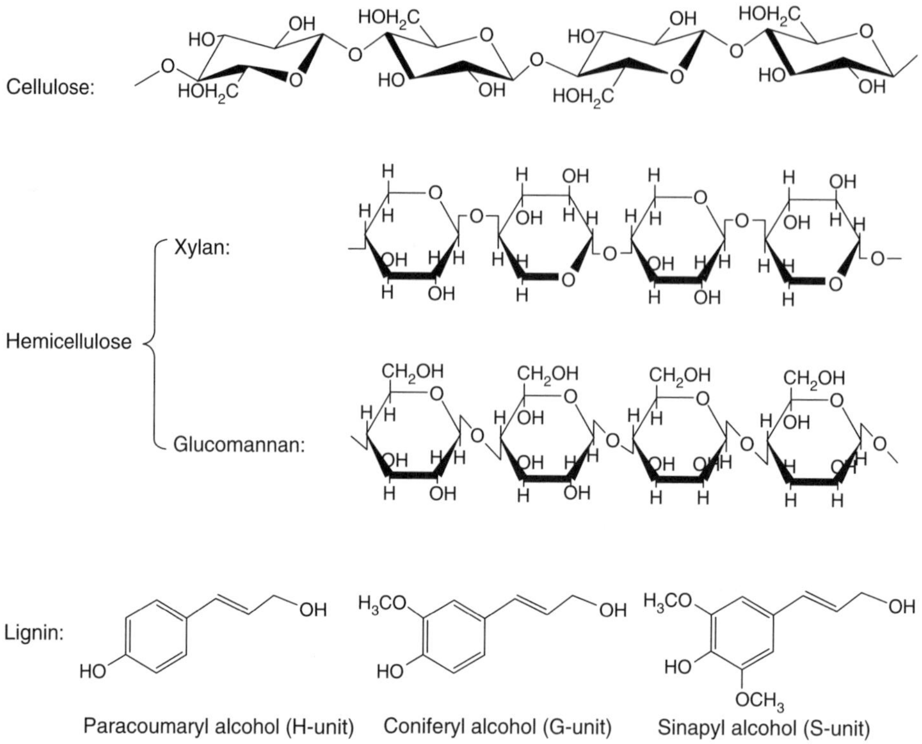

Figure 1.2 Three main components of lignocellulosic biomass: cellulose, hemicellulose, and lignin [15].

allowing lignocellulosic biomass to be used for food and fuel production. Biochemical processing of lignocellulosic biomass employs a variety of microorganisms that secrete enzymes that catalyze the hydrolysis of glycosidic bonds in either cellulose or hemicellulose. Many animals, such as cattle and other ruminants, have developed symbiotic relationships with these microorganisms to allow them to digest cellulose. Thermal energy and catalysts can also break glycosidic bonds, usually more inexpensively but less selectively than enzymes.

Lignin, a complex cross-linked phenolic polymer, is indigestible by most animals and microorganisms. In fact, it protects the carbohydrate against biological attack. Thus, even ruminant animals that have evolved on diets of lignocellulosic biomass, such as grasses and forbs, can only extract 50–80% of the energy content of this plant material because some of the polysaccharides and all of the lignin pass through the gut undigested. Biochemical processing has many similarities to the digestive system of ruminant animals. Physical and chemical pretreatments release cellulose fibers from the composite matrix, making them more susceptible to enzymatic hydrolysis, which releases simple sugars that can be fermented or otherwise metabolized [17].

Thermochemical processing occurs at temperatures that are several hundred degrees Celsius and sometimes over 1000 °C. At these temperatures, thermochemical processes occur very rapidly whether catalysts are present or not. In contrast, biochemical processes occur at only a few tens of degrees Celsius above ambient temperature, with the result that they can take hours or even days to complete even in the presence of biocatalysts. Thermal depolymerization of cellulose in the absence of alkali or alkaline earth metals produces predominately levoglucosan, an anhydrosugar of the monosaccharide glucose [18]. Under certain conditions, it appears that lignin depolymerizes to monomeric phenolic compounds [19]. Under conditions of high-temperature combustion and gasification, chemical equilibrium among products is attained. Thus, thermochemical processing offers opportunities for rapid processing of diverse feedstocks, including recalcitrant materials and unique intermediate feedstocks, for production of fuels, chemicals, and power.

1.2 Thermochemical Conversion Technologies

Thermochemical conversion can be categorized as combustion, gasification, pyrolysis, and solvent liquefaction, as shown in Figure 1.3. The key operating parameters governing these routes are degree of oxidation, temperature, heating rate, and residence time. End products from various technologies include electric power, heat, fuels, and chemicals [20].

1.2.1 Direct Combustion

Direct combustion of biomass produces moderate- to high-temperature thermal energy (800–1600 °C) suitable for heat and power applications. This is realized by rapid reaction of fuel and oxygen to obtain thermal energy and flue gas, consisting primarily of carbon dioxide and water. Depending on heating value and moisture content of the fuel, the fuel-to-air ratio, and the construction of the furnace, flame temperatures can exceed 1650 °C. Direct combustion of biomass has the advantage of employing well-developed and commercially available technology. Combustion is the foundation of much of the electric power generation around the world. Direct combustion of biomass is burdened by three prominent disadvantages. These include penalties associated with burning high-moisture fuels, agglomeration and ash fouling due to alkali compounds in biomass, and difficulty of

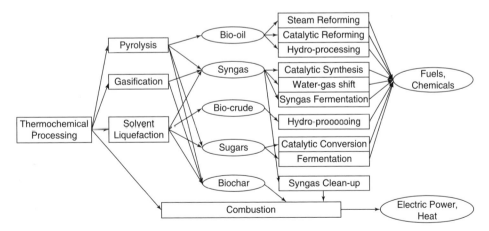

Figure 1.3 Thermochemical options for production of fuels, chemicals, and power. Text in rectangles indicate technologies and text in ovals indicate products or intermediate products.

providing and safeguarding sufficient supplies of bulky biomass to modern electric power plants.

While most of the focus on bioenergy has been on the production of liquid fuels, it has been argued that a better use of biomass would be to burn it for the generation of electricity to power battery electric vehicles (BEVs) [21, 22]. However, this is contingent on further development of batteries that can store sufficient electricity to match the power, range, and cost of ICE vehicles.

1.2.2 Gasification

Thermal gasification is the conversion of carbonaceous solids at elevated temperatures (700–1000 °C) and under oxygen-starved conditions into syngas, which is a flammable gas mixture of carbon monoxide, hydrogen, methane, nitrogen, carbon dioxide, and smaller quantities of hydrocarbons [23]. The produced syngas can be used either to generate electric power or to synthesize fuels or other chemicals using catalysts or using even microorganisms (syngas fermentation) [24]. Gasification has been under development for almost 200 years, beginning with the gasification of coal to produce so-called "manufactured gas" or "town gas" for heating and lighting. Coal gasification has also been used for large-scale production of liquid transportation fuels, first in Germany during World War II and then later in South Africa during a period of worldwide embargo as a result of that country's apartheid policies.

Gasification can be used to convert any carbonaceous solid or liquid to low molecular weight gas mixtures. In fact, the high volatile matter content of biomass allows it to be gasified more readily than coal. Biomass gasification has found commercial application where waste wood was plentiful or fossil resources were scarce. An example of the former was Henry Ford's gasification of wood waste derived from shipping crates at his early automotive plants. An example of the latter was the employment of portable wood gasifiers in Europe during World War II to power automobiles. With a few exceptions, gasification in all its forms gradually declined over the twentieth century due to the emergence of electric lighting, the development of the natural gas industry, and the success of the petroleum industry in continually expanding proven reserves of petroleum. In the twenty-first century, as natural gas and petroleum become more expensive, gasification of both coal and biomass is likely to be increasingly employed.

One of the most attractive features of gasification is its flexibility of application, including thermal power generation, hydrogen production, and synthesis of fuels and chemicals. This offers the prospect of gasification-based energy refineries, producing a mix of energy and chemical products or allowing the staged introduction of technologies as they reach commercial viability.

The simplest application of gasification is production of heat for kilns or boilers. Often the syngas can be used with minimal clean-up because tars or other undesirable compounds are consumed when the gas is burned and process heaters are relatively robust to dirty gas streams. Syngas can be used in ICEs if tar loadings are not too high and after removal of the greater part of particulate matter entrained in the gas leaving the gasifier. Gas turbines offer prospects for high-efficiency integrated gasification–combined-cycle power, but they require more stringent gas cleaning [25]. As the name implies, syngas can also be used to synthesize a wide variety of chemicals, including organic acids, alcohols, esters, and hydrocarbon fuels, but the catalysts for these syntheses are even more sensitive to contaminants than are gas turbines.

1.2.3 Pyrolysis

Pyrolysis is a thermal conversion technology that can either be considered an initial step for other thermal conversion processes, such as gasification and combustion, or a conversion process in its own right for production of biofuels. Pyrolysis decomposes biomass in the absence of oxygen, within a temperature range of 300–900 °C [26] and a heating rate that varies greatly from less than 0.005 °C/s to more than 10 000 °C/s [27]. Depending on operating conditions, pyrolysis can be classified as slow, intermediate, fast, or flash pyrolysis. Slow pyrolysis operates at relatively low heating rate, low temperature, and long residence times, with the main product being solid char. Fast pyrolysis is characterized by high heating rate, high temperature, and short residence time compared to slow pyrolysis. Intermediate pyrolysis is a technology with moderate operating temperature and heating rate. Flash pyrolysis has the highest heating rate and shortest residence time, which requires special reactors to achieve [20].

Bio-oil is an energy-rich liquid recovered from the condensable vapors and aerosols produced during fast pyrolysis. It is a complex mixture of oxygenated organic compounds, including carboxylic acids, alcohols, aldehydes, esters, saccharides, phenolic compounds, and lignin oligomers. Other products include flammable gas (syngas) and biochar [28]. However, bio-oil is the majority product – with yields as high as 70–80 wt% [20]. Under suitable processing conditions, fast pyrolysis can also yield significant quantities of sugars and anhydrosugars [29]. These "thermolytic sugars" can either be fermented or catalytically upgraded to fuel molecules.

The great virtues of fast pyrolysis are the simplicity of the process and the attractiveness of a liquid as the intermediate product for upgrading to finished fuels and chemicals compared to either syngas from gasification or raw biomass. Early attempts to use bio-oil as fuel for both boilers and gas turbine engines were hindered by its cost, corrosiveness, and instability during storage. More recent strategies upgrade bio-oil to either heavy fuel oil substitutes or transportation fuels. For example, light oxygenates in bio-oil can be steam reformed to provide hydrogen [30] while the heaviest fraction of bio-oil can be cracked to gasoline and diesel fuel [31]. Techno-economic analysis [32] indicating bio-oil could be upgraded to gasoline and diesel for $2–$3/gal (about $0.53–$0.79/l) gasoline equivalent has spurred interest in fast pyrolysis and bio-oil upgrading.

Hydroprocessing bio-oil into hydrocarbons suitable for use as transportation fuel is similar to the process for refining petroleum. Hydroprocessing was originally developed to convert petroleum into motor fuels by reacting it with hydrogen at high pressures in the presence of catalysts. Hydroprocessing includes two distinct processes. Hydrotreating is designed to remove sulfur, nitrogen, oxygen, and other contaminants from petroleum. When adapted to bio-oil, the main contaminant to be removed is oxygen. Thus, hydrotreating bio-oil from pyrolysis of lignocellulosic biomass is primarily a process of deoxygenation. Hydrocracking is the reaction of hydrogen with organic compounds to break long-chain molecules into lower molecular weight compounds. Although fast pyrolysis attempts to depolymerize plant molecules, a number of oligomers (especially from lignin) are found in bio-oil, which hydrocracking can convert into more desirable paraffin or naphthene molecules. Some researchers have employed catalysts in pyrolysis reactors to directly produce hydrocarbons. Similar to the process of fluidized catalytic cracking used in the petroleum industry, the process occurs at atmospheric pressure over acidic zeolites. Yields of C_5–C_{10} hydrocarbons as high as 17% have been reported for catalytic pyrolysis of poplar wood [33]. Although superior to conventional bio-oil, this product still needs refining to gasoline and diesel fuel.

1.2.4 Solvent Liquefaction

Solvent liquefaction is the thermal decomposition of biomass in the presence of a solvent at moderate temperatures and pressures, typically 105–400 °C and 2–20 MPa, to produce predominately liquid or solubilized products with smaller amounts of gaseous and solid co-products. Like pyrolysis, solvent liquefaction can produce sugars from carbohydrate and phenolic compounds from lignin. A wide range of solvents can be employed, including non-polar solvents, such as toluene and tetralin, polar aprotic solvents, such as gamma-valerolactone and tetrahydrofuran, protic solvents, such as water and ethanol, and ionic liquids, such as 1-ethyl-3-methylimidazolium chloride.

Solvent liquefaction in water, referred to as hydrothermal processing, is particularly attractive for wet feedstock, which can be handled as slurries with solids loadings in the range of 5 to 20 wt%. Hydrothermal processing occurs at elevated pressures of 50–250 atm (~5–25 MPa) to prevent boiling of the water in the slurry and at temperatures ranging from 200 to 500 °C, depending upon whether the desired products are fractionated and hydrolyzed plant polymers [34], partially deoxygenated liquid product known as biocrude [35], or syngas [36]. As illustrated in Figure 1.4, processing pressure must be increased as reaction temperature increases to prevent boiling of water in the wet biomass. At temperatures around 100 °C, extraction of high-value plant chemicals such as resins, fats, phenolics, and phytosterols is possible. At 200 °C and 20 atm (~2 MPa), fibrous biomass undergoes a fractionation process to yield cellulose, lignin, and hemicellulose degradation products such as furfural. Further hydrothermal processing can hydrolyze the cellulose to glucose. At 300–350 °C and 120–180 atm (~12.2–18.2 MPa), biomass undergoes more extensive chemical reactions, yielding a hydrocarbon-rich liquid known as biocrude. Although superficially resembling bio-oil, it has lower oxygen content and is less miscible in water, making it more amenable to hydrotreating. At 600–650 °C and 300 atm (30.4 MPa) the primary reaction product is gas, including a significant fraction of methane.

Continuous feeding of biomass slurries into high-pressure reactors, efficient energy integration, and product separation from solvent are significant engineering challenges to be overcome before solvent liquefaction results in a commercially viable technology.

1.3 Diversity of Products: Electric Power, Transportation Fuels, and Commodity Chemicals

Theoretically, biomass is a resource that can be used to produce all types of products: heat, electricity, transportation fuels, and commodity chemicals. Currently, heat is the major

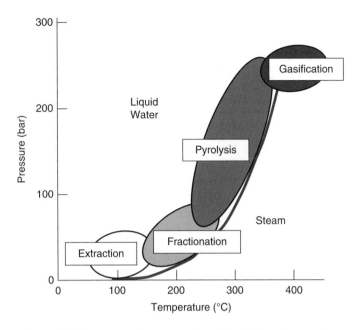

Figure 1.4 *Temperature/pressure regimes of hydrothermal processing.*

product of biomass utilization, with roughly 60% of worldwide bioenergy comprising traditional applications for cooking and heating in developing countries [37]. With regard to electricity production from biomass, dedicated biomass power plants or biomass co-firing plants are feasible means of reducing CO_2 emissions and producing green energy. Much of the recent research focus worldwide on bioenergy has been liquid transportation fuels in an effort to displace imported petroleum.

1.3.1 Biopower

Biopower can help reduce the use of coal; coal is expected to decline from 40% of world net electricity generation in 2015 to only 31% by 2040 [38]. Several examples can be cited. Drax, one of the largest coal-fired power plant in the UK, started co-firing biomass in 2003, eventually achieving 100% replacement of coal. Atikokan Generating Station in Canada achieved 100% conversion to biomass in 2014. DONG Energy in Denmark announced in February 2017 that their thermal power plants (all of which employ co-generation) would completely replace coal with biomass by 2023 [39].

Plug-in electric vehicles utilizing biopower provides a promising option as WTWs analyses indicate that BEVs are superior to biofuel-powered ICE vehicles in terms of primary energy consumed, GHG emissions, life-cycle water usage, and cost when evaluated on the basis of kilometers driven [22].

Combined-cycle power based on gasification of biomass is another route to biopower. Although gasification can efficiently convert a wide range of feedstocks into a flammable mixture of carbon monoxide and hydrogen, it also contains contaminants including tar, solid

10 *Thermochemical Processing of Biomass*

particulates, alkali compounds, sulfur, nitrogen, and chlorine that must be removed before the gas can be burned for power generation to avoid in-plant corrosion and air pollution emissions. Currently, syngas clean-up is the key barrier for the reliable and cost-efficient operation of power plants based on biomass gasification [20].

1.3.2 Biofuels

Biofuels are defined as transportation fuels derived from biomass. Second-generation biofuels are illustrated in Figure 1.5 [20]. These are predominantly liquids at ambient conditions, to be compatible with the existing infrastructure for the transportation fuels, but also includes methane and hydrogen, which are gases at ambient conditions but which can be compressed or liquefied for use as transportation fuels. Liquid biofuels can generally be categorized as alcohols, drop-in biofuels, or fuel additives.

Methanol, a C_1 alcohol, is traditionally synthesized from syngas (CO and H_2) derived from biomass gasification. Due to undesirable properties such as toxicity, water solubility, low vapor pressure, and phase separation, methanol has received less attention than ethanol as a substitute for gasoline. Ethanol has been widely integrated in transportation fuel infrastructures of Brazil, Europe, and the Unites States, although it is currently derived from

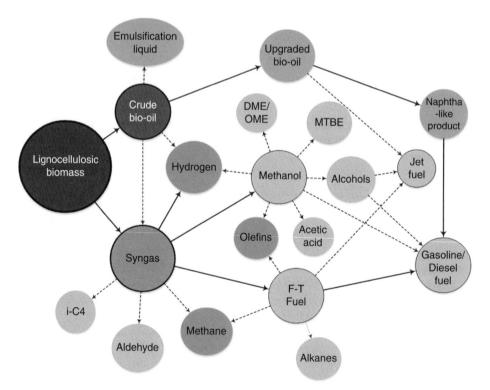

Figure 1.5 *Second-generation biofuels. F-T, Fischer-Tropsch; i-C4, isobutene and isobutane; DME, dimethyl ether; OME, oxymethylene ethers; MTBE, Methyl tert-butyl ether; HRJs, hydro-processed renewable jet fuels [20]. See the plate section for a color representation of this figure.*

biochemical processing of sugars and starches from sugar cane and grains [40]. Ethanol is typically blended with gasoline with ratios of 10% (E10), 15% (E15), and 25% (E25). Flexible-fuel vehicles are able to use up to 100% ethanol fuel (E100).

Drop-in biofuels are fully compatible with existing fuel utilization systems. Thus, they are liquids containing very little oxygen and are completely miscible in petroleum-derived gasoline and diesel fuel. Drop-in biofuels that are pure hydrocarbons are usually referred to as renewable or green gasoline and diesel. An example of a low-oxygen-content drop-in biofuel is butanol. Drop-in biofuels would address concerns about first-generation biofuels damaging fuel systems and causing phase separation of water in gasoline pipelines.

Biomass can also be processed into fuel additives to improve engine performance. Examples include methanol, ethanol, butanol, dimethyl ether (DME), and oxymethylene ethers (OME). Blended with diesel fuel, these oxygenated compounds help reduce soot and NO_x emissions [41]. Blended with gasoline, fuel additives serve as octane boosters.

Hydrogen has been of interest as a carbon-free energy carrier for several decades. Although it is currently produced mostly from fossil fuels, it could also be produced from biomass gasification. The resulting syngas could be converted into a mixture of carbon dioxide and hydrogen by the water-gas shift reaction ($CO + H_2O \rightarrow CO_2 + H_2$) and the carbon dioxide removed to yield a pure stream of hydrogen. An alternative route for hydrogen production from biomass is pyrolysis followed by catalytic steam reforming of the bio-oil. The overall yield of hydrogen from biomass via bio-oil reforming is lower than direct gasification of biomass feedstock, although the combination of decentralized pyrolysis of biomass and centralized gasification of bio-oil lends itself to distributed processing of biomass.

1.3.3 Bio-Based Chemicals

Although biofuels represent much larger volume products from biomass, commodity chemicals from biomass (bio-based chemicals) potentially represent much larger revenue from biomass. The production of a wide range of chemicals from biomass has been demonstrated [20]. For example, olefins, which can be produced from syngas, are platform molecules to produce plastics and detergents; the water-insoluble fraction of bio-oil, referred to as pyrolytic lignin, can be converted into resin; the biomass-derived carboxylic acids can be used to produce calcium salts as road deicers (the conversion has not been commercialized; however, it has the potential to be scaled up) [20]. Additionally, bio-methanol is a widely used platform material to produce bio-based chemicals such as acetic acid and formaldehyde [42].

Despite the high value of many bio-based chemicals compared to biofuels, they must be produced in high yields and efficiently separated at high purity. The growing interest in replacing fossil fuels with renewable resources should encourage the further research and development needed move society toward a bioeconomy [43].

1.4 Economic Considerations

Among the thermochemical technologies, combustion is the most widely deployed in commercial practice. Gasification was commercially deployed for heating and lighting over 200 years ago, but mostly using coal rather than biomass as feedstock. Development of more convenient and cleaner petroleum and natural gas resources has mostly displaced gasification, with a few exceptions around the world. Gasification has gained renewed interest as a way to efficiently utilize solid biomass resources. However, most schemes to convert syngas into fuel and power require gasifiers to operate at elevated pressures and produce syngas with very low contaminant levels, both of which contribute to high capital costs [42].

Solvent liquefaction has found few commercial applications, in part due to the capital costs associated with high-pressure reactors but also due to the relatively high operating costs associated with the use of large amounts of solvent and recovering products from the solvent.

Fast pyrolysis is a relatively simple process that can be operated at small scale, lending itself to distributed processing of biomass and reducing investment risk compared to more capital intensive investments of gasification or biochemical processing of cellulosic biomass. Nevertheless, it has not achieved cost parity with petroleum-derived fuels.

The commercialization of second-generation biofuels is widely limited by their high cost of production, which can be as much as two to three times higher than for fossil fuels [20]. As might be expected for recalcitrant feedstock, processing costs are much higher for lignocellulosic biomass than sugars and starches. Feedstocks also show considerable variability across species and production regions, including the relative amounts of carbohydrate and lignin, differences in composition of lignin, and variations in ash composition and amount. All of these factors influence the ability to process the biomass, with attendant influences on capital and operating costs.

1.5 Environmental Considerations

Biofuels produced from sustainably grown biomass have several environmental benefits compared to petroleum-derived gasoline and diesel [44]. Among the most important of these is net reductions in life-cycle GHG emissions, which arise from the fact that the carbon in biofuels is taken up from the atmosphere through photosynthesis of growing biomass. GHG emission and the relevant net energy assessment are the main focus of life-cycle analysis (LCA) studies. Typical GHG emissions for several pathways are shown in Figure 1.6 [20]. First-generation biofuels reduce GHG emissions on the order of 40–80% compared to petroleum-derived fuels while second-generation biofuels can even achieve negative GHG emissions [20]. Biofuels have a larger variation in GHG emissions compared to conventional fuels, mainly due to the different maturity levels of these technologies and the several potential routes for producing biofuels.

The use of fossil fuels in the production of biomass (application of fertilizer and use of farm machinery) and the processing of biomass (natural gas burned for process heat to support drying and distillation operations) contributes to net GHG emissions associated with the use of biofuels. Furthermore, conversion of farmland from food production to biofuel production can indirectly encourage the conversion of forests and grasslands in other parts of the world to cropland, which often is associated with release of GHG to the atmosphere due to burning of standing biomass or oxidizing soil carbon during tillage of the land [45]. The extent of this so-called "indirect land-use change" to the net emissions of GHG from biofuels production is unclear and is much debated [46, 47]. Similarly, the effect of biofuel production on water resources and biodiversity and socioeconomic impacts such

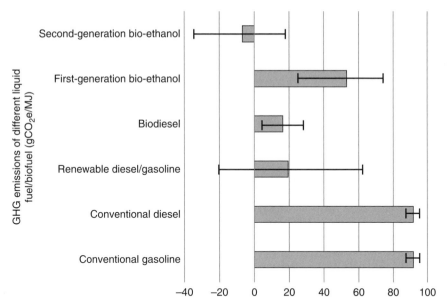

Figure 1.6 *GHG emissions (gCO$_2$e/MJ) associated with production of several kinds of biofuels and comparison with conventional diesel and gasoline [20].*

as income and employment should also be considered in comparing impacts of biofuels to fossil fuels [48].

1.6 Organization of This Book

Thermochemical processing is distinguished by a large number of approaches to converting lignocellulosic biomass into fuels, chemicals, and power. This book is a compilation of review articles on distinct approaches to thermochemically deconstructing biomass into intermediates and their upgrading into final products. Chapter 1 (this chapter) is an overview of these technologies. Chapter 2 reviews the science of thermal deconstruction of biomass, both from experimental and modeling perspectives. Chapter 3 is devoted to biomass combustion – including fundamentals, combustor types, operational issues, and power options. Chapter 4 focuses on gasification, including fundamentals, gasifier types, and the state of the technology development and commercialization. Chapter 5 explains the issues associated with syngas cleaning and upgrading, with detailed descriptions of the kinds of contaminants found in syngas, the unit operations associated with removal of each of these contaminants, and the synthesis of fuels and chemicals from syngas. Chapter 6 covers fast pyrolysis of biomass, including current understanding of the chemistry of pyrolysis, different kinds of pyrolyzers, and the state of development and commercialization. Chapter 7 explores stabilization and upgrading of bio-oil by physical, chemical, and catalytic means to transportation fuels. Chapter 8 discusses solvent liquefaction of carbohydrate and lignin to liquids and solubilized products. Chapter 9 describes hybrid processing of biomass, which is defined as the integration of a thermochemical process to deconstruct lignocellulosic biomass into an intermediate suitable as substrate for biochemical upgrading to finished products. Chapter 10 provides cost estimates for a wide range of thermochemical processes, ranging from electric power generation to the production of liquid biofuels and other chemicals. These analyses provide a useful starting point for exploring the feasibility of different approaches to thermochemical processing. Chapter 11 reviews the literature on LCA of thermochemical processes, including air and water impacts and GHG emissions.

References

1. NSF (2008). *Breaking the Chemical and Engineering Barriers to Lignocellulosic Biofuels: Next Generation Hydrocarbon Biorefineries* (ed. G.W. Huber). Washington D.C: National Science Foundation. Chemical, Bioengineering, Environmental, and Transport Systems Division.
2. Wright, M. and Brown, R.C. (2007). Establishing the optimal sizes of different kinds of biorefineries. *Biofuels, Bioproducts and Biorefining* **1**: 191–200.
3. Renewable Fuels Association. Industry Statistics, World Fuel Ethanol Production. 2016. https://ethanolrfa.org/resources/industry/statistics/#1537559649968-e206480c-7160 (accessed November 17, 2018).
4. Statista. Leading biodiesel producers worldwide in 2016, by country (in billion liters). https://www.statista.com/statistics/271472/biodiesel-production-in-selected-countries/ (accessed November 17, 2018).
5. Farrell, A.E., Plevin, R.J., Turner, B.T. et al. (2006). Ethanol can contribute to energy and environmental goals. *Science* **311**: 506.
6. Gebremariam, S.N. and Marchetti, J.M. (2017). Biodiesel production technologies: review. *AIMS Energy* **5**: 425–457.
7. Hill, J., Nelson, E., Tilman, D. et al. (2006). Environmental, economic, and energetic costs and benefits of biodiesel and ethanol biofuels. *Proceedings of the National Academy of Sciences of the United States of America* **103**: 11206.
8. Shields-Menard, S.A., Amirsadeghi, M., French, W.T., and Boopathy, R. (2018). A review on microbial lipids as a potential biofuel. *Bioresource Technology* **259**: 451–460.
9. Leiva-Candia, D.E., Pinzi, S., Redel-Macías, M.D. et al. (2014). The potential for agro-industrial waste utilization using oleaginous yeast for the production of biodiesel. *Fuel* **123**: 33–42.
10. Klass, D.L. (1998). Chapter 10 – Natural biochemical liquefaction. In: *Biomass for Renewable Energy, Fuels, and Chemicals* (ed. D.L. Klass), 333–382. San Diego: Academic Press.
11. Trabucco, A., Achten, W.M.J., Bowe, C. et al. (2010). Global mapping of Jatropha curcas yield based on response of fitness to present and future climate. *GCB Bioenergy* **2**: 139–151.
12. Basiron, Y. (2007). Palm oil production through sustainable plantations. *European Journal of Lipid Science and Technology* **109**: 289–295.
13. Sheehan, J., Dunahay, T., Benemann, J., and Roessler, P. 1998. Look Back at the U.S. Department of Energy's Aquatic Species Program: Biodiesel from Algae; US Department of Energy's Office of Fuels Development. https://www.nrel.gov/docs/legosti/fy98/24190.pdf (accessed 22 November 2018).
14. Bwapwa, J.K., Anandraj, A., and Trois, C. (2017). Possibilities for conversion of microalgae oil into aviation fuel: a review. *Renewable and Sustainable Energy Reviews* **80**: 1345–1354.
15. Zhang, X., Yang, W., and Blasiak, W. (2011). Modeling study of woody biomass: interactions of cellulose, hemicellulose, and lignin. *Energy and Fuels* **25**: 4786–4795.
16. Sjöström, E. (ed.) (1993). *Wood Chemistry*, 2e, 1–20. San Diego: Academic Press.
17. Brown, R.C. and Brown, T.R. (2014). *Biorenewable Resources*, 171–194. Wiley.
18. Patwardhan, P.R., Satrio, J.A., Brown, R.C., and Shanks, B.H. (2009). Product distribution from fast pyrolysis of glucose-based carbohydrates. *Journal of Analytical and Applied Pyrolysis* **86**: 323–330.
19. Patwardhan, P.R., Brown, R.C., and Shanks, B.H. (2011). Understanding the fast pyrolysis of lignin. *ChemSusChem* **4**: 1629–1636.

20. Zhang, X. (2016). Essential scientific mapping of the value chain of thermochemically converted second-generation bio-fuels. *Green Chemistry* **18**: 5086–5117.
21. Campbell, J.E., Lobell, D.B., and Field, C.B. (2009). Greater transportation energy and GHG offsets from bioelectricity than ethanol. *Science* **324**: 1055.
22. Gifford, J.D. and Brown, R.C. (2011). Four economies of sustainable automotive transportation. *Biofuels, Bioproducts and Biorefining* **5**: 293–304. https://doi.org/10.1002/bbb.287.
23. Rezaiyan, J. and Cheremisinoff, N.P. (2005). *Gasification Technologies: A Primer for Engineers and Scientists*. Boca Raton: Taylor & Francis.
24. Brown, R.C. (2008). Biomass refineries based on hybrid thermochemical/biological processing – an overview. In: *Biorefineries, Bio-Based Industrial Processes and Products* (ed. B. Kamm, P.R. Gruber and M. Kamm), 227–252. Wiley-VCH.
25. Cummer, K.R. and Brown, R.C. (2002). Ancillary equipment for biomass gasification. *Biomass and Bioenergy* **23**: 113–128.
26. Guo, M., Song, W., and Buhain, J. (2015). Bioenergy and biofuels: history, status, and perspective. *Renewable and Sustainable Energy Reviews* **42**: 712–725.
27. Collard, F. and Blin, J. (2014). A review on pyrolysis of biomass constituents: mechanisms and composition of the products obtained from the conversion of cellulose, hemicelluloses and lignin. *Renewable and Sustainable Energy Reviews* **38**: 594–608.
28. Bridgwater, A.V. and Peacocke, G.V.C. (2000). Fast pyrolysis processes for biomass. *Renewable and Sustainable Energy Reviews* **4**: 1–73.
29. Brown, R.C., Radlein, D., and Piskorz, J. (2001). Pretreatment processes to increase pyrolytic yield of levoglucosan from herbaceous feedstocks. In: *Chemicals and Materials from Renewable Resources*, 123–132. Washington, DC: American Chemical Society.
30. Czernik, S., French, R., Feik, C., and Chornet, E. (2002). Hydrogen by catalytic steam reforming of liquid byproducts from biomass thermoconversion processes. *Industrial and Engineering Chemistry Research* **41**: 4209–4215.
31. Elliott, D.C. (2007). Historical developments in hydroprocessing bio-oils. *Energy and Fuels* **21**: 1792–1815.
32. Wright, M.M., Daugaard, D.E., Satrio, J.A., and Brown, R.C. (2010). Techno-economic analysis of biomass fast pyrolysis to transportation fuels. *Fuel* **89**: S2–S10.
33. Carlson, T., Vispute, T., and Huber, G. (2008). Green gasoline by catalytic fast pyrolysis of solid biomass derived compounds. *ChemSusChem* **1**: 397–400.
34. Allen, S.G., Kam, L.C., Zemann, A.J., and Antal, M.J. (1996). Fractionation of sugar cane with hot, compressed, liquid water. *Industrial and Engineering Chemistry Research* **35**: 2709–2715.
35. Elliott, D.C., Beckman, D., Bridgwater, A.V. et al. (1991). Developments in direct thermochemical liquefaction of biomass: 1983–1990. *Energy and Fuels* **5**: 399–410.
36. Elliott, D.C., Neuenschwander, G.G., Hart, T.R. et al. (2004). Chemical processing in high-pressure aqueous environments. 7. Process development for catalytic gasification of wet biomass feedstocks. *Industrial and Engineering Chemistry Research* **43**: 1999–2004.
37. IPCC (Intergovernmental Panel on Climate Change) (2012). *Renewable Energy Sources and Climate Change Mitigation*. Cambridge University Press.
38. U.S. Energy Information Administration 2017. International Energy Outlook 2017. https://www.eia.gov/outlooks/ieo/pdf/0484(2017).pdf (accessed 22 November 2018).
39. Aikawa, T. 2017. Biomass co-firing: For the reduction of coal-fired power plants. Tokyo, Japan: Renewable Energy Institute.
40. Renewable Fuels Association (2012). *Accelerating Industry Innovation: 2012 Ethanol Industry Outlook*. Renewable Fuels Association.
41. Zhang, X., Oyedun, A.O., Kumar, A. et al. (2016). An optimized process design for oxymethylene ether production from woody-biomass-derived syngas. *Biomass and Bioenergy* **90**: 7–14.

42. Spath, P.L. and Dayton, D.C. 2003. Preliminary Screening – Technical and Economic Assessment of Synthesis Gas to Fuels and Chemicals with Emphasis on the Potential for Biomass-Derived Syngas (NREL/TP-510-34929). National Renewable Energy Laboratory. https://www.nrel.gov/docs/fy04osti/34929.pdf (accessed 22 November 2018).
43. Sheldon, R.A. (2014). Green and sustainable manufacture of chemicals from biomass: state of the art. *Green Chemistry* **16**: 950–963.
44. Brown, R.C. and Brown, T.R. (2014). Environmental impact of the bioeconomy. In: *Biorenewable Resources*, 2e, 261–287. Wiley.
45. Kim, S. and Dale, B.E. (2011). Indirect land use change for biofuels: testing predictions and improving analytical methodologies. *Biomass and Bioenergy* **35**: 3235–3240.
46. Tokgoz, S. and Laborde, D. (2014). Indirect land use change debate: what did we learn? *Current Sustainable/Renewable Energy Reports* **1**: 104–110.
47. Finkbeiner, M. (2014). Indirect land use change – Help beyond the hype? *Biomass and Bioenergy* **62**: 218–221.
48. Searchinger, T. and Heimlich, R. 2015. Avoiding Bioenergy Competition for Food Crops and Land. Working Paper, Installment 9 of Creating a Sustainable Food Future. Washington, DC: World Resources Institute. https://www.wri.org/publication/avoiding-bioenergy-competition-food-crops-and-land (accessed 22 November 2018).

2

Condensed Phase Reactions During Thermal Deconstruction

Jake K. Lindstrom[1], Alexander Shaw[2], Xiaolei Zhang[2] and Robert C. Brown[1,3]

[1]*Department of Mechanical Engineering, Iowa State University, Ames, IA, USA*
[2]*School of Mechanical and Aerospace Engineering, Queen's University Belfast, Belfast, UK*
[3]*Bioeconomy Institute, Iowa State University, Ames, IA, USA*

2.1 Introduction to Condensed Phase Reactions During Thermal Deconstruction of Biomass

All biomass thermochemical processes begin with depolymerization and decomposition reactions within a solid substrate that release volatile products and typically leave behind carbonaceous solid products. Despite their universal importance, condensed phase (liquid and solid) reactions during thermal deconstruction have received relatively little attention due to the difficulty in directly analyzing these reactions. While external conditions vary widely among thermochemical processes, reactions within the condensed phase are largely similar. Understanding condensed phase reactions can provide insight into the wide range of thermochemical processes.

Six thermochemical processes are discussed in this chapter: torrefaction, slow pyrolysis, solvent liquefaction, fast pyrolysis, gasification, and combustion. With the exception of solvent liquefaction, the operating temperature range is distinct for each process, as shown in Figure 2.1, although these limits can be flexible.

Figure 2.2 shows typical yields of solid, liquid, and gas for these thermochemical processes. (Solvent liquefaction is excluded because yields are widely variable depending upon operating conditions.) The trend is clear: higher temperatures favor gaseous products over solid products.

Understanding thermal deconstruction of biomass requires familiarity with biomass composition. Lignocellulosic biomass, the common feedstock for thermochemical processes,

---

Thermochemical Processing of Biomass: Conversion into Fuels, Chemicals and Power, Second Edition.
Edited by Robert C. Brown.
© 2019 John Wiley & Sons Ltd. Published 2019 by John Wiley & Sons Ltd.

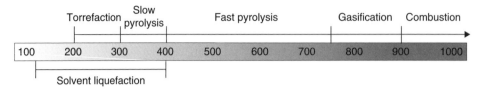

Figure 2.1 Approximate operating temperatures for torrefaction [1], slow pyrolysis [2], solvent liquefaction [3], fast pyrolysis [2], gasification [2], and combustion range from 105 °C to greater than 1000 °C. Temperatures shown in degrees Celsius.

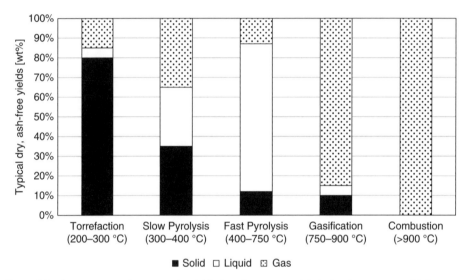

Figure 2.2 As temperature increases, the yield of gas increases and that of carbonaceous solid decreases for a wide range of thermochemical processes. Source: figure produced using data from Bridgwater [2].

typically contains 50–60 wt% structural carbohydrates (namely cellulose and hemicellulose) and 15–40 wt% lignin, with the remaining mass composed of extractives, proteins, ash, etc. [4]. This composition can vary significantly among different plant varieties, plant parts (e.g. stalk versus leaves), and cell wall layers (e.g. primary versus secondary cell wall). Regardless of composition, these components intertwine to form a complex composite structure (Figure 2.3) [5–7].

This intricate structure complicates chemical and kinetic analysis. As a result, thermal deconstruction of the three main biopolymers is often studied individually. Isolated versions of these polymers and smaller model compounds, such as monomers and dimers, are commonly used despite marked differences from their native forms. There is value in this approach, but correctly applying the results to lignocellulosic biomass can be challenging. For example, thermochemical conversion analysis usually accounts for biomass varieties and, to some extent, plant components, but plant cellular structure is mostly neglected. See Harris and Stone [6] for a comprehensive review of plant cell wall structure and chemistry.

Cellular microstructure plays an underappreciated role in thermochemical processes. Plants evolved to resist physical, chemical, and biological attack, making them recalcitrant

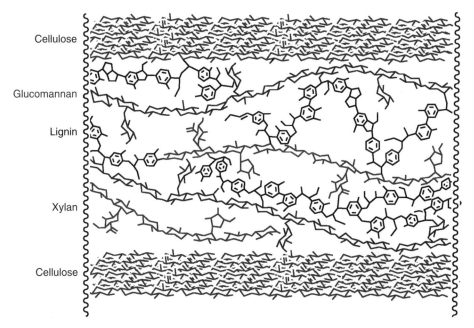

Figure 2.3 *This depiction of cellulose, hemicellulose (glucomannan, xylan), and lignin shows some of the difficulty of using an extracted version of the polymer as a model compound. All the components in biomass are interconnected to form a complicated three-dimensional structure.* Source: figure reproduced from Reference [5], with permission from Elsevier.

to deconstruction. Cell structure and arrangement also affect thermal deconstruction. In particular, heat and mass transfer within biomass particles affects product yields. Section 2.3.2 "The Role of Cell Wall Structure in Thermal Deconstruction" examines these effects in more detail.

Accounting for the effects of cellular structure may improve the understanding of thermal deconstruction, but its importance may depend upon scale. For example, random error at the macroscale may overwhelm minute – but real – differences at the micro or nano domains. This error propagation indicates that efforts to model deconstruction at multiple scales are inherently imprecise. Their primary usefulness is providing a qualitative understanding of the overall process.

This chapter explores condensed phase reactions important to thermal deconstruction, first by noting common physical and chemical transformations among different thermochemical processes. This chapter attempts to unify a wide spectrum of thermochemical processes in terms of the fundamental thermal deconstruction mechanisms that occur in the condensed phase of biomass.

2.2 Thermochemical Processes

Thermal deconstruction of biomass is integral to several kinds of processes for which heat is the driving force of physical and chemical transformations. Many of the most important transformations occur in the condensed phase – with solid biomass depolymerizing and decomposing into liquids, vapors, gases, and solid residue – although secondary reactions of vaporized species can also be important in determining the ultimate yields of products. Despite similarities, thermochemical processes can be categorized according to whether the principal products are gases (combustion and gasification), liquids (pyrolysis and solvent liquefaction), or solids (torrefaction and slow pyrolysis), although they all produce smaller amounts of the other products as well.

2.2.1 Processes Yielding Chiefly Solids

Torrefaction and slow pyrolysis are characterized as producing solids as the principal products, specifically torrefied biomass and charcoal. These carbonaceous solids have applications as solid fuels. Charcoal has also been used as a soil amendment, in which case it is referred to as biochar.

2.2.1.1 Torrefaction

Torrefaction is the low temperature (200–300 °C) partial decomposition of biomass in the absence of oxygen. See Tumuluru et al. [1] for an extensive review of biomass torrefaction. Its Latin root word *torrefacio* means *parch* but the transformation involves more than dehydration. Torrefaction converts biomass into a carbonaceous solid fuel, superficially resembling peat or charcoal but with important chemical differences. The resulting fuel, known as torrefied biomass, is typically used as a solid combustion fuel for heat or electricity generation.

Generating torrefied biomass involves slowly heating biomass, which drives out water and other volatile chemicals. These products primarily derive from hemicellulose, which completely decomposes during torrefaction. Lignin and cellulose degrade to a lesser extent because they are more stable than hemicellulose [1]. The reaction temperature most strongly dictates the extent of this degradation [8].

The products of thermal deconstruction of hemicellulose account for most of the mass and energy loss during torrefaction, including the reduction of a significant fraction of the hydrogen and oxygen present in the biomass. Partially removing these elements increases the higher heating value (HHV) of torrefied biomass (Figure 2.4) by transforming it into a more carbonaceous solid fuel (Figure 2.5).

Torrefaction represents relatively limited thermal deconstruction compared to other thermochemical processes as only hemicellulose completely decomposes. Cellulose and lignin decompose to a very limited extent compared to higher temperature processes.

2.2.1.2 Slow Pyrolysis

Slow pyrolysis is the low temperature (300–400 °C) thermal deconstruction of biomass under oxygen-starved conditions over the course of hours or even days to form biochar and relatively low molecular weight vapors and non-condensable gases. Derived from the Greek roots *pyro* and *lysis* meaning *fire* and *splitting* respectively, pyrolysis involves all components of the plant cell wall. Slow pyrolysis differs from fast pyrolysis only in heating rate and final reaction temperature, although sometimes research on slow pyrolysis is conducted

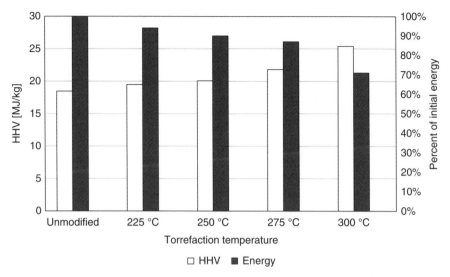

Figure 2.4 The higher heating value (HHV) of torrefied biomass (pine) increases with torrefaction temperature. Source: figure produced using data from Phanphanich and Mani [9].

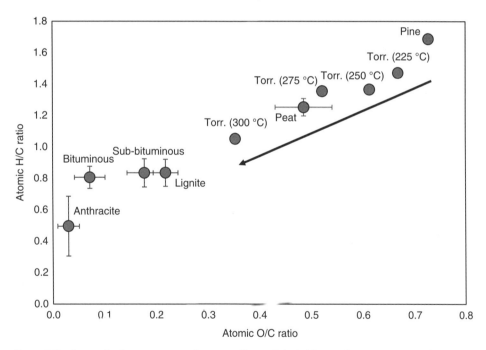

Figure 2.5 As torrefaction temperature increases, lignocellulosic biomass becomes more carbonaceous and similar to peat and coal, as illustrated in this van Krevelen diagram of pine torrefaction [9]. Peat averages and standard deviations were determined from 21 samples in the Phyllis2 database maintained by the Energy Research Centre of the Netherlands [10]. Coal averages and standard deviations were computed after classifying 6573 coal samples from the U.S. Geological Survey COALQUAL [11] database according to ASTM D388-18 [12].

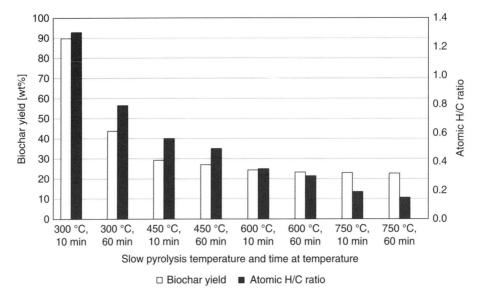

Figure 2.6 *Yield and atomic hydrogen to carbon (H/C) ratio for biochar produced during slow pyrolysis both decrease with increasing temperature and reaction time. Source: figure produced using data from Ronsse et al. [16].*

at higher temperatures. The principal product of slow pyrolysis is a solid carbonaceous residue that is commonly called charcoal or biochar.

Humans began producing charcoal from slow pyrolysis thousands of years ago [13]. Cave art is the most enduring and famous evidence of charcoal use, but charcoal also found use as a soil amendment [14] as well as the main component in gunpowder [15], among other uses. Nevertheless, historically and today, charcoal is primarily used as solid fuel.

Slow reaction rate favors high charcoal yields. The exact chemical mechanisms are not fully understood but low temperatures, gas ventilation rates, and solids residence times are known to increase char yield (Figure 2.6) [16, 17].

Heating rate obscures the relationship between temperature and reaction time (Figure 2.7) [18]. Two otherwise identical biomass samples heated at different rates to the same final temperature have distinct product yields as a result of the very different spatial and temporal temperature profiles experienced by the samples.

Regardless of heating profile, cellulose and lignin are depolymerized along with hemicellulose during slow pyrolysis. Slow pyrolysis deconstructs biomass to a considerably greater extent than does torrefaction, with charcoal containing a higher carbon content than torrefied biomass and on par with some coals.

2.2.2 Processes Yielding Chiefly Liquids

Processes that generate primarily liquid products – solvent liquefaction and fast pyrolysis – occupy a middle ground between lower temperatures promoting char formation and higher temperatures that promote cracking of vapors into permanent gases. Solvent liquefaction and fast pyrolysis both rapidly depolymerize and deconstruct biomass to

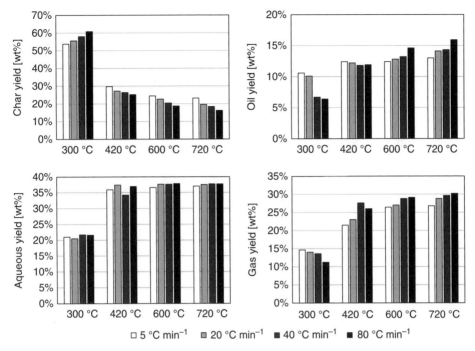

Figure 2.7 *Heating rate alters product yields differently at low and high temperatures. Despite this incongruous variable, temperature predicts yields better than heating rate. Source: figure produced using data from Williams and Besler [18].*

yield primarily liquid products of similar chemical composition. The intent of solvent liquefaction is to dissolve or otherwise disperse liquid products into the solvent medium for subsequent recovery as liquid known as bio-oil or biocrude. Fast pyrolysis attempts to vaporize liquid products into an inert gas stream as quickly as they are formed, which is subsequently condensed to a liquid known as bio-oil or pyrolysis liquid.

2.2.2.1 Solvent Liquefaction

Solvent liquefaction is the thermal deconstruction of biomass in a solvent at moderate temperatures (105–400 °C) and high pressure (2–20 MPa) [3]. For more information about solvent liquefaction, see Chapter 8. The process generates primarily solubilized products but also gases and solids. Solvent plays several roles, including transporting heat to the biomass, dissolving select components of the biomass, dispersing products (preventing their condensation to larger molecules), and changing the thermodynamic environment in favor of certain chemical reactions.

In solvent liquefaction, water, organic solvents, or non-aqueous inorganic solvents (such as concentrated mineral acids) are used as the reaction medium. These solvents perturb the thermodynamic properties of reactants and reaction intermediates, profoundly influencing the final products of thermal deconstruction [19]. In contrast, the gaseous environment of other thermochemical processes only influence heat and mass transport among reactants and products.

Depending on the solvent, intermolecular forces between solvent and biomass affect kinetic parameters but do not necessarily change the reaction mechanism [20, 21]. For example, Ghosh et al. [21] determined the apparent activation energies for cellulose depolymerization in γ-valerolactone, acetonitrile, and tetrahydrofuran, and showed that the activation energies were reduced up to 67% compared to fast pyrolysis of cellulose. In these polar aprotic solvents, product distributions were also strongly dependent on choice of solvent although the kinds of products remained the same [21].

Furthermore, many protic solvents chemically react with either biomass or its products to form products not otherwise expected from solvent liquefaction. Alcohols, for example, can alkylate solubilized carbohydrates [22]. Water, notably, can hydrolyze glycosidic bonds, forming smaller oligosaccharides or monosaccharides [23]. Shuai et al. [24] exploited this phenomena by stabilizing lignin deconstruction products via reactions with formaldehyde, significantly reducing secondary reactions of these products that form intractable carbon–carbon bonds.

The unique opportunity for solvent liquefaction is the potential of solvents to perturb the chemical kinetics of thermal deconstruction and improve product selectivity under mild reaction conditions. This distinct advantage compared to other thermochemical processes is somewhat countervailed by the expense of solvents and difficulties of solvent recovery.

2.2.2.2 Fast Pyrolysis

Fast pyrolysis is the moderately high temperature (400–750 °C) thermal deconstruction of organic matter in the absence of oxygen. For more information about fast pyrolysis, see Chapter 6. It produces mainly bio-oil – a viscous, acidic liquid – plus non-condensable gases and char. Depending on process conditions, fast pyrolysis can produce approximately 75 wt% bio-oil [2]. Bio-oil originates from the evaporation or thermal ejection of liquids formed during the thermal deconstruction of biomass polymers. These vapors and aerosols are subsequently condensed or otherwise separated from the non-condensable gas stream, thus exiting the pyrolysis reactor as liquid bio-oil. For more information about bio-oil upgrading, see Chapter 7.

As for all thermochemical processes, temperature is a major determinant of the yield of fast pyrolysis products, as shown in Figure 2.8 [25]. The existence of an optimal temperature for maximum bio-oil yield demonstrates the balance between char formation at lower temperatures and gas generation at higher temperatures.

Fast pyrolysis is fundamental to combustion and gasification, producing the vapors that are ultimately cracked and/or oxidized into flue gas and producer gas, respectively. Thus, an understanding of fast pyrolysis as a thermal deconstruction process provides insights into the physical and chemical mechanisms of combustion and gasification, as subsequently described.

2.2.3 Processes Yielding Chiefly Gases

Gasification and combustion generate primarily gaseous products. Biomass is aggressively deconstructed into vapors and gases followed by gas-phase cracking and oxidation reactions to form ideally only permanent gases. The product distributions of gasification and combustion are distinct. The theoretical products of gasification, determined by thermodynamic

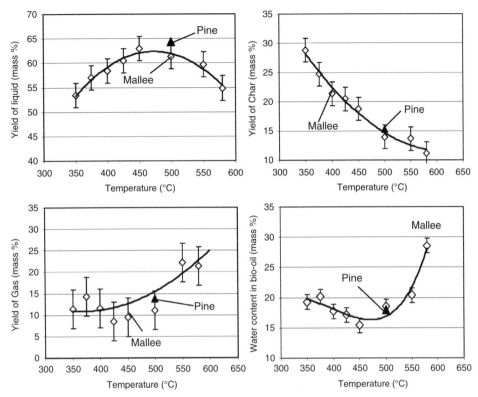

Figure 2.8 The effect of temperature on the yield of solids (char), liquids (bio-oil), gases, and water in the bio-oil from fast pyrolysis of mallee wood. Most notably the liquid yield decreases as secondary reactions begin to play a larger role at higher temperatures. Source: figure reproduced with permission from Reference [25].

equilibrium, are mixtures of carbon monoxide (CO), carbon dioxide (CO_2), hydrogen (H_2), methane (CH_4), and small amounts of low molecular weight alkanes and alkenes, with virtually no char or tar present. In practice, chemical equilibrium is difficult to attain and the products include tar and char, the amounts depending upon temperature and reaction time, as well as contaminants such as hydrogen sulfide and hydrogen cyanide. The product stream is sometimes called producer gas although more often referred to as syngas (an abbreviation of *synthesis gas*, indicating its use in catalytic synthesis of fuels and chemicals). The theoretical products of combustion are only CO_2 and water, although in practice they usually include small amounts of soot, tar, nitrogen oxides, and sulfur oxides.

Some gasifiers are indirectly heated, using heat exchangers or heat carriers to transport thermal energy into the gasifier. More commonly gasifiers are directly heated, admitting oxygen or air into the reactor where it reacts with biomass or pyrolysis products to provide thermal energy to deconstruct the biomass. Air or oxygen is always supplied to a combustor. Despite operation at different equivalence ratios, the devolatilization products of gasification and combustion are similar before being converted into final products by oxidation reactions (Figure 2.9).

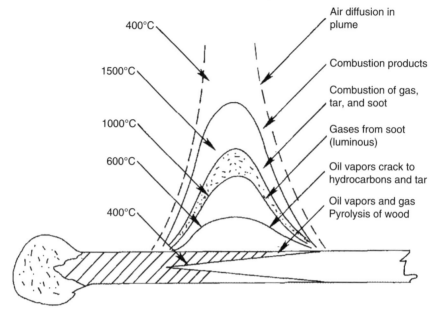

Figure 2.9 This depiction of a wooden match burning illustrates how biomass undergoes pyrolysis, gasification, and then combustion. Source: figure reproduced from the Solar Energy Research Institute report Handbook of Biomass Downdraft Gasifier Engine Systems [26].

2.2.3.1 Gasification

Gasification occurs at temperatures greater than 650 °C, sometimes in the presence of steam and/or oxygen, yielding primarily producer gas but also tar and char. For more information about gasification, see Chapter 4. Most biomass gasifiers operate in the range of 750–900 °C to prevent ash fouling although in a few instances entrained flow, slagging reactors have been developed, which operate at considerably higher temperatures [27, 28]. Devolatilization rapidly releases gases and liquids from the pyrolyzing biomass, which is followed by more gradual secondary reactions in the gas phase.

Gasification undergoes four major stages of thermal deconstruction: heating and drying, pyrolysis, gas-phase reactions, and gas–solid reactions. All thermal deconstruction processes include drying and pyrolysis, which release vapors and gases. The extent of gas-phase and solid–gas reactions are important in determining the ultimate gas composition leaving the gasifier.

The ability to accurately predict product distributions of producer gas, tar, and char from a gasifier is dependent as much on the condensed phase reactions of biomass as the secondary reactions of tar (condensable vapors), gas, and char. Devolatilization of biomass is very fast compared to the gas–solid reactions of char and the gas-phase reactions of producer gas and tars; thus the immediate products of biomass devolatilization can be considered as reactants for subsequent gas–solid and gas-phase reactions. Whereas most of the relevant gas–solid and gas-phase reaction kinetics are well known, the elementary reactions responsible for condensed phase reactions are poorly understood and usually represented by semi-empirical global reaction mechanisms.

2.2.3.2 Combustion

Combustion is the high temperature (typically greater than 900 °C) exothermic oxidation of fuel, producing flue gas. For more information about combustion, see Chapter 3. The earliest form of bioenergy, humans have used fires from wood combustion as a source of energy for hundreds of thousands of years [29]. Today, this thermal energy is used for a wide range of applications including process heat and electricity generation.

Combustion oxidizes organic compounds into carbon dioxide and water while leaving behind ash from the mineral content in the biomass. Combustion follows four main steps: heating and drying, pyrolysis, flaming combustion, and char combustion. Flaming combustion occurs in a thin flame front surrounding the fuel particle where volatiles diffusing away from the biomass and oxygen from the surrounding air reach a critical equivalence ratio (Figure 2.10). As long as volatiles are being expelled from the biomass, essentially no oxygen reaches the biomass surface. Once devolatilization is complete – usually on the order of a few seconds or less at typical combustion temperatures – oxygen is able to penetrate to the particle surface and commence the gas–solid reaction of char oxidation.

These last two oxidation steps are similar to the gas-phase and gas–solid reactions that occur during gasification, except that they occur at higher equivalence ratios and temperatures. Ideally, the products of combustion are carbon dioxide and water, and the chemical energy of the reactants has been wholly converted into high temperature thermal energy, which can be used for process heating, steam production, or electric power generation.

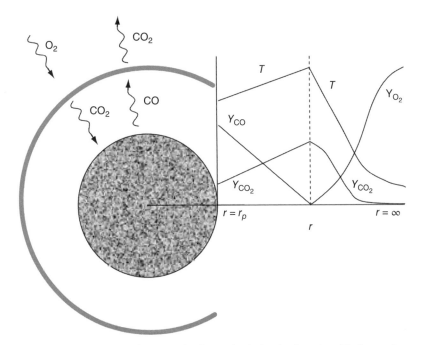

Figure 2.10 Biomass combustion begins with release of volatiles that burn in a thin flame where oxygen diffusing toward the biomass particle reaches a critical equivalence ratio. Source: figure reproduced from Reference [30] with permission from Elsevier.

2.3 Understanding Condensed Phase Reactions

The fact that researchers have broadly neglected the study of the condensed phase reactions that occur during thermal deconstruction reflects the difficulties in directly probing them; however, understanding condensed phase reactions is vital to improving the performance of thermochemical processes. As one example, fast pyrolysis of biomass is well known to produce anhydro-monosaccharides but at much lower yields than expected based on experiments with pure cellulose and hemicellulose. This discrepancy was traced to condensed phase reactions in the pyrolyzing biomass catalyzed by naturally occurring alkali and alkaline earth metals (AAEM) that fragmented pyranose rings. Understanding this cause led to the development of a biomass pretreatment that passivated AAEM, leading to higher sugar yields [31, 32]. (These reactions are discussed further at the end of Section 2.3.3.2.5 "Effects of alkali and alkaline earth metals".) In fact, there has been recent progress in modeling elementary reactions of complex thermal deconstruction reactions [33–37], an important advance over the use of global reaction mechanisms [38–40] to describe these complex processes. This progress should encourage future experimental studies of condensed phase reactions to validate these models.

2.3.1 Challenges in Investigating Condensed Phase Reactions

Much of the research on thermal deconstruction of biomass has focused on volatile products since these are much easier to access and analyze than the intermediate (and sometimes transient) products that form in the condensed phase. In general, biopolymers are not readily dissolved or volatilized, but this is the basis of many analytical techniques such as by liquid or gas chromatography. Progress in evaluating the thermal deconstruction of biopolymers requires approaches in which large oligomeric products can be detected.

Capturing intermediate products can present significant challenges. Low volatility and short reaction times paired with high temperatures make interrogating condensed phase products difficult. Some experiments have used entrained flow reactors [41, 42] or concentrated radiation [43–45] to heat samples rapidly. Both kinds of apparatus are able to rapidly terminate heating by removing the samples from the heated zone or turning off the radiation source although this does not necessarily quench thermal deconstruction reactions. It is very likely that reactions continue as the intermediate products are slowly cooled. Few fast pyrolysis experiments have been able to both rapidly heat biopolymer samples and rapidly cool condensed phase products, whether studying cellulose [46–48] or lignocellulose [47, 49, 50].

Although much of the research on biomass thermal deconstruction has actually employed extracted polysaccharides or lignin, thermal deconstruction of the ex situ polymer may depart significantly from the in situ polymer because interactions between the biomass components are lost. Tiarks [51] has explicitly demonstrated these differences in experiments with a filament pyrolyzer enclosed within an optically accessible chamber. A powdered sample of technical lignin obtained from enzymatic hydrolysis of corn stover was observed to melt, coalesce into a hemispherical shape, and lose mass through both devolatilization and liquid droplet ejection. However, it was clear that droplet ejection was an artifact of the experimental arrangement, arising from the coalescence of individual lignin particles into a single liquid mass and non-uniform heating of the sample. Rather than forming at the free surface between melted lignin and gas atmosphere, volatiles were generated at the interface of the hot filament and the bottom of the melt. These vapors could only escape by flowing upward through the melt as bubbles, exploding at the free surface, and ejecting liquid droplets into the gas flow. In contrast, practical reactors achieve more uniform heating of particles, releasing volatiles directly to the surrounding gas. Furthermore, lignin polymers within actual lignocellulosic biomass are dispersed among the cellulose microfibrils, which is expected to prevent their surface tension-driven coalescence. In fact, although Tiarks [51] observed some evidence of lignin migrating to the surface of biomass fibers during pyrolysis, it was not sufficient to cause coalescence as observed for extracted lignin. When extracted lignin was mixed with a fumed silica matrix, the lignin was sufficiently dispersed to prevent melted particles from agglomerating during pyrolysis. As shown in Figure 2.11, the resulting film of melted lignin showed only minor ejection phenomena, with most of the mass loss due to vaporization from the melt phase. Tiarks [51] also observed little thermal ejection when pyrolyzing cellulose that was dispersed as small particles along the filament heater of the pyrolyzer, arguing that previous studies reporting this phenomenon suffered from non-uniform heating of the sample [52, 53].

Even if extracted polymers behaved like their in situ counterparts [54], most reactors and chemical instruments are ill-suited to examine condensed phase reactions. For instance,

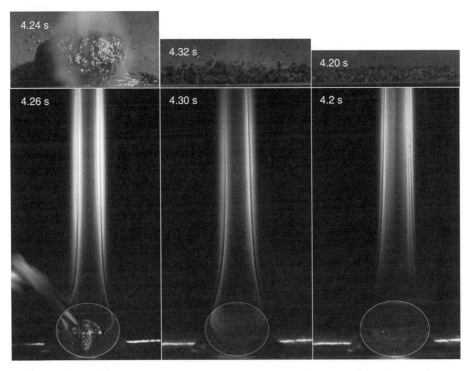

Figure 2.11 Comparing pyrolysis of extracted lignin (left), extracted lignin mixed with silica matrix (center), and red oak (right) demonstrates how thermal ejection is an artifact of particle coalescence and non-uniform heating from below the sample. Source: figure adapted with permission from Tiarks [51]. See the plate section for a color representation of this figure.

extracted cellulose is well known to progress through a liquid intermediate during fast pyrolysis. (See Lédé's review for a thorough account [55].) These liquid intermediate products have been identified as anhydro-oligosaccharides [56]. Their maximum degree of polymerization (DP) is typically measured to be around seven [41, 57]; however, Lindstrom et al. [48] recently identified anhydro-oligosaccharides up to DP 60 from partially pyrolyzed cellulose. More importantly, this work concluded that larger oligomers probably exist but have evaded detection by conventional water-based analyses because they are insoluble in water. This subtle instrumentation issue resulted in theories and models biased toward water-soluble anhydro-oligosaccharides, which is not the full picture.

Present methods for the analysis of biopolymers or their oligomers are relatively unwieldy. To probe specific reactions within biopolymers, model compounds with specific moieties or bonds are often chosen to represent particular areas of interest. Model compound experiments often focus on yields or types of products [58, 59], but more advanced methods, such as isotope labeling [60–63], are needed for more detailed insights. However, model compound experiments can be confounded by heat and mass transfer limitations, among other problems, so researchers often resort to the well-controlled conditions offered by computational chemistry to provide insights into polymer decomposition.

Computational chemistry and its many approximations are discussed in Section 2.3.3.1, but a few assumptions are worth noting here. These calculations are often performed as gas-phase reactions with the results assumed to hold true for condensed phase reactions. Furthermore, there is significant uncertainty in computed values of kinetic parameters, depending on the nature of the computations. Depending on the level of theory used, these constants can vary significantly. For example, the commonly used Becke-three parameter-Lee-Yang-Parr (B3LYP) functional is thought to predict pre-exponential factors [33, 34] within an order of magnitude and activation energies [64] within $\pm 4.8\,\text{kcal mol}^{-1}$. Taken together, errors in rate coefficients can be orders of magnitude. Lastly, computational power limitations bias modeling toward smaller molecules.

Biomass thermal deconstruction is not simply the depolymerization of the three main biopolymers in lignocellulosic biomass. It also entails the deconstruction of the lignocellulosic matrix, a structure involving the interaction of cellulose, hemicellulose, and lignin to produce a composite material with its own unique physical properties. Too often, experiments and modeling overlook these interactions, which likely has hindered progress in understanding thermal deconstruction of biomass.

2.3.2 The Role of Cell Wall Structure in Thermal Deconstruction

Cell wall structure plays an important role in the thermal deconstruction of biomass, determining the rate at which heat is conducted into the composite structure of lignocellulosic biomass and the rate at which mass is transported out of the disintegrating plant material. Thus, two key areas discussed in this section are heat transfer and mass transfer during thermal deconstruction, and the structural breakdown of the plant cell wall.

Lignocellulosic biomass has unusual heat transfer characteristics for a solid material. Conventional heat transfer calculations, mainly the Biot number, demonstrate that the relatively low thermal conductivity of biomass creates thermal gradients in particles (Figure 2.12). However, detailed studies indicate this simple calculation significantly underestimates the actual thermal gradients that occur [67–70].

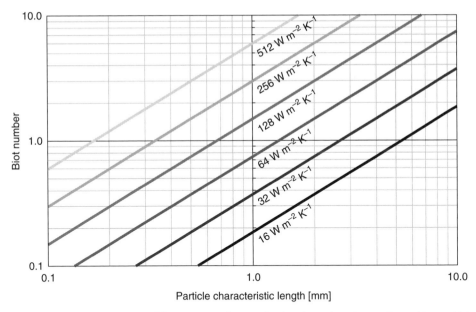

Figure 2.12 Even relatively small biomass particles can develop thermal gradients when subject to high heat transfer. Produced with assistance from Chad A. Peterson using the thermal conductivity of pine [65, 66].

Within biomass particles, asymmetry in the microstructure of biomass dramatically influences heat transfer. Biomass has the highest thermal conductivity through its cell walls axial to their lumina [71] (the interiors of the cells); however, lumina, as well as larger tubes such as xylem, inhibit radial thermal conduction. To examine intraparticle heat and mass transfer more accurately, Ciesielski et al. [68] modeled fast pyrolysis of pine and aspen particles with realistic particle, cell wall, and cell lumina dimensions, as measured by multiple microscopy and spectroscopy methods. Accounting for the lumina predicts significantly slower heat transfer, particularly into the cores of particles (Figure 2.13).

Different biomass species can exaggerate this effect. Using the same realistic biomass models, Pecha et al. [70] calculated aspen to have a heat transfer coefficient roughly 20% greater than pine in a laminar flow regime because of their differing microstructures. Understanding these heat transfer effects is essential for realistic modeling. Thermal gradients can alter product yields because reaction rates increase exponentially with temperature [72, 73]. Large particles, in particular, are more susceptible to these gradients – as described by the Biot number, which is directly related to particle characteristic length. There are limits to reducing particle size in industrial practice, however, because comminution costs grow exponentially with decreasing particle size [74].

Other factors have also received limited attention but deserve more thorough investigation. As biomass is heated, moisture is driven out of the particles, and the high specific heat of water and high enthalpy of water vaporization likely exaggerates thermal gradients. Similarly, volatile product formation and vaporization may slow heat transfer, especially as most condensed phase thermal deconstruction reactions are endothermic.

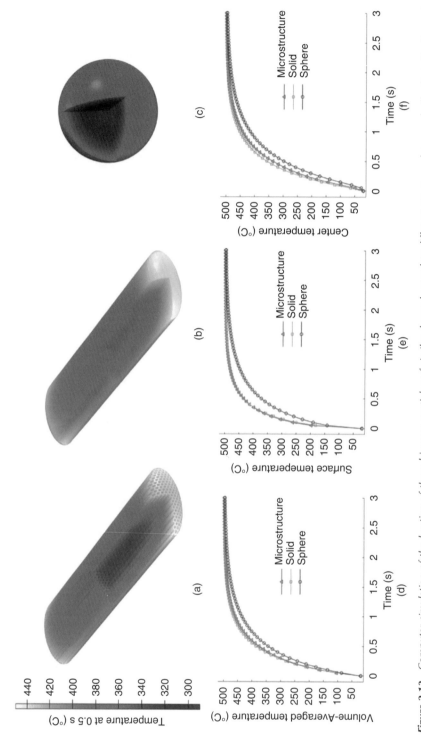

Figure 2.13 Computer simulation of the heating of three biomass particles of similar thermal mass but different structure show markedly dissimilar thermal gradients. Realistic morphology (a) develops larger temperature gradients than particles with the same proportions but without cell lumina (b). Spherical particles (c), often used for modeling simplicity, have dramatic thermal gradients that do not accurately represent biomass. Source: figure reproduced with permission from Reference [68]. See the plate section for a color representation of this figure.

The diffusion time for volatiles released from biomass particles likely plays a strong role in secondary reactions, which often are responsible for molecules decomposing into non-condensable gases, thus reducing yields of liquid products. While the structure of cell lumina can impede radial heat transfer through particles, it assists mass transfer of volatile products out [68], so further analysis of this transport phenomena may contribute toward improving liquid yields.

Inclusion of cell morphology in the analysis of transport phenomena is insufficient if it is static. The complex cell wall microstructure breaks down and changes during thermal deconstruction [42]. Haas, Nimlos, and Donohoe saw significant morphological changes when they heated small, thin sections of poplar at approximately $2.5\,\mathrm{K\,s^{-1}}$ under a light microscope with an inert atmosphere [75]. The poplar expanded only in the radial direction (Figure 2.14). (Videos of this expansion are available in the supplemental material of Reference [75].) Cell walls swelled, grew taut, partially converted into volatile products, and then contracted as these products escaped. The final configuration had about 10–15% more lumina area compared to unmodified cell walls due to this expansion and contraction, but more so from the volatile product mass transfer. These changes likely slow heat transfer and increase volatile product mass transfer. Clearly, cell wall structure is important to both the rate and final products of thermal deconstruction of biomass.

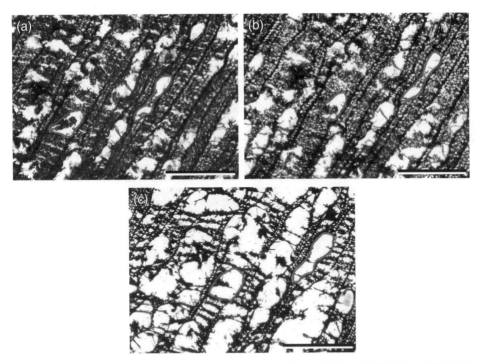

Figure 2.14 Light microscope images of a cross section of poplar at 26 °C (a), 299 °C (b), and 501 °C (c) show how cell walls expand when heated before contracting as volatile products leave. The red outlines surround xylem. Scale bars equal 1 mm. Source: figure reproduced with permission from Reference [75]. See the plate section for a color representation of this figure.

2.3.3 Use of Computational Chemistry to Understand Thermal Deconstruction

Computational chemistry has increasingly helped overcome the dearth of experiments on condensed phase reactions. Starting in the mid-2000s, researchers have used computational chemistry in combination with experimental measurements to examine condensed phase reactions. The rise of theoretical methods has allowed researchers to better understand how species evolve during the thermal deconstruction of various kinds of biomass components. Furthermore, the results of theoretical investigations are helping guide experimentalists to tailor reaction conditions, with the aim of achieving greater control over product distribution.

Experiments have revealed hundreds of compounds produced from thermal deconstruction of biomass. A key advantage of computational methodologies is that they allow rapid simulation of a wide range of experimental parameters which would take significantly longer to test in a laboratory. In this way, computational chemistry can be of great assistance to the experimentalist, both as a guide to potential research avenues and as a time-saving tool.

The following portion of text serves as a primer on computational chemistry methods. This overview is followed with a description of key condensed phase reactions that occur during biomass thermal deconstruction, and concludes with a brief discussion of current deficiencies in these methods and suggestions for future computational studies.

2.3.3.1 Computational Methodology

Computational investigations of biomass deconstruction are used to estimate how the energy of a system of reacting molecules evolves over the course of the reaction. For instance, the difference in energy between the pyranose and furanose forms of glucose indicates which structure is more stable. Calculating the energy of a lignin model compound and its decomposition products can determine whether the reaction proceeds endothermically or exothermically. The prediction of reaction mechanisms and their associated enthalpies is a common undertaking and allows for researchers to understand the reasons why certain products are favored under different experimental conditions.

The Hartree-Fock (HF) method, an early computational approach to describing molecular systems, was never widely applied toward the study of biomass deconstruction. HF does not account explicitly for the Coulombic repulsion of individual electrons, termed the electron correlation, and instead works only with an averaged repulsion. Owing to this fact, the energy obtained using the HF method converges to a value that is always above the true ground state energy, at a point known as the HF limit. Due to this overestimation, improvements dubbed "post-HF" were introduced to include electron correlation. One such example shown to be highly accurate is the coupled cluster method. This method is useful in biomass deconstruction investigations in areas such as modeling the decomposition of glucose and fructose molecules [76].

An alternative to HF-based methods is density functional theory (DFT), which has seen widespread use in the field of biomass conversion. In DFT, the energy of the system is calculated as a function of electronic density. For purposes of this description, it is sufficient to note that DFT makes approximations to account for electronic quantum mechanics, namely the Born-Oppenheimer approximation. This has resulted in a hierarchy of methods ranging

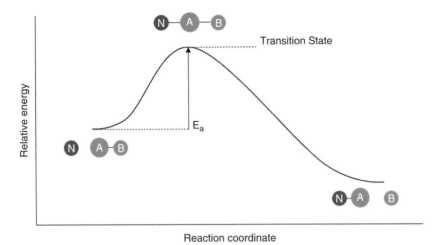

Figure 2.15 This reaction coordinate diagram demonstrates a generic substitution reaction, highlighting the position of the transition state.

in complexity from a local density approximation (LDA) to the double hybrid methods. Some of the most popular methods fall into the hybrid and meta-hybrid categories. One that has seen significant application for biomass deconstruction studies over the course of the last decade is the B3LYP functional [77–80]. The speed of the B3LYP functional and its prediction of accurate geometries has led to its use in modeling a range of phenomena including degradation of cellulose, hemicellulose, and lignin [81–83]. More recently, the Minnesota family of functionals from Zhao and Truhlar [84, 85] has gained popularity in the biomass field and is regularly encountered in the literature, while use of the aforementioned B3LYP method is increasingly less common.

DFT is used to calculate the energy of reactants and products for a reaction. Transition state theory (TST) is used to understand the mechanism by which a reaction occurs, and traditional TST posits that a reaction proceeds from reactant to product via a transition state complex. Using TST, a complicated mechanism can be divided into a number of elementary reactions, with each step proceeding through a transition state complex. Figure 2.15 illustrates the energy change for a hypothetical substitution reaction.

As this simplified example shows, computational chemistry has the potential to provide insights not possible through experimentation techniques; however, despite accounting for quantum details, the information and conclusions derived are imperfect. The reaction intermediates, mechanisms, and slew of products predicted by computational chemistry seem definite, but like all computer models, the accuracy of results depend on the assumptions about mechanisms and boundary conditions.

2.3.3.2 Formation of Liquid Products

Thermal deconstruction of solid biomass can produce solids, liquids, vapors, and gases as primary products. The elevated temperatures of thermal deconstruction quickly vaporize most liquid products although there is evidence that a small amount may also escape as aerosol [51, 52, 86]. In the case of fast pyrolysis, the desire is to quickly sweep these vapors from the reactor and cool them into liquid products. At sufficiently high temperatures and long residence times, these vapors crack or dehydrate by secondary reactions into gases and char, although this is not the focus of this section.

2.3.3.2.1 Cellulose Conversion Reactions

Levoglucosan, as the major product of the thermal deconstruction of cellulose, has attracted substantial attention in computational studies of biomass pyrolysis. Experimental studies have shown fast pyrolysis of pure cellulose produces approximately 60 wt% levoglucosan under optimal conditions [87], although the presence of AAEM in lignocellulosic biomass catalyzes pyranose ring fragmentation which can dramatically reduce this yield [88].

Cellulose thermal deconstruction experiments, however, like others using isolated biopolymers, are fraught with confounding variables. The cellulose source and isolation method impacts the material significantly. Drying the cellulose irreparably alters the structure [89], and the resulting cellulose crystalline allomorphs, as well as the degree of crystallinity, may influence experimental outcomes [90, 91]. However, attempting to model reactions of cellulose consisting of many thousands of connected glucose monomers would be impractical. Instead, early computational researchers supposed that glucopyranose, the monomeric unit of cellulose, was a suitable model compound for their computational studies. In particular, several groups investigated how levoglucosan could arise from glucopyranose through a dehydration and ring-forming mechanism based on a single transition state [81, 92–94].

The problem with modeling the formation of levoglucosan from glucopyranose is the assumption that other monomers in the polysaccharide chain do not influence the process. In fact, depolymerization of cellulose to form levoglucosan entails breaking glycosidic bonds between adjacent pyranose rings, so these bonds should be included in simulations of thermal deconstruction. An example is the study of Zhang et al. [95], who determined energy requirements for homolytic and heterolytic breakage of the glycosidic bond in cellobiose, a glucose dimer. Although the homolytic route was found to be significantly more favorable in terms of bond dissociation energy, the energy requirement of 79 kcal mol^{-1} was still large, suggesting this may not be the most likely route for levoglucosan formation.

An alternative route to cellulose depolymerization proposed by Mayes and Broadbelt employed a low-energy concerted mechanism [96], illustrated in Figure 2.16. The glycosidic bonds between pyranose units in the cellulose chain are cleaved to produce anhydro-oligosaccharides, as has been observed experimentally [41, 48, 57]. Subsequent cleavage at the terminal glycosidic linkage of an anhydro-oligosaccharide yields levoglucosan. This route provides the lowest-energy mechanism for levoglucosan formation from cellulose. Zhang et al. [97] compared these three possible levoglucosan formation mechanisms in a dry gaseous atmosphere and verified that homolytic cleavage has a substantially higher energy barrier than their proposed levoglucosan chain-end mechanism.

Cellulose, and its subsequent anhydro-oligosaccharides, break into increasingly small anhydro-oligosaccharides. Because levoglucosan forms from terminal monomers, the relative abundance of these monomers controls the rate of levoglucosan formation. As a result, cellulose produces very little levoglucosan before it has substantially depolymerized

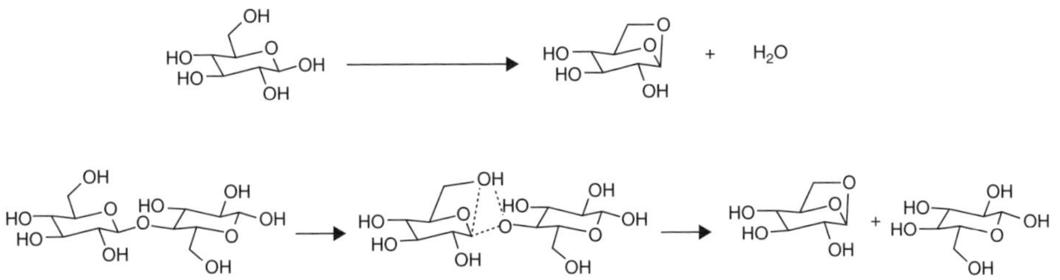

Figure 2.16 Top: Formation of levoglucosan from glucopyranose in a single step reaction. Bottom: Simplified representation of Mayes and Broadbelt's chain-end mechanism for levoglucosan formation [96].

[35, 48]. Levoglucosan and many less valuable, lower molecular weight products primarily derive from low DP anhydro-oligosaccharides.

After levoglucosan is formed, it can decompose into lower molecular weight molecules, especially as reaction temperature increases [98]. Such secondary reactions may help explain differences observed experimentally in liquid product yields. Zhang et al. [99] evaluated the decomposition pathways for levoglucosan and concluded that dehydration has a lower energy barrier compared to C–C or C–O bond breaking, suggesting it dominates secondary reactions of levoglucosan decomposition.

Assary and Curtiss suggested that levoglucosenone (LGO) can be formed from the double dehydration of levoglucosan; however, the thermal stability of levoglucosan and the high activation energy required for the reaction discourage this as a feasible route [92]. Lu et al. suggest an alternative mechanism in which LGO forms from cellulose through a chain-end mechanism similar to that for levoglucosan formation [100]. This hypothesis is supported by experimental work in which pyrolysis of cellobiose produced more LGO than did pyrolysis of glucose [87].

Hydroxymethyl furfural (HMF) is another important product of biomass thermal deconstruction. Based on its chemical formula, HMF requires two dehydration reactions to derive from glucopyranose. A concerted mechanism of formation from cellulose has been considered, but the high energy requirements render this route unfeasible [101]. Correspondingly, two formation mechanisms from glucopyranose have been proposed, both initiated through an opening of the pyran ring and subsequent dehydrative cyclization to yield the furanose form of the sugar [81, 93]. Between these two proposed mechanisms, the order of the elementary steps differ but the energy barriers are very similar.

Furthermore, experimental work has shown that pyrolysis of fructose produced greater amounts of HMF than did glucose, indicating that glucose–fructose tautomerization proposed by previous work may be more likely. Mayes et al. [102] compared multiple routes for HMF formation in a comprehensive study and found that tautomerization prior to dehydration was the most energetically favorable pathway.

While it is reasonable to expect furfural (FF) to form following removal of the hydroxymethyl group from HMF, the energy barrier for this reaction is very high [103]. Alternatively, it has been suggested that the formation of FF and HMF are competitive with one another and that a Grob fragmentation could be a significant low-energy step for removal of the hydroxymethyl group [104].

The multitude of reactions forming these small molecules are well suited for computational chemistry; however, they still require firm experimental evidence. For example, thermohydrolysis, a depolymerization reaction that generates a glucose molecule, has been proposed [33, 35]. By this theory, the glucose produced via thermohydrolysis degrades into many lower molecular weight products. This reaction has not been experimentally validated for cellulose, even though it can predict final product yields very accurately [33, 35]. Experiments examining the solid and liquid-phase intermediate products during cellulose pyrolysis have not found anywhere close to the amount of glucose these models predict [48, 105, 106]. Reactions, such as thermohydrolysis, should be experimentally validated if possible.

2.3.3.2.2 Hemicellulose Conversion Reactions

Thermal deconstruction of hemicellulose typically leads to a variety of small, oxygenated products with relatively low sugar and anhydrosugar yields, such as xylose and anhydroxylopyranose [107]. Studies on the deconstruction of hemicellulose are less numerous than cellulose, most likely owing to the greater variability in products and monosaccharides within the polymer. Experimental studies are typically performed on extracted hemicelluloses [107–109], but computational chemistry is perhaps better suited [36]. Hemicellulose, unlike cellulose, is a branched heteropolymer composed of pentoses, hexoses, hexuronic acids, and acetyl groups [110]. A standard structure or composition cannot be defined as it differs among biomass species [110], stages of plant growth [111, 112], and even cell wall layers [113]. Nevertheless, the prevalence of xylose in hemicelluloses – and not in cellulose – means that it is often chosen as a model compound for investigation.

The formation of FF from xylose appears to be more straightforward than its formation from glucose. Wang et al. [114] suggest the process involves a pyranose ring opening, removing two hydroxy groups, and then forming a five-membered furan ring. The energy barriers in this conversion can be greatly reduced by explicitly including a water molecule in the computations [114]. Just as Seshadri and Westmoreland [94] found that hydroxyl groups promote glucose deconstruction, xylose conversion to FF may be assisted by water as well as other small, oxygenated chemicals surrounding hemicellulose. Similar routes have been proposed for other monosaccharides found in hemicelluloses, such as arabinose [82].

By molecular weight, FF is one of the larger products produced from hemicellulose deconstruction, but many small-chain carboxylic acid, ketone, and aldehyde-type species are also derived through degradation of hemicellulose saccharides [82, 115–117]. Hemicellulose thermally degrades more easily than cellulose and lignin [1], which explains the pervasiveness of these small chemicals in the liquid fractions of thermochemical processing methods. Likely in part due to relative instability and low economic value of its products, hemicellulose has not received as much study as cellulose or lignin, although it has recently garnered more attention [36, 109].

2.3.3.2.3 Lignin Conversion Reactions

Lignin thermal deconstruction generally produces large amounts of phenolic monomers and oligomers, as well as non-condensable gases and char [118]. Studying these lignin deconstruction reactions is perhaps more difficult than for cellulose or hemicellulose because of the large number of potential products. Like the other main biopolymers, extracting lignin from biomass can significantly modify the structure [119–121]. Lignin structure varies dramatically among plant species [122] and even within the same plant at different times during its growth [123, 124]. Unlike cellulose or hemicellulose, lignin is highly cross-linked with an apparently random structure, leading to its well-known recalcitrance [122, 125].

To avoid these complications, researchers often employ model compounds to study lignin deconstruction. Models of lignin deconstruction are rarely based on elementary reaction mechanisms, instead employing simple lumped parameter models. With lignin's complexity, experiments typically only provide product yields from isolated lignin [118, 126] or changes in bond prevalence before and after heating [127], not specific reaction pathways or mechanisms. Model compounds are necessary for more detailed reaction and kinetic study toward the eventual goal of a comprehensive model [37].

Phenethyl phenyl (PPE) is frequently employed as a model compound to investigate fragmentation of β-O-4 ether linkages, one of the most prevalent bonds in lignin. The relatively simple structure of PPE is convenient for computational studies while accounting for the influence of aromatic groups on important structural effects in lignin deconstruction. More comprehensive information on lignin deconstruction reactions can be found in Kawamoto [128].

Homolytic reactions are thought to dominate lignin deconstruction, but concerted mechanisms may also play a role. Jarvis et al. [129] found strong experimental evidence for concerted mechanisms at temperatures below 1000 °C but at higher temperatures homolytic pathways appear to be more thermodynamically favorable. Huang et al. [83] narrowed the likely concerted reactions to retro-ene fragmentation [130] and Maccoll elimination [131]. Elder and Beste [132] further analyzed these reactions with computational chemistry, determining that retro-ene fragmentation presents the most favorable low-temperature pathway. Many other lignin model compounds are also likely to degrade via concerted reactions at low temperatures and homolysis at higher temperatures [133], although there are exceptions. For example, 1-(4-methoxyphenyl)-2-(2-methoxyphenoxy) ethanol, a synthesized lignin dimer model compound, likely reverses this temperature trend [134].

These findings highlight the complexity of condensed phase reactions during lignin thermal deconstruction. Small changes in the substituents attached to aromatic rings within lignin can greatly alter the degradation reaction rates and even the type of decomposition reactions. Owing to the diversity of aromatic substituent groups in lignin, these reactions warrant further investigation.

2.3.3.2.4   Evolution of Water

The production of water during biomass deconstruction impacts the properties of the products, frequently necessitating its removal prior to utilizing the products in specific applications (for instance, if they are to be used as liquid fuels). The concentration of water in liquid products is often very high (typically 20 wt%) [25], so it is important to understand how biomass deconstruction reactions generate water. Many works show water arising from the removal of hydroxyl groups from cellulose and hemicellulose monomers [93, 102, 103, 115, 117], suggesting that the presence of water in deconstruction products is most likely unavoidable.

2.3.3.2.5 Effects of Alkali and Alkaline Earth Metals

Biomass contains notable amounts of AAEM, with potassium and calcium being noteworthy elements. The presence of AAEM can have a catalytic effect on deconstruction reactions [32, 88]. Notably, the yield of sugars decreases with increasing presence of AAEM because other degradation reactions dominate. In a comprehensive study, Mayes et al. [135] examined the effect of sodium cations on a wide range of reactions that occur during glucose pyrolysis. Their findings suggested that rate coefficients were increased considerably for a majority of the reaction pathways, though not all [135]. Importantly, the presence of a sodium ion did not alter the mechanisms of the various reaction pathways, leading to generation of the same products but in different concentrations. These computational findings agree with their experimental results. In another work, potassium was modeled for comparison to sodium and similar catalytic trends were observed for both ions [136].

Notably, this trend can be reversed by passivating the AAEM with stoichiometric quantities of mineral acids [32, 88]. Most likely, thermally stable salts form between the conjugate bases and AAEM, preventing their catalytic action.

2.3.3.3 Formation of Gaseous Products

Compared to the hundreds of chemicals in the liquid products of thermochemical processing, gas composition is fairly simple [18, 137–139]. It has been proposed that CO and CO_2 arise through decarbonylation and decarboxylation reactions, respectively [140]. CO can be formed from lignin through decarbonylation of benzaldehyde-type moieties, while decarboxylation of benzoic acid derivatives leads to CO_2 formation. The formation of CO has a significantly higher energy barrier than for the formation of CO_2. This is in agreement with pyrolysis experiments, where CO_2 is formed in greater quantities than CO [18, 34, 137–139].

Methane is thought to evolve from homolytic cleavage of methyl groups from the ends of methoxy functionalities in lignin, as shown in Figure 2.17 [141]. The resulting methyl radical is then converted to methane by combining with an unbound hydrogen atom. This route, while feasible, is highly speculatively and requires significant energy to overcome the homolysis barrier. An alternative route requires the methoxy group be located ortho to a hydroxy group, enabling a concerted reaction to occur in which the methyl group sequesters the neighboring phenolic hydrogen, liberating methane and leaving two adjacent carbonyl type groups attached to the ring. This route is considerably more energetically facile than the homolytic route.

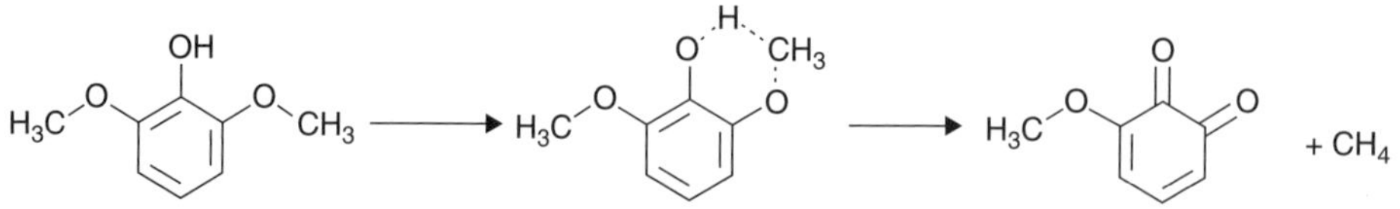

Figure 2.17 One potential route for methane formation from lignin is homolytic cleavage of methyl groups from the ends of methoxy functionalities as illustrated here for the model lignin compound syringol [141].

2.3.3.4 Char Formation

Theoretical investigations into the formation of char and carbonaceous materials during thermal deconstruction of biomass are limited. Many theories have been proposed as to why these compounds form, but studies do not provide a direct link to mechanisms of polycyclic aromatic species formation. In high-temperature thermal deconstruction processes, such as fast pyrolysis and gasification, char is often an unwanted side product formed from the dehydration of more desirable product molecules. A greater understanding of char formation mechanisms is essential to reducing its formation. Once understood, these mechanisms can be used to help tailor thermal decomposition to minimize char production and increase the output of more valuable organic compounds.

2.3.3.5 Computational Studies of Large-Scale Polymers

Over the past decade, the application of computational methods to study thermal deconstruction of biomass has expanded greatly. Ever advancing computational power has led to an increase in the complexity of model systems, and, as a result, a greater range of mechanisms may be investigated to account for the experimentally observed products.

In some instances, computational work is expanding beyond small model compounds and investigating large-scale polymers. These studies are beginning to bridge the gap between chemistry and materials science – a vital step [142–146]. A continuation of this trend may lead to detailed research on larger polymeric species of lignin, or perhaps facilitate the modeling of intermolecular interactions between polymer chains to more accurately capture the chemistry of decomposition reactions.

2.4 Conclusions

All thermochemical processes are closely related. Often distinctions among them are the result of imposed definitions, not the reactions themselves. Examining condensed phase biomass thermal deconstruction reactions shows that the processes – with the exception of solvent liquefaction – merely occupy different positions on a continuum of reaction times and temperatures. Despite varying temperatures, heating rates, and reaction times, the different methodologies for thermochemical processing of biomass are fundamentally similar.

This relationship can be exploited to probe difficult to analyze reactions. For example, the relative strength of bonds can be determined with slow heating rate and low temperature processes, such as torrefaction and slow pyrolysis. These conclusions should be verified at higher heating rates and temperatures but they provide a sound basis for further work. Additionally, volatile products from fast pyrolysis could be used as a starting point for gasification and combustion studies.

This exploitation is possible because the similarities are more than parallels; they are often the same reactions. On the other hand, overreliance on these comparisons invites error. Relative reaction rates change with temperature, and these differences can grow as intermediate product concentrations and subsequent condensed phase reaction rates shift. Within reason, the condensed phase reactions during thermal deconstruction can act as powerful tools for suggesting answers to research questions or point to new avenues of study.

References

1. Tumuluru, J.S., Sokhansanj, S., Hess, J.R. et al. (2011). A review on biomass torrefaction process and product properties for energy applications. *Industrial Biotechnology* **7** (5): 384–401.
2. Bridgwater, A.V. (2012). Review of fast pyrolysis of biomass and product upgrading. *Biomass and Bioenergy* **38**: 68–94.
3. Haverly, M.R. (2016). *An Experimental Study on Solvent Liquefaction*. Iowa State University.
4. Kenney, K.L., Smith, W.A., Gresham, G.L., and Westover, T.L. (2013). Understanding biomass feedstock variability. *Biofuels* **4** (1): 111–127.
5. Oinonen, P., Zhang, L., Lawoko, M., and Henriksson, G. (2015). On the formation of lignin polysaccharide networks in Norway spruce. *Phytochemistry* **111**: 177–184.
6. Harris, P.J. and Stone, B.A. (2008). Chemistry and molecular organization of plant cell walls. In: *Biomass Recalcitrance: Deconstructing the Plant Cell Wall for Bioenergy* (ed. M.E. Himmel), 61–93. Blackwell Publishing.
7. Henriksson, G. (2009). Lignin. In: *Wood Chemistry and Wood Biotechnology* (ed. M. Ek, G. Gellerstedt and G. Henriksson), 121–145. Berlin, New York: Walter de Gruyter.
8. Tumuluru, J. S., Kremer, T., Wright, C. T., and Boardman, R. D. Proximate and Ultimate Compositional Changes in Corn Stover during Torrrefaction Using Thermogravimetric Analyzer and Microwaves. American Society of Agricultural and Biological Engineers Annual Meeting, Dallas, Texas. 2012, doi:. https://www.osti.gov/servlets/purl/1056026.
9. Phanphanich, M. and Mani, S. (2011). Impact of torrefaction on the grindability and fuel characteristics of forest biomass. *Bioresource Technology* **102** (2): 1246–1253.
10. Phyllis2, *Database for Biomass and Waste*. https://phyllis.nl (accessed 8 January 2019).
11. Palmer, C. A., Oman, C. L., Park, A. J., and Luppens, J. A. 2015. The U.S. Geological Survey Coal Quality (COALQUAL) Database Version 3.0. U.S. Geological Survey Data Series 975. http://dx.doi.org/10.3133/ds975
12. ASTM (2018). *Standard D388, 2018, "Standard Classification of Coals by Rank"*. West Conshohocken, PA: ASTM International.
13. Antal, M.J. and Grønli, M. (2003). The art, science, and technology of charcoal production. *Industrial and Engineering Chemistry Research* **42** (8): 1619–1640.
14. Denevan, W.M. (1996). A bluff model of riverine settlement in prehistoric Amazonia. *Annals of the Association of American Geographers* **86** (4): 654–681.
15. Gray, E., Marsh, H., and McLaren, M. (1982). A short history of gunpowder and the role of charcoal in its manufacture. *Journal of Materials Science* **17** (12): 3385–3400.
16. Ronsse, F., van Hecke, S., Dickinson, D., and Prins, W. (2013). Production and characterization of slow pyrolysis biochar: influence of feedstock type and pyrolysis conditions. *GCB Bioenergy* **5** (2): 104–115.
17. Brown, R.C., del Campo, B., Boateng, A.A. et al. (2015). Fundamentals of biochar production. In: *Biochar for Environmental Management: Science, Technology and Implementation* (ed. J. Lehmann and S. Joseph), 39–61. London: Taylor & Francis Group.
18. Williams, P.T. and Besler, S. (1996). The influence of temperature and heating rate on the slow pyrolysis of biomass. *Renewable Energy* **7** (3): 233–250.
19. Shuai, L. and Luterbacher, J. (2016). Organic solvent effects in biomass conversion reactions. *ChemSusChem* **9** (2): 133–155.
20. Kawamoto, H., Hatanaka, W., and Saka, S. (2003). Thermochemical conversion of cellulose in polar solvent (sulfolane) into levoglucosan and other low molecular-weight substances. *Journal of Analytical and Applied Pyrolysis* **70**: 303–313.
21. Ghosh, A., Brown, R.C., and Bai, X. (2016). Production of solubilized carbohydrate from cellulose using non-catalytic, supercritical depolymerization in polar aprotic solvents. *Green Chemistry* **18** (4): 1023–1031.

22. Ishikawa, Y. and Saka, S. (2001). Chemical conversion of cellulose as treated in supercritical methanol. *Cellulose* **8**: 189–195.
23. Mellmer, M.A., Martin Alonso, D., Luterbacher, J.S. et al. (2014). Effects of γ-valerolactone in hydrolysis of lignocellulosic biomass to monosaccharides. *Green Chemistry* **16** (11): 4659–4662.
24. Shuai, L., Amiri, M.T., Questell-Santiago, Y.M. et al. (2016). Formaldehyde stabilization facilitates lignin monomer production during biomass depolymerization. *Science* **354** (6310): 329–333.
25. Garcia-Perez, M., Wang, X.S., Shen, J. et al. (2008). Fast pyrolysis of oil mallee woody biomass: effect of temperature on the yield and quality of pyrolysis products. *Industrial and Engineering Chemistry Research* **47** (6): 1846–1854.
26. Reed, T.B. and Das, A. (1988). *Handbook of Biomass Downdraft Gasifier Engine Systems*. Colorado: Golden.
27. van der Drift, A., Boerrigter, H., Coda, B. et al. (2004). *Entrained Flow Gasification of Biomass: Ash Behaviour, Feeding Issues, and System Analyses*. ECN (Energy Research Centre of the Netherlands). https://www.ecn.nl/docs/library/report/2004/c04039.pdf (accessed 27 November 2018).
28. Chhiti, Y. and Kemiha, M. (2013). Thermal conversion of biomass, pyrolysis and gasification: a review. *International Journal of Engineering Science* **2** (3): 75–85.
29. Gowlett, J.A.J. (2016). The discovery of fire by humans: a long and convoluted process. *Philosophical Transactions of the Royal Society B: Biological Sciences* **371** (1696): 20150164.
30. Glassman, I., Yetter, R.A., and Glumac, N.G. (2014). Combustion of nonvolatile fuels. In: *Combustion*, 477–536. Elsevier Inc.
31. Kuzhiyil, N., Dalluge, D., Bai, X. et al. (2012). Pyrolytic sugars from cellulosic biomass. *ChemSusChem* **5** (11): 2228–2236.
32. Dalluge, D.L., Daugaard, T., Johnston, P. et al. (2014). Continuous production of sugars from pyrolysis of acid-infused lignocellulosic biomass. *Green Chemistry* 4144–4155.
33. Vinu, R. and Broadbelt, L.J. (2012). A mechanistic model of fast pyrolysis of glucose-based carbohydrates to predict bio-oil composition. *Energy and Environmental Science* **5** (12): 9808–9826.
34. Zhou, X., Nolte, M.W., Mayes, H.B. et al. (2014). Experimental and mechanistic modeling of fast pyrolysis of neat glucose-based carbohydrates. 1. Experiments and development of a detailed mechanistic model. *Industrial and Engineering Chemistry Research* **53** (34): 13274–13289.
35. Zhou, X., Nolte, M.W., Shanks, B.H., and Broadbelt, L.J. (2014). Experimental and mechanistic modeling of fast pyrolysis of neat glucose-based carbohydrates. 2. Validation and evaluation of the mechanistic model. *Industrial and Engineering Chemistry Research* **53** (34): 13290–13301.
36. Zhou, X., Li, W., Mabon, R., and Broadbelt, L.J. (2018). A mechanistic model of fast pyrolysis of hemicellulose. *Energy and Environmental Science* **11** (5): 1240–1260.
37. Yanez, A.J., Natarajan, P., Li, W. et al. (2018). Coupled structural and kinetic model of lignin fast pyrolysis. *Energy and Fuels* **32** (2): 1822–1830.
38. Diebold, J.P. (1994). A unified, global model for the pyrolysis of cellulose. *Biomass and Bioenergy* **7** (1–6): 75–85.
39. Burnham, A.K., Zhou, X., and Broadbelt, L.J. (2015). Critical review of the global chemical kinetics of cellulose thermal decomposition. *Energy and Fuels* **29** (5): 2906–2918.
40. Ranzi, E., Debiagi, P.E.A., and Frassoldati, A. (2017). Mathematical modeling of fast biomass pyrolysis and bio-oil formation. Note I: kinetic mechanism of biomass pyrolysis. *ACS Sustainable Chemistry and Engineering* **5** (4): 2867–2881.
41. Piskorz, J., Majerski, P., Radlein, D. et al. (2000). Flash pyrolysis of cellulose for production of anhydro-oligomers. *Journal of Analytical and Applied Pyrolysis* **56** (2): 145–166.
42. Thompson, L.C., Ciesielski, P.N., Jarvis, M.W. et al. (2017). Estimating the temperature experienced by biomass particles during fast pyrolysis using microscopic analysis of biochars. *Energy and Fuels* **31** (8): 8193–8201.
43. Boutin, O., Ferrer, M., and Lédé, J. (2002). Flash pyrolysis of cellulose pellets submitted to a concentrated radiation: experiments and modelling. *Chemical Engineering Science* **57**: 15–25.

44. Lédé, J., Blanchard, F., and Boutin, O. (2002). Radiant flash pyrolysis of cellulose pellets: products and mechanisms involved in transient and steady state conditions. *Fuel* **81** (10): 1269–1279.
45. Liu, Q., Wang, S., Wang, K. et al. (2008). Mechanism of formation and consequent evolution of active cellulose during cellulose pyrolysis. *Acta Physico-Chimica Sinica* **24** (11): 1957–1963.
46. Krumm, C., Pfaendtner, J., and Dauenhauer, P.J. (2016). Millisecond pulsed films unify the mechanisms of cellulose fragmentation. *Chemistry of Materials* **28**: 3108–3114.
47. Lindstrom, J. K., Ciesielski, P. N., Johnston, P. A., Peterson, C. A., Gable, P., and Brown, R. C. 2017. Thermal Deconstruction Opens Biomass for Acid Hydrolysis to Monosaccharides. In: AIChE Annual Meeting, Minneapolis, MN.
48. Lindstrom, J.K., Proano-Aviles, J., Johnston, P.A. et al. (2019). Competing reactions limit levoglucosan yield during fast pyrolysis of cellulose. *Green Chemistry*. **21**: 178–186. https://doi.org/10.1039/C8GC03461C
49. Gable, P. and Brown, R.C. (2016). Effect of biomass heating time on bio-oil yields in a free fall fast pyrolysis reactor. *Fuel* **166**: 361–366.
50. Maduskar, S., Facas, G.G., Papageorgiou, C. et al. (2018). Five rules for measuring biomass pyrolysis rates: pulse-heated analysis of solid reaction kinetics of lignocellulosic biomass. *ACS Sustainable Chemistry and Engineering* **6** (1): 1387–1399.
51. Tiarks, J.A. (2018). *Investigation of Fundamental Transport and Physicochemical Phenomena in Lignocellulosic Fast Pyrolysis*. Iowa State University.
52. Teixeira, A.R., Mooney, K.G., Kruger, J.S. et al. (2011). Aerosol generation by reactive boiling ejection of molten cellulose. *Energy and Environmental Science* **4**: 4306.
53. Teixeira, A.R., Gantt, R., Joseph, K.E. et al. (2016). Spontaneous aerosol ejection: origin of inorganic particles in biomass pyrolysis. *ChemSusChem* **9** (11): 1322–1328.
54. Zhang, J., Choi, Y.S., Yoo, C.G. et al. (2015). Cellulose–hemicellulose and cellulose–lignin interactions during fast pyrolysis. *ACS Sustainable Chemistry and Engineering* **3** (2): 293–301.
55. Lédé, J. (2012). Cellulose pyrolysis kinetics: an historical review on the existence and role of intermediate active cellulose. *Journal of Analytical and Applied Pyrolysis* **94**: 17–32.
56. Radlein, D.S.A.G., Grinshpun, A., Piskorz, J., and Scott, D.S. (1987). On the presence of anhydro-oligosaccharides in the sirups from the fast pyrolysis of cellulose. *Journal of Analytical and Applied Pyrolysis* **12** (1): 39–49.
57. Lin, Y.-C., Cho, J., Tompsett, G.A. et al. (2009). Kinetics and mechanism of cellulose pyrolysis. *Journal of Physical Chemistry C* **113** (46): 20097–20107.
58. Chu, S., Subrahmanyam, A.V., and Huber, G.W. (2013). The pyrolysis chemistry of a β-O-4 type oligomeric lignin model compound. *Green Chemistry* **15** (1): 125–136.
59. Kim, K.H., Bai, X., and Brown, R.C. (2014). Pyrolysis mechanisms of methoxy substituted α-O-4 lignin dimeric model compounds and detection of free radicals using electron paramagnetic resonance analysis. *Journal of Analytical and Applied Pyrolysis* **110**: 254–263.
60. Paine, J.B., Pithawalla, Y.B., Naworal, J.D., and Thomas, C.E. (2007). Carbohydrate pyrolysis mechanisms from isotopic labeling part 1: the pyrolysis of glycerin: discovery of competing fragmentation mechanisms affording acetaldehyde and formaldehyde and the implications for carbohydrate pyrolysis. *Journal of Analytical and Applied Pyrolysis* **80** (2): 297–311.
61. Paine, J.B., Pithawalla, Y.B., and Naworal, J.D. (2008). Carbohydrate pyrolysis mechanisms from isotopic labeling part 2. The pyrolysis of d-glucose: general disconnective analysis and the formation of C1 and C2 carbonyl compounds by electrocyclic fragmentation mechanisms. *Journal of Analytical and Applied Pyrolysis* **82** (1): 10–41.
62. Degenstein, J.C., Murria, P., Easton, M. et al. (2015). Fast pyrolysis of 13C-labeled cellobioses: gaining insights into the mechanisms of fast pyrolysis of carbohydrates. *Journal of Organic Chemistry* **80** (3): 1909–1914.

63. Hutchinson, C.P. and Lee, Y.J. (2017). Evaluation of primary reaction pathways in thin-film pyrolysis of glucose using 13C labeling and real-time monitoring. *ACS Sustainable Chemistry and Engineering* **5** (10): 8796–8803.
64. Lynch, B.J., Fast, P.L., Harris, M., and Truhlar, D.G. (2000). Adiabatic connection for kinetics. *The Journal of Physical Chemistry A* **104** (21): 4811–4815.
65. Austin, L.W. and Eastman, C.W. (1900). *On the Relation between Heat Conductivity and Density in Some of the Common Woods*, 539–543. Wisconsin Academy of Sciences, Arts, and Letters.
66. Mason, P.E., Darvell, L.I., Jones, J.M., and Williams, A. (2016). Comparative study of the thermal conductivity of solid biomass fuels. *Energy and Fuels* **30** (3): 2158–2163.
67. Bahng, M.K., Donohoe, B.S., and Nimlos, M.R. (2011). Application of an Fourier transform-infrared imaging tool for measuring temperature or reaction profiles in pyrolyzed wood. *Energy and Fuels* **25** (1): 370–378.
68. Ciesielski, P.N., Crowley, M.F., Nimlos, M.R. et al. (2015). Biomass particle models with realistic morphology and resolved microstructure for simulations of intraparticle transport phenomena. *Energy and Fuels* **29** (1): 242–254.
69. Wiggins, G.M., Ciesielski, P.N., and Daw, C.S. (2016). Low-order modeling of internal heat transfer in biomass particle pyrolysis. *Energy and Fuels* **30** (6): 4960–4969.
70. Pecha, M.B., Garcia-Perez, M., Foust, T.D., and Ciesielski, P.N. (2017). Estimation of heat transfer coefficients for biomass particles by direct numerical simulation using microstructured particle models in the laminar regime. *ACS Sustainable Chemistry and Engineering* **5** (1): 1046–1053.
71. Forest Products Laboratory (2010). *Wood Handbook: Wood as an Engineering Material*, vol. **FPL–GTR–19**. Madison, WI: Forest Products Laboratory, part of the US Department of Agriculture.
72. Pan, Y. and Kong, S.-C. (2017). Simulation of biomass particle evolution under pyrolysis conditions using lattice Boltzmann method. *Combustion and Flame* **178**: 21–34.
73. Pan, Y. and Kong, S.-C. (2017). Predicting effects of operating conditions on biomass fast pyrolysis using particle-level simulation. *Energy and Fuels* **31** (1): 635–646.
74. Himmel, M.E., Tucker, M., Baker, J. et al. (1986). Comminution of biomass: hammer and knife mills. *Biotechnology and Bioengineering Symposium* **15** (15): 39–58.
75. Haas, T.J., Nimlos, M.R., and Donohoe, B.S. (2009). Real-time and post-reaction microscopic structural analysis of biomass undergoing pyrolysis. *Energy and Fuels* **23** (7): 3810–3817.
76. Assary, R.S. and Curtiss, L.A. (2012). Comparison of sugar molecule decomposition through glucose and fructose: a high-level quantum chemical study. *Energy and Fuels* **26** (2): 1344–1352.
77. Vosko, S.H., Wilk, L., and Nusair, M. (1980). Accurate spin-dependent electron liquid correlation energies for local spin density calculations: a critical analysis. *Canadian Journal of Physics* **58** (8): 1200–1211.
78. Becke, A.D. (1993). Density-functional thermochemistry. III. The role of exact exchange. *The Journal of Chemical Physics* **98** (7): 5648–5652.
79. Lee, C., Yang, E., Parr, R.R.G. et al. (1988). Development of the Colle-Salvetti correlation energy formula into a functional of the electron density. *Physical Review B* **37** (2): 785–789.
80. Stephens, P.J., Devlin, F.J., Chabalowski, C.F., and Frisch, M.J. (1994). Ab initio calculation of vibrational absorption and circular dichroism spectra using density functional force fields. *The Journal of Physical Chemistry* **98** (45): 11623–11627.
81. Huang, J., Liu, C., Wei, S. et al. (2010). Density functional theory studies on pyrolysis mechanism of β-d-glucopyranose. *Journal of Molecular Structure: THEOCHEM* **958** (1–3): 64–70.
82. Wang, S., Ru, B., Lin, H., and Luo, Z. (2013). Degradation mechanism of monosaccharides and xylan under pyrolytic conditions with theoretic modeling on the energy profiles. *Bioresource Technology* **143**: 378–383.
83. Huang, X., Liu, C., Huang, J., and Li, H. (2011). Theory studies on pyrolysis mechanism of phenethyl phenyl ether. *Computational and Theoretical Chemistry* **976** (1–3): 51–59.

84. Zhao, Y. and Truhlar, D.G. (2008). The M06 suite of density functionals for main group thermochemistry, thermochemical kinetics, noncovalent interactions, excited states, and transition elements: two new functionals and systematic testing of four M06-class functionals and 12 other function. *Theoretical Chemistry Accounts* **120** (1–3): 215–241.
85. Zhao, Y. and Truhlar, D.G. (2006). Density functional for spectroscopy: no long-range self-interaction error, good performance for Rydberg and charge-transfer states, and better performance on average than B3LYP for ground states. *The Journal of Physical Chemistry A* **110** (49): 13126–13130.
86. Dedic, C., Tiarks, J. A., Sanderson, P. D., Brown, R. C., Michael, J. B., and Meyer, T. R. 2015. Optical Diagnostic Techniques for Investigation of Biomass Pyrolysis. In Fourth International Conference on Thermochemical Biomass Conversion Science; Chicago, IL.
87. Patwardhan, P.R., Satrio, J.A., Brown, R.C., and Shanks, B.H. (2009). Product distribution from fast pyrolysis of glucose-based carbohydrates. *Journal of Analytical and Applied Pyrolysis* **86** (2): 323–330.
88. Patwardhan, P.R., Satrio, J.A., Brown, R.C., and Shanks, B.H. (2010). Influence of inorganic salts on the primary pyrolysis products of cellulose. *Bioresource Technology* **101** (12): 4646–4655.
89. Atalla, R.H., Brady, J.W., Matthews, J.F. et al. (2008). Structures of plant cell wall celluloses. In: *Biomass Recalcitrance: Deconstructing the Plant Cell Wall for Bioenergy* (ed. M.E. Himmel), 188–210. Wiley.
90. Zhang, J., Nolte, M.W., and Shanks, B.H. (2014). Investigation of primary reactions and secondary effects from the pyrolysis of different celluloses. *ACS Sustainable Chemistry and Engineering* **2** (12): 2820–2830.
91. Mukarakate, C., Mittal, A., Ciesielski, P.N. et al. (2016). Influence of crystal allomorph and crystallinity on the products and behavior of cellulose during fast pyrolysis. *ACS Sustainable Chemistry and Engineering* **4** (9): 4662–4674.
92. Assary, R.S. and Curtiss, L.A. (2012). Thermochemistry and reaction barriers for the formation of levoglucosenone from cellobiose. *ChemCatChem* **4** (2): 200–205.
93. Wang, S., Guo, X., Liang, T. et al. (2012). Mechanism research on cellulose pyrolysis by Py-GC/MS and subsequent density functional theory studies. *Bioresource Technology* **104**: 722–728.
94. Seshadri, V. and Westmoreland, P.R. (2012). Concerted reactions and mechanism of glucose pyrolysis and implications for cellulose kinetics. *The Journal of Physical Chemistry A* **116** (49): 11997–12013.
95. Zhang, X., Li, J., Yang, W., and Blasiak, W. (2011). Formation mechanism of levoglucosan and formaldehyde during cellulose pyrolysis. *Energy and Fuels* **25** (8): 3739–3746.
96. Mayes, H.B. and Broadbelt, L.J. (2012). Unraveling the reactions that unravel cellulose. *The Journal of Physical Chemistry A* **116** (26): 7098–7106.
97. Zhang, X., Yang, W., and Dong, C. (2013). Levoglucosan formation mechanisms during cellulose pyrolysis. *Journal of Analytical and Applied Pyrolysis* **104**: 19–27.
98. Nimlos, M.R. and Evans, R.J. (2002). Levoglucosan pyrolysis. *Fuel Chemistry Division Preprints* **47** (1): 393.
99. Zhang, X., Yang, W., and Blasiak, W. (2012). Thermal decomposition mechanism of levoglucosan during cellulose pyrolysis. *Journal of Analytical and Applied Pyrolysis* **96**: 110–119.
100. Lu, Q., Zhang, Y., Dong, C. et al. (2014). The mechanism for the formation of levoglucosenone during pyrolysis of β-d-glucopyranose and cellobiose: a density functional theory study. *Journal of Analytical and Applied Pyrolysis* **110** (1): 34–43.
101. Zhang, Y., Liu, C., and Chen, X. (2015). Unveiling the initial pyrolytic mechanisms of cellulose by DFT study. *Journal of Analytical and Applied Pyrolysis* **113**: 621–629.
102. Mayes, H.B., Nolte, M.W., Beckham, G.T. et al. (2014). The alpha-bet(a) of glucose pyrolysis: computational and experimental investigations of 5-hydroxymethylfurfural and levoglucosan formation reveal implications for cellulose pyrolysis. *ACS Sustainable Chemistry and Engineering* **2** (6): 1461–1473.

103. Zhang, Y., Liu, C., and Xie, H. (2014). Mechanism studies on β-d-glucopyranose pyrolysis by density functional theory methods. *Journal of Analytical and Applied Pyrolysis* **105**: 23–34.
104. Wang, M., Liu, C., Xu, X., and Li, Q. (2016). Theoretical investigation on the carbon sources and orientations of the aldehyde group of furfural in the pyrolysis of glucose. *Journal of Analytical and Applied Pyrolysis* **120**: 464–473.
105. Liu, D., Yu, Y., and Wu, H. (2013). Evolution of water-soluble and water-insoluble portions in the solid products from fast pyrolysis of amorphous cellulose. *Industrial and Engineering Chemistry Research* **52** (36): 12785–12793.
106. Gong, X., Yu, Y., Gao, X. et al. (2014). Formation of anhydro-sugars in the primary volatiles and solid residues from cellulose fast pyrolysis in a wire-mesh reactor. *Energy and Fuels* **28** (8): 5204–5211.
107. Patwardhan, P.R., Brown, R.C., and Shanks, B.H. (2011). Product distribution from the fast pyrolysis of hemicellulose. *ChemSusChem* **4** (5): 636–643.
108. Werner, K., Pommer, L., and Broström, M. (2014). Thermal decomposition of hemicelluloses. *Journal of Analytical and Applied Pyrolysis* **110**: 130–137.
109. Zhou, X., Li, W., Mabon, R., and Broadbelt, L.J. (2017). A critical review on hemicellulose pyrolysis. *Energy Technology* **5** (1): 216.
110. Holtzapple, M.T. (2003). Hemicelluloses. In: *Encyclopedia of Food Sciences and Nutrition* (ed. B. Caballero), 3060–3071. Elsevier.
111. Bikova, T. and Treimanis, A. (2002). Solubility and molecular weight of hemicelluloses from Alnus incana and Alnus glutinosa. Effect of tree age. *Plant Physiology and Biochemistry* **40** (4): 347–353.
112. Kurata, Y., Mori, Y., Ishida, A. et al. (2018). Variation in hemicellulose structure and assembly in the cell wall associated with the transition from earlywood to latewood in Cryptomeria japonica. *Journal of Wood Chemistry and Technology* **38** (3): 254–263.
113. Meier, H. (1964). General chemistry of cell walls and distribution of the chemical constituents across the walls. In: *The Formation of Wood in Forest Trees* (ed. M.H. Zimmerman), 137–151. Elsevier.
114. Wang, M., Liu, C., Li, Q., and Xu, X. (2015). Theoretical insight into the conversion of xylose to furfural in the gas phase and water. *Journal of Molecular Modeling* **21** (11): 296.
115. Huang, J., He, C., Wu, L., and Tong, H. (2017). Theoretical studies on thermal decomposition mechanism of arabinofuranose. *Journal of the Energy Institute* **90** (3): 372–381.
116. Lu, Q., Tian, H.Y., Hu, B. et al. (2016). Pyrolysis mechanism of holocellulose-based monosaccharides: the formation of hydroxyacetaldehyde. *Journal of Analytical and Applied Pyrolysis* **120**: 15–26.
117. Huang, J., He, C., Wu, L., and Tong, H. (2016). Thermal degradation reaction mechanism of xylose: a DFT study. *Chemical Physics Letters* **658**: 114–124.
118. Patwardhan, P.R., Brown, R.C., and Shanks, B.H. (2011). Understanding the fast pyrolysis of lignin. *ChemSusChem* **4** (11): 1629–1636.
119. Kim, J.Y., Oh, S., Hwang, H. et al. (2013). Structural features and thermal degradation properties of various lignin macromolecules obtained from poplar wood (Populus albaglandulosa). *Polymer Degradation and Stability* **98** (9): 1671–1678.
120. Katahira, R., Elder, T.J., and Beckham, G.T. (2018). A brief introduction to lignin structure. In: *Lignin Valorization: Emerging Approaches* (ed. G.T. Beckham), 1–20. London: Royal Society of Chemistry.
121. Houtman, C. (2018). Lessons learned from 150 years of pulping wood. In: *Lignin Valorization: Emerging Approaches* (ed. G.T. Beckham), 62–73. London: Royal Society of Chemistry.
122. Ragauskas, A.J., Beckham, G.T., Biddy, M.J. et al. (2014). Lignin valorization: improving lignin processing in the biorefinery. *Science* **344** (6185): 1246843.
123. Rencoret, J., Gutierrez, A., Nieto, L. et al. (2011). Lignin composition and structure in young versus adult Eucalyptus globulus plants. *Plant Physiology* **155** (2): 667–682.
124. Grabber, J.H. (2005). How do lignin composition, structure, and cross-linking affect degradability? A review of cell wall model studies. *Crop Science* **45** (3): 820.
125. Li, M., Pu, Y., and Ragauskas, A.J. (2016). Current understanding of the correlation of lignin structure with biomass recalcitrance. *Frontiers in Chemistry* **4**: 45.

126. Jiang, G., Nowakowski, D.J., and Bridgwater, A.V. (2010). Effect of the temperature on the composition of lignin pyrolysis products. *Energy and Fuels* **24** (8): 4470–4475.
127. Kim, J.-Y., Hwang, H., Oh, S. et al. (2014). Investigation of structural modification and thermal characteristics of lignin after heat treatment. *International Journal of Biological Macromolecules* **66**: 57–65.
128. Kawamoto, H. (2017). Lignin pyrolysis reactions. *Journal of Wood Science* **63** (2): 117–132.
129. Jarvis, M.W., Daily, J.W., Carstensen, H.-H. et al. (2011). Direct detection of products from the pyrolysis of 2-phenethyl phenyl ether. *The Journal of Physical Chemistry A* **115** (4): 428–438.
130. Hoffmann, H.M.R. (1969). The ene reaction. *Angewandte Chemie International Edition in English* **8** (8): 556–577.
131. Maccoll, A. (1969). Heterolysis and the pyrolysis of alkyl halides in the gas phase. *Chemical Reviews* **69** (1): 33–60.
132. Elder, T. and Beste, A. (2014). Density functional theory study of the concerted pyrolysis mechanism for lignin models. *Energy and Fuels* **28** (8): 5229–5235.
133. Huang, J., Liu, C., Wu, D. et al. (2014). Density functional theory studies on pyrolysis mechanism of β-O-4 type lignin dimer model compound. *Journal of Analytical and Applied Pyrolysis* **109**: 98–108.
134. Chen, L., Ye, X., Luo, F. et al. (2015). Pyrolysis mechanism of β-O-4 type lignin model dimer. *Journal of Analytical and Applied Pyrolysis* **115**: 103–111.
135. Mayes, H.B., Nolte, M.W., Beckham, G.T. et al. (2015). The alpha-bet(a) of salty glucose pyrolysis: computational investigations reveal carbohydrate pyrolysis catalytic action by sodium ions. *ACS Catalysis* **5** (1): 192–202.
136. Zhang, Y. and Liu, C. (2014). A new horizon on effects of alkalis metal ions during biomass pyrolysis based on density function theory study. *Journal of Analytical and Applied Pyrolysis* **110** (1): 297–304.
137. Prins, M.J., Ptasinski, K.J., and Janssen, F.J.J.G. (2006). Torrefaction of wood. Part 2. Analysis of products. *Journal of Analytical and Applied Pyrolysis* **77** (1): 35–40.
138. Mullen, C.A., Boateng, A.A., Goldberg, N.M. et al. (2010). Bio-oil and bio-char production from corn cobs and stover by fast pyrolysis. *Biomass and Bioenergy* **34** (1): 67–74.
139. Abdoulmoumine, N., Kulkarni, A., and Adhikari, S. (2014). Effects of temperature and equivalence ratio on pine syngas primary gases and contaminants in a bench-scale fluidized bed gasifier. *Industrial and Engineering Chemistry Research* **53** (14): 5767–5777.
140. Liu, C., Huang, J., Huang, X. et al. (2011). Theoretical studies on formation mechanisms of CO and CO_2 in cellulose pyrolysis. *Computational and Theoretical Chemistry* **964** (1): 207–212.
141. Huang, J., Liu, C., Tong, H. et al. (2014). A density functional theory study on formation mechanism of CO, CO_2 and CH_4 in pyrolysis of lignin. *Computational and Theoretical Chemistry* **1045**: 1–9.
142. Beckham, G.T., Matthews, J.F., Peters, B. et al. (2011). Molecular-level origins of biomass recalcitrance: decrystallization free energies for four common cellulose polymorphs. *Journal of Physical Chemistry B* **115** (14): 4118–4127.
143. Matthews, J.F., Himmel, M.E., and Crowley, M.F. (2012). Conversion of cellulose Iα to Iβ via a high temperature intermediate (I-HT) and other cellulose phase transformations. *Cellulose* **19** (1): 297–306.
144. Bu, L., Himmel, M.E., and Crowley, M.F. (2015). The molecular origins of twist in cellulose I-beta. *Carbohydrate Polymers* **125**: 146–152.
145. Matthews, J.F., Bergenstråhle, M., Beckham, G.T. et al. (2011). High-temperature behavior of cellulose I. *Journal of Physical Chemistry B* **115** (10): 2155–2166.
146. Payne, C.M., Himmel, M.E., Crowley, M.F., and Beckham, G.T. (2011). Decrystallization of oligosaccharides from the cellulose Iβ surface with molecular simulation. *Journal of Physical Chemistry Letters* **2** (13): 1546–1550.

3

Biomass Combustion

Bryan M. Jenkins[1], Larry L. Baxter[2] and Jaap Koppejan[3]

[1] Department of Biological and Agricultural Engineering, University of California, Davis, CA, USA
[2] Department of Chemical Engineering, Brigham Young University, Provo, UT, USA
[3] Procede Biomass BV, Enschede, Netherlands

Nomenclature

a, b	constant coefficients of linearized enthalpy functions
C_p	mass or molar specific heat (J/kg/K, J/mol K)
e	excess oxidant or air in reactants (kg/kg)
$f_{i,j}$	mass fraction of fuel constituent i in product j (kg/kg)
H	enthalpy, total enthalpy (kJ/kmol, kJ)
h	molar enthalpy (kJ/kmol)
h_{fg}	enthalpy of vaporization (kJ/kg)
M_{db}	dry basis moisture content of feedstock (kg/kg)
$m_{p,j}$	mass of the product species j (kg)
$m_{r,i}$	mass of the reactant species i (kg)
m_w	mass of water as moisture in feedstock (kg)
M_{wb}	wet basis moisture content of feedstock (kg/kg)
Q_h	higher heating value (constant pressure or volume) (kJ/kg)
$Q_{h,o}$	higher heating value, dry (constant pressure or constant volume) (kJ/kg)
Q_l	heat transfer to reaction system (negative for heat loss) (kJ)
$Q_{p,h}$	higher heating value at constant pressure (kJ/kg)
$Q_{residual}$	residual feedstock heating value after evaporating feedstock moisture (%)

Thermochemical Processing of Biomass: Conversion into Fuels, Chemicals and Power, Second Edition.
Edited by Robert C. Brown.
© 2019 John Wiley & Sons Ltd. Published 2019 by John Wiley & Sons Ltd.

$Q_{v,h}$ higher heating value at constant volume (kJ/kg)
$Q_{v,h,o}$ higher heating value at constant volume, dry (kJ/kg)
$Q_{v,l}$ lower heating value at constant volume (kJ/kg)
T absolute temperature (K)
u_{fg} internal energy of vaporization (kJ/kg)
W_i molar mass of species i (kg/kmol)
ϕ fuel/air equivalence ratio
λ air/fuel equivalence ratio, air factor
ν stoichiometric coefficient

3.1 Introduction

Since humans first learned to manage fire a quarter of a million years or more ago [1], the burning of fuels has served as a defining phenomenon for the development of societies. Releasing the energy needed for large-scale land clearing and agricultural expansion, combustion also provided the means for industrial growth, rapid transportation, the increase and concentration of populations, the waging of world wars, and the globalization of trade and culture. As the world population continues to expand and increasing emphasis is placed on renewable resources for transportation, power generation, and other sectors, the environmental impacts of current fuel-burning practices cannot be sustained into the future. Continuing evolution of heat and power generation is likely to see dramatic transformations toward low- and zero-emission alternatives, and the future design of combustion systems will be heavily challenged to adapt to more stringent regulations affecting environmental performance while maintaining economic competitiveness.

Biomass resources supported early industrialization efforts until largely supplanted by fossil energy resources – coal, petroleum, and natural gas – and hydroelectric and nuclear power. Ancient uses of fire are still employed by a large fraction of the world's population that is without access to more expensive fuels or electricity. Firewood gathering constitutes a significant burden of work and environmental harm, and uncontrolled emissions are responsible for respiratory and other diseases mostly among women and children [2]. Firewood use in fireplaces and woodstoves for heating purposes is a major demand sector for biomass. These uses of biomass are typically associated with low conversion efficiency and high pollutant emissions. Biomass use in cookstoves, however, serves a vital role around the world and methods continue to be investigated to reduce pollutant exposures while maintaining acceptable performance for users, although there is much controversy surrounding the subject [3].

Although the sustainability of biomass production and conversion to fuels and power has recently seen increasing scrutiny due to indirect land use change and other effects associated with global food and energy markets [4], as a well-managed renewable resource biomass has the potential to contribute more substantially to the development of a sustainable economy. The combined processes of plant photosynthesis and respiration produce in biomass a chemically complex resource supporting a wide range of uses. Emulating these processes in manufacturing fuels and chemicals from sunlight but without the need of life processes is now viewed as one of the scientific grand challenges [5]. The energy storage in biomass enables its use as a renewable resource for baseload power generation, an integral component in managing electricity distribution systems as generating capacity increases among more variable solar and wind energy resources. The stored energy in biomass also allows its use in generation under dispatch, however, and integrated renewable systems deploying multiple resources including biomass to meet variable demand and supply conditions are of considerable interest for optimized microgrid and other grid applications [6].

Historically, and still so today, the most widely applied conversion method for biomass is combustion. The chemical energy of the fuel is converted via combustion into heat. Heat is useful in and of itself, and it may be transformed by heat engines of various types into mechanical and, hence, electrical energy. Direct conversion of biomass to electricity by magnetohydrodynamic energy conversion has been considered due to certain similarities to coal, particularly for potassium in ash as an easily ionizable material, but no practical use of the technology has occurred [7, 8].

Electricity generation using biomass fuels expanded rapidly in the USA following legislation changing utility regulatory policy in 1978, but stalled for economic and environmental reasons after the mid-1990s. U.S. generating capacity at present is approximately $10\,GW_e$ of electricity, with global capacity about five times that amount [9]. Combustion plays a major role in waste disposal, complementing other waste management practices. Incentives for future expansion exist in the form of renewable portfolio standards, such as that enacted in California in 2002 and in subsequent legislation calling for 33% renewable electricity by 2020 and 60% by 2030 with 100% of all retail electricity sales in the state supplied from renewable and zero-carbon resources by 2045 [10]. Integration of power and heat generation in biorefinery operations will also lead to capacity expansions for biomass combustion and related systems. Meeting environmental and economic performance requirements into the future will prove challenging, however, and there continues to be the need for targeted research in advanced system design.

This chapter outlines technologies and performance issues in biomass combustion, summarizing system designs, feedstock properties, and environmental impacts. Combustion fundamentals are also briefly reviewed, including combustion stoichiometry, equilibrium, and kinetics. As highlighted by the simple burning of logwood, combustion is a complex process involving multiple simultaneous phenomena. More detailed predictive capability facilitating analysis, design, operation, control, and regulation remains a goal for continued research and development.

3.2 Combustion Systems

3.2.1 Fuels

Combustor design and selection are dictated both by fuel type and end use. Within the class of biomass fuels are solids, gases, and liquids, the latter two being derived by physical, chemical, or biological conversion of the parent feedstock. Comparative properties of selected fuel types are listed in Table 3.1.

3.2.1.1 Solids

Solids constitute the primary class of biomass fuels, including woody and herbaceous materials such as wood and bark, lumber mill residues, grasses, cereal straws and stovers, other agricultural and forest residues, and energy crops such as switchgrass, *Miscanthus*, poplar, willow, and numerous others. Manures and other animal products include a fraction of solids that are also used as fuels. Municipal solid waste (MSW) also includes a biomass

Table 3.1 Properties of selected fuels.

Fuel	Approximate molar mass	Mass density (kg/l)	Mass energy density (MJ/kg)	Volume energy density (MJ/l)	Stoichiometric air/fuel ratio (kg/kg)	LHV[a] of stoichiometric mixture (MJ/kg)	Octane no.	Cetane no.
Gasoline (l)	110	0.750	44.0	33.0	14.6	2.8	87–94	
Methane (g)	16	0.00065	50.0	0.033	17.2	2.7	120	
LNG (l)	16	0.424	49.5	21.0	~17	~2.7	~120	
Methanol (l)	32	0.792	20.0	15.8	6.5	2.7	99	
Ethanol (l)	46	0.785	26.9	21.1	9.0	2.7	98	
Butanol (l)	74	0.810	33.0	26.7	11.2	2.7	104	
Hydrogen (g)	2	0.00008	120.0	0.010	34.3	3.4	>130	
Hydrogen (l)	2	0.070	120.0	8.4	34.3	3.4	>130	
Diesel no. 2 (l)	200	0.850	42.8	36.4	14.4	2.8		40–55
Biodiesel (soy B100) (l)	292	0.880	38.3	33.7	13.8	2.6		48–60

[a]LHV: lower heating value.

fraction and is used in waste-to-energy (WTE) systems to provide volume reduction along with useful heat and electricity. Depending on location and local policies, WTE units may employ mass burning of unseparated wastes or combustion of separated wastes in which recyclables and other constituents have first been sorted from the waste stream. Properties of biomass feedstocks are reviewed in Section 3.3.1.

Other solids derived from biomass include torrefied materials and charcoal. Torrefaction is a light pyrolysis of the feedstock and results in a partially carbonized fuel with a lower moisture and volatile content than the original feedstock. Charcoal production is an ancient technology in which a large fraction of the volatile matter in biomass is first driven off by heating and pyrolysis. Charcoal yields from traditional processes are often below 10% of the biomass dry matter, with yields from industrial charcoal making in the range up to about 30%, although more modern techniques can increase this substantially [11]. Charcoal is widely used throughout the world as a "smokeless" cooking and heating fuel, although pollutant emissions are still high in most applications using open fires and simple stoves. Traditional charcoal making as practiced in many countries is a heavily polluting process due to uncontrolled venting of volatiles to the atmosphere. In some applications, charcoal has advantages over crude biomass in terms of handling, storage, gasification, and combustion, but unless the manufacturing process includes energy recovery, a large fraction of the energy in biomass goes unutilized.

3.2.1.2 Gases

Gaseous fuels can be produced from biomass by anaerobic digestion, pyrolysis, thermal gasification, and various fuel synthesis pathways using intermediates from these processes. The biological conversion of biomass through anaerobic digestion generates a biogas consisting primarily of methane (CH_4) and carbon dioxide (CO_2) with much smaller amounts of hydrogen sulfide (H_2S), ammonia, and other products. The CH_4 concentration typically ranges from 40% to 70% by volume depending on the types of feedstock and reactor. Anaerobic digesters are employed for conversion of animal manures, MSW, food wastes, and many other feedstocks, and have long been used in wastewater treatment operations.

Incentives such as feed-in tariffs for renewable power have stimulated wider use of digesters for grain, energy crop, and other agricultural biomass in addition to wastes, especially in Europe. The anaerobic conditions in landfills also result in the production of a similar biogas. Biogas or landfill gas can be burned directly or treated to remove moisture and contaminants such as H_2S to improve fuel value for reciprocating engines, microturbines, fuel cells, boilers, and other devices. Sulfur removal is important to avoid catalyst deactivation where stringent nitrogen oxide (NO_x) emission limits must be met and post-combustion catalysts employed, a common problem for reciprocating engines used for power generation. Scrubbing of the biogas to remove CO_2 and contaminants generates biomethane (or renewable natural gas), which in some cases is suitable for injection into utility natural gas pipelines.

Pyrolysis and gasification produce fuel gases, although pyrolysis is more generally optimized for solids (char) or liquids (bio-oil) production. Gasifiers generate fuel gases of variable composition depending on the type of feedstock and oxidant used and the reactor design. Air-blown units make a producer gas consisting of carbon monoxide (CO), H_2, CO_2 and H_2O, along with hydrocarbons (HCs) and a large fraction of N_2. Oxygen-blown units incur the cost of oxygen separation from air but eliminate nitrogen dilution in the gas to produce a synthesis quality gas, or syngas, useful for burning as well as chemical synthesis or electrochemical conversion via fuel cells after reforming to hydrogen (some fuel cells, such as the solid-oxide type, are internally reforming). Steam gasifiers also produce low nitrogen syngas, and several dual reactor designs have been developed to provide heat demand and energy for steam-raising through residual char combustion. Syngas can be used to make substitute natural gas (SNG), another source of biomethane, and reformed to produce hydrogen. Details on gasification processes are described in Chapter 4.

3.2.2 Types of Combustors

Biomass combustion involves a range of technologies from primitive open fires and traditional cooking stoves to highly controlled furnaces used for power generation and combined heat and power (CHP) applications. These span a wide range of scales, from kilowatt-size stoves to multi-megawatt furnaces and boilers. Current estimates of the energy in biomass used annually for traditional and modern combustion applications are 33.5 and 16.6 EJ, respectively [12]. The largest use of biomass by combustion is still in traditional cooking, heating, and lighting applications, mostly in developing nations. Pollutant emissions from these systems are a major health concern [2, 13] and contribute to greenhouse gas emissions. More modern uses for power generation and CHP are roughly equally deployed around the world among developed and developing nations. Co-firing of biomass with coal and other fuels is also expanding the industrial use of biomass for power and heat.

3.2.2.1 Small-Scale Systems

Considerable effort is focused on the development of clean and efficient wood-burning and other biomass combustion appliances for heating and cooking, to reduce both fuel demand and emissions. Developments in stove design for these types of application are the subject of active discussion and debate around the world [14]. More sophisticated stoves have been developed for residential and small commercial and industrial heating applications.

These often involve automatic control and the use of preprocessed fuels, such as pelletized biomass, to maintain good control over combustion and reduce emissions. Despite many improvements in combustor design, biomass remains one of the most difficult heating fuels to burn cleanly [15]. Small biomass systems typically emit considerable amounts of CO, particulate matter (PM), polycyclic aromatic hydrocarbons (PAHs), and other products of incomplete combustion. These emissions are exacerbated by heat control schemes that limit air supply to reduce the rate of heat output and the frequency of manually stoking new fuel to the stove. The ability to automatically fire more uniform fuels such as pellets provides substantially greater control over heat output rates while maintaining adequate air supply with reduced emissions compared to stick- or log-wood-fueled furnaces. The inclusion of a catalytic combustor in some designs improves emissions performance by continuing to react combustion products to lower temperatures (around 260 °C) than would occur otherwise outside the primary firebox. Reductions in emissions have accompanied improvements in stove design, test standards, flue gas cleaning systems, system installation, and better education of users on stove operations [13]. Average emissions of CO, for example, have been reduced by half or more over the last two decades. PM emissions from advanced pellet stoves now range from 15 to 25 mg/MJ compared with log-wood boilers and stoves that commonly exceed 300 mg/MJ. Electrostatic precipitators and cloth baghouses are now being deployed for emission control on small systems in addition to their more conventional use on large-scale biomass combustors. The International Energy Agency (IEA) coordinates research and outreach on these and other combustion technologies through its Task 32, *Biomass Combustion and Co-firing* [13, 15, 16].

3.2.2.2 Large-Scale Systems for Power and Heat Generation

Total installed capacity in biomass power generation around the world has been expanding at a rate of approximately 5% per year, growing from about 50 GWe in 2007 to more than 106 GW_e by 2015 [16, 17]. In many regions of the world, Asia being a possible exception, biomass utilization is below the sustainable resource capacity and potential exists to increase uses for fuels, heat, and power [18].

The most common type of biomass-fueled power plant today utilizes the Rankine or steam cycle (Figure 3.1). The fuel is burned in a boiler, which consists of a combustor with one or more heat exchangers used to make steam. Typical medium-efficiency units designed for biomass fuels utilize steam temperatures and pressures of up to 540 °C and 6–10 MPa, although installed systems include pressures up to 17 MPa [19]. The steam is expanded through one or more turbines (or multistage turbines) that drive an electrical generator. In smaller systems up to a few MW, reciprocating and screw-type steam engines are sometimes used in place of the steam turbine. The steam from the turbine exhaust is condensed and the water recirculated to the boiler through feedwater pumps. Combustion products exit the combustor, are cleaned, and discharged to the atmosphere. Typical cleaning devices include wet or dry scrubbers for control of sulfur and chlorine compounds, especially with WTE units burning MSW, cyclones (or other inertial separation devices), baghouses (high-temperature cloth filters), and/or electrostatic precipitators for PM removal. Selective catalytic reduction (SCR) or selective non-catalytic reduction (SNCR) of NO_x may also be included. Low CO and HC emissions are generally maintained by proper control of air/fuel

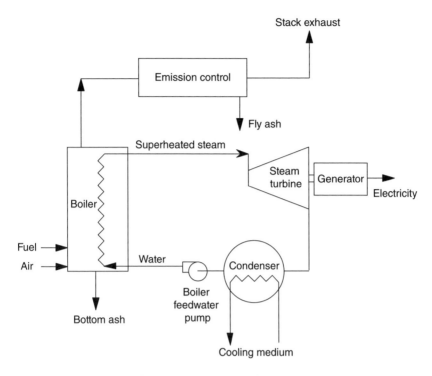

Figure 3.1   Schematic Rankine cycle.

(AF) ratio in the furnace and boiler. Organic working fluids can also be used with the Rankine cycle instead of water, in which case the system is referred to as an organic Rankine cycle (ORC). These are typically applied to lower temperature operations such as waste heat recovery or solar thermal systems.

Fireside fouling of steam superheaters and other heat exchange equipment in boilers by ash is a particular concern with biomass fuels (Figure 3.2), and larger boiler designs frequently incorporate soot-blowing capacity for intermittent cleaning. Severe fouling may require an outage (shutdown) of the plant to remove deposits. Such outages reduce operating time and thus increase the cost of delivered energy. Corrosion is also an issue with many biomass types, especially those containing higher concentrations of chlorine. Feedstock pretreatment to remove chlorine prior to firing has distinct advantages in reducing corrosion and fouling, but also increases cost.

Rankine cycle power plants using biomass fuels typically range up to about 50 MW$_e$ electrical generating capacity, which is relatively small in comparison with coal-fired power plants that in the U.S. range up to 1426 MW$_e$ for individual units and to about twice that capacity for sites deploying multiple units [20]. Larger sizes are possible and optimal size selection is accomplished through an analysis of fuel resource availability, plant design and economy, electricity and heat markets, and local regulations. The distributed nature of biomass fuels and the limited economy of scale associated with plants of this type have kept the size of individual facilities relatively small in comparison with coal or nuclear generating stations. The Polaniec power station in Poland, for example,

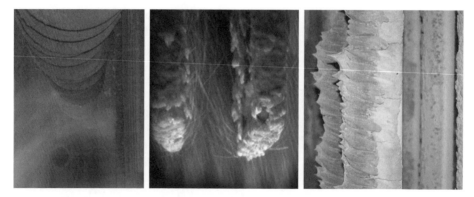

Figure 3.2 Ash fouling on superheaters in biomass-fueled boilers. Left: flame impingement on a superheater pendant and incipient ash deposition during co-firing of energy cane biomass and coal. Center: ash deposits on a superheater in a wood-fired power boiler. Right: characteristic deposits along the leading edges of superheaters in a power boiler fueled with agricultural residues (wood, shells, and pits).

has a biomass-fueled (wood and agricultural residue) capacity of 205 MW$_e$ [21]. The Ironbridge B power station in the UK, converted from coal to operate on wood in 2012, was perhaps the world's largest biomass power plant at 740 MW$_e$ but was decommissioned in 2015 under EU regulation after achieving maximum total lifetime operating hours [22]. Optimization studies have more generally suggested larger sizes are feasible than most biomass units currently built [23, 24]. The efficiencies of biomass power plants are generally lower than comparable fossil-fueled units because of higher fuel moisture content, lower steam temperatures and pressures to control fouling at higher combustion gas temperatures, and to some extent the smaller sizes, with a proportionately higher parasitic power demand needed to run the pumps, fans, and other components of the power station. Biomass integrated gasification combined cycles (BIGCCs) have been projected to exceed 35% electrical efficiency [25]. Co-firing biomass in higher capacity, higher efficiency fossil (e.g. coal) stations can also lead to higher efficiency in biomass-fueled power generation [12]. CHP applications realize much higher efficiencies (80% or better) due to the use of much of the heat that is otherwise rejected in power-only applications.

Biomass-fueled power boilers utilize three principal types of combustors: grate burners, suspension burners, and fluidized beds [26]. The differences in these units relate primarily to the relative velocities of fuel particles and gas and the presence of an intervening heat-transfer medium as in the fluidized bed. They also differ in their abilities to handle fouling-type fuels, their levels of emissions, and a number of other operating considerations.

3.2.2.3 Co-firing

Co-firing is the simultaneous burning of two or more fuels in the same combustor or furnace [9]. Co-firing of biomass is an attractive option for reducing greenhouse gas emissions associated with the combustion of coal and for utilizing biomass at higher efficiencies than in most biomass-dedicated power plants. Co-firing advanced rapidly in a number of countries, with more than 230 operations around the world reported by 2004 [9]. Roughly half of these were in pulverized coal facilities, with the rest principally in bubbling and circulating fluidized beds along with a few grate-fired units. Typical biomass co-firing rates without derating (reducing rated power) of the plant are 5–10% of total fuel input energy. Even at these fractions, co-firing in a large coal plant requires a substantial biomass supply, similar to a 25–50 MW$_e$ biomass plant or larger. Moderate investments are needed for storage and handling equipment until the fraction of biomass begins to exceed about 10%. Beyond this, or if the biomass is fired separately into the furnace, changes are needed to mills, burners, dryers, and other equipment, which increase cost. New pulverized coal units can co-fire up to about 40% biomass [12].

Biomass can be added in a co-firing application by pre-mixing with the coal prior to injection into the furnace, by direct injection with the coal, or by burning separately in the same furnace [27]. A special case is that of firing biomass in the upper levels of the furnace as a reburning fuel to help control NO_x emissions. Reburning is a multistage (normally two) fuel-injection technique which uses fuel as a reducing agent to react with and remove NO_x [28]. Metal oxide promoters (Na-, K-, Ca-containing additives) can be injected with or downstream of the reburning fuel to enhance the NO_x reduction, although modeling studies have shown that neither of these locations is as effective as co-injection with the main fuel [29]. Gasification of biomass and co-firing of producer gas have also been tested for reburning purposes [30].

Various technical concerns associated with biomass co-firing in coal facilities include fuel preparation, storage, handling, and supply, ash deposition, fuel conversion, pollutant formation, corrosion, ash utilization, impacts on SCR systems for emission control, and formation of striated flows in the boiler, and research is directed toward understanding and mitigating adverse impacts [31].

Comingling of biomass ash with coal ash may influence the latter's value in construction and other applications, such as concrete additives [16], although there appear to have been few quality concerns with co-fired wood fuels [15]. Higher concentrations of alkali metals and chlorine in straw and other herbaceous fuels may be of more concern, but most impacts appear to be manageable [31]. Restrictions on comingled ash in various standards for concrete admixtures pose significant penalties on co-firing operations, and research is continuing on this issue [32, 33]. New standards have been developed that include testing procedures to ensure quality of comingled ash is not reduced compared with coal ash for these applications.

3.2.2.4 *Alternative Combustion and Power Generation Concepts*

Steam Rankine cycles are the dominant power generation concept employed at present for solid biomass fuels. Alternatives to the conventional steam cycle include various enhancements, such as supercritical Rankine cycles that operate at higher temperatures and pressures (above the critical point of water at 22.1 MPa, 647 K), and are more commonly used with coal and other fossil fuels. Ash fouling and superheater corrosion are a primary concern with higher temperature systems, so most biomass-dedicated plants remain subcritical.

ORCs operate on the same cycle as steam power stations, but they replace water with another working fluid such as ammonia or a hydrocarbon like propane or butane. Operating temperatures are generally lower for ORC units that are deployed for biomass at scales of around 400–1500 kW$_e$ with efficiencies of up to 20% [34]. ORC units can also be used to take advantage of waste heat in cogeneration applications to improve overall efficiency.

Conversion of biomass by gasification to make producer gas is another alternative for power generation using conventional spark-ignited or dual-fuel compression-ignited reciprocating engines. The technology has a long history of use and development, and typically suffers from inadequate gas purification for small distributed and transportation-related applications. Dry scrubbing has been employed by at least one company in attempting to resolve issues of liquids (tars) condensation in engine intake systems and other combustion-related devices, although this approach may incur penalties in the form of power derating due to lower fuel gas densities at elevated temperatures [35].

Other cycles used with solid biomass fuels include Brayton (gas turbine) and Stirling engines. Attempts have also been made to direct-fire powdered biomass into Diesel engines [36], but scoring of the cylinder walls can occur in addition to problems of handling solids and ash. Of these, the most advanced for direct-firing of solid fuels is the Stirling engine, although this class of engines remains mostly developmental. Solid-fuel direct-fired gas-turbine engines have received considerable attention, but cleaning the combustion products sufficiently to run through the turbine blading has proved difficult [37]. Indirect-fired (hot air) turbines have also been investigated, but in these engines high-temperature heat exchange is limited by materials. Compression-ignited (Diesel), spark-ignited (Otto), and Brayton engines, especially microturbines, are currently used with biogas and landfill gas, biomethane, biodiesel, and alcohols, and should also be compatible with HCs, mixed alcohols, SNG, and other clean fuels made from syngas by Fischer–Tropsch synthesis and other techniques. Pyrolysis oils (bio-oils) and vegetable oils can also be used after hydrotreating or other refining to improve viscosity and stability, remove oxygen, and reduce corrosivity. BIGCCs, in which biomass is gasified to generate a fuel gas (producer gas or syngas) that can be used in a combined gas-turbine–steam cycle (combined cycle), similar to the use of coal in an integrated gasification combined cycle, have been under active investigation for improving power plant efficiency and potentially repowering existing steam plants [38, 39]. The Värnamo BIGCC plant demonstrated in Sweden was designed for a net electrical efficiency of 32% (lower heating value) while simultaneously generating 6 MW$_e$ of electricity and 9 MW$_t$ of heat for district heating. Overall efficiency was rated at 83% in cogeneration mode [38].

An alternative to direct combustion of biomass for power generation is conversion to gas for electrochemical oxidation in fuel cells. Five major types of fuel cell have been developed, with all practical fuel cells at present using hydrogen as the energy carrier. Alkaline, acid, and the solid polymer (or polymer electrolyte membrane)-type fuel cells require any HC, syngas, or biomass feed to first be reformed to hydrogen. Molten carbonate and solid oxide fuel cells are higher temperature types and internally reforming, so that a reforming stage upstream of the fuel cell may not be needed when using biomethane, biogas, or syngas from biomass as long as purity is high. The solid oxide fuel cell operates at temperatures in the range of 600–1000 °C and could be used to replace the gas turbine in a combined cycle operation. In such cases, the peak net electrical efficiency might be improved from about 50% to close to 70% at low loads, including parasitic demands of the cell operation, and from 30% to about 55% at high loads [40]. Significant research remains for biomass integrated fuel cell applications and many other advanced options.

3.3 Fundamentals of Biomass Combustion

Combustion is a complex phenomenon involving simultaneous coupled heat and mass transfer with chemical reaction and fluid flow. For the purposes of design and control, thorough knowledge is required of fuel properties and the manner in which these properties influence the outcome of the combustion process. Combustion conditions must also be specified, including type of oxidant (air, oxygen, oxygen-enrichment), oxidant-to-fuel ratio (stoichiometry), type of combustor (e.g. pile, grate, suspension, fluidized bed), emission limits, and many other factors. Fully detailed models of the combustion process include pyrolysis and gasification of solid feedstock along with homogeneous and heterogeneous oxidation involving a substantial number of reactions and reaction intermediates. Comprehensive models have been developed for combustion of fuels such as hydrogen and CH_4, but have not so far been completed for more complex fuels such as biomass. Fortunately, simpler approaches involving more global reaction processes can be used to account for specific feedstock properties and combustion conditions.

3.3.1 Combustion Properties of Biomass

Combustion of biomass is heavily influenced by the properties of the feedstock and the reaction conditions (e.g. air/fuel ratio). The amount of heat released during combustion depends on the energy content of the fuel along with the conversion efficiency of the reaction. The organic matter assembled by photosynthesis and plant respiration contains the majority of the energy in biomass, but knowledge of the inorganic fraction is also important in the design and operation of the combustion system, particularly in regards to ash fouling, slagging, and in the case of fluidized bed combustors, agglomeration of the bed medium.

3.3.1.1 Composition of Biomass

Photosynthesis and plant respiration result in the production of a diverse and chemically complex array of structural and nonstructural carbohydrates and other compounds, including cellulose, hemicellulose, lignin, lipids, proteins, simple sugars, starches, HCs, and ash, which along with water comprise the majority of the biomass. The concentration of each class of compounds varies depending on species, type of plant tissue, stage of growth, and growing conditions. Cellulose is a linear polysaccharide of β-D-glucopyranose units linked with (one to four) glycosidic bonds. Hemicelluloses are polysaccharides of variable composition including both five- and six-carbon monosaccharide units. The lignin is an irregular polymer of phenylpropane units [41–45]. Plants producing large amounts of free sugars, such as sugar cane and sweet sorghum, are attractive as feedstock for fermentation, as are starch crops such as maize (corn) and other grains. Lignins are not yet generally considered fermentable, although research is active in this regard, and thermochemical means such as combustion are frequently proposed for their conversion. Combustion can be applied either to the direct conversion of the whole biomass or to portions remaining following biochemical separation, such as by fermentation. Combustion, unlike the biochemical and some other thermochemical conversion strategies, is essentially nonselective in its conversion of the biomass and reduces the whole fuel to products of CO_2 and water along with generally smaller amounts of other species, depending on feedstock composition and process efficiency. However, the complex structure of biomass still has a significant influence on its combustion behavior.

Owing to its carbohydrate structure, biomass is highly oxygenated in comparison with conventional fossil fuels. Typically, 30–40 wt% of the dry matter in biomass is oxygen (Table 3.2). The principal constituent of biomass is carbon, making up from 30 to 60 wt% of dry matter depending on ash content, and most woods are about half carbon when dry. Of the organic component, hydrogen is the third major constituent, comprising typically 5–6% dry matter. Nitrogen, sulfur, and chlorine can also be found in quantity, usually in concentrations less than 1% dry matter, but occasionally well above this. These are important in the formation of criteria and hazardous air pollutant emissions and in other design and operating

Table 3.2 Properties of selected biomass feedstocks [55].

Property	Hybrid poplar wood	RDF[a]	Rice straw	Sugar cane bagasse	Willow wood
Proximate analysis (% dry matter)					
Fixed carbon	12.49	0.47	15.86	11.95	16.07
Volatile matter	84.81	73.4	65.47	85.61	82.22
Ash	2.7	26.13	18.67	2.44	1.71
Total	100	100	100	100	100
Ultimate analysis (% dry matter)					
Carbon	50.18	39.7	38.24	48.64	49.9
Hydrogen	6.06	5.78	5.2	5.87	5.9
Oxygen (by difference)	40.43	27.24	36.26	42.82	41.8
Nitrogen	0.6	0.8	0.87	0.16	0.61
Sulfur	0.02	0.35	0.18	0.04	0.07
Chlorine	0.01		0.58	0.03	< 0.01
Ash	2.7	26.13	18.67	2.44	1.71
Total	100	100	100	100	100
Elemental composition (% ash)					
SiO_2	5.9	33.81	74.67	46.61	2.35
Al_2O_3	0.84	12.71	1.04	17.69	1.41
TiO_2	0.3	1.66	0.09	2.63	0.05
Fe_2O_3	1.4	5.47	0.85	14.14	0.73
CaO	49.92	23.44	3.01	4.47	41.2
MgO	18.4	5.64	1.75	3.33	2.47
Na_2O	0.13	1.19	0.96	0.79	0.94
K_2O	9.64	0.2	12.3	0.15	15
SO_3	2.04	2.63	1.24	2.08	1.83
P_2O_5	1.34	0.67	1.41	2.72	7.4
CO_2	8.18				18.24
Total	100	100	100	100	100
Undetermined	1.91	12.58	2.68	1.39	8.38
Higher heating value (MJ/kg)	19.02	15.54	15.09	18.99	19.59
Alkali index (kg/GJ)	0.14	0.23	1.64	0.06	0.14
Stoichiometric air:fuel (kg/kg)	6.10	5.38	4.61	5.75	5.96
Enthalpy of formation (kJ/kg)	−5995	−5662	−4796	−5266	−5113
Adiabatic flame temperature[b] (K)	2211	2016	2192	2301	2313

[a] Refuse-derived fuel.
[b] Linearized estimate.
Source: reprinted from Jenkins et al. [55], with permission from Elsevier.

considerations, including materials selection. Nitrogen is a macronutrient for plants, and critical to growth, but is also involved in NO_x and nitrous oxide (N_2O) formation during combustion. Sulfur and chlorine contribute to fouling, slagging, corrosion, and emissions of major air pollutants. Inorganic elements can also be found in high concentrations. In annual growth tissues, concentrations of the macronutrient potassium frequently exceed 1% dry matter. Like sodium, another alkali metal, potassium is involved in slag formation and ash fouling. Sodium is toxic to most plants other than the halophytes, and so is usually found in lower concentrations than potassium. In at least trace amounts, virtually every element can be found in biomass, with important consequences in the design, operation, and environmental performance of combustion facilities [46–48]. Elemental properties of biomass have been determined for a wide range of fuel types [15, 49–54]. Databases of biomass properties are also maintained online (e.g. https://phyllis.nl). These properties include moisture content, heating value, elemental composition, bulk density, specific gravity, thermal conductivity, and mechanical, acoustic, and electrical properties.

Biomass is similar to other fuel types in the need for standardized methods of analysis leading to accurate and consistent evaluations of fuel properties. Standards have been developed for many properties, although this is still an area of active development. ASTM (http://www.astm.org), ISO (http://www.iso.org), and other organizations maintain standards for analyzing chemical composition, heating value, density, ash fusibility, and other properties, many of which are summarized in a number of more detailed references [15, 55]. Standards for similar types of analysis, but developed for coal and other fuels, should be used only when biomass-specific standards are not available, and then they should be used with caution, recognizing that differences in the chemistry of the feedstock may lead to difficulties in analysis.

The molar ratios of oxygen and hydrogen to carbon are very similar for a wide range of biomass feedstock. Glucose ($C_6H_{12}O_6$), a primary product of photosynthesis, can be written in terms of elemental molar composition as CH_2O. Cellulose ($C_6H_{10}O_5)_n$, a polymer built from glucose, yields $CH_{1.67}O_{0.83}$ with H:O in the ratio of 2. Averaging across many types of biomass, the elemental molar composition can be expressed as $CH_{1.41}O_{0.64}$, indicating the presence of lignin and other less-oxygenated species [50]. This is also seen in properties such as the theoretical air/fuel ratio for complete combustion, equal to 6.0 kg of dry air per kilogram of dry organic (CHO) matter for the generalized biomass, 5.1 kg/kg for cellulose, and 4.6 kg/kg for glucose. By comparison, the theoretical air/fuel ratio for carbon is 11.4 kg/kg. Increasing carbonization occurs in the transformation of biomass to coals of increasing rank [56]. Black liquor produced during pulping also exhibits increased lignification in the reduction of the O:C ratio while retaining a H:C ratio roughly equal to the parent biomass.

Inorganic material in biomass can be divided into two fractions: ash inherent in the feedstock and ash inadvertently added to the fuel through collection and processing. The latter, adventitious material, such as soil accumulated during harvesting, often makes up a major fraction of the ash content of wood fuels used in power plants and originates from skidding and other operations used to move trees and slash (branches and tops) from the forest. Its composition is typically different from that of the inherent materials, as is the mode of occurrence of the elements (e.g. crystalline silicates and aluminum arising from the incorporation of sands, clays, and other soil particles and potassium incorporated in feldspars with relatively little contribution to the alkali reactions that lead to fouling other than by inertial impaction and sticking of particles). The inherent inorganic matter is more intimately distributed throughout the fuel, and is sometimes referred to as atomically dispersed material. Elements including Si, K, Na, S, Cl, P, Ca, Mg, and Fe are involved in reactions leading to ash fouling and slagging, and the principal mechanisms describing these phenomena in biomass combustors are now reasonably well understood [15, 52–54, 56, 57]. Descriptions of the detailed chemistry and means to control or mitigate these processes other than by fuel selection are the subjects of much research [47, 58–65]. The impacts on the fouling and slagging behavior of biomass by removing certain elements, such as potassium and chlorine, confirm much of what is perceived about the mechanisms involved [66–70].

Herbaceous fuels, such as grasses and straws, contain silicon and potassium as their principal ash-forming constituents. They are also commonly high in chlorine relative to other biomass fuels. These properties portend potentially severe ash deposition problems at high or moderate combustion temperatures. The primary sources of these problems arise from the reaction of alkali with silica to form alkali silicates that melt or soften at low temperatures, and the reaction of alkali with sulfur to form alkali sulfates on combustor heat transfer surfaces. Sugar cane bagasse is substantially depleted in potassium relative to the parent material due to the washing of the cane that occurs during sugar extraction. Similar results have been obtained for solid–liquid extraction of biomass [66]. The composition of rice straw ash, for example, is remarkably similar in respect to alkali metal and silica concentrations in an ordinary soda-lime glass except that potassium is the major element rather than the sodium predominantly used for glass making [55]. A substantial fraction (>80% typically) of potassium can be extracted from straw and other biomass by simple water leaching, yielding an inorganic fraction enriched in silica with a much higher melting temperature. Chemical fractionation of the feedstock using different solvents to detect the form of the ash constituents indicates the mobility of inorganic elements during combustion [53, 55].

3.3.1.2 Moisture Content

Moisture content is highly variable in biomass and has a large influence on the combustion chemistry and energy balance. Moisture concentration is defined in two principal ways: dry basis and wet basis. The dry basis moisture M_{db} expresses the mass of water in the feedstock per unit mass of dry matter. The wet basis M_{wb} relates the water to the total wet weight, dry matter plus water.

$$M_{db} = \frac{m_w}{m_{f,d}} \quad M_{wb} = \frac{m_w}{m_{f,d} + m_w} \quad M_{wb} = \frac{M_{db}}{1 + M_{db}} \quad M_{db} = \frac{M_{wb}}{1 - M_{wb}} \quad (3.1)$$

The oven-dry (also known as bone-dry) state achieved after heating in an air oven at temperatures below about 104 °C is generally defined to be the zero moisture datum, although bound moisture may still be present and other light volatiles such as alcohols may be lost during drying. The dry basis moisture has no upper bound; the wet basis is bounded by 100%. Feedstock will vary in moisture when in contact with air of variable humidity. Equilibrium moisture in biomass ranges typically above 25% wet basis as the relative humidity exceeds 90–95%.

3.3.1.3 Heating Value

The energy content or heating value of biomass is defined as the heat released by combustion under specific conditions. For measurement purposes, the reaction is carried out either at constant pressure or constant volume, the latter being the more commonly reported for solid biomass feedstock when analyzed using a bomb calorimeter. For feedstocks that contain hydrogen, the water formed by oxidation (refer to Section 3.3.2) may be either in vapor or liquid phase when the reaction completes. If in the vapor phase, the amount of heat released as measured by the calorimeter will be less by the heat of vaporization than if the water is condensed. The higher heating value (also referred to as the gross calorific value), either at constant pressure or constant volume, is measured with product water condensed. The lower heating value (or net calorific value) measures the heat released with water in the vapor phase. Reporting the basis of the measurement is important for the purposes of comparing analyses and system performance. For example, boiler or power plant efficiency is by convention commonly reported using the lower heating value of the fuel in some countries, while in others the higher heating value is used. For the same facility, the efficiency will be greater when reported on the basis of the lower heating value of the fuel.

The as-fired higher heating value of a moist fuel is directly related to the dry basis higher heating value and the moisture content as

$$Q_h = Q_{h,o}(1 - M_{wb}) \tag{3.2}$$

The as-fired lower heating value can be determined from the higher heating value, hydrogen concentration H (kg/kg) in the dry feedstock, and the moisture content:

$$Q_{V,l} = (1 - M_{wb})\left[Q_{v,h,o} - u_{fg}\left(M_{db} + \frac{W_{H_2O}}{2W_H}H\right)\right] \tag{3.3}$$

The internal energy of vaporization u_{fg} (kJ/kg), in Eq. (3.3) is used for the case of constant-volume combustion (no flow work). If the reaction is instead carried out at constant pressure, u_{fg} is replaced by h_{fg} (kJ/kg), the enthalpy of vaporization. The W_i are the molar masses (molecular weights, kg/kmol) of water and hydrogen.

The maximum temperature achieved by combustion is dependent in part on the amount of sensible heat (that capable of raising temperature in a system) available for heating reaction products and vaporizing feedstock moisture. The greater the moisture content of the feedstock, the greater the fraction of heat needed to evaporate water (Figure 3.3):

$$Q_{residual}(\%) = 100\left[1 - \frac{u_{fg}M_{wb}}{(1-M_{wb})Q_{v,h,o}}\right] \tag{3.4}$$

Beyond about 90% moisture for most biomass, the energy needed to vaporize water exceeds the heating value of the feedstock. The autothermal limit below which the fire is self-sustaining is typically in the vicinity of 70–80% [71], and flame stability becomes poor in most combustion systems above 50–55% moisture wet basis. Natural gas and other fuels are often co-fired for flame stabilization when fuel moisture is high.

The heating value can be partially correlated with ash concentration. Woods with less than 1% ash typically have heating values near 20 MJ/kg dry weight Each 1% increase in

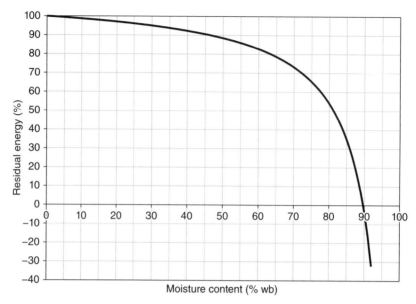

Figure 3.3 Sensible energy available from combustion after evaporating moisture in feedstock ($Q_{v,h,0} = 20$ MJ/kg, $u_{fg} = 2.3$ MJ/kg).

ash translates roughly into a decrease of 0.2 MJ/kg [49] because the ash does not contribute substantially to the overall heat released by combustion, although elements in ash may be catalytic to the thermal decomposition. Heating values can also be correlated to carbon concentration, with each 1% increase in carbon elevating the heating value by approximately 0.39 MJ/kg dry weight [49], a result that is identical to that found by Shafizadeh [72] for woods and wood pyrolysis products. The heating value relates to the amount of oxygen required for complete combustion, with 14.0 MJ released for each kilogram of oxygen consumed [72]. Cellulose has a smaller heating value (17.3 MJ/kg) than lignin (26.7 MJ/kg) because of its higher degree of oxidation.

Various efforts have also been made to correlate heating value with the ultimate elemental composition of biomass, similar to correlations developed for coal. One correlation is that of Gaur and Reed [73] for $Q_{v,h,0}$ (MJ/kg):

$$Q_{v,h,o} \text{ (MJ kg}^{-1}) = 0.3491 \times C + 1.1783 \times H + 0.1005 \times S - 0.1034 \times O \\ - 0.0151 \times N - 0.0211 \times \text{Ash} \tag{3.5}$$

where C, H, S, O, N, and Ash are the weight percent (dry weight basis) of carbon, hydrogen, sulfur, oxygen, nitrogen, and ash in the feedstock. For the generalized biomass ($CH_{1.41}O_{0.64}$) noted above, the correlation of Eq. (3.5) yields a heating value equal to 20.31 MJ/kg dry weight. As with other models of this type, the agreement with analytical data is variable and often poor, overestimating by up to about 2 MJ/kg dry weight in some cases. Estimations should not be used in place of measurements for critical design considerations.

3.3.1.4 Density, Particle Size, and Other Properties

Biomass density and particle size influence feedstock handling and combustion rates and efficiencies. Density is affected by moisture content due to both the change in feedstock mass and the volume change (e.g. swelling) accompanying changes in moisture content. Densification, such as briquetting or pelletizing, increases the bulk density, but may also alter the apparent density of particles (excluding interstitial volumes between particles). True density (excluding pore volumes in particles) for cellulose from wood has been reported as 1580 kg/m^3, with lignin in the range 1380–1410 kg/m^3 [74]. Bulk densities are highly variable and range from about 20–80 kg/m^3 for loose or loose chopped materials to above 700 kg/m^3 for pelleted materials. The latter can in some cases exceed the density of water. Particle size, size distribution, and particle morphology are also highly variable, depending on the type of processing used to prepare biomass for firing, if any. These influence aerodynamic properties, surface area, and conversion rates along with a number of other factors. In general, the process energy needed to reduce the particle size increases as mean particle size decreases.

Many other chemical, electrical, mechanical, and thermal properties of the feedstock influence the handling, processing, and combustion of biomass [15, 49, 50]. Detailed design should rely on specific analyses of intended feedstocks rather than more general literature data, although the latter are useful for preliminary assessments.

3.3.2 Combustion Stoichiometry

3.3.2.1 Simplified Global Reaction

A simplified global combustion reaction for biomass is represented by the mass balance of Eq. (3.6), in which the mass coefficients $m_{r,i}$ designate the masses of the reactant species and the $m_{p,j}$ designate the product species. In this case the biomass is divided into two fractions: an organic phase and an ash phase. The feedstock elemental mass fractions described by the coefficients y_i arise from the analysis of the dry feedstock for C, H, O, N, S, and Cl. Moisture in biomass is represented by a separate liquid water fraction, even though water may be held differently in the biomass. Air as oxidant is considered to consist of the gases O_2, CO_2, H_2O, and N_2. The reaction results in 11 main gas-phase products and a residual mass containing ash and unreacted portions of the other element masses from the biomass. The residual can consist of solids such as PM, carbon in ash as charcoal or carbonates, and chlorides and sulfates in furnace deposits. In practice, liquids and gases can also result, which can be represented in the reaction given appropriate property data. Equation (3.6) is reasonably general, in that it can equally be used to describe the combustion of other fuels such as (bio)methane, biodiesel, Fischer–Tropsch liquids and other HCs, bio-oils from pyrolysis, and many others.

$$m_{r,1}\left(\sum_i yi\bigg|_{C,H,O,N,S,Cl,Ash}\right) + m_{r,2}H_2O(l) + m_{r,3}O_2 + m_{r,4}CO_2 + m_{r,5}H_2O(g) + m_{r,6}N_2$$
$$= m_{p,1}CH_4 + m_{p,2}CO + m_{p,3}CO_2 + m_{p,4}H_2 + m_{p,5}H_2O + m_{p,6}HCl$$
$$+ m_{p,7}N_2 + m_{p,8}NO + m_{p,9}NO_2 + m_{p,10}O_2 + m_{p,11}SO_2 + m_{residual} \qquad (3.6)$$

The reactant masses in Eq. (3.6) are specified by reaction conditions and properties of the feedstock or fuel. Except in the case of complete oxidation, where a number of the product masses are taken as zero, Eq. (3.6) is not determinate for the product masses by elemental mass balances alone, as there are more unknowns than element balances for C, H, O, N, S, and Cl along with the overall mass balance. To determine the product masses, other information must be given or another solution technique employed, such as an equilibrium solution. If the fractions of feedstock elements, $f_{i,j}$, converted to certain product species, such as fuel N to NO and NO_2, sulfur to SO_2, and Cl to HCl are known a priori or otherwise assumed, the reaction is then determinate. Nitrogen is problematical in this regard, as for any oxidant containing nitrogen and oxygen, such as air, thermal and prompt oxides of nitrogen (NO_x) can also be produced in addition to those from fuel nitrogen. In such a case the conveniently prescribed fraction of fuel N converted to NO_x could exceed unity, although the mass balance of Eq. (3.6) remains applicable. In addition to the product composition, Eq. (3.6) allows for the estimation of other properties of the reaction, such as the flame temperature, although additional information may be needed, such as the heat loss from the reaction if not adiabatic.

3.3.2.2 Air/Fuel Ratio

Equation (3.6) also specifies the oxidant-to-fuel ratio, or in the case of air as the source of oxygen, the air/fuel ratio, along with the equivalence ratios often used to specify the combustion conditions. The air/fuel ratio defines the mass of air added relative to the mass of feedstock, expressed on either a wet or dry basis. The stoichiometric value (AF_s) defines the special case in which only the amount of air theoretically needed to completely burn the feedstock is added to the reaction. The fuel/air equivalence ratio ϕ, the air/fuel equivalence ratio or air factor λ, and the excess air e are related as

$$\phi = \frac{AF_s}{AF} \qquad \lambda = \frac{1}{\phi} \qquad e = \frac{1}{\phi} - 1 \qquad (3.7)$$

The combustion regimes are defined by the value of ϕ with $\phi = 1$ ($e = 0$) the stoichiometric combustion, $\phi > 1$ the fuel-rich regime (insufficient air), and $\phi < 1$ the fuel-lean regime (excess air). For fuel-rich conditions, concentrations increase for products of incomplete combustion, such as HCs, CO, and PM. Other pollutant species not included in Eq. (3.6) are also produced in varying amounts under all three combustion regimes.

3.3.2.3 Flame Temperature

The reaction temperature associated with Eq. (3.6) is found through application of the energy balance. For the constant-pressure reaction, the enthalpy of the products is equal to the sum of the reactant enthalpies and any heat transfer to the system:

$$H_p = H_r + Q_1 \qquad (3.8)$$

or

$$\sum_j \frac{m_{p,j}}{W_j} h_{p,j} - \sum_i i \frac{m_{r,i}}{W_i} h_{r,i} - Q_1 = 0 \qquad (3.9)$$

where H_p is the total enthalpy of the products, H_r the total enthalpy of the reactants, and Q_1 the external heat transfer to the reaction, which is negative for heat loss. The enthalpies of formation for all species other than the biomass are known, and the latter can be solved by energy balance on the reaction in oxygen carried out in determining the heating value. The other species' enthalpies h_i and h_j are in most cases nonlinear functions of temperature. Reference values of the molar enthalpies (kJ/mol) have been compiled in the NIST database for $298.15 < T$ (K) < 6000 (for steam, the lower temperature bound is 500 K [75]).

Owing to the nonlinear behavior of the enthalpies with temperature, a solution for the flame temperature generally involves a root-finding or similar technique. As noted by Jenkins and Ebeling [76], the enthalpy functions are nearly linear over a fairly large range of temperature, including the flame temperatures in air for many fuels. Fitting the enthalpies to linear functions allows for a direct approximation of the flame temperature. Based on the data of Linstrom and Mallard [75], functions of the form $h = aT - b$ were fit over the temperature range 500–3000 K, where a and b are regression coefficients and T (K) is the absolute temperature. Coefficients of the functions are listed in Table 3.3, along with the correlation coefficients indicating quality of fit. For higher temperatures, such as achieved in oxy-fuel combustion, dissociation reactions must also be considered.

Enthalpies for the solid phases including ash and char, among other products, also appear in Eq. (3.9). For ash, the correlation of Kirov [77], as attributed by Hanrot et al. [78], gives the specific heat (J/kg K) as a function of temperature ($0 < T < 1200$) for T in Celsius:

$$C_{p,ash} = 752 + 0.293T \qquad (3.10)$$

From the definition of the enthalpy and extrapolating and linearizing over the temperature range 500–3000 K (227–2727 °C), the ash enthalpy is approximated as above with $a = 1.5510$ and $b = 856$, but for h on a mass basis (kJ/kg). For carbon black as a residual solid, NIST specifies a constant $C_{p,c} = 10.68$ J/mol K over the temperature range $300 < T$ (K) < 1800. The enthalpy of product carbon at the flame temperature T_f is calculated from $h = C_{p,c}T - h_{298.15}$, where the specific heat is referenced to the temperature 298.15 K and extrapolated for temperatures above 1800 K. Enthalpies of other solid species depend on the form in which they appear in the products, although if the masses are small they may be simply lumped with the ash. Solid products may achieve temperatures considerably different

Table 3.3 Linearized enthalpies (kJ/kmol) of selected gases, (500–3000 K), H = aT − b, T in kelvin.

	a	b	r^2
CH_4	87.701 7	123 160	0.994 542 9
CO	35.330 1	123 988	0.999 381 7
CO_2	58.528 8	418 206	0.999 054 1
H_2	33.179 5	12 508	0.998 743 1
H_2O	48.340 9	263 954	0.996 427 3
HCl	34.474 1	105 620	0.998 966 9
N_2	34.855 5	13 041	0.999 217 1
NO	35.855 1	−76 847	0.999 529 9
NO_2	55.104 0	−10 755	0.999 432 4
O_2	36.890 6	13 996	0.999 346 8
SO_2	57.082 6	319 166	0.999 573 2

from the flame temperature in actual combustors, so an assumption of a uniform temperature for all products constitutes a special case. Note that the range in specific heat computed for ash from 25 to 1200 °C (0.76–1.10 kJ/kg K) also includes the specific heat for carbon black (0.89 kJ/kg K).

Using the linearized enthalpy equations, $h = aT - b$, the product enthalpy can be written thus:

$$H_p = \sum_j m_j aT - b_j = \sum_j a_j T m_j - \sum_j b_j m_j \quad (3.11)$$

with the reactant inlet temperatures specified (e.g. with air preheat), the flame temperature can be directly estimated:

$$T = \frac{H_r + Q_1 + \sum_j b_j m_j}{\sum_j a_j m_j} \quad (3.12)$$

The use of this equation is illustrated in Figure 3.4 for the case of hybrid poplar wood combustion in air for 0–100% excess air and the same feedstock moisture range (dry basis). Adiabatic flame temperatures are compared for inlet air temperatures of 298 and 500 K. Heat loss equal to 10% of fuel heating value is sufficient to drop the peak flame temperature by approximately 200 K.

3.3.3 Equilibrium

Only in the case of simple reactions or where conversion fractions are already known will element balances be sufficient to determine the reaction products. When more species and phases are present, the element balances alone may not yield a sufficient number of equations to make the system determinate. More detailed kinetic analyses may be required, or if the reactions proceed rapidly enough, an assumption of equilibrium can be employed to solve the stoichiometry. A number of species are not adequately predicted from equilibrium, however, including important pollutant species such as NO_x.

The equilibrium composition can be determined from the state at which the Gibbs free energy of the system reaches a minimum at a given temperature T and pressure P. A number of sophisticated computer-based equilibrium solvers are available, and the thermodynamic basis for determining the equilibrium is detailed elsewhere (e.g. [79]).

3.3.4 Rates of Reaction

The rate of combustion is also important to the design of combustion systems. Underdesigned furnaces, boilers, and other combustion units lead to reduced capacity and poor economy of operation. Ash fouling and slagging can also be exacerbated in underdesigned units when forced to meet design capacity due principally to the need to operate above design temperature.

Typical design heat release rates (expressed per unit grate area) for stoker-fired traveling grate combustors are in the range 2–4 MW_t/m^2. Design for a conceptual whole-tree combustion concept using a deep bed of fuel estimated a heat release rate of 6 MW_t/m^2, but this was not confirmed [80]. Some circulating fluidized bed furnaces firing biomass have heat release rates approaching 10 MW_t/m^2 [55].

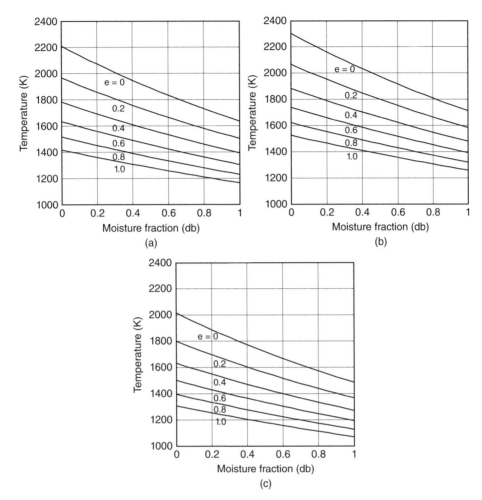

Figure 3.4 Estimated flame temperatures for combustion of hybrid poplar wood in air (e is the fraction of excess air, moisture given on a dry basis). (a) Adiabatic, inlet air temperature is 298 K. (b) Adiabatic, inlet air temperature is 500 K. (c) With heat loss, inlet air temperature is 298 K, heat loss is 10% of fuel higher heating value.

The rate at which biomass burns depends predominantly on the rate of heat transfer and the kinetic rates of reaction [81]. Particle size and morphology dominate the influence of heat transfer. Combustion occurs both in the gas phase, with the burning of volatile materials released through pyrolysis and gasification of the fuel, and heterogeneously in the solid phase as char oxidation. The burning of volatiles is generally quite rapid and follows as fast as volatiles are released; the oxidation of the char occurs much more slowly. The residence time of the particle in the furnace and the environment of the particle are important, therefore, to the total conversion attained through combustion, as well as the emissions from the combustor.

Fundamental to the combustion rate are the rates of fuel pyrolysis and char oxidation. The standard method of measuring these rates is via dynamic thermogravimetric (TG) analysis,

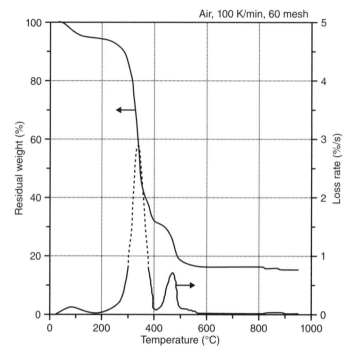

Figure 3.5 Thermogram produced under a constant heating ramp of 100 K/min for 60-mesh particle size rice straw in air [82]. Source: reprinted with permission from Bining and Jenkins [82].

whereby a small sample of the fuel is heated at a controlled rate in a controlled atmosphere while simultaneously recording weight, time, and temperature. Evolved-gas analysis also monitors the composition of products during reaction and provides additional information. A new technique coupling a TG analyzer with an inductively coupled plasma mass spectrometer (ICPMS) has been developed for this purpose [48]. The thermogram has a characteristic shape for biomass heated and burned in air (Figure 3.5). Starting from room temperature with a dry sample, feedstock drying is first observed (even if initially oven dried, biomass being hygroscopic) with a small weight loss up to about 150 °C. Between about 200 and 400 °C there is a rapid loss of weight due to the evolution of volatile material, which in an oxidative environment (e.g. air) at high enough temperature will ignite and burn. A slow loss of weight accompanies the residual char burning after the release of volatiles. Kinetic parameters can be determined from the thermogram, allowing prediction of the overall rate of reaction. Although there is no fundamental basis for the kinetic model used, Arrhenius kinetics are frequently employed [82], although isoconversional or model-free kinetic estimation can also be employed.

Metals in biomass are known to have an effect on reaction rates and are thought to be catalytic to pyrolysis. Leaching biomass in water results in a distinct effect on the conversion kinetics [66]. The result is consistent with what is known about the effects of alkali chlorides on the pyrolysis rates of biomass [83]. Under isothermal heating, the emission rate for volatiles has been observed to terminate earlier with the leached material than for the untreated material [84], and leached straw has been observed to ignite more readily than untreated straw. Chlorine is known to retard flame propagation by terminating free-radical chain reactions. Chlorine is leached from biomass by water in solid–liquid extraction [67], which, therefore, has an effect on ignition and burning in addition to formation of hazardous emissions.

Biomass particle morphology varies widely depending on type of feedstock (e.g. straw, wood) and level and type of processing (e.g. knife milling, hammer milling). Biomass particles generally exhibit large aspect ratios and more commonly resemble cylinders or plates than spheres, with corresponding differences in surface area-to-volume ratio that influence the burning rates. Larger sized biomass particles also develop thicker boundary layers, with the result that some or all of the flame may be contained within the thermal and mass transfer boundary layers. The particle, therefore, experiences an increase in surface temperature during most of the oxidation stage, although this constitutes a small fraction of the total reaction time [85]. Isothermal particle models normally employed for predicting combustion of small coal particles do not capture the effects of large temperature gradients that develop in biomass particles, an important feature in modeling biomass co-firing in pulverized coal boilers. More comprehensive models that account for drying, devolatilization, recondensation, char gasification and oxidation, and gas-phase combustion have been developed for predicting single particle combustion of variable geometry and properties [86].

3.4 Pollutant Emissions and Environmental Impacts

Environmental impacts are of primary importance for the design of combustion systems at any scale. Modern biomass combustion power plants are often zero discharge for wastewater (treated or evaporated on site) or reduce wastewater discharges to levels that are accommodated within the local municipal sewer system. Water supply and water quality are, however, critical issues for plant siting and operations, and most boiler facilities include a water treatment plant to produce high-purity water for the boiler. The largest fraction of solid waste is typically the bottom and fly ashes from the furnace and emission control equipment. With the exception of metals-contaminated feedstock, such as some urban and industrial wastes, ash often has value in secondary markets, such as land application as agricultural fertilizer, as admixtures for concrete, and in the steel-making industry. Air emissions generally constitute the largest environmental concern for most combustion systems and have become a principal inhibitor to expanded development in many regions with poor air quality due to the high cost of stringent emission control or emission offsets. Smoke emissions from uncontrolled fires and stoves have long been recognized as major contributors to human respiratory disease and other disease. Increasing use of small biomass combustion systems for distributed power generation will similarly be of concern for increased risk of exposure to pollutant emissions among other environmental factors, such as noise, odors, and fugitive emissions (e.g. dust and debris). Despite these concerns, progress has been made in reducing emissions from combustion systems and improving emissions measurement and monitoring. Environmental issues in biomass combustion, including emission control devices, are reviewed in greater detail by van Loo and Koppejan [15].

Primary pollutants formed during biomass combustion include PM, CO, HCs, oxides of nitrogen (NO_x, principally NO and NO_2), and oxides of sulfur (SO_x, principally as SO_2). Other acid gases, such as HCl, accompany the use of halogenated feedstock, such as MSW and straw. Elevated halogen concentrations can also lead to the formation and emission of hazardous air pollutants in addition to accelerated corrosion in the combustion system. Of particular concern for chlorinated fuels is the formation of dioxins and furans, especially with waste fuels, although new emission standards for WTE facilities have greatly reduced dioxin contamination from this source [87, 88]. The presence of heavy metals, such as lead, cadmium, selenium, and zinc, in the feedstock leads to their concentration in ash during burning, sometimes to levels above the hazardous waste thresholds as measured by the toxicity characteristic leaching procedure and related methods [47, 89], or, as in the case of mercury (Hg), released as vapor to the atmosphere with subsequent downwind condensation or reaction and deposition. Heavy metals can be present in high concentration in certain urban and refuse-derived fuels, especially if treated or painted woods are present [90]. Control of heavy metals is generally by exclusion of contaminated feedstock, unless the facility is specifically designed and permitted for such use. Research into oxy-fuel combustion processes that replace air with oxygen holds promise for reducing emissions of greenhouse gases, especially CO_2, in addition to products of incomplete combustion and thermal NO_x, but still must contend with environmental concerns over fuel NO_x and metals, in addition to added costs for oxygen production. Advanced systems such as BIGCC integrated with downstream oxy-fuel, electrochemical, and other process options suggest the need for expanded research in this area. Greenhouse gas emissions are of less concern for biomass conversion than with fossil fuels when biomass is produced on a renewable basis. Fossil fuel use, especially diesel fuel used in biomass production, harvesting, collection, and transport, implies that the use of biomass is not completely carbon neutral unless an additional increment of biomass is continuously grown to offset the fossil greenhouse gas emissions, or carbon from the biomass is sequestered. Other greenhouse gases, such as CH_4 and N_2O emitted during biomass processing or combustion, also make biomass systems less than carbon neutral due to the higher global warming potential (GWP) associated with these species in comparison with the CO_2 taken up by plants and used in photosynthesis. Combustion of biomass and waste in various small systems has been found to lead to higher emissions of N_2O in addition to CO [91]. Indirect effects leading to deforestation and agricultural expansion with high greenhouse gas emissions elsewhere in the world when biomass is produced as an energy crop may further reduce the sustainability of bioenergy and potentially increase exposures to criteria and other pollutants as evidenced, for example, by considerations of indirect land use change effects as part of fuel carbon intensities in the California low carbon fuel standard (LCFS) although this area is remains controversial and subject to revision in application of standards and regulations [92]. Nonetheless, the use of biomass in properly designed systems has the potential to radically reduce the release of greenhouse gases relative to the use of fossil fuels.

3.4.1 Oxides of Nitrogen and Sulfur

One of the largest concerns for many areas in the USA and elsewhere around the world is emissions of NO_x due to already high emissions from existing industries and vehicles. Increasingly stringent NO_x emission standards in California, for example, reduce the economic feasibility of dairy biogas systems with on-site power generation and make siting of new solid fuel and other biomass systems more difficult [93, 94]. NO_x emissions have also been of concern in the use of biodiesel as vehicle fuel, where a number of studies have yielded NO_x increases compared with petroleum diesel [95], although this is an issue of continuing research.

Emissions of sulfur oxides are associated with sulfur in the fuel unless some other source of sulfur is available, such as H_2S in a co-fired waste or process gas stream. Sulfur oxides are respiratory irritants and their effects are enhanced in the presence of PM due to transport deep within the lung. Both sulfur oxides and nitrogen oxides contribute to acid precipitation. In addition to sulfur oxides, some sulfur remains in the ash or is deposited in the furnace. Some may also be released as salts or H_2S [15]. SO_x emissions can be reduced through dry or wet scrubbing. Removal of sulfur upstream of post-combustion catalysts is often important to avoid catalyst fouling and deactivation. Inadequate sulfur control has been a consistent problem with engines burning digester or landfill gas and using catalytic after-treatment of combustion gases for NO_x control [96, 97]. Limestone or dolomite injection in fluidized bed combustors has been a primary control measure for SO_x, in addition to reducing fireside fouling from sulfates in boilers [15, 98].

Emissions of NO_x and other nitrogenous species, such as N_2O (a strong greenhouse gas), in biomass combustion systems arise mostly from fuel N concentrations. NO is the primary species (>90%) formed during combustion and is later oxidized to NO_2 in the atmosphere [15]. The latter compound typically serves as the reporting basis for NO_x emissions. NO_x formation is associated with three principal mechanisms: fuel NO_x stemming from nitrogen in the fuel, thermal NO_x resulting from high-temperature reactions between nitrogen and oxygen mostly from the air, and prompt NO_x due to reactions of nitrogen in air with HC radicals to form hydrogen cyanide (HCN) that then follows the fuel NO_x formation route [15, 98]. For biomass, the fuel NO_x mechanism is presently the most important in most commercial systems due to relatively low flame temperatures. NO is also formed from residual N during char combustion, but is reduced via a fast heterogeneous char reaction to N_2 [15]. NO formation from fuel nitrogen occurs on time scales comparable to HC oxidation, and is known to be sensitive to equivalence ratio, with fuel-lean conditions producing higher yields and fuel-rich conditions producing lower yields [99, 100]. Under fuel-rich conditions, the relatively fast conversion of fuel carbon to CO competes for oxygen, leading to a reduced availability of oxygen for NO_x formation. The fractional conversion of fuel nitrogen to NO_x has been shown to decrease with increasing fuel N concentration for hydrocarbon fuels and coal [99–101]. Data obtained from commercial biomass-fueled fluidized bed combustors [86] and laboratory experiments with fir and birch wood [102] also suggest declining fuel N conversion with increasing fuel N concentration. N_2O also results from fuel N oxidation, but is generally produced in much lower amounts. The high GWP of this species – a factor 310 greater than CO_2 with N_2O having a similar lifetime to CO_2 in the atmosphere [103] – makes even small emissions important from a climate change perspective [15]. NO_x photochemically reacts with volatile organic compounds (VOCs) to form ozone. While stratospheric ozone is critically important to life on earth by absorbing much of the ultraviolet (UV) radiation from the sun, ozone (when found at ground level) is a lung and eye irritant and a major air pollutant in many urban environments and is often heavily regulated in terms of allowable concentration. Uncontrolled tropospheric or ground-level ozone formation is therefore undesirable. NO_x also participates in the formation of fine particles in the presence of other species, such as ammonia to form ammonium nitrate [104]. Although most of the NO_x involved in ozone formation in many regions originates from other sources, including vehicles, stringent NO_x emission regulations increase the cost of control for any new generators, including biomass power plants which might provide other benefits, such as greenhouse gas emission reductions. No satisfactory regulatory process yet exists for aggregating multiple emission impacts to achieve optimal overall environmental performance, an area in need of substantial research.

3.4.2 Products of Incomplete Combustion

CO and HCs, including VOCs and PAHs, are products of incomplete combustion. These species are largely controlled by proper management of combustion stoichiometry and feedstock moisture to ensure sufficient temperature and more complete burning [105, 106]. Increased emission of these species is linked to inadequate mixing in the combustion chamber, overall lack of oxygen, low combustion temperatures, short residence times, or low radical concentrations, especially in the final stages of batch combustion processes [15].

The PAHs include a large number of environmentally persistent aromatic compounds of variable toxicity [105]. The U.S. Department of Health and Human Services has classified a number of these as known animal carcinogens. Benzo[a]pyrene is listed by the International Agency for Research on Cancer (IARC) as a known human carcinogen [106, 107]. PAHs, including benzo[a]pyrene and others, are found in smoke from cigarettes and open burning of biomass [108–110] and are partitioned to both the gas and particle phase [111]. PAHs, like most other products of incomplete combustion, can be reduced by proper management of combustion conditions to provide adequate oxygen, residence time, and temperature.

3.4.3 Particulate Matter

PM includes soot, ash, condensed fumes (tars/oils), and sorbed materials such as VOCs and PAHs [99, 112, 113]. Most combustion-generated particles are in the size range below 1 μm aerodynamic particle diameter. Respirable particles 10 μm or smaller (PM10) are breathing hazards, due to their retention deep in the alveoli of the lung. Mechanically generated PM, including carry-over fines from the feedstock and ash particles, tend to be fairly large compared with combustion aerosols and are more readily controlled by filters and other emission-control equipment. Biogenic silica in some materials, such as rice straw, is partly released as fibrous particles of high aspect ratio and has become of concern recently for lung disease [114]. Mutagenicity of PM extracts has been found to increase from engines burning biodiesel blends, with a maximum occurring at around 20% biodiesel (B20) in the blend [115]. PM can be controlled by a combination of proper management of combustion conditions, to ensure more complete combustion, and post-combustion control equipment, such as cyclones, baghouses, scrubbers, and electrostatic precipitators, in some cases specified by regulation [15, 98, 116–120].

3.4.4 Dioxin-Like Compounds

The class of dioxin-like compounds includes a large number of chemicals having similar structure, physicochemical properties, and common toxic responses and includes the polychlorinated dibenzo-p-dioxins (CDDs), polychlorinated dibenzofurans (CDFs), polybrominated dibenzo-p-dioxins, polybrominated dibenzofurans, and polychlorinated biphenyls (PCBs) [88]. Polybrominated biphenyls are likely to have similar properties to the PCBs. Of the 629 congeners of these various species, 47 are thought to exhibit dioxin-like toxicity. The toxicity varies in extent relative to the most toxic and widely studied of these compounds, 2,3,7,8-tetrachlorodibenzo-*p*-dioxin (TCDD), for which sufficient evidence now exists that the IARC classifies it as a known human carcinogen [107]. Toxicity of the other compounds is referred to that of TCDD in terms of toxic equivalence (TEQ), with values ranging from 10^{-5} to 1.0, the latter equal to the toxicity of TCDD. Different schemes for measuring the TEQ have emerged including the international scheme adopted by the US EPA and those of the World Health Organization [88]. Mixtures of dioxin-like compounds are assigned TEQ values on the basis of a mass-weighted average.

Prior to 1995, MSW combustion was listed as the leading source of dioxin emissions in the USA [87, 88]. Since 1995, regulations have been promulgated by the US EPA to limit CDD/CDF emissions from numerous sources. Installation of maximum achievable control technology (MACT) on MSW combustors with aggregate combustion capacities greater than 227 Mg/day (250 short tons/day) and retirement of facilities for which retrofit was infeasible have led to large reductions in emissions of dioxins and furans from this source [87, 88, 121]. Total environmental releases of dioxin-like compounds declined 90% in the USA between 1987 and 2000 (Table 3.4). Reductions in emissions from MSW combustion exceed 99% since 1987. Emissions from industrial wood burning increased 36% over the same period due to a 58% increase in firing capacity and wood demand without a concurrent requirement to meet MACT standards but with enhancements in best available control technology requirements. Total air emissions from industrial burning of 51×10^6 t of wood and 0.8×10^6 t of salt-laden wood in 2000 were about half those from the 31×10^6 t of MSW burned in the USA during the same year [121]. Backyard refuse burning is one of the largest sources of dioxins [121] and no federal standards exist to regulate this source, although state and local standards and prohibitions apply in some locations. Additional sources not incorporated in the total environmental release listed in Table 3.4 include wildfire and prescribed

Table 3.4 Releases of dioxin-like compounds (g TEQ/year) in the USA 1987–2000 [121].

Release category	1987	1995	2000
Total environmental releases of dioxin-like compounds[a]	13 965	3 444	1 422
Municipal waste combustion	8 905	1 394	84
Medical waste/pathological incineration	2 570	487	378
Wastewater treatment sludge (land application and incineration)	85	133	90
Open (backyard) refuse burning	604	628	499
Residential wood burning	22	16	11
Industrial wood combustion	27	26	42
Coal-fired utility boilers	51	60	70
Hazardous waste incineration	5	6	3
Hazardous waste combustion in cement kilns	118	156	19
Cigarette burning	1.0	0.8	0.4

[a] Not including wildland and prescribed fires.
Source: reproduced from EPA [121].

burning for land clearing or residue disposal. Estimates for releases from wildfires range up to 4560 g TEQ/year depending on year and area burned, but are highly uncertain.

Although chlorine must be present in the system in order to form CDDs and CDFs, chlorine concentration in the feedstock has not been observed to be a dominant factor in the emission of dioxin-like compounds from commercial waste incinerators [88]. Other factors appear to have more control over their formation and emission, including overall combustion efficiency, flue gas residence times and temperatures in emission control equipment downstream of the furnace, and opportunistic catalysis supporting CDD/CDF synthesis. Emissions of dioxins from chlorine-containing herbaceous feedstock, such as straw, have generally been low, possibly as a result of the simultaneous formation of KCl, NaCl, and other salts that reduce the levels of chlorine available for reaction. Sufficient oxygen is also required for the destruction of dioxins, as is enhanced turbulence to promote good mixing, to eliminate cold spots in the furnace, and to provide adequate residence time at high temperature [120]. Individual combustor designs may vary substantially in terms of the dependence of dioxin emissions on chlorine concentration in the feed, and chlorine concentration may be more of an issue for open or uncontrolled burning of biomass [15, 88]. Wildfires in wildland–urban interface zones, such as recent fires in northern California, likely increase emissions of these and other toxics due to the burning of manufactured products such as polyvinylchloride (PVC) plastic, vehicles, and other materials in addition to vegetation [122].

3.4.5 Heavy Metals

Mercury is strongly released to the flue gas during combustion and can be classified as a highly volatile element together with Cl, F, and Se [47, 120]. Some Hg removal occurs during the control of other species, such as SO_x and NO_x, and Hg can be strongly concentrated in fly ash, but for fuels containing higher concentrations of Hg, active control such as activated carbon injection as a sorbent may be required [123]. Water leaching of the feedstock to remove alkali metals and chlorine does not appear to alter Hg mobility in biomass [47]. Other less volatile metals, such as copper and zinc, are enriched in slag and bottom ash, while cadmium, selenium, and more volatile elements vaporize in the combustion zone, condense downstream on fine particles, and are either captured in flyash or escape to the atmosphere [120]. Lead and chromium exhibit enrichment factors of three or more in incinerator bottom ash [46], but are also enriched in flyash [120].

3.4.6 Radioactive Species

Radioactive constituents may also be present in biomass and become enriched in ash or released to the atmosphere. Incineration of medical waste and human or animal tissues containing radioisotopes is a direct source [120]. In addition to naturally occurring radionuclides found in soil, artificial radionuclides occur in wood and other biomass as a result of deposition from atmospheric testing of nuclear weapons, the Chernobyl nuclear reactor accident, and other radioactive contamination [124, 125]. Wood samples collected in Croatia after the Chernobyl accident contained levels of ^{137}Cs, ^{214}Bi, and ^{40}K in the range 1.6–37.3, 0.2–27.1, and 21.5–437.1 Bq/kg respectively (1 Bq = 1 nuclear decay per second = 2.7×10^{-11} Ci) [124]. Both ^{214}Bi and ^{40}K are naturally occurring and so are always found in biomass, ^{214}Bi being a decay product of uranium contained in soil and ^{40}K a natural isotope of potassium in the environment. ^{137}Cs is an artificial isotope and indicative of radioactive contamination. Combustion results in enrichment in ash as radionuclides evaporate and recondense on flyash particles.

3.4.7 Greenhouse Gas Emissions

Using sustainably grown biomass as feedstock is one means of displacing fossil fuel to aid in stabilizing atmospheric greenhouse gas concentrations. Carbon dioxide released to the atmosphere by biomass conversion is offset by that taken up by plants in producing new biomass. Although the use of biomass for energy is often said to be carbon neutral – that is, as much carbon is taken up in the growth of new biomass as is released in its conversion – on a lifecycle basis this is not at present entirely true. The use of diesel fuel, gasoline, natural gas, coal-fired electricity, and other fossil-based energy in the production, harvesting, processing, and conversion of biomass is not offset when biomass is grown on a purely replacement basis. An additional amount of biomass would need to be grown and the carbon sequestered to offset the fossil carbon release from biomass handling. CH_4, N_2O, and other pollutants released during biomass production and conversion also have larger GWPs than CO_2. Combustion of biomass, therefore, results in an imbalance in greenhouse gas emission and uptake. The large displacement of fossil-derived CO_2 through the substitution of biomass in well-controlled systems, however, still results in a substantial decrease in net greenhouse gas emissions. Advanced biomass conversion technologies reduce the emission of CO_2 per unit of product energy by increasing efficiency and reducing the feedstock demand. For biomass utilization to be effective in managing atmospheric CO_2, it must be produced on a renewable and sustainable basis, including indirect effects associated with global market-mediated impacts [126]. Carbon capture and storage, currently under development for use with fossil fuels, and a number of other techniques to sequester carbon (including crop and land management to enhance soil carbon) have the potential to achieve a net reduction in atmospheric carbon when applied to sustainable biomass conversion systems. Achieving substantially lower levels of pollutant emissions and realizing potential carbon benefits remain important objectives for biomass combustion research and development.

References

1. Niele, F. (2005). *Energy – Engine of Evolution*. Amsterdam: Elsevier.
2. WHO (2009). Indoor Air Pollution Health Effects. http://www.who.int/indoorair/health_impacts/disease/en
3. Coffey, E.R., Muvandimwe, D., Hagar, Y. et al. (2017). New emission factors and efficiencies from in-field measurements of traditional and improved cookstoves and their potential implications. *Environmental Science & Technology* **51** (21): 12508–12517.
4. Searchinger, T., Heimlich, R., Houghton, R.A. et al. (2008). Use of U.S. croplands for biofuels increases greenhouse gases through emissions from land-use change. *Science* **319** (5867): 1238–1240.
5. DOE (2008). *New Science for a Secure and Sustainable Energy Future*. Washington, DC: US Department of Energy. http://www.er.doe.gov/bes/reports/files/NSSSEF_rpt.pdf (accessed 10 November 2010).

6. Zheng, Y.Y., Jenkins, B.M., Kornbluth, K., and Traeholt, C. (2018). Optimization under uncertainty of a biomass-integrated renewable energy microgrid with energy storage. *Renewable Energy* **123**: 204–217.
7. English, P.E. and Rantell, T.D. (1969). The determination of the electrical conductivity of the gases in a pilot-scale coal-fired magnetohydrodynamic combustor. *Journal of Physics D: Applied Physics* **2**: 1215–1225.
8. Hoffman, M.A. (1977). *Personal Communication*. Davis: University of California.
9. DOE (2004). *Biomass Cofiring in Coal-Fired Boilers*. US Department of Energy, DOE/EE-0288. http://www.techtp.com/Cofiring/BiomassCofiring in Coal-Fired Boilers.pdf (accessed 10 November 2010).
10. Jenkins, B.M., Williams, R.B., Parker, N. et al. (2009). Sustainable use of California biomass resources can help meet state and national bioenergy targets. *California Agriculture* **63** (4): 168–177. See also California Senate Bill 350, https://leginfo.legislature.ca.gov/faces/billNavClient.xhtml?bill_id=201720180SB100 (accessed 19 November 2018).
11. Antal, M.J. and Gronli, M.G. (2003). The art, science, and technology of charcoal production. *Industrial & Engineering Chemistry Research* **42**: 1919.
12. Koppejan, J. (2009). Challenges in Biomass Combustion and Co-firing: The Work of IEA Bioenergy Task 32. IEA Bioenergy Conference, Vancouver, British Columbia.
13. Edwards, R.D., Smith, K.R., Zhang, J., and Ma, Y. (2003). Models to predict emissions of health-damaging pollutants and global warming contributions of residential fuel/stove combinations in China. *Chemosphere* **50**: 201–215.
14. Miles, T.R. Jr. (ed.) (2009). Improved Biomass Cooking Stoves. http://www.bioenergylists.org (accessed 6 November 2018).
15. Van Loo, S. and Koppejan, J. (eds.) (2008). *The Handbook of Biomass Combustion and Co-Firing*. London: Earthscan.
16. IEA (2007). Biomass for Power Generation and CHP, International Energy Agency. https://www.iea.org/publications/freepublications/publication/essentials3.pdf (accessed 20 November 2018).
17. REN21 (2016). Renewables 2016 Global Status Report, Renewable Energy Policy Network. Paris: REN21 Secretariat. http://www.ren21.net/wp-content/uploads/2016/06/GSR_2016_Full_Report.pdf (accessed 20 November 2018).
18. Parikka, M. (2004). Global biomass resources. *Biomass and Bioenergy* **27**: 613–620.
19. Wiltsee, G. (2000). *Lessons Learned from Existing Biomass Power Plants*, NREL/SR-570-26946, . Golden, CO: National Renewable Energy Laboratory.
20. EIA (2018). Preliminary Monthly Electric Generator Inventory. https://www.eia.gov/electricity/data/eia860m (accessed 6 November 2018).
21. Power Technology (2018). Polaniec Biomass Power Plant. https://www.power-technology.com/projects/polaniec-biomass-power-plant-poland (accessed 6 November 2018).
22. Voegele, E. (2015). E.ON to close Ironbridge Power Station. *Biomass Magazine*, 24 November 2015. http://biomassmagazine.com/articles/12608/e-on-to-close-ironbridge-power-station (accessed 6 November 2018).
23. Jenkins, B.M. (1997). A comment on the optimal sizing of a biomass utilization facility under constant and variable cost scaling. *Biomass and Bioenergy* **13** (1–2): 1–9.
24. Searcy, E. and Flynn, P. (2009). The impact of biomass availability and processing cost on optimum size and processing technology selection. *Applied Biochemistry and Biotechnology* **154**: 271–286.
25. Craig, K.R. and Mann, M.K. (1996). Cost and Performance Analysis of Biomass-based Integrated Gasification Combined-cycle (BIGCC) Power Systems, NREL/TP-430-21657. https://www.nrel.gov/docs/legosti/fy97/21657.pdf (accessed 20 November 2018).
26. Thorn, W.F., Hoskins, R.L., and Wilson, D. (1980). A study of the Feasibility of Cogeneration Using Wood Waste as Fuel, EPRI AP-1483. Electric Power Research Institute, Palo Alto, CA.

27. EUBIA (2007). Biomass Co-combustion, European Biomass Industry Association. http://www.eubia.org/cms/wiki-biomass/co-combustion-with-biomass/ (accessed 20 November 2018).
28. Lissianski, V.V., Zamansky, V.M., Maly, P.M., and Sheldon, M.S. (2001). Reburning chemistry-mixing model. *Combustion and Flame* **125**: 1310–1319.
29. Lissianski, V.V., Zamansky, V.M., and Maly, P.M. (2001). Effect of metal-containing additives on NO_x reduction in combustion and reburning. *Combustion and Flame* **125**: 1118–1127.
30. Meister, B.C., Williams, R.B., and Jenkins, B.M. (2004). Utilization of Waste Renewable Fuels in Boilers with Minimization of Pollutant Emissions: Laboratory Scale Gasification Screenng Experiments, Final report CEC/UCD 500-99-013, California Energy Commission, Sacramento, CA.
31. Damstedt, B., Pederson, J.M., Hansen, D. et al. (2007). Biomass cofiring impacts on flame structure and emissions. *Proceedings of the Combustion Institute* **31**: 2813–2820.
32. Wang, S., Llamazos, E., Baxter, L., and Fonseca, F. (2008). Durability of biomass fly ash concrete: freezing and thawing and rapid chloride permeability tests. *Fuel* **87**: 359–364.
33. Rissanen, J., Ohenoja, K., Kinnunen, P., and Illikainen, M. (2017). Partial replacement of Portland-composite cement by fluidized bed combustion fly ash. *Journal of Materials in Civil Engineering* **29** (8), 04017061-1-10. https://doi.org/10.1061/(ASCE)MT.1943-5533.0001899.
34. Dong, L., Liu, H., and Riffat, S. (2009). Development of small-scale and micro-scale biomass fuelled CHP systems – a literature review. *Applied Thermal Engineering* **29**: 2119–2126.
35. Williams, R.B. and Kaffka, S. (2015). Biomass Gasification, Technical Report CEC-500-11-020, California Energy Commission, Sacramento, California. https://biomass.ucdavis.edu/wp-content/uploads/Task7-Report_Biomass-Gasification_DRAFT.pdf (accessed 6 November 2018).
36. Pawlikowski, R. (1929). Powdered fuels for engines (brief note). *Chemistry & Metallurgy* **36**: 287. http://pubs.acs.org/doi/abs/10.1021/ed006p2180.2 (accessed 10 November 2010).
37. Cocco, D., Deiana, P., and Cau, G. (2006). Performance evaluation of small size externally fired gas turbine (EFGT) power plants integrated with direct biomass dryers. *Energy* **31**: 1459–1471.
38. Stahl, K. and Neergaard, M. (1998). IGCC power plant for biomass utilization, Varnamo, Sweden. *Biomass and Bioenergy* **15** (3): 205–211.
39. Valero, A. and Uson, S. (2006). Oxy-co-gasification of coal and biomass in an integrated gasification combined cycle (IGCC) power plant. *Biomass and Bioenergy* **31**: 1643–1655.
40. Colson, C.M., Nehrir, M.H., Deibert, M.C. et al. (2009). Efficiency evaluation of solid-oxide fuel cells in combined cycle operations. *Journal of Fuel Cell Science and Technology* **6**: 021006.
41. Chum, H.L. and Baizer, M.M. (1985). *The Electrochemistry of Biomass and Derived Materials*, ACS Monograph 183, . Washington, DC: American Chemical Society.
42. Schultz, T.P. and Taylor, F.W. (1989). Wood. In: *Biomass Handbook* (ed. O. Kitani and C.W. Hall), 133–141. New York: Gordon and Breach.
43. Sudo, S., Takahashi, F., and Takeuchi, M. (1989). Chemical properties of biomass. In: *Biomass Handbook* (ed. O. Kitani and C.W. Hall), 892–933. New York: Gordon and Breach.
44. Lynd, L.R. (1990). Large-scale fuel ethanol from lignocellulose. *Applied Biochemistry and Biotechnology* **24–25**: 695–719.
45. Wyman, C.E. and Hinman, N.D. (1990). Ethanol, fundamentals of production from renewable feedstocks and use as a transportation fuel. *Applied Biochemistry and Biotechnology* **24–25**: 735–753.
46. Ludwig, C., Hellweg, S., and Stucki, S. (eds.) (2003). *Municipal Solid Waste Management: Strategies and Technologies for Sustainable Solutions*. Berlin: Springer.
47. Thy, P. and Jenkins, B.M. (2009). Mercury in biomass feedstock and combustion residuals. *Water, Air, and Soil Pollution* **209** (1–4): 429–437. https://doi.org/10.1007/s11270-009-0211-9.
48. Thy, P., Barfod, G.H., Cole, A.M. et al. (2017). Trace metal release during wood pyrolysis. *Fuel* **203**: 548–556.
49. Jenkins, B.M. (1989). Physical properties of biomass. In: *Biomass Handbook* (ed. O. Kitani and C.W. Hall), 860–891. New York: Gordon and Breach.

50. Jenkins, B.M. (1993). Properties of biomass in *Biomass Energy Fundamentals*, EPRI TR-102107, Electric Power Research Institute, Palo Alto, CA.
51. Domalski, E.S., Jobe, T.L. Jr., and Milne, T.A. (1987). *Thermodynamic Data for Biomass Materials and Waste Components*. New York: American Society of Mechanical Engineers.
52. Tillman, D.A. (1994). *Trace Metals in Combustion Systems*. San Diego, CA: Academic Press.
53. Miles, T.R., Miles, T.R. Jr., Baxter, L.L. et al. (1995). *Alkali Deposits Found in Biomass Power Plants: A Preliminary Investigation of their Extent and Nature*. Golden, CO: National Renewable Energy Laboratory.
54. Baxter, L.L., Miles, T.R., Miles, T.R. Jr. et al. (1996). The behavior of inorganic material in biomass-fired power boilers – field and laboratory experiences. *Fuel Processing Technology* **54**: 47–78.
55. Jenkins, B.M., Baxter, L.L., Miles, T.R. Jr., and Miles, T.R. (1998). Combustion properties of biomass. *Fuel Processing Technology* **54**: 17–46.
56. Baxter, L.L. (1993). Ash deposition during biomass and coal combustion: a mechanistic approach. *Biomass and Bioenergy* **4** (2): 85–102.
57. Jenkins, B.M., Baxter, L.L., Miles, T.R. et al. (1994). Composition of ash deposits in biomass fuelled boilers: results of full-scale experiments and laboratory simulations, ASAE Paper No. 946007, ASAE, St Joseph, MI.
58. Misra, M.K., Ragland, K.W., and Baker, A.J. (1993). Wood ash composition as a function of furnace temperature. *Biomass and Bioenergy* **4** (2): 103–116.
59. Baxter, L.L., Miles, T.R., Miles, T.R. Jr. et al. (1998). The behavior of inorganic material in biomass-fired power boilers: field and laboratory experiences. *Fuel Processing Technology* **54**: 47–78.
60. Thy, P., Jenkins, B.M., and Lesher, C.E. (1999). High temperature melting behavior of urban wood fuel ash. *Energy and Fuels* **13** (4): 839–850.
61. Thy, P., Lesher, C.E., and Jenkins, B.M. (2000). Experimental determination of high temperature elemental losses from biomass fuel ashes. *Fuel* **79**: 693–700.
62. Thy, P., Grundvig, S., Jenkins, B.M. et al. (2005). Analytical controlled losses of potassium from straw ashes. *Energy and Fuels* **19** (6): 2571–2575.
63. Thy, P., Jenkins, B.M., Grundvig, S. et al. (2006). High temperature elemental losses and mineralogical changes in common biomass ashes. *Fuel* **85**: 783–795.
64. Thy, P., Jenkins, B.M., Williams, R.B., and Lesher, C.E. (2006). Compositional constraints on slag formation and potassium volatilization from rice straw blended wood fuel. *Fuel Processing Technology* **87** (5): 383–408.
65. Thy, P., Jenkins, B.M., Williams, R.B. et al. (2010). Bed agglomeration in a fluidized bed combustor fueled by wood and rice straw blends. *Fuel Processing Technology* **91** (11): 1464–1485.
66. Jenkins, B.M., Bakker, R.R., and Wei, J.B. (1995). Removal of inorganic elements to improve biomass combustion properties, in Proceedings Second Biomass Conference of the Americas, Portland, OR, pp. 483–492.
67. Jenkins, B.M., Bakker, R.R., and Wei, J.B. (1996). On the properties of washed straw. *Biomass and Bioenergy* **10** (4): 177–200.
68. Dayton, D.C., Jenkins, B.M., Turn, S.Q.R.R. et al. (1999). Release of inorganic constituents from leached biomass during thermal conversion. *Energy and Fuels* **13** (4): 860–870.
69. Turn, S.Q., Kinoshita, C.M., Jakeway, L.A. et al. (2003). Fuel characteristics of processed, high-fiber sugar cane. *Fuel Processing Technology* **81**: 35–55.
70. Turn, S.Q., Jenkins, B.M., Jakeway, L.A. et al. (2006). Test results from sugar cane bagasse and high fiber cane co-fired with fossil fuels. *Biomass and Bioenergy* **30**: 565–574.
71. Shelton, R.D. (1978). Stagewise gasification in a multiple-hearth furnace. In: *Solid Wastes and Residues – Conversion by Advanced Thermal Processes* (ed. J.L. Jones and S.B. Radding), ACS Symposium Series 76, , 165–190. Washington, DC: American Chemical Society.

72. Shafizadeh, F. (1981). Basic principles of direct combustion. In: *Biomass Conversion Processes for Energy and Fuels* (ed. S.S. Sofer and O.R. Zaborsky), 103–124. New York: Plenum.
73. Gaur, S. and Reed, T.B. (1995). *An Atlas of Thermal Data for Biomass and Other Fuels*, NREL/TP-433-7965, . Golden, CO: National Renewable Energy Laboratory.
74. Kollman, F.F.P. and Cote, W.A. (1968). *Principles of Wood Science and Technology, I, Solid Wood*. New York: Springer Verlag.
75. Linstrom, P.J. and Mallard, W.G. (eds.) (2009). *NIST Chemistry WebBook*, NIST Standard Reference Database Number 69, . Gaithersburg, MD: National Institute of Standards and Technology. http://webbook.nist.gov (accessed 10 November 2010).
76. Jenkins, B.M. and Ebeling, J.M. (1985). Correlation of physical and chemical properties of terrestrial biomass with conversion. In: *Proceedings of IGT Energy from Biomass and Wastes IX*. Chicago, IL: Institute of Gas Technology.
77. Kirov, N.Y. (1965). Specific heats and total heat contents of coals and related materials at elevated temperatures. *British Coal Utilisation Research Association Monthly Bulletin* **29**: 33.
78. Hanrot, F., Ablitzer, D., Houzelot, J.L., and Dirand, M. (1994). Experimental measurement of the true specific heat capacity of coal and semicoke during carbonization. *Fuel* **73** (2): 305–309.
79. Smith, J.M. and Van Ness, H.C. (1975). *Introduction to Chemical Engineering Thermodynamics*. New York: McGraw-Hill.
80. Wiltsee, G.A. (1993). *Biomass Energy Fundamentals*, vol. **1** , EPRI TR-102107, . Palo Alto, CA: Electric Power Research Institute.
81. Kanury, A.M. (1994). Combustion characteristics of biomass fuels. *Combustion Science and Technology* **97**: 469–491.
82. Bining, A.S. and Jenkins B.M. (1992). Thermochemical reaction kinetics for rice straw from an approximate integral technique, ASAE Paper 926029, ASAE, St Joseph, MI.
83. Williams, P.T. and Horne, P.A. (1994). The role of metal salts in the pyrolysis of biomass. *Renewable Energy* **4** (1): 1–13.
84. Jenkins, B.M., Bakker, R.R., Baxter, L.L. et al. (1996). Combustion characteristics of leached biomass. In: *Developments in Thermochemical Biomass Conversion* (ed. A.V. Bridgwater and D.G.B. Boocock), 1316–1330. London: Blackie Academic and Professional.
85. Lu, H., Ip, E., Scott, J. et al. (2010). Effects of particle shape and size on devolatilization of biomass particle. *Fuel* **89**: 1156–1168.
86. Lu, H., Robert, W., Peirce, G. et al. (2008). Comprehensive study of biomass particle combustion. *Energy and Fuels* **22**: 2826–2839.
87. Williams, R.B. (2006). Biomass in solid waste in California: utilization and policy alternatives, California Biomass Collaborative, CEC Contract 500-01-016, California Energy Commissione, Sacramento, CA. https://biomass.ucdavis.edu/wp-content/uploads/09-20-2013-2006-cbc-biomass-in-solid-waste-in-california-utilization-and-policy-alternatives.pdf (accessed 20 November 2018).
88. US EPA (2004). Exposure and Human Health Reassessment of 2,3,7,8-Tetrachlorodibenzo-P-Dioxin (Tcdd) and Related Compounds. National Academy Sciences (External Review Draft). Washington, DC: US Environmental Protection Agency. EPA/600/P-00/001Cb. https://cfpub.epa.gov/ncea/risk/recordisplay.cfm?deid=87843 (accessed 20 November 2018).
89. EPA (1992). Method 1311, Toxicity Characteristic Leaching Procedure. US Environmental Protection Agency. https://www.epa.gov/sites/production/files/2015-12/documents/1311.pdf (accessed 20 November 2018). See also https://www.epa.gov/hw-sw846/sw-846-test-method-1311-toxicity-characteristic-leaching-procedure (accessed 20 November 2018).
90. C.T. Donovan Associates, Inc.. (1995). Air emissions and ash disposal at wood-burning facilities: a sourcebook and case studies for the Great Lakes region, Great Lakes Regional Biomass Energy Program.
91. Gutierrez, M.J.F., Baxter, D., Hunter, C., and Svoboda, K. (2005). Nitrous oxide (N_2O) emissions from waste and biomass to energy plants. *Waste Management & Research* **23**: 133–147.

92. Leland, A., Hoekman, S.K., and Liu, X. (2018). Review of modifications to indirect land use change modeling and resulting carbon intensity values within the California Low Carbon Fuel Standards regulations. *J. Cleaner Production* **180**: 698–707.
93. Williams, R.B. (2005). Environmental issues for biomass development in California, PIER Collaborative Report, Contract 500-01-016, California Energy Commission, Sacramento, CA. https://biomass.ucdavis.edu/wp-content/uploads/10-11-2013-2005-cbc-environmental-issues-biomass-development-in-california.pdf (accessed 20 November 2018).
94. Jenkins, B.M., Aldas, R.E., Gildart, M. et al. (2005). Biomass in California: challenges, opportunities, and potentials for sustainable management and development, PIER Collaborative Report, Contract 500-01-016, Sacramento, CA: California Energy Commission. https://biomass.ucdavis.edu/wp-content/uploads/10-16-2013-2005-cbc-biomass-in-ca-white-paper.pdf (accessed 20 November 2018).
95. Canakci, M. (2009). NOx emissions of biodiesel as an alternative diesel fuel. *International Journal of Vehicle Design* **40** (1–4): 213–228.
96. Marsh, M. and LaMendola, T. (2005). Dairy power production program: dairy methane digester system 90-day evaluation report, Cottonwood Dairy (Joseph Gallo Farms), CEC-500-2005-116, California Energy Commission, Sacramento, CA.
97. Abatzoglou, N. and Boivin, S. (2008). A review of biogas purification processes. *Biofuels, Bioproducts and Biorefining* **3**: 42–71.
98. Grass, S.W. and Jenkins, B.M. (1994). Biomass fueled fluidized bed combustion: atmospheric emissions, emission control devices and environmental regulations. *Biomass and Bioenergy* **6** (4): 243–260.
99. Seinfeld, J.H. (1986). *Atmospheric Chemistry and Physics of Air Pollution*. New York: Wiley.
100. Bowman, C.T. (1991). Chemistry of gaseous pollutant formation and destruction. In: *Fossil Fuel Combustion: A Source Book* (ed. W. Bartok and A.F. Sarofim), 215–260. New York: Wiley.
101. Miller, J.A. and Bowman, C.T. (1989). Mechanism and modeling of nitrogen chemistry in combustion. *Progress in Energy and Combustion Science* **15**: 287–338.
102. Leckner, B. and Karlsson, M. (1993). Gaseous emissions from circulating fluidized bed combustion of wood. *Biomass and Bioenergy* **4** (5): 379–389.
103. EPA (2009). Inventory of U.S. greenhouse gas emissions and sinks: 1990–2006, US Environmental Protection Agency. https://www.epa.gov/ghgemissions/inventory-us-greenhouse-gas-emissions-and-sinks-1990-2006 (accessed 20 November 2018). See also updated inventory at https://www.epa.gov/ghgemissions/inventory-us-greenhouse-gas-emissions-and-sinks (accessed 20 November 2018).
104. Stockwell, W.R., Watson, J.G., Robinson, N.F. et al. (2000). The ammonium nitrate particle equivalent of NO_x emissions for wintertime conditions in Central California's San Joaquin Valley. *Atmospheric Environment* **34**: 4711–4717.
105. EPA (2008) Polycyclic aromatic hydrocarbons (PAHs). https://www.epa.gov/sites/production/files/2014-03/documents/pahs_factsheet_cdc_2013.pdf (accessed 20 November 2018).
106. ATSDR (1995). *Toxicological Profile for Polycyclic Aromatic Hydrocarbons*. Atlanta, GA: US Department of Health and Human Services/Public Health Service/Agency for Toxic Substances and Disease Registry, http://www.atsdr.cdc.gov/phs/phs.asp?id=120&tid=25 (accessed 10 November 2010).
107. IARC (2009). A review of human carcinogens – Part F: chemical agents and related occupations. *The Lancet Oncology* **10** (12): 1143–1144.
108. Hecht, S.S. (1999). Tobacco smoke carcinogens and lung cancer. *Journal of the National Cancer Institute* **91** (14): 1194–1210.
109. Jenkins, B.M., Turn, S.Q., Williams, R.B. et al. (1996). Atmospheric pollutant emission factors from open burning of agricultural and forest biomass by wind tunnel simulations, Final Report (3 vols), CARB Project A932-126, California Air Resources Board, Sacramento, CA.

110. Jenkins, B.M., Jones, A.D., Turn, S.Q., and Williams, R.B. (1996). Emission factors for polycyclic aromatic hydrocarbons (PAH) from biomass burning. *Environmental Science and Technology* **30** (8): 2462–2469.
111. Jenkins, B.M., Jones, A.D., Turn, S.Q., and Williams, R.B. (1996). Particle concentrations, gas–particle partitioning, and species intercorrelations for polycyclic aromatic hydrocarbons (PAH) emitted during biomass burning. *Atmospheric Environment* **30** (22): 3825–3835.
112. Finlayson-Pitts, B.J. and Pitts, J.N. Jr. (1986). *Atmospheric Chemistry: Fundamentals and Experimental Techniques*. New York: Wiley.
113. CARB (2009). Ambient air quality standards (AAQS) for particulate matter. https://www.arb.ca.gov/research/aaqs/pm/pm.htm (accessed 20 November 2018).
114. Lawson, R.J., Schenker, M.B., McCurdy, S.A. et al. (1995). Exposure to amorphous silica fibers and other particulate matter during rice farming operations. *Applied Occupational and Environmental Hygiene* **10** (8): 677–684.
115. Munack, A., Krahl, J., and Bunger, J. (2009). Political framework and tail pipe emissions for rapeseed oil based fuels, in IEA Bioenergy Conference, Vancouver, British Columbia.
116. Hasler, P. and Nussbaumer, T. (1999). Gas cleaning for IC engine applications from fixed bed biomass gasification. *Biomass and Bioenergy* **16**: 285–295.
117. Lin, G.Y., Tsai, C.J., Chen, S.C. et al. (2010). An efficient single-stage wet electrostatic precipitator for fine and nanosized particle control. *Aerosol Science and Technology* **44**: 38–45.
118. Ni, Y., Zhang, H., Fan, S. et al. (2009). Emissions of PCDD/Fs from municipal solid waste incinerators in China. *Chemosphere* **75**: 1153–1158.
119. World Bank Group (2018). Environmental, health, and safety general guidelines. https://www.ifc.org/wps/wcm/connect/topics_ext_content/ifc_external_corporate_site/sustainability-at-ifc/policies-standards/ehs-guidelines (accessed 20 November 2018).
120. Lisk, D.J. (1988). Environmental implications of incineration of municipal solid waste and ash disposal. *The Science of the Total Environment* **74**: 39–66.
121. EPA (2006). An inventory of sources and environmental releases of dioxin-like compounds in the United States for the years 1987, 1995, and 2000. EPA/600/P-03/002F. http://cfpub.epa.gov/ncea/cfm/recordisplay.cfm?deid=159286 (accessed 6 November 2018).
122. Johnson, K. (2017). Cleanup from California fires poses environmental and health risks, *New York Times*, 16 October 2017. https://www.nytimes.com/2017/10/16/us/california-fires-cleanup.html (accessed 6 November 2018).
123. EPA (2004). *Control of Mercury Emissions from Coal-Fired Electric Utility Boilers*. Research Triangle Park, NC: Air Pollution Prevention and Control Division, http://www.epa.gov/ttn/atw/utility/hgwhitepaperfinal.pdf (accessed 10 November 2010).
124. Grammelis, P., Skodras, G., Kakaras, E. et al. (2006). Effects of biomass co-firing with coal on ash properties – part II: leaching, toxicity, and radiological behavior. *Fuel* **85**: 2316–2322.
125. Hus, M., Kosutic, K., and Lulic, S. (2001). Radioactive contamination of wood and its products. *Journal of Environmental Radioactivity* **55**: 179–186.
126. Tilman, D., Socolow, R., Foley, J.A. et al. (2009). Beneficial biofuels – the food, energy, and environment trilemma. *Science* **325** (5938): 270–271.

4

Gasification

Karl M. Broer[1] and Chad Peterson[2]

[1]*Gas Technology Institute, Des Plaines, IL, USA*
[2]*Department of Mechanical Engineering, Iowa State University, Ames, IA, USA*

4.1 Introduction

Gasification is defined as high temperature (>650 °C) conversion of carbonaceous material into a combustible gas mixture under reducing conditions. The enthalpy for gasification can be provided either by partial oxidation of reactants or products, or by high temperature heat transfer into the reactor. In both cases, water, in the form of steam, may be added to promote additional production of hydrogen via the water gas shift reaction. Through gasification of biomass, a heterogeneous solid material is converted into a gaseous fuel intermediate that can be used for heating, industrial process applications, electricity generation, and liquid fuels production. Biomass gasification is significantly discussed in modern contexts by Reed [1], Rezaiyan and Cheremisinoff [2], Probstein and Hicks [3], Higman and van der Burgt [4], Knoef [5], and Basu [6].

4.1.1 History of Gasification

Gasification evolved from efforts to understand the physics and chemistry of gases, particularly flammable and toxic gases encountered in coal mining. Beginning around 1650, through the historical field of "pneumatic chemistry," humankind began to investigate these gases and how to produce, manipulate, purify, and store them. Pneumatic chemists began preparing flammable gases for their research via heating solids in retorts, which had long been in use by alchemists for distillation and heating applications. These research activities

Thermochemical Processing of Biomass: Conversion into Fuels, Chemicals and Power, Second Edition.
Edited by Robert C. Brown.
© 2019 John Wiley & Sons Ltd. Published 2019 by John Wiley & Sons Ltd.

soon demanded improvements to the retorts, such as making them out of iron to withstand higher temperatures, and fabricating them to more exacting standards. The technological demands of pneumatic chemistry also led to the invention of the "pneumatic trough" by Stephen Hales and the gasometer by Antoine Lavoisier. These devices allowed gas from the retorts to be effectively captured and stored. Storage capability was important for enabling the gas to be cooled down, purified, and utilized later at controllable flow rates [7].

The pneumatic chemists also conceived the first applications for the gases produced by their retorts. Primitive gas lamps were developed by Alessandro Volta in the 1770s, and Jan-Pieter Minckelers experimented with lighting his lecture hall at the University of Louvain in 1785 using coal gas [7]. The first commercial gasification plants emerged in the early 1800s. These plants used equipment scaled up from the laboratory instruments, and produced gas mixtures rich in H_2 and CO for machine and textile factories in Great Britain [7]. As the industry took off, complementary knowledge and experience from the coal distillation process for making coke from coal was gained, and companies such as the London "Gas Light and Coke Company" were formed to distribute the gas throughout cities for lighting and heating. Gas lighting began in the United States in 1816 in Baltimore, MD [6]. Technological improvements continued rapidly, and by 1826, nearly every town in Britain with a population over 10 000 had coal gas lighting [4, 7]. Gasification was widely used in the United States as well, with over 1200 plants operating by the late 1920s [2]. Although the need for coal gas declined some after deployment of electric lighting in ca. 1900, it continued to be important for heating and cooking [4]. Natural gas production and distribution dramatically increased from 1935 and 1960 due to installation of intercontinental pipelines, made possible by wartime needs, improvements in steel, and the high energy density of natural gas. Due to wide availability and low costs, natural gas quickly displaced coal gas [1, 6, 8].

Since the end of coal gas for heating and lighting purposes, there have also been several special circumstances where gasifiers were deployed as a means of dealing with shortages of liquid vehicle fuels. During World War II, severe shortages of petroleum in Germany and other European countries led to rapid development of small wood stove-like gasifiers that were strapped onto vehicles to allow the civilian population to continue use of agricultural tractors, cars, delivery trucks, buses, and even boats [9]. These gasifiers were small, designed specifically for vehicle application, and usually fueled by either charcoal or wood cut into consistent chunks [10]. The produced gas was burned in internal combustion engines (ICEs). By the end of World War II, an impressive 700 000 of these gasification units were in use [9]. In Sweden, for example, 90% of all vehicles were powered by gasifiers [9]. It was not particularly convenient to power a vehicle in this way. Power output from the engine was reduced [8], fuel had to be manually loaded, maintenance needs were high, and reliability was inferior to gasoline. Despite this, these gasifiers enabled society to carry on in the face of crippling petroleum shortages. As soon as World War II was over, these small gasifiers were rapidly abandoned. Much can be learned from the original documentation about the designs and technology that went into these gasifiers, as the designs were meant to be affordable, practical, simple, and portable, although this documentation is rare and difficult to access. Fortunately, some has been recovered, translated, and republished into various works [8–14].

War economics and petroleum supply restrictions have also pressed nations to make synthetic liquid fuels from the CO and H_2 produced from gasifiers. In contrast to vehicle mounted gasifiers, these stationary installations were very large in scale, and usually fueled by coal. Germany used coal gasification coupled with Fischer–Tropsch (FT) synthesis to produce as much as 2.3 million liters of gasoline per day by the late 1930s. Plants based on similar technology were also built by Japan, Britain, and France. By the early 1940s, Germany transitioned to using mainly coal hydrogenation to make fuel, because Fischer–Tropsch technology was still relatively underdeveloped, reactor size small, and overall production modest [3]. Both kinds of synthetic fuels were halted at the end of World War II in favor of conventional petroleum. A larger scale gasification and Fischer–Tropsch synthesis operation was commissioned by SASOL in South Africa in the mid-1950s as a response to minimal domestic petroleum resources. Plant capacity was initially about 1.3 million l/d, but then continued and expanded as a means of dealing with petroleum embargos implemented against South Africa during the Apartheid era [3, 15, 16]. There was also strong interest in gasification technology development in the United States to take advantage of plentiful domestic coal and biomass resources during times of high crude oil prices (1974–1985 and 2005–2014).

Gasification has renewed prospects as a way to convert biomass into transportation fuels or electric power. Bioenergy is attractive for its low net CO_2 emissions compared to fossil fuels. Biomass gasification coupled to carbon capture and storage can actually remove carbon dioxide from the atmosphere, a climate mitigation strategy recently endorsed by IPCC [17]. Gasification also poses a solution to increasing costs of solids waste disposal and decreasing availability of landfill facilities, while at the same time offsetting fossil fuel use and energy costs.

4.1.2 Gasification Terminology

Perhaps due to the complexity of gasification history, the nomenclature of gasification is sometimes not clearly defined.

Producer gas refers to the low heating value gas mixture of carbon monoxide (CO), hydrogen (H_2), carbon dioxide (CO_2), methane (CH_4), other low molecular weight hydrocarbons, and nitrogen (N_2) produced from gasification of feedstocks in air. Historical applications of producer gas have included heat and electricity production, as well as the production of synthetic liquid fuels.

Synthesis gas (syngas) refers to a gas mixture of predominantly CO and H_2 produced from gasification of carbonaceous feedstocks in oxygen and steam followed by gas separation to remove CO_2. This H_2-rich mixture was developed for synthesis of fuels and chemicals. Although not strictly correct, the term syngas is widely used to describe the gaseous product from any kind of gasification process, and will be employed here.

Cold gas efficiency (CGE) is the ratio of the chemical energy in the produced gas to the chemical energy of the incoming feedstock. CGE is calculated via Eq. (4.1), where LHV_g and $\dot{m}_g$ are the lower heating value and mass flow rate of the produced gas, and LHV_f and $\dot{m}_f$ are the lower heating value and mass flow rate of the incoming feedstock [6].

$$CGE = \frac{LHV_g \dot{m}_g}{LHV_f \dot{m}_f} \quad (4.1)$$

Typical CGE values range from 60% to 80%. Because gasification occurs at high temperatures, syngas as produced contains significant sensible energy. In applications where this sensible energy can be exploited, hot gas efficiency (HGE) should be referenced instead. HGE is calculated via Eq. (4.2), where the term $(h_{prod} - h_{ref})\dot{m}_g$ has been added to the numerator to account for sensible heat energy. The quantity h_{prod} represents the enthalpy of the produced gas at its temperature and pressure, and h_{ref} is the enthalpy of the gas at reference temperature and pressure. HGE values of 80–95% are typical for gasifiers. The difference between CGE and HGE tends to increase as syngas exit temperature increases. This occurs because CGE is negatively impacted by increased temperatures, as the chemical energy of the biomass fuel must contribute more toward sensible heat of the produced gas. This comes at the expense of the chemical energy embodied in the produced gas.

$$\text{HGE} = \frac{\text{LHV}_f \dot{m}_g + (h_{prod} - h_{ref})\dot{m}_g}{\text{LHV}_f \dot{m}_f} \quad (4.2)$$

Gasifier performance can also be characterized by carbon conversion. Carbon conversion is the percentage of carbon in the feed that is converted to gaseous products, including tar vapors. Typical carbon conversion values range from 60% to 99%. Increasing operating temperature is generally the most reliable way to maximize carbon conversion efficiency. Very high operating temperatures (>1000 °C) generally lead to nearly 100% carbon conversion, but the high temperature ultimately detracts from the CGE, promotes ash slagging, and increases capital costs of the plant. Other factors exert influence on carbon conversion, including fuel particle heating rates, residence time of particles and gases in the reactor, oxygen exposure patterns, fuel properties, and fuel pretreatment. It is not always advantageous to operate at high carbon conversion rates. Lower carbon conversion rates may be acceptable for operations where biochar is a desirable product, or temperatures need to be kept low to avoid metal corrosion and fatigue.

Air-to-fuel ratio (A/F) is the amount of air (mass or moles) provided to the reactor relative to the amount of fuel. From this can be calculated the gasification equivalence ratio (ER), which is the actual A/F divided by the A/F required for stoichiometric combustion, as given by Eq. (4.3). Note that this equivalence ratio is distinct from the equivalence ratio used by the combustion community, which is based on fuel-to-air ratio (F/A) [18].

$$\text{ER} = \frac{(A/F)_{actual}}{(A/F)_{stoic}} \quad (4.3)$$

4.2 Fundamentals of Gasification

During biomass gasification, the solid fuel goes through a progression of distinct steps. As shown in Figure 4.1, these steps include heating and drying, pyrolysis, gas–solid reactions, and gas-phase reactions [19]. Unlike combustion, the gas-phase reactions produce only limited amounts of CO_2 and H_2O due to the restricted availability of oxygen. The pace at which the gasification process proceeds to completion for an individual particle of biomass fuel entering a gasification reactor is dependent on reactor design and fuel particle size, and can range from less than one second to a few tens of minutes.

Gasification 89

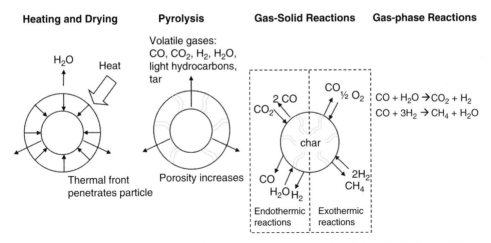

Figure 4.1 The process of thermal gasification [19]. Source: reproduced from R. C. Brown, 2014, *Biorenewable Resources: Engineering New Products from Agriculture*, 2nd Edition, with permission from John Wiley and Sons.

4.2.1 Heating and Drying

Heating and drying is the first step in gasification, transforming biomass moisture content (MC) from 10–50 wt% to bone dry [19]. Both sensible and latent energy must be supplied to heat the biomass and evaporate the moisture, respectively. Although it is possible to gasify wet feedstocks such as livestock manure and greenwood, the energy required to remove this extra moisture comes at the expense of the chemical enthalpy of the product gas, manifesting itself as a syngas with increased CO_2, H_2O, and N_2 content, and reduced CO, H_2, and CH_4 content. This reduced syngas quality hinders downstream applications and decreases plant efficiency. Thus, some drying of biomass before feeding into a gasifier is highly desirable. In the case of difficult fuels that have both high moisture content and low energy content (e.g. sewage sludge cake), the penalty to the chemical enthalpy of the product gas may become large enough that drying becomes essential, otherwise either gasification temperatures cannot be achieved, the nonflammable content of the syngas becomes prohibitively high, or both.

The process of heating and drying begins on the outside surface of a biomass particle and then progresses toward the center. A thermal front forms, with the temperature of the outside surface rising immediately after entering the reactor, and the temperature at the center of the particle lagging behind. The larger the particle of biomass inserted into the reactor, the more pronounced this lag becomes.

4.2.2 Pyrolysis

Rapid thermal decomposition of biomass in the absence of oxygen is known as pyrolysis [8]. Although some reactions commence at temperatures as low as 225 °C, the process becomes progressively more rapid and complete as temperatures reach 400–500 °C. The process is accompanied by the release of volatiles, which includes produced water (arising from chemical reactions rather than from moisture content of the biomass), permanent gases (those that do not condense upon cooling), and tarry vapors (those that do condense upon cooling), and formation of a porous, carbonaceous solid known as char. Permanent gases include CO, CO_2, H_2, and light hydrocarbons, particularly methane. Tarry vapors immediately released upon pyrolysis consist of anhydrosugars and other highly oxygenated compounds (from the decomposition of cellulose and hemicellulose) and phenolic monomers and oligomers (from the depolymerization of lignin). Exposed to high temperatures, these compounds can crack to smaller compounds or condense to larger compounds, including polyaromatic hydrocarbons, which can be a major constituent of condensed tar. Pyrolysis converts around 70–90 wt% of the biomass into vapors and gases, depending upon the type of biomass fuel and the process conditions [20]. This proportion is roughly double that of coals, which is an advantage that biomass has over coal as a gasification fuel. Once pyrolysis is complete, a porous solid called char remains. The char contains carbon and inorganic compounds (ash), and will not volatilize further at gasification temperatures; however, it can continue to interact with gaseous species in the gasification environment.

4.2.3 Gas–Solid Reactions

Following pyrolysis, chemical reactions continue to occur between the char and the surrounding gas species. There are four major gas–solid reactions responsible for this conversion [19]: the carbon–oxygen reaction, the Boudouard reaction, the carbon–water reaction, and the carbon hydrogenation reaction.

Carbon–oxygen reaction:

$$C + \tfrac{1}{2}O_2 \rightleftharpoons CO; \quad \Delta H_R = -110.5 \text{ MJ kmol}^{-1}$$

Boudouard reaction:

$$C + CO_2 \rightleftharpoons 2CO; \quad \Delta H_R = 172.4 \text{ MJ kmol}^{-1}$$

Carbon–water reaction:

$$C + H_2O \rightleftharpoons H_2 + CO; \quad \Delta H_R = 131.3 \text{ MJ kmol}^{-1}$$

Carbon hydrogenation reaction:

$$C + 2H_2 \rightleftharpoons CH_4; \quad \Delta H_R = -74.8 \text{ MJ kmol}^{-1}$$

The highly exothermic carbon–oxygen reaction is important for supplying energy to drive the endothermic processes of heating, drying, and pyrolysis. It also provides thermal energy to drive the Boudouard and carbon–water reactions, which are important in gasifying char into CO and H_2. The hydrogenation reaction also contributes energy to support endothermic reactions, although its rate is much slower than the carbon–oxygen reaction.

If chemical equilibrium is attained, char would convert to gaseous products. In practice, the contact time between char and gaseous reactants at elevated temperatures is usually insufficient to achieve equilibrium, and significant amounts of char are produced, amounting to as much as 10% of the weight of the incoming biomass.

4.2.4 Gas-Phase Reactions

Volatiles released during pyrolysis participate in gas-phase reactions as long as they remain at elevated temperatures. Two of the most important reactions in determining final gas composition are the water gas shift reaction and the methanation reaction [19].

Water gas shift reaction:

$$CO + H_2O \rightleftharpoons H_2 + CO_2; \quad \Delta H_R = -41.1 \text{ MJ kmol}^{-1}$$

Methanation:

$$CO + 3H_2 \rightleftharpoons CH_4 + H_2O; \quad \Delta H_R = -206.1 \text{ MJ kmol}^{-1}$$

The water gas shift reaction is important in increasing the H_2 content of syngas, while the methanation reaction strongly influences the CH_4 content of syngas. Both of these reactions are exothermic, and thus thermodynamically favored at low temperatures. An effective approach to promoting hydrogen formation is to add steam, while methane is promoted by increasing the partial pressure of hydrogen in the gasifier.

4.3 Feed Properties

Feedstock properties are frequently evaluated with proximate analysis, ultimate analysis, ash analysis, and enthalpy of combustion. These analyses provide information on moisture content, volatility, relative amounts of major constituent elements, heating value, and tendency for the ash in the fuel to slag or foul the gasifier.

Proximate analysis classifies the fuel in terms of moisture content, MC, volatile matter, VM, ash content, and fixed carbon, FC. In the test procedure, the volatile material is driven off in an inert atmosphere at high temperatures (950 °C) using a slow heating rate (rapid heating yields more volatile matter). Moisture measured by proximate analysis represents physically bound water only; water released by chemical reactions is classified as part of the VM. The ash content is determined by combustion of the volatile and fixed-carbon fractions. The resulting ash fraction is not representative of the original ash, more appropriately termed mineral matter, because the mineral matter in the biomass is oxidized. In the most exact analyses, small corrections to the ash weight are necessary to correct it to a mineral matter basis. The fixed-carbon content of an as-received sample is calculated by material balance. Thus:

$$FC = 1 - MC - VM - ash \tag{4.4}$$

Proximate analysis is useful in screening fuels to use in particular kinds of gasifiers. For example, fuels with low volatile content can be more effectively processed by partial-oxidation gasification, while fuels with high volatile content may be processed by indirect gasification. Proximate analyses for a variety of solid fuels are given in Table 4.1, from which it is evident that common biomass materials are more readily devolatilized (pyrolyzed) than lignite and bituminous coals, yielding considerably less fixed-carbon residue. This is due to the highly aromatic structure of the coals. Proximate analyses also reveal that most kinds of biomass have lower ash content than coals.

Ultimate analysis reports the major elemental composition of a solid fuels, and typically includes carbon, hydrogen, nitrogen, sulfur, and oxygen (the latter often by difference) [30].

Table 4.1 Proximate analysis data for selected solid fuels and biomass.

	Analysis (wt%, dry basis)			
	Volatile matter (VM)	Fixed carbon (FC)	Ash	Ref.
Coals and peat				
Lignite	43.0	46.6	10.4	[21]
Pittsburgh seam coal	33.9	55.8	10.3	[21]
Wyoming Elkol coal	44.4	51.4	4.2	[21]
Peat	68.4	26.4	5.2	[22]
Woods				
Cedar	77.0	21.0	2.0	[23]
Douglas fir	86.2	13.7	0.1	[23]
Hybrid poplar	84.81	12.49	2.70	[24]
Ponderosa pine	82.54	17.17	0.29	[19]
Red oak sawdust	86.22	13.47	0.31	[24]
Redwood	83.5	16.1	0.4	[23]
Western hemlock	84.8	15.0	0.2	[23]
White fir	84.4	15.1	0.5	[23]
Willow	85.23	13.82	0.95	[24]
Barks				
Cedar	86.7	13.1	0.2	[23]
Douglas fir	70.6	27.2	2.2	[23]
Ponderosa pine	73.4	25.9	0.7	[23]
Redwood	71.3	27.9	0.8	[23]
Western hemlock	74.3	24.0	1.7	[23]
White fir	73.4	24.0	2.6	[23]
Herbaceous biomass				
Alfalfa stems	78.92	15.81	5.27	[24]
Corn stover	75.17	19.25	5.58	[19]
Miscanthus, silberfeder	84.40	12.55	3.05	[24]
Rice hulls	63.52	16.22	20.26	[24]
Sugar cane bagasse, HI	85.61	11.95	2.44	[24]
Switchgrass, IA	82.3	14.5	6.3	[25]
Wheat straw, Denmark	81.24	14.87	3.89	[24]
Wastes and by-products				
Black liquor	55.60	8.80	35.60	[26]
DDGS[a]	78.2	14.7	7.1	[27]
Demolition wood	74.56	12.32	13.12	[24]
Railroad ties	79.2	18.7	2.6	[28]
RDF[b], Tacoma, WA	73.40	0.47	26.13	[24]
Sewage sludge	55.1	7.1	37.8	[29]
Walnut shells	78.28	21.16	0.56	[19]
Yard waste (woody)	69.63	13.87	16.50	[24]

[a]Distillers dried grains with solubles.
[b]Refuse-derived fuel.
Source: adapted from [30].

Examples of ultimate analyses for several kinds of biomass and coals are given in Table 4.2. Care must be exercised when interpreting ultimate analysis data for high moisture content fuels because any moisture content will be included in the ultimate analysis as additional hydrogen and oxygen. To avoid this confusion, ultimate analyses are generally reported on a dry basis.

Table 4.2 Ultimate analysis (dry basis) for selected solid fuels and biomass.

	C	H	N	S	O	Cl	Ash	HHV[a] (MJ/kg)	Ref.
Coals and peat									
Lignite	64.0	4.2	0.9	1.3	19.2		10.4	24.9	[21]
Pittsburgh seam coal	75.5	5.0	1.2	3.1	4.9		10.3	31.7	[31]
Wyoming Elkol coal	71.5	5.3	1.2	0.9	16.9		4.2	29.5	[21]
Peat	54.3	5.7	1.5	0.2	33.1		5.2		[22]
Woods									
Hybrid poplar	50.18	6.06	0.60	0.02	40.44	0.01	2.70	19.0	[24]
Ponderosa pine	49.25	5.99	0.06	0.03	44.36	0.01	0.29	20.02	[19]
Red oak sawdust	49.96	5.92	0.03	0.01	43.77	<0.01	0.31	19.4	[24]
Willow	47.94	5.84	0.63	0.06	44.43	<0.01	1.10	19.3	[24]
Barks									
Douglas fir	56.2	5.9	0.0	0.0	36.7		1.2	22.0	[31]
Pine	52.3	5.8	0.2	0.0	38.8		2.9	20.4	[31]
Herbaceous biomass									
Alfalfa stems	47.17	5.99	2.68	0.20	38.69	0.50	5.27	18.6	[24]
Corn stover	43.65	5.56	0.61	0.01	43.31	0.60	6.26	17.65	[19]
Miscanthus, silberfeder	47.29	5.75	0.33	0.06	43.52	0.06	3.05	28.7	[24]
Rice hulls	38.83	4.75	0.52	0.05	35.59	0.12	19.2	15.8	[24]
Sugar cane bagasse, HI	48.64	5.87	0.16	0.04	42.85		2.44	19.0	[24]
Switchgrass, IA	46.7	4.99	0.86	0.06	41.0	1.49	6.3		[25]
Wheat Straw, Denmark	47.55	5.86	0.59	0.09	42.02	0.17	3.89	18.3	[24]
Wastes and by-products									
Black Liquor	25.70	5.20	0.09	4.80	38.10	0.09	35.60	11.54	[26]
DDGS[b]	49	6.3	4.5	0.4	33.6		7.1	19.8	[27]
Demolition wood	46.3	5.39	0.57	0.12	34.5	0.05	13.12	18.4	[24]
Railroad ties	53.0	5.63	0.26	0.10	38.4	0.016	2.6	21.1	[28]
RDF[c], Tacoma, WA	39.70	5.78	0.80	0.35	27.24		26.13	15.5	[24]
Sewage sludge	34.7	4.6	4.6	1.2	17.0	0.087	37.8	14.9	[29]
Walnut shells	49.98	5.71	0.21	0.01	43.35	0.03	0.71	20.18	[19]
Yard waste (woody)	41.54	4.79	0.85	0.24	32.21	0.3	20.37	16.3	[24]

[a] Higher heating value.
[b] Distillers dried grains with solubles.
[c] Refuse-derived fuel.
Source: adapted from [30].

Nitrogen, chlorine, and sulfur are usually minor constituents of a feedstock, but during gasification they convert to trace species with undesirable impacts on utilization of the syngas. Recent investigation has revealed that gasification converts the nitrogen content of the fuel into at least five significant forms (Figure 4.2), with their relative proportions changing in response to operating conditions [32]. Out of these, ammonia (NH_3) and hydrogen cyanide (HCN) are of particular concern because they are known to cause catalyst poisoning and NO_X air pollution in downstream syngas applications. HCN is also highly toxic; concentrations of merely 50 ppm are considered immediately hazardous to life and health [33]. On the other hand, significant fuel nitrogen appears in the char product or is transformed into N_2, suggesting gasifier operation under certain conditions may favor these nonproblematic forms [32].

The chlorine and sulfur in fuels also form undesirable products; the most prevalent being hydrochloric acid (HCl), hydrogen sulfide (H_2S), and carbonyl sulfide (COS), all of which

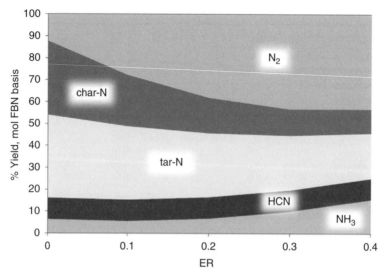

Figure 4.2 *As the ER of switchgrass gasification was increased from 0 to 0.4, yields of NH_3 and N_2 increased, HCN held steady, and char-N and tar-N yields both declined. Reactor pressure and temperature for all tests were 1 atm and 750 °C. Source: reprinted from* The Effect of Equivalence Ratio on Partitioning of Nitrogen during Biomass Gasification, *Copyright (2015), with permission from American Chemical Society.*

can poison catalysts used in downstream applications or convert to air pollutants if the syngas is directly combusted. Chlorine compounds and sulfur compounds are also well known for accelerating corrosion of the metals and refractories used to construct gasification and combustion equipment.

The composition of ash in feedstocks is also important to gasification performance. Inorganic compounds include silica, aluminum, titanium, iron, calcium, potassium, magnesium, sodium, and phosphorus. These components are reported as the highest oxide form and normalized as a weight percent of ash. Sulfur trioxide (SO_3), Cl, and CO_2 may also be reported. In evaluating ash composition for gasification, alkali and phosphorus contents are particularly important because of their potential to cause slagging and fouling in the gasifier and in downstream operations, such as gas coolers and tar reformers. Alkali and phosphorus are particularly associated with herbaceous and manure fuels. When investigating slagging and fouling, total alkali is typically reported on a mass per energy content basis (e.g. kg/GJ). Although most of the historical analyses of ash composition have been performed for combustion systems, the same general guidelines hold true in evaluating gasification feedstock suitability. Miles et al. [24] noted the tendency of alkali in biomass to form deposits or slag as alkali level increases from 0.17 to 0.34 kg/GJ. Potassium is of special concern because, in combination with silica, it can form a low-melting point eutectic at gasification temperatures. The impact may be more severe for partial-oxidation gasifiers, especially fixed-bed gasifiers. In many cases the fouling may be alleviated by an additive that changes the solid–liquid phase behavior. A common additive is magnesia. Partial-oxidation gasifiers that use limestone as a bed medium may also see sequential conversion from carbonate to oxide to carbonate that may lead to fouling.

Recently there has been increased interest in gasifying waste products such as refuse-derived fuel (RDF) and sewage sludge. Some relevant waste feedstocks have been included in Tables 4.1 and 4.2 to facilitate comparison with conventional solid fuels.

Despite these challenges, gasification is a promising solution for solid waste disposal. Waste fuels are sometimes called negative value feedstocks, since there can be value associated with simply destroying them to avoid their disposal costs. This provides an additional revenue stream to a gasification plant processing them, one that is mostly independent of energy and fuel prices, which can be volatile. The high temperatures of gasification also make it more capable of destroying recalcitrant wastes such as plastic, which resists biological conversion processes and lower temperature thermochemical conversions such as pyrolysis. Even toxic organic compounds present in some wastes can often be converted to syngas along with the rest of the material. Gasification temperatures are also sufficient to render biologically hazardous wastes harmless.

4.4 Classifying Gasifiers According to Method of Heating

Although there are many ways to classify gasifiers, the manner in which energy is provided to the process is probably the most common and useful classification scheme. The conversion of carbonaceous solids into flammable gas mixtures is a high-temperature, endothermic process. Thus, energy must be generated within or conveyed to the gasification vessel at high temperatures. Partial-oxidation gasifiers, also referred to as air-blown, oxygen-blown, or directly heated gasifiers, use the exothermic reaction between oxygen and carbonaceous feedstocks to provide the heat necessary for gasification. When air is used as the oxidant, the resulting product gas is diluted with nitrogen and typically has a relatively low dry-basis calorific value of 4–6 MJ/N-m^3 (where N-m^3 is normal cubic meters). The calorific value of the product gas can be increased to 13–14 MJ/N-m^3 by using pure oxygen gas instead of air, though obtaining the oxygen requires operation of an air separation system, which increases the plant capital and operating expenses. In contrast to partial-oxidation gasifiers, indirectly heated gasifiers convey heat to the gasifier through heat transfer surfaces or heat transfer media. Since neither oxygen nor air is added, the product gas is not diluted with nitrogen or significant carbon dioxide. As a result, the gas has a significantly higher heating value, such as 17–19 MJ/N-m^3 [34].

4.4.1 Air-Blown Gasifiers

Different combinations of air, nitrogen, steam, and oxygen have been used for partial-oxidation gasification of biomass, but the simplest and cheapest option is air. As a result, air-blown biomass gasification has been the most thoroughly explored and has been most frequently employed for commercial-scale gasification projects to date.

Air-blown gasification can be conducted in fixed-bed, bubbling fluidized-bed (BFB), and circulating fluidized-bed (CFB) reactors. Because of the significant quantity of air required to achieve favorable equivalence ratio, the final syngas composition produced by an air-blown gasifier is dominated by molecular nitrogen, which significantly reduces the heating value of the syngas. Despite this disadvantage, syngas from air-blown gasification has been successfully used in furnaces, boilers, and ICEs. The nitrogen becomes more problematic if the syngas is to be used for chemical or fuel synthesis applications, because processing vessels and purge gas volumes must be significantly increased to accommodate the large gas volume. In the case of Fischer–Tropsch synthesis, nitrogen has a negative effect on the production of hydrocarbons of suitable chain length for liquid fuel production [35].

Other major constituents of syngas are CO, H_2, CO_2, and CH_4. Ethylene, acetylene, ethane, and other light hydrocarbons are commonly found in syngas in smaller amounts, on the order of 3% molar basis or less. Syngas usually contains some tar and char, which are undesirable, and often must be removed before the syngas can be used. Table 4.3 illustrates syngas composition for a number of air-blown gasification projects, which used a variety of biomass feedstocks, equivalence ratios, and reactor types. Factors influencing syngas composition for air-blown gasification include reaction temperature, equivalence ratio, residence time, type of reactor, and feedstock composition.

4.4.2 Oxygen/Steam-Blown Gasifiers

The diluent effect of the nitrogen content of the air can be avoided by using pure oxygen as a gasifying agent instead. Often steam is applied with the oxygen to encourage the water gas shift reaction to maximize hydrogen gas production, assist with bed fluidization, and moderate reactor temperatures. The resulting syngas contains higher concentrations of H_2 and CO than from air-blown gasification, giving it higher energy content and making it more suitable for downstream chemical processes such as Fischer–Tropsch synthesis. Syngas composition data for representative oxygen/steam gasification projects are shown in Table 4.4.

4.4.3 Indirectly Heated Gasifiers

Although directly heated gasifiers are relatively simple in construction and operation, they have the disadvantage of diluting the syngas with oxidation products (CO_2 and H_2O) and inert N_2 (if air is used). Indirectly heated gasifiers avoid this disadvantage by transporting heat into the reactor from an external energy source instead of burning part of the feedstock within the reactor. The primary reactions of indirect gasification are devolatilization of the feedstock to produce permanent gases, condensable vapors, and char. Depending on the reaction medium and residence time, secondary gas-phase reactions such as the water gas shift reaction may also influence final gas composition. The range of operating temperatures for indirectly heated gasifiers is broader than for partial-oxidation reactors, with temperature ranging from 650 to 1500 °C, although the majority of systems operate in the 650–850 °C range because of the difficulty of transferring high-temperature heat into the reactor. Gas compositions typical of a generic indirect gasifier (National Renewable Energy Laboratory [NREL] Process Development Unit [PDU]) are given in Table 4.5.

A number of indirectly heated gasification systems have been developed that use heat transfer from a hot divided solid, such as crushed olivine or sand. Notable systems of this type have been developed by Repotec (Austria) [47], the Silvagas gasifier technology (USA), which is now owned by Kaidi [48, 49], and the MILENA gasifier (The Netherlands) [50]. These biomass gasifiers are based on fluid coking technology developed in the petroleum industry for processing heavy refinery residuals and employ a combination of fluidized-bed and CFB technologies for the indirectly heated gasifier and the associated combustor that typically burns the residual char.

Table 4.3 Gas composition from various air-blown gasifiers.

Operator	Gasifier type	Fuel(s)	Gas species (% v/v dry basis)						Ref.
			N_2	CO_2	CO	H_2	CH_4	C_xH_y	
SAFI	CFB	RDF	50.8	17.3	9.7	9.5	7.2	5.4	[36]
Muni Distr Htg, Harboore, Denmark	Updraft	Wood chips	40.7	11.9	22.8	19.0	5.3		[36]
Sydkraft AB	CFB	Wood, straw, RDF	48–52	14.4–17.5	16–19	9.5–12	5.8–7.5		[36]
Tech. U. of Denmark	2-stage downdraft	Wood chips	33.3	15.4	19.6	30.5	1.2		[36]
CIEMAT	CFB	Orujillo	59.5–63.2	19.0–21.7	6.9–8.6	5.4–9.3	1.8–3.0	0.9–1.9	[37]
U. of Maine	BFB	Black liquor solids		5.1–14.9	6.9–17.7	5.2–8.3	0.95–1.35		[38]
U. of Seville	BFB	Orujillo		16–20	9–14	8–12	5–7		[39]
U. of Complutense	BFB	Pine sawdust	45.0–66.5	12.0–15.0	10.0–18.0	7.0–9.5	2.4–4.5		[40]
ECN	CFB	Various	51.4	16.4	14.2	11.9	4.0	1.45	[41]
EPI	BFB	Wood	51.9	15.8	17.5	5.8	4.65	2.58	[42]
Iowa State U.	BFB	Wood	55.9	12.8	23.9	4.1	3.1		[42]
Southern Electric Intl.	BFB	Wood	47.9	15.9	15.5	12.7	5.72	2.27	[42]
Tampella Power, Inc.	BFB	Wood	48.8	15.7	16.4	13.7	5.8		[42]

Table 4.4 Composition of typical syngas from pure oxygen-blown gasification.

Organization	Gasifier type	Fuel	Gas species (% v/v dry basis)					Ref.
			CO_2	CO	H_2	CH_4	C_xH_y	
Guangzhou Inst.	Downdraft	Pine wood	24	39	29	6.0	1.3	[43]
U. of Hawaii	BFB	Sawdust	32	30.	25	9.5	2.5	[44]
U. of Saragossa	BFB	Pine wood	14–37	30–50	13–29	5.0–7.5	2.3–3.8	[45]
Union Carbide	Updraft	MSW	24	39	23	5.5	4.9	[42]
Iowa State U.	BFB	Switchgrass	30.–40.	23–33	14–21	7.1–12	0.88–4.8	[25]

Table 4.5 Average gas compositions for NREL indirectly-heated gasification pilot plant.

	Corn stover	Vermont wood	Wheat straw	Switchgrass
H_2	27.3	29.1	25.8	23.9
CO	25.1	23.9	27.9	33.7
CO_2	23.7	23.7	23.7	23.7
CH_4	15.5	15.8	16.6	17.3
C_2H_4	4.27	3.96	4.37	5.18
C_2H_2	0.46	0.39	0.32	0.35
C_3H_8	0.41	0.62	0.82	0.83
C_3H_6	0.12	0.09	0.10	0.10

Steam-to-biomass ratio of 1.0, fluidized bed reactor temperature of 650 °C, and thermal cracker temperature of 875 °C. Gas compositions are on a vol/vol N_2-free dry basis.
Source: adapted from [46].

Other systems are being developed that introduce heat through the reactor wall. The ThermoChem Recovery International (TRI) gasifier in the USA combusts a portion of the product gas in a pulse combustor and transfers heat to a fluidized-bed gasifier through tubular heat exchangers suspended in the bed [51]. Other systems, such as the Pearson gasifier, have attempted to use producer gas or natural gas to heat a fire box enclosing an entrained-flow gasifier [52]. These gasifiers typically have longer residence times to minimize char production. The Range Fuels gasifier was a variation on this concept. In the case of the Pyromex technology, now owned by PowerHouse Energy Group Plc, electrical heating of the reactor walls is used to produce temperatures representative of slagging gasification [53].

Indirect gasification utilizing a heated reactor wall is challenging to scale up because separate gasification and combustion reactor compartments are required, and heat transfer rates between the two can become limiting. An interesting solution was proposed and tested by Lysenko et al. [54], where a fluidized bed was outfitted with a thermal ballast of lithium fluoride and then operated in alternating modes of combustion and gasification. The ballast underwent cycles of melting and solidifying, transferring latent heat from the exothermic combustion phase to the endothermic gasification phase.

4.5 Classifying Gasifiers According to Transport Processes

Most gasifiers are designed as steady-flow processes rather than batch operations. Flow of feedstock through a reactor and mixing it with air and oxygen (for partial-oxidation

Table 4.6 Approximate tar content for four different common types of gasifiers.

Reactor type	Ref.	Tar content (mg/N-m^3)
Updraft	[55]	50 000
Indirect	[55]	48 000–83 000
BFB and CFB	[55]	10 000
Downdraft	[46]	1000

gasifiers) or with a heat carrier (for indirectly heated gasifiers) can be accomplished in many ways. Four distinct gasifier designs based on transport phenomena are generally recognized: fixed-bed reactors, BFB reactors, CFB reactors, and entrained-flow reactors. Among the most prominent differences among these gasifiers is the tar content of the syngas, as shown in Table 4.6.

4.5.1 Fixed Bed

In fixed-bed gasifiers, biomass is processed in bulk, flowing through consecutive zones of drying, pyrolysis, gasification, and char combustion. Fixed-bed gasifiers are the oldest types of gasifier and have historically been developed for smaller scale applications. Developed before modern computer control systems, they are relatively simple systems to design, operate, and maintain. Because of the high temperatures attained in the char combustion zones, fixed-bed gasifiers have a high potential for ash slagging, which impacts their reliability. The slag is most likely to form around where the oxidizing gas is admitted and the highest temperatures are attained. Another challenge for fixed-bed gasifiers is their sensitivity to the size and consistency of the incoming fuel particles since the oxidizing agent and syngas must move evenly and freely through the deep bed of fuel inside the gasifier in order to avoid hot spots and achieve smooth operation. Fixed-bed gasifiers include updraft and downdraft designs, each with distinctive operating characteristics.

Updraft gasifiers were the first and simplest gasifiers. As illustrated in Figure 4.3, fuel enters the top of an updraft gasifier by means of a lock hopper or rotary valve. As it moves in counterflow to air or oxygen, the biomass goes through stages of drying, devolatilizing, and char combustion, with unburned char and ash exiting via a grate at the bottom. Air or oxygen entering the bottom of the gasifier reacts with char in the combustion zone to form CO, CO_2, and H_2O at temperatures as high as 1200 °C. This hot gas provides the energy to drive heating, drying, and pyrolysis of the biomass. In the pyrolysis zone, these gases contact dry biomass in the temperature range 400–800 °C and devolatilize the biomass to produce pyrolysis products and residual char. Above this zone, the gases and pyrolytic vapors dry the entering biomass. Typical product exit temperatures are relatively low (80–100 °C). The counterflow design of the updraft gasifier results in large quantities of tars in the product gas, making the exit pipe susceptible to plugging. For this reason, syngas from updraft gasifiers is usually directed straight into a furnace or boiler to produce steam or hot water, since this setup is relatively tolerant of tars.

The high tar content of the syngas produced by updraft gasifiers can be remedied by admitting oxidant higher up the gasification vessel and allowing the syngas to

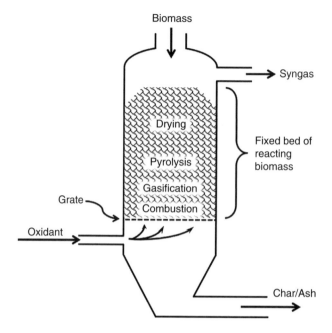

Figure 4.3 Updraft gasifier.

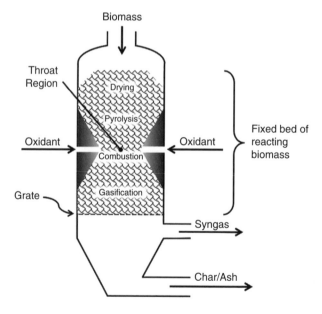

Figure 4.4 Downdraft gasifier.

exit at the bottom, leading to concurrent flow of gases and solids (Figure 4.4). This configuration is known as downdraft gasification. Syngas tar content is dramatically reduced compared to updraft gasification because the tar-rich volatiles released during the gradual heating of the biomass must pass through a high-temperature char combustion zone (800–1200 °C) where tars are efficiently cracked. Tar conversion rates of 99% or greater can be achieved [9]. The hot char also reacts with CO_2 and H_2O released during combustion to produce CO and H_2. Exit gas temperatures are generally high (~700 °C). Feedstock with low ash content and high ash fusion temperature is important to prevent slagging in the high-temperature combustion zone, making wood fuels best for downdraft gasifiers. Downdraft gasifiers must also have fuel with moisture content less than about 20% in order to achieve temperatures high enough to crack tars. The crossflow gasifier is a variation of the downdraft gasifier, with air introduced tangentially into a throat located near the bottom of the gasifier to form a char oxidation zone.

4.5.2 Bubbling Fluidized Bed (BFB)

In a BFB gasifier, illustrated in Figure 4.5, gas flows upward through a bed of free-flowing granular material at a gas velocity sufficient to agitate the material into a churning emulsion of levitated particles and gas bubbles [56]. A gas distribution manifold or series of sparge tubes is used to introduce fluidization gas to the bed [57]. The fluidized bed itself resembles a boiling liquid and has many of the same physical properties as a fluid. Commonly used bed materials include sand, olivine, limestone, dolomite, or alumina. Beds can be fluidized with the gasification agent, typically air, oxygen, and/or steam. The superficial velocity (volumetric flow rate/cross-sectional area) of the gas in the bottom zone is controlled to

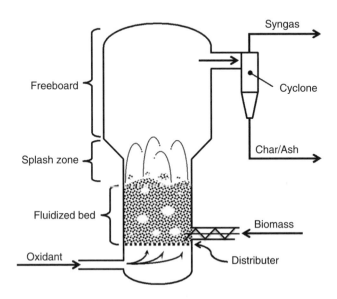

Figure 4.5 Bubbling fluidized bed gasifier.

maintain the bed in a fluidized state. The upper portion of the gasifier is called the freeboard. Its cross-sectional area is often enlarged to lower the superficial gas velocity, helping return particles to the bed to maintain solids inventory. The enlargement also helps to increase gas-phase residence time, giving more time for gasification reactions to convert solids and tars to gases.

Biomass is either injected into the bed from the side by an auger, or sometimes fed from above, where it falls into the bed [58]. In-bed introduction is advantageous because it provides residence time for fines which would otherwise be entrained in the fluidizing gas and leave the bed without complete conversion. In-bed introduction also promotes more uniform biomass heating through better mixing of the biomass and bed material. Biomass entering the hot BFB is almost instantly devolatilized. Tars may partially crack and char may be partially gasified, but residence time for both the gases and the char particles is relatively short and the gas composition does not closely approach equilibrium. In partial-oxidation fluidized-bed gasifiers, combustion of char in the bed provides the heat to maintain the bed temperature and devolatilize biomass. In indirectly heated BFB gasifiers, heat is introduced by heat exchange through the walls of the gasifier or through heat exchanger tubes in the bed.

Fluidized-bed gasifiers have the advantage of being extremely well mixed and having high rates of heat transfer, resulting in very uniform bed conditions. For partial-oxidation systems, gasification is very efficient with 95–99% carbon conversion. For indirectly heated fluidized-bed steam-blown gasifiers, carbon conversion is typically lower, in the range 60–75%. In many indirectly heated gasifiers, the residual carbon is combusted in an external char combustor which generates heat that is returned to the gasifier. This effectively results in carbon conversion efficiencies comparable to partial-oxidation systems. The notable advantage is that the CO_2 formed in char combustion is not included in the syngas. BFB gasifiers are normally designed for complete char/ash carryover, necessitating the use of cyclones or other types of inertial separators for particulate control. In general, fluidized bed gasifiers produce syngas with moderate tar and high char loadings compared to fixed-bed systems. The high char loading occurs because the fluidized bed gasifier must rely on pneumatic conveyance to eliminate all of its char and ash, leading to the cyclone essentially playing the same role as the grate in a fixed-bed reactor.

BFB reactors are readily scaled to large sizes, a notable advantage when designing commercial gasification plants based on data and operating experience from small pilot plants. Bed fluidization behavior remains predictable across a great range of sizes, and reactor size is only limited by the ability to evenly distribute incoming feedstock across the fluidized bed cross-section. This can be addressed by the use of multiple feeding locations, or by increasing reactor pressure, which allows more biomass throughput to occur for the same fluidized-bed cross-sectional area.

BFB gasifiers can be operated between about 700 and 925 °C [59]. Higher bed temperatures are desirable because they generally lead to higher carbon conversion and increased tar cracking, but the fluidized bed temperature must be kept well below the ash-fusion temperature of the biomass ash, otherwise the ash particles soften, become slightly sticky, and begin to adhere bed particles to each other. This quickly leads to defluidization of the entire bed, a condition known as "agglomeration." Once the bed has agglomerated and stopped fluidizing, the condition generally cannot be reversed without shutting the gasifier down and manually replacing the bed material.

4.5.3 Circulating Fluidized Bed (CFB)

As gas flow increases through a fluidized bed, the bed voidage increases and solids loading in the freeboard increases. As the interface between the fluidized bed and the freeboard becomes difficult to discern, the reactor is said to be operating in the turbulent bed regime [60]. As gas flow is further increased, elutriation of particles becomes significant enough that the bed is quickly depleted of particles, and a cyclone becomes essential for returning particles via a downcomer to the bottom of the reactor. This operating regime is known as a CFB, as illustrated in Figure 4.6.

CFBs can be used for both partial-oxidation and indirectly heated gasifiers. In partial-oxidation systems, either air or pure oxygen is admitted to the bottom of the gasifier in a fashion similar to a BFB reactor, or sometimes in the lower portion of the solids return leg coming from the cyclone. In indirect CFB gasifiers, the char is combusted externally, flue gas is separated from the solids, and heated solids are returned to the gasifier. This changes the composition of the syngas, in that the oxidation products are not mixed back into the syngas. The solids circulation rate in such a CFB is governed by the amount of energy needed to gasify the biomass and to maintain the maximum temperature below slagging conditions. Typical ratios of bed media solids to biomass are $15:1-30:1$. Circulating bed gasifiers have the advantages of high throughput, good

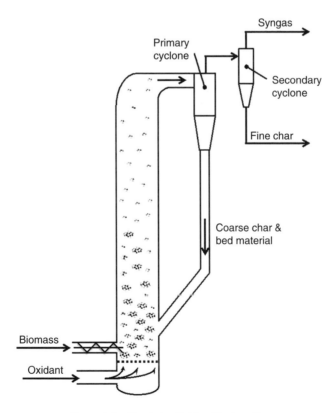

Figure 4.6 Circulating fluidized bed gasifier.

4.5.4 Entrained Flow

An entrained-flow gasifier suspends finely ground fuel particles or atomized liquid in a gas flow (Figure 4.7). A flame front usually exists at the entrance to the gasifier, achieving very high temperatures (1300–1800 °C), which then decreases as the gas flow approaches equilibrium composition. While most gasifiers avoid ash slagging, entrained-flow gasifiers are designed to melt the ash, which runs down the sides of the gasifier to exit at the bottom where it is cooled with water. The extremely high temperatures result in almost complete destruction of tars, and the gas composition typically approaches an equilibrium composition and contains very low levels of CH_4 and other light hydrocarbons [61].

The high temperatures and molten slag pose materials selection challenges, but designers have come up with clever solutions. In an entrained-flow gasifier design by Siemens, tubes with cooling water line the inside of the reactor. Slag initially condenses on the tubes, quickly forms a solid, protective, and thermally insulating layer, and then all further molten slag flows downward over the slag layer and into a slag pot at the bottom of the gasifier without damaging the cooling tubes. This design allows the reactor to withstand the extreme temperatures without the use of refractory material to protect the reactor walls. The absence of refractory means that the reactor can be started up and shut down quickly. The entrained-flow design also tolerates high ash content fuels well [61].

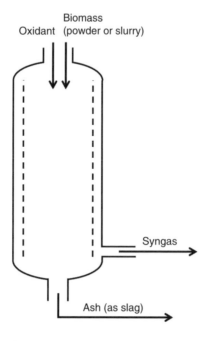

Figure 4.7 Entrained-flow slagging gasifier.

Entrained-flow gasifiers have been developed by Shell, Siemens, Texaco (GE), Conoco Phillips, and others, mainly for coal fuel, but with some biomass experimentation. There are a number of reasons for the lack of application to biomass, but the high cost of feed preparation to reduce moisture content to low levels and reduce the particle size (to 100 μm to 1 mm) are the primary concerns [62]. There are some interesting solutions that have been proposed to address these feedstock prep concerns, such as the Carbo-V technology developed by Choren, and now owned by Linde. The Carbo-V process uses a multistage gasifier, which first pyrolyzes the biomass at low temperatures, and then uses an entrained-flow reactor to process the tar-rich pyrolysis vapors alongside the char residues, further converting both to additional syngas. Pyrolysis oil and pyrolysis oil/char slurries are also being investigated as feed for entrained-flow gasifiers, with the pyrolysis being carried out at small, decentralized facilities as an energy densification strategy, and then the crude pyrolysis oil being fed to the entrained-flow slagging reactor at a centralized location to make high-quality fuels at large scale [5, 34].

Although usually directly heated, an indirectly heated, entrained-flow gasifier has also been developed, as shown in Figure 4.8. Finely divided biomass (less than 3 mm diameter) is entrained in steam or steam/nitrogen mixtures and injected into tubular reactors operated at 700–950 °C. The biomass rapidly gasifies into syngas. The tubular reactors are installed in furnaces fired with product gas or natural gas to provide the energy to pyrolyze the biomass. These indirectly heated entrained-flow gasifiers have considerable flexibility in operating conditions, which allows better control of char and tar yields. The size of these reactors is

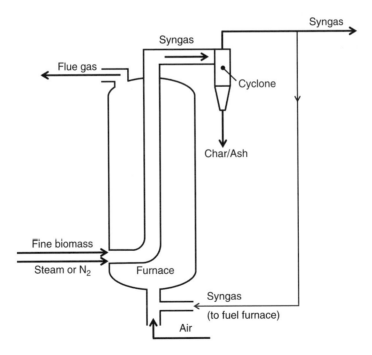

Figure 4.8 Indirectly heated entrained-flow gasifier.

4.6 Pressurized Gasification

Gasification under elevated pressures has several advantages. For a given mass throughput, gas volumes are smaller, resulting in smaller process vessels and piping. Many reactions are accelerated under pressurized conditions, which can more closely approach the equilibrium products. A pressurized gasifier can also have strong advantages from a systems perspective. The reactor pressure can be used directly to drive the produced syngas through cleanup equipment, and then into downstream applications, eliminating the need for syngas compression equipment.

One of the major challenges of pressurized gasifiers is feeding of biomass. Gases and liquids can be readily pressurized for injection into a pressure vessel, but solids need to be moved through lock hoppers, rotary valves, or other suitable mechanical devices which are susceptible to plugging, back-flowing, or leaking. Many of these devices are suitable only for modest pressures, on the order of tens of kPa. Hydrophobic solids such as coal have been successfully slurried to produce a pumpable mixture. Unfortunately, biomass is extremely hydrophilic, absorbing large amounts of water before forming a stable slurry. Wetted biomass will still retain the general appearance, feel, and mechanical handling characteristics of a dry solid, even after adding an equal mass of water.

4.7 Products of Gasification

Gasification converts solid, carbonaceous feedstocks into primarily gaseous products but also liquids (tar) and solid residue (ash). Anticipating syngas composition is important for evaluating suitability for downstream applications. Liquids and carbonaceous residues are not expected if gasification achieves chemical equilibrium, but this is rarely attained in practice. Accordingly, a knowledge of conditions that encourage the formation of tar and ash are important in the design of gasification systems.

4.7.1 Gaseous Products

Gas composition is a function of the biomass fuel properties, inlet gas composition (air, oxygen, or steam), gasifier design, and operating conditions, especially the oxygen–fuel ratio applied. Simple chemical equilibrium calculations that consider only elemental ratios, temperature, and pressure can be used to obtain first estimates of syngas composition. Equilibrium results should be interpreted with some caution, as only high-temperature slagging gasifiers operate at temperatures high enough to closely approach equilibrium; nevertheless, the results are insightful for all types of gasifiers.

For example, the chemical equilibrium composition of syngas for steam-blown, indirect gasification of biomass was calculated using STANJAN [63] modeling software. The elemental composition of the biomass was assumed to be ($CH_{1.4}O_{0.6}$). It was reacted with an equal mass of steam at 1 atm with temperature varied from 600 to 900 °C. The results on a dry gas basis are shown in Figure 4.9. The trends in Figure 4.9 demonstrate that CH_4

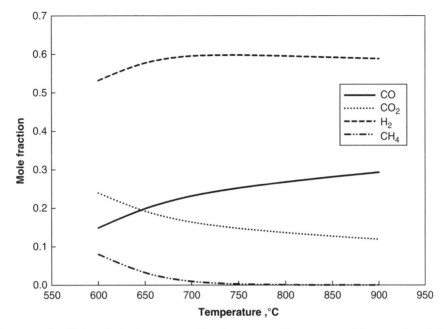

Figure 4.9 *Equilibrium dry basis gas composition for steam gasification (steam/biomass ratio = 1.0). A constant pressure of 1 atm was assumed, and specified constant temperatures from 600 to 900°C. Biomass elemental composition was assumed to be $CH_{1.4}O_{0.6}$.*

concentrations can be expected to decline as reactor temperatures are increased. They also predict high H_2 and CO concentrations because there is limited availability of oxygen atoms to further oxidize the hydrogen and carbon atoms, and also no diluent effect from N_2, as there is with air gasification.

For comparison, the syngas composition for air-blown gasification is presented in Figure 4.10, assuming a steam-to-biomass ratio of 0.5, an ER of 0.3, and pressure of 1 atm. Figure 4.11 gives a comparable dry basis equilibrium makeup for pure oxygen-blown gasification, also with an ER of 0.3 and a pressure of 1 atm.

Figure 4.12 shows the effect of steam/biomass ratio on produced gas composition. The H_2 to CO ratio is a strong function of the steam/biomass ratio, where H_2/CO ratios from <1 to >2 could in theory be obtained by varying steam rate. High H_2/CO ratios are desirable for fuel synthesis applications, such as Fischer–Tropsch synthesis.

The composition of syngas from real gasifiers is more complex than suggested by these calculations because chemical equilibrium is rarely approached. In addition to CO, H_2, CO_2, H_2O, and CH_4, real gasifiers produce significant light hydrocarbons, tar, and char. The gasification mechanism begins with the initial pyrolysis step, which produces primary pyrolysis products (permanent gases, condensable aromatics, and char). Secondary reactions then further convert the aromatics and char into gases, though at modest rates. Char conversion is typically hindered by mass transfer limitations, while the aromatics have relatively high stability and convert via slow kinetics, leaving substantial amounts unconverted as tar. Reaching equilibrium (permanent gases with no char or tar remaining)

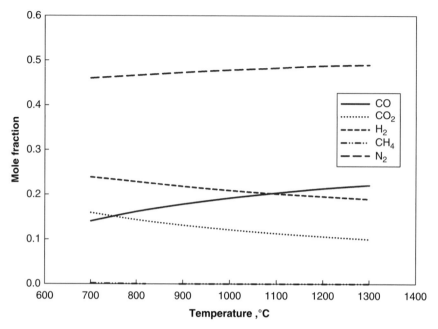

Figure 4.10 Equilibrium dry basis gas composition for air gasification at 30% of stoichiometric oxygen, a steam-to-biomass ratio of 0.5, a constant pressure of 1 atm, and specified constant temperatures from 700 to 1300°C. Biomass elemental composition was assumed to be $CH_{1.4}O_{0.6}$.

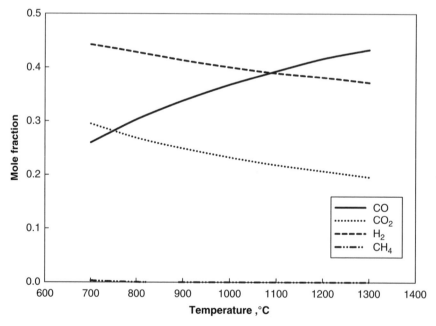

Figure 4.11 Equilibrium dry basis gas composition for oxygen gasification at 30% of stoichiometric oxygen, a steam-to-biomass ratio of 0.5, a constant pressure of 1 atm, and specified constant temperatures from 700 to 1300°C. Biomass elemental composition was assumed to be $CH_{1.4}O_{0.6}$.

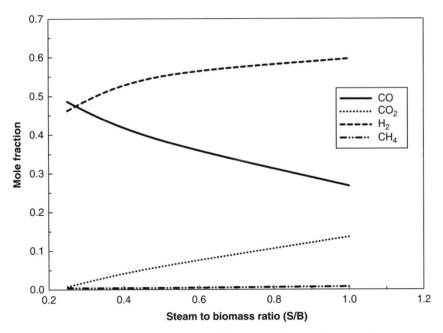

Figure 4.12 *Equilibrium dry basis gas composition for steam gasification. Pressure and temperature were held constant at 1 atm and 800 °C. Biomass elemental composition was assumed to be $CH_{1.4}O_{0.6}$.*

is theoretically possible with sufficient time and/or temperatures, but such conditions are usually not practical. Entrained-flow gasifiers approach equilibrium most closely by using very small particle sizes to overcome heat and mass transfer limitations, and high temperatures to accelerate kinetics.

4.7.2 Char and Tar

Syngas produced by most types of air-blown, oxygen-blown, and indirectly heated gasifiers can be expected to contain significant amounts of tar and char. Table 4.6 provides approximate tar content of syngas for partial-oxidation gasifiers and indirect gasifiers [46, 55]. Table 4.7 provides approximate char content for partial-oxidation gasifiers [28, 64, 65]. These concentrations are high enough to justify gas cleaning for most downstream applications, including ICEs, gas turbines, and liquid fuel synthesis (Table 4.8).

Table 4.7 *Examples of char concentrations in producer gas reported by three air-blown gasification studies.*

Reactor type	Char content (g/N-m³)	Ref.
CFB	1.7–13.1	[28]
BFB	1.0–44	[64]
Downdraft	9.3–30	[65]

Table 4.8 Approximate maximum acceptable levels of particulate (mainly char) and tar in syngas for various applications [42, 66].

Application	Particulate	Tar
Internal comb. engine	50 mg/N-m^3	10–100 mg/N-m^3
Gas turbine	30 mg/N-m^3	0.5–5 mg/N-m^3
FT synthesis	n.d.[a]	0.01–1 µl/l[b]
Methanol synthesis	0.02 mg/N-m^3	0.1 mg/N-m^3

[a] Not detectable.
[b] Maximum acceptable is 0.01 µl/l for FT catalyst inhibiting tars; 1 µl/l otherwise.

Many studies have investigated reducing char and tar through changes to gasifier operating conditions. Increasing the temperature of the reactor, using calcined dolomite as bed material in a fluidized-bed gasifier, and adding steam have all demonstrated some benefit. High-temperature operation (above 1200–1300 °C) is very effective at reducing char and tar yields [42], but the materials of construction become quite demanding. Increasing the equivalence ratio reliably achieves higher temperatures, but also results in more oxidized products, reducing syngas heating value and CGE of the gasification plant. Dolomite can be used as an ingredient in the bed of fluidized-bed gasifiers to catalytically crack tar, but it rapidly breaks into fines that elutriate from a fluidized bed. The use of steam may impact plant efficiency if heat load for raising the steam is larger than the heat recovery available from syngas coolers. Nevertheless, these strategies can be utilized to an appropriate extent, and used in combination to reduce tar production.

Char and tar produced during gasification can be removed by a number of gas-cleaning technologies, often used in series. Gas cyclones are able to remove most particulate matter above about 10 µm in size [19]. The remaining particulate matter and tars must be removed by barrier filters, electrostatic precipitators, or wet scrubbers. It is also possible to catalytically crack tars into noncondensable gases at high temperatures. This is advantageous compared to tar removal as the tar cracking reactions then contribute additional syngas yield, but catalyst expense, fouling, and deactivation can be challenging [66].

4.8 System Applications

Applications of gasification plants can be categorized as process heat, combined heat and power (CHP), or liquid fuel/chemical synthesis (see Table 4.9).

4.8.1 Process Heat

Process heat is used for a variety of industrial applications, including drying, distilling, or steam raising. It is among the lowest value applications of syngas because it often substitutes for coal, natural gas, or other relatively low-cost fuels. However, it is among the least capital intense gasification applications to implement because syngas quality requirements are very modest. Nexterra of British Columbia has completed projects where syngas is burned directly for wood drying and lime kilns. The syngas produced by the gasifier is used in place of, or alongside, natural gas in otherwise conventional wood drying kilns [88]. Gasifiers have also been used in developing countries such as India and China to provide

Table 4.9 Representative gasification technologies.

Company name	Kaidi [48, 49, 67–71]	Alter NRG	Taylor biomass energy [72, 73]	TRI [74]	Repotech GmbH [75]
Headquarters	Wuhan, China		Montgomery, NY	Baltimore, MD	Güssing, Austria
Gasifier technologies	Silvagas	ClearFuels		The Taylor Gasification Process	Steam Reforming Gasification
Gasifier type	Indirectly heated twin CFB	HTR	Fixed bed with plasma-assisted cupola	Indirectly heated twin CFB	Fluidized bed steam reforming with pulse combustors followed by a char reduction stage	Indirectly heated two-stage fluidized bed
Temperature (absolute)	815 °C	982 °C	1100 °C	815 °C	600 °C	850 °C
Pressure (absolute)	Near atmospheric	28 bar	Near atmospheric	Near atmospheric	Bed 4.4 bar Freeboard 2 bar	Near atmospheric
Intended use	CHP v a boiler, also biofuels	Automotive fuels from biomass via FT synthesis	Power plant retrofits, ethanol production	CHP via boiler	Power generation and biorefineries	Power generation via ICE, pilot scale production of SNG, and FT research
Fuel(s) processed	Wood and agricultural residues	Wood and sugar cane bagasse	MSW, sewage sludge, medical waste, hazardous waste	MSW and construction waste		
Feed preparation	Cleaning, screening to 5 cm minus, drying to 20% MC	Cleaning, screening, size reduction to ~0.5 cm, drying to ~9–15% MC		Cleaning, screening to 5 cm minus, drying to 20% MC	Screening to 5 cm minus, drying optional	Hammer milling and screening
Feed introduction	Lock hoppers and rotary valves with additive		Fuel fed with metallurgic coke and lime (ash fluxing agent)	Lock hoppers and rotary valves with additive	Hydraulic piston feeder	

(Continued)

Table 4.9 *Continued*

Gas cleanup	Gas cyclones followed by gas quench scrubber	Multiple step gas cleaning to prepare for catalytic conversion	Cooling, particulate-, mercury-, sulfur filters, and carbon capture	Integrated gas conditioning reactor, heat recovery, polishing scrubber	HRSG, venturi scrubber, tar capture & recycling, water removal, H_2S & NH_3 scrubbing	Biodiesel scrubbing, fabric filter, tar capture and combustion
Advantages	Demonstrated at commercial scale, medium calorific value gas		Significant reduction of mercury, SO_2, NO_x, and CO_2, feedstock flexibility, low tars	Demonstrated at commercial scale, medium calorific value gas, and conditioning reactor to reduce tars	Medium calorific value gas, customized composition, feedstock flexibility	14 years of commercial operation at 2 MW_e
Size Demonstration	25–50 MW_e 8 MW_e in Burlington, VT	Demonstration facility in Commerce city, CO, sponsored by DOE grant	Operating 40 000 gal/year ethanol plant at Madison, PA with Coskata	20–25 MW_e Project underdevelopment at Montgomery, NY (2017)	Up to 60 MW_e Extensive commercial demonstration on spent pulping liquor. Norampac Cascades Paper, Ontario Canada. Process demonstration unit at TRI center in Durham, NC. 500 ton/d Fulcrum Biofuels under development (2017) Reno NV	2 MW_e and 4.5 MW_t Commercial operation in Güssing, Austria

	Linde [76]	Foster wheeler [77]	Carbona-andritz [78]		Air liquide [79, 80]
Company name Headquarters Gasifier Name(s)	Linde [76] Munich, Germany Carbo-V	Foster wheeler [77] London, U.K.	Carbona-andritz [78] Graz, Austria High Pressure	Low Pressure	Air liquide [79, 80] Paris, France Lurgi Multi Purpose Gasifier
Gasifier Type	Low temperature Pyrolyzer, Pyrolysis vapors to an entrained-flow gasifier	CFB	BFB	BFB	Oxygen-blown entrained-flow with cooling shield
Temperature Pressure (absolute) Intended Use	1400–1700°C 5 bar Automotive fuels from biomass via FT synthesis	650–900°C Near atmospheric Cofiring/repowering 60 MW$_t$ unit, SNG, FT	850–900°C 21 bar Combined cycle power generation stand alone or repowering condensing plant	850°C 2 bar CHP via ICE	1200°C 30–100 bar Liquid synthetic fuels
Fuel(s) Processed		Wood Biomass, coal, petroleum coke	Coal, wood chips, forest residue, straw	Coal, wood chips, forest residue, straw	Biosyncrude from pyrolysis, coal slurry, waste slurry
Feed Preparation	Drying to 15–20% MC, hammer mill and chipping	Cleaning and screening	Cleaning, crushing to 5 cm minus, and drying to ~20% MC via belt dryer	Cleaning, crushing to 5 cm minus, and drying to % MC via belt dryer	Feed prepared remotely at small satellite pyrolysis units close to the source, then transported as a slurry
Feed Introduction	Biomass metered into mechanically mixed pyrolyzer, pyrolysis gases flow into gasifier	Metering bins followed by screw feeders	Atmospheric storage/weigh silo, lock hoppers, surge hopper, metering screw, and feeding screw	Two lock hoppers followed by feeding screws	Fuel in liquid form is pumped through a spray nozzle into the gasifier

(Continued)

Table 4.9 Continued

Gas cleanup	Cooling, fabric filter, water scrubbing, water washing, and gas conditioning for fuel production	Gas coolers and hot gas filters, catalytic and thermal cracking, oil scrubbing	Hot gas filtering at 250–400 °C (allows alkali metals to condense on dust particles and tars to remain in gas phase)	Tar reformer (catalyst), two gas coolers, fly ash filter, and water scrubber	Cold, high pressure methanol washing used to remove nitrogen species, sulfur species, and CO_2
Advantages	Very clean gas, high carbon efficiency (99.5% by mass) but process complex and expensive. Mechanical feeding problems at pilot scale stalled progress	Years of experience with plants operating at Lahti, Finland, Ruien, Belgium, and Varkaus, Finland	Technology based on extensive piloting and engineering design	Long history of solving scale up and operating problems. Commercial-scale plant operating.	Remote pyrolysis could solve feedstock issues.
Size		>60 MW_t, 100 Nm^3/hour SNG	50–150 MW_t	1–15 MW_t	5 MW_t
Demonstration	Commercial experience, 1000+ hour of pilot scale operation	Commercial experience, 12 gasifiers between 1981 and 2015		110 dt/d biomass to produce 5.4 MW_e and 11.5 MW_t at Skive, Denmark	Commercial experience

Company name	Frontline BioEnergy [81–83]	Outotec [84]	Enerkem [85, 86]	Aries clean energy [87]	
Headquarters	Ames, IA	Helsinki, Finland	Montreal, Canada	Nashville, TN	
Gasifier name(s)	TarFreeGas	Advance Staged Gasifier	BIOSYN Process	LF64	
Gasifier type	Multi-Modal BFB/CFB	BFB	BFB, O_2-enriched air and steam	Downdraft	BFB
Temperature	675–900 °C	540–980 °C	750 °C	Varies between zones	705 °C
Pressure (absolute)	2–11 bar	Near atmospheric	2 bar	1.1–1.4 bar	Near atmospheric
Intended use	Boiler, ICE, and catalytic synthesis	Repowering of coal-fired boilers, other applications that can utilize hot syngas	Green fuels and chemicals	Electricity generation and solid waste disposal	Electricity generation and solid waste disposal
Fuel(s) processed	Wood residues, Ag Residues, RDF	Woody biomass, forest residue, tires	Municipal solid waste	Waste wood, supplemented with tires and sewage sludge	Sewage sludge and other wastes
Feed preparation	Sizing to 5.0 or 7.5 cm minus, depending on gasifier scale	Inert removal, sizing, drying to 15–25% MC advised	Drying, sorting, shredding to 5 cm minus, then drying to 20% MC	Drying	Drying, size reduction for some fuels
Feed introduction	Dry hopper storage, lock hopper introduction	Metering bin and rotary feeder	"Front-end feeding system." No need for pelletization	Lock hoppers and augers	Lock hoppers and augers

(Continued)

Table 4.9 Continued

Gas cleanup	Proprietary PMFreeGas technology, additional cleanup as required	Refractory lined cyclones, gas cleanup with dry sorbent, baghouse filters	Cyclones, carbon/tar conversion, heat recovery, carbon/tar reinjection		Cyclones, developing biochar media filtration, microwave tar cracking, wet scrubbing
Advantages	Feedstock flexible, commercial experience	Proven solutions at scale, over 100 plants in operation	Feedstock flexible, integrated tar capture and destruction	No size reduction needed, ORC generation of electricity is robust	Feedstock flexible
Size	Up to 900 Mg dry/day		Demonstration plant (Westbury, Quebec) 1.2 million gal/year of methanol/ethanol	Up to 58 Mg/day	22 Mg/day at 10% MC
Demonstration	Commercial unit operated in Benson, MN (>68 Mg/day); another under construction in Wednesbury, UK (36 Mg/day).	100 gasifier plants in operation, with over 250 feedstocks processed	Projects 10 million gal/year methanol/ethanol in Edmonton, Canada	Operating units in Covington, TN and Lebanon, TN	Gasifier technology from MaxWest, operated in Sanford, FL

CFB, circulating fluidized bed; BFB, bubbling fluidized bed; CHP, combined heat and power; SNG, synthetic natural gas; FT, Fischer–Tropsch; MSW, municipal solid waste; ICE, internal combustion engine; MC, moisture content; HRSG, heat recovery steam generator; HTR, hydrothermal reformer; RDF, refuse-derived fuel; ORC, organic Rankine cycle.

syngas for use in cooking and heating as an alternative to open fires in the home, although with some difficulties in reliability [89].

4.8.2 Combined Heat and Power (CHP)

Biomass gasification can be used to generate electricity and waste heat suitable for steam for industrial process heat or for district heating. Projects of this type tend to be relatively small scale (less than 10 MW electric). For these applications, the syngas is typically combusted in a stationary ICE equipped with a generator and provisions for heat recovery. This mode of electricity generation is simple, and ICEs are moderately robust to syngas impurities (Table 4.8), allowing the syngas cleanup equipment to be relatively simple compared to other power generation options. Disadvantages of power generation using ICEs include the possibility of unacceptable levels of air pollutants and high engine maintenance costs [5].

Examples of companies around the world that have developed CHP projects via use of ICEs include Syntech Bioenergy, LLC of Englewood, CO, which markets 145 kW$_e$ modular gasifiers originally developed by Community Power Corporation for CHP applications [90]. Babcock and Wilcox have a demonstration plant in Harboøre, Denmark, where wood chips are gasified to fuel ICEs, generating heat and electrical power for a municipality [91]. Similar projects have also been carried out by ANDRITZ Carbona in Skive, Denmark [92], and REPOTEC in Güssing, Austria [93].

For larger-scale operations, electricity and heat can be generated by integrated gasification/combined cycle (IGCC) power [94]. An IGCC system consists of an air separation plant to generate pressurized oxygen, a pressurized gasifier, a high-temperature gas-cleaning system, a gas turbine topping cycle, a heat recovery steam generator (HRSG), and a steam turbine bottoming cycle. A demonstration plant of an IGCC fired with biomass was successfully operated in Värnamo, Sweden, from 1996 to 2000, and had an output of 6 MW electric [95]. The economic viability of the production of power via IGCC is currently weak, but could become stronger if electricity prices increase, or if the price of carbon emission allowances becomes a significant factor in selecting power options [95, 96].

4.8.3 Fuel and Chemical Synthesis

Synthesis of fuels and chemicals is perhaps the most technically challenging application for biomass gasification. Despite these technical challenges, companies such as CHORAN in Freiburg, Germany, have created demonstration synthetic fuel plants. The ability to create synfuels from biomass using gasification is important because it provides a way to convert all components of biomass, including the lignin fraction, into the final fuel product. The high temperatures utilized by gasification also tend to make it more agnostic to fuel properties compared to pyrolysis and enzymatic conversion, making it especially suitable for conversion of wastes and other low-quality feedstocks. Provided that syngas can be purified into a clean mixture of CO and H_2, it can be processed catalytically to make a wide range of fuels and chemicals including diesel fuel, gasoline, jet fuel, methanol, ethanol, larger alcohols, pure hydrogen, aldehydes, and dimethyl ether [97]. The difficulty lies in cleaning the raw syngas emerging from the gasification reactor to a standard of cleanliness high enough to not poison the chemical catalysts, which are highly sensitive to the nitrogen, sulfur, and halides present in the syngas (Table 4.10).

Table 4.10 Nitrogen, sulfur, and halide species limits for synthetic fuels production [66].

	Sulfur (H_2S, COS)	Nitrogen (NH_3, HCN)	Halides (mainly HCl)
Methanol synthesis	<1 mg/m^3	<0.1 mg/m^3	<0.1 mg/m^3
FT synthesis	<0.01 ml/l	<0.02 ml/l	<0.01 ml/l

It is also possible to use microorganisms to synthesize fuels and chemicals from syngas [98]. In a process known as syngas fermentation, bacteria metabolize CO and H_2 to a wide variety of products. Companies like LanzaTech [99], Jupeng Bio [100], and Synata Bio [101] are exploring ethanol production via syngas fermentation. LanzaTech is using syngas fermentation to produce chemical products such as acetic acid and 2,3-butanediol, commodity chemicals that are useful for making a variety of fuels and products [102]. Syngas fermentation-based production of fuels and chemicals provides a similar overall transformation as enzymatic hydrolysis, but with some interesting advantages. By using gasification, deconstruction of biomass does not require expensive enzymes, gasification is robust to difficult feedstocks such as waste products, both carbohydrate and lignin are utilized, and preliminary economic assessments estimate that production costs may be lower [98]. Syngas fermentation also has advantages compared with Fischer–Tropsch synthesis or other catalytic synthesis routes: the microorganisms are insensitive to sulfur in the syngas and the process does not require high pressures or temperatures. Disadvantages of syngas fermentation include relatively slow gas–liquid exchange rates in aqueous fermentation media, toxicity of HCN and some tar constituents to microorganisms, and more modest research investment compared to other pathways to advanced biofuels [98].

Acknowledgement

This chapter is an update of *Gasification* prepared by Richard L. Bain and Karl M. Broer for the first edition of this book (*Thermochemical Processing of Biomass: Conversion into Fuels, Chemicals and Power*, edited by Robert C. Brown, © 2011 John Wiley & Sons, Ltd.).

References

1. Reed, T.B. (ed.) (1981). *Biomass Gasification: Principals and Technology*. Park Ridge, NJ: Noyes Data Corporation.
2. Rezaiyan, J. and Cheremisinoff, N.P. (2005). *Gasification Technologies: A Primer for Engineers and Scientists*. Boca Raton, FL: CRC Press.
3. Probstein, R.F. and Hicks, R.E. (2006). *Synthetic Fuels*. Mineola, NY: Dover Publications, Inc.
4. Higman, C. and van der Burgt, M. (2008). *Gasification*, 2e. Burlington, MA: Elsevier, Inc.
5. Knoef, H.A.M. (ed.) (2012). *Handbook Biomass Gasification*, 2e. Enschede: BTG Biomass Technology Group BV.
6. Basu, P. (2013). *Biomass Gasification, Pyrolysis and Torrefaction*, 2e. London: Elsevier, Inc.
7. Tomory, L. (2012). *Progressive Englightenment: The Origins of the Gaslight Industry, 1780–1820*. Cambridge, MA: The MIT Press.
8. Goss, J.R., Coward, L.D.G., Desrosiers, R.E. et al. (1983). *Producer Gas: Another Fuel for Motor Transport*. Washington, DC: National Academy Press.

9. Reed, T.B. and Das, A. (1988). *Handbook of Biomass Downdraft Gasifier Engine Systems.* Golden.
10. Swedish Academy of Engineering (1998). *Generator Gas: The Swedish Experience – Gas 1939–1945.* Golden, CO: Biomass Energy Foundation.
11. Williams, T.R. (1939). *Producer Gas Vehicles.* Melbourne: Department of Defence.
12. LaFontaine, H. and Zimmerman, F.P. 1989. Construction of a simplified wood gas generator for fueling internal combustion engines in a petroleum emergency, Oak Ridge, 1989.
13. Cash, J.D. and Cash, M.G. (1940). *Producer Gas for Motor Vehicles.* Sydney: Angus and Robertson Ltd.
14. Kaupp, A. (1984). *State of the Art for Small Scale Producer-engine Systems.* Golden, CO: Biomass Energy Foundation Press.
15. Sichinga, J., Jordaan, N., Govender, M. and van de Venter, E. 2005. Sasol Coal-to Liquids Developments, a presentation to the Gasification Technologies Council Conference, San Francisco, 2005.
16. Murphy, C. 1979. To cope with embargoes, S. Africa converts coal into oil, The Washington Post, 27 April 1979.
17. IPCC 2014. Climate Change 2014: Mitigation of Climate Change. Contribution of Working Group III to the Fifth Assessment Report of the Intergovernmental Panel on Climate Change. Cambridge University Press, Cambridge and New York.
18. Turns, S.R. (2000). *An Introduction to Combustion*, 2e, 19. New York, NY: McGraw-Hill.
19. Brown, R.C. and Brown, T.R. (2014). *Biorenewable Resources: Engineering New Products from Agriculture*, 2e. Ames, IA: Wiley.
20. de Wiebren, J. (2005). *Nitrogen Compounds in Pressurised Fluidised Bed Gasification of Biomass and Fossil Fuels.* Rotterdam: Optima Grafische Communicatie.
21. Bituminous Coal Research, Inc. 1974. Gas Generator Research and Development, Phase II: Process and Equipment Development.
22. Leppälahti, J. and Kurkela, E. (1991). Behaviour of nitrogen compounds and tars in fluidized bed air gasification of peat. *Fuel* **70**: 491–497.
23. Howlett, K. and Gamache, A. 1977. Forest and mill residues as potential source of biomass, McLean.
24. Miles, T.R., Miles, T.R. Jr., Baxter, L.L., et al. 1995. Alkali deposits found in biomass power plants: A preliminary investigation of their extent and nature, Golden.
25. Broer, K.M., Woolcock, P.J., Johnson, P.A., and Brown, R.C. (2015). Steam/oxygen gasification system for the production of clean syngas from switchgrass. *Fuel* **140**: 282–292.
26. ECN 2012. Black Liquor. https://phyllis.nl/Biomass/View/1396 (accessed 9 November 2018).
27. Giuntoli, J., de Jong, W., Arvelakis, S. et al. (2009). Quantitative and kinetic TG-FTIR study of biomass residue pyrolysis: dry distiller's grains with solubles (DDGS) and chicken manure. *Journal of Analytical and Applied Pyrolysis* **85**: 301–312.
28. van der Drift, A., van Doorn, J., and Vermeulen, J.W. (2001). Ten residual biomass fuels for circulating fluidized-bed gasification. *Biomass and Bioenergy* **20**: 45–56.
29. Pinto, F., Andre, R.N., Lopes, H. et al. (2008). Effect of experimental conditions on gas quality and solids produced by sewage sludge cogasification. 2. Sewage sludge mixed with biomass. *Energy and Fuels* **22**: 2314–2325.
30. Graboski, M. and Bain, R. (1981). Properties of biomass relevant to gasification. In: *Gasification: Principles and Technology* (ed. T.B. Reed), 41–71. Park Ridge, NJ: Noyes Data Corporation.
31. Tillman, D.A. (1978). *Wood as an Energy Resource.* New York, NY: Academic Press.
32. Broer, K.M. and Brown, R.C. (2016). Effect of equivalence ratio on partitioning of nitrogen during biomass gasification. *Energy and Fuels* **30** (1): 407–413.
33. The National Institute for Occupational Safety and Health (NIOSH) 2014. Hydrogen Cyanide. https://www.cdc.gov/niosh/idlh/74908.html (accessed 7 November 2018).
34. Heidenreich, S., Müller, M., and Foscolo, P.U. (2016). *Advanced Biomass Gasification: New Concepts for Efficiency Increase and Product Flexibility*, 69. Ann Arbor, MI: Elsevier.

35. Tijmensen, A., Faaij, A., Hamelinck, C., and van Hardeveld, M. (2002). Exploration of the possibilities for production of Fischer–Tropsch liquids and power via biomass gasification. *Biomass and Bioenergy* **23**: 129–152.
36. Hofbauer, H. and Knoef, H. (2005). Success stories on biomass gasification. In: *Handbook Biomass Gasification*, 1e (ed. H.A.M. Knoef), 115–161. Enschede: BTG Biomass Technology Group BV.
37. Garcia-Ibañez, P., Cabanillas, A., and Sánchez, J.M. (2004). Gasification of leached orujillo (olive oil waste) in a pilot plant circulating fluidised bed reactor. Prelminary results. *Biomass and Bioenergy* **27**: 183–194.
38. Zeng, L. and van Heiningen, A.R.P. (2000). Carbon gasification of kraft black liquor solids in the presence of TiO_2 in a fluidized bed. *Energy and Fuels* **14**: 83–88.
39. Gomez-Barea, A., Arjona, R., and Ollero, P. (2005). Pilot-plant gasification of olive stone: a technical assessment. *Energy and Fuels* **19**: 598–605.
40. Narvaez, I., Orio, A., Aznar, M.P., and Corella, J. (1996). Biomass gasification with air in an atmospheric bubbling fluidized bed. Effect of six operational variables on the quality of the produced raw gas. *Industrial and Engineering Chemistry Research* **35**: 2110–2120.
41. Kersten, S., Prins, W., van der Drift, A., and van Swaaij, W.P.M. (2003). Experimental fact-finding in CFB biomass gasification for ECN's 500 kWth pilot plant. *Chemical Engineering Science* **58** (3–6): 725–731.
42. Ciferno, J. and Marano, J. 2002. Benchmarking biomass gasification technologies for fuels, chemicals, and hydrogen production.
43. Lv, P., Yuan, Z., Ma, L. et al. (2007). Hydrogen-rich gas production from biomass air and oxygen/steam gasification in a downdraft gasifier. *Renewable Energy* **32**: 2173–2185.
44. Wang, Y. and Kinishita, C. (1992). Experimental analysis of biomass gasification with steam and oxygen. *Solar Energy* **49** (3): 153–158.
45. Gil, J., Aznar, M.P., and Caballero, M.A. (1997). Biomass gasification in a fluidized bed at pilot scale with steam-oxygen mixtures. Product distribution for very different operating conditions. *Energy and Fuels* **11** (6): 1109–1118.
46. Carpenter, D.L., Bain, R.L., and Davis, R.E. (2010). Pilot-scale gasification of corn stover, switchgrass, wheat straw, and wood: 1. Parametric study and comparison with literature. *Industrial and Engineering Chemistry Research* **49** (4): 1859–1871.
47. Repotec. Biomass gasification – steam fluidized bed. http://www.repotec.at/index.php/biomass-gasification-steam-fluidized-bed.html (accessed 7 November 2018).
48. Rentech, Inc., Gasification. http://www.rentechinc.com/gasification.php (accessed 21 October 2017).2017].
49. RES Kaidi, Silvagas Process. http://reskaidi.com/technology/silvagas-process (accessed 3 November 2017).
50. van der Meijden, C.M., Sierhuis, W., and van der Drift, A. 2011. Waste wood fueled gasification demonstration project. Presented at the Renewable Energy World Europe Conference and Exhibition, Milan, 2011.
51. ThermoChem Recovery International, Gasification. http://tri-inc.net/steam-reforming-gasification (accessed 7 November 2018).
52. TSS Consultants 2005. Gridley Ethanol Demonstration Project Utilizing Biomass Gasification Technology: Pilot Plant Gasifier and Syngas Conversion Testing.
53. PowerHouse Energy Group Plc. https://www.powerhouseenergy.net (accessed 7 November 2018).
54. Lysenko, S., Sadaka, S., and Brown, R.C. (2012). Comparison of mass and energy balances for air blown and thermally ballasted fluidized bed gasifiers. *Biomass and Bioenergy* **45**: 95–108.
55. Milne, T., Evans, R., and Abatzoglou, N., 1998. Biomass Gasifier "Tars": Their Nature, Formation, and Conversion, National Renewable Energy Laboratory: Golden, CO. https://www.nrel.gov/docs/fy99osti/25357.pdf (accessed 7 November 2018).

56. Perry, R.H. and Chilton, C.H. (1973). *Chemical Engineers' Handbook*, 5e. New York, NY: McGraw-Hill.
57. Basu, P. (2006). *Combustion and Gasification in Fluidized Beds*, 359–380. Boca Raton, FL: CRC Press.
58. Corella, J., Herguido, J., and Alday, F.J. (1988). Pyrolysis and steam gasification of biomass in fluidized beds. Influence of the type and location of the biomass feeding point on the product distribution. In: *Research in Thermochemical Biomass Conversion*, 384–398. Elsevier Science Publishers Ltd.
59. Timmer, K.J. 2008. Carbon conversion during bubbling fluidized bed gasification of biomass. *Retrospective Theses and Dissertations*. 15822. https://lib.dr.iastate.edu/rtd/15822 (accessed 7 November 2018).
60. Babcock & Wilcox (2005). Fluidized-bed combustion. In: *Steam: Its Generation and Use*, 41e (ed. J.B. Kitto and S.C. Stultz), 17–12. Barberton, OH: Babcock & Wilcox.
61. Siemens AG 2014. Siemens Fuel Gasification Technology. https://www.siemens.com/press/pool/de/pressemitteilungen//2014/energy/power-generation/EP201405047e.pdf (accessed 23 November 2018).
62. Larson, E.D. and Katofsky, R.E. (1993). Production of hydrogen and methanol from biomass. In: *Advances in Thermochemical Biomass Conversion* (ed. A.V. Bridgwater). Dordrecht: Springer.
63. Colorado State University. Chemical Equilibrium Calculation. http://navier.engr.colostate.edu/code/code-4/index.html (accessed 23 November 2018).
64. P. Meehan 2009. Investigations into the fate and behavior of selected inorganic compounds during biomass gasification. Graduate Theses and Dissertations. 10713. https://lib.dr.iastate.edu/etd/10713 (accessed 7 November 2018).
65. Wander, P.R., Altafini, C.R., and Barreto, R.M. (2004). Assessment of a small sawdust gasification unit. *Biomass and Bioenergy* **27**: 467–476.
66. Woolcock, P.J. and Brown, R.C. (2013). A review of cleaning technologies for biomass-derived syngas. *Biomass and Bioenergy* **52**: 54–84.
67. SkyFuel, Inc. 2018. Sunshine Family Companies. http://www.skyfuel.com/about/sunshine-family-companies (accessed 23 November 2018).
68. Alter NRG Corp 2016. AlterNRG. http://www.alternrg.com (accessed 7 November 2017).
69. H. Wright 2011. Rentech-ClearFuels IBR Project. http://www.gasification-syngas.org/uploads/eventLibrary/32WRIGHT.pdf (accessed 7 November 2018).
70. Yeh, B. 2011. Independent Assessment of Technology Characterizations to Support the Biomass Program Annual State-of-Technology Assessments. https://doi.org/10.2172/1010864 (accessed 23 November 2018).
71. Juniper Consultancy Services Limited 2008. The Alter NRG/Westinghouse Plasma Gasification Process. http://energy.cleartheair.org.hk/wp-content/uploads/2013/09/Westinghouse_Plasma_Gasification.pdf (accessed 7 November 2018).
72. Taylor Biomass Energy. Taylor Technology. http://www.taylorbiomassenergy.com/taylorbiomass03_tayl_ttm.html (accessed 7 November 2018).
73. Paisley, M.A. 2007. Advanced Biomass Gasification for the Production of Biopower, Fuels, and Chemicals. Presentation at the 2007 AIChE Annual Meeting, Salt Lake City, 2007. https://www.aiche.org/conferences/aiche-annual-meeting/2007/proceeding/paper/498c-advanced-biomass-gasification-production-biopower-fuels-and-chemicals (accessed 7 November 2018).
74. ThermoChem Recovery International TRI Gasification. http://tri-inc.net (accessed 7 November 2018).
75. Repotec Technology. http://www.repotec.at/index.php/technology.html (accessed 7 November 2018).
76. CHOREN. CHOREN Coal Gasification. http://www.choren.com/en/company/choren-coal-gasification/specifications.html (accessed 7 November 2018).

77. Mancuso, L. and Ruggeri, F. 2015. Biomass gasification for the production of SNG: a practical route through available and new technologies, in ANIMP Conference: Renewables, Grid, Energy Storage, 2 July 2015. http://animp.it/prodotti_editoriali/materiali/convegni/pdf/energia_2015/Mancuso_Ruggeri_AMEC_CnANIMP_ATI_2015.pdf (accessed 7 November 2018).
78. ANDRITZ. ANDRITZ Carbona biomass gasifiers for clean energy. https://www.andritz.com/products-en/group/environmental-solutions/power-generation/gasification (accessed 7 November 2018).
79. Koss, U. and Schlichting, H., Lurgi's MPG Gasfication plus Rectisol Gas Purification – Advanced Process Combination for Reliable Syngas Production. Presentation at Gasification Technologies 2005, San Francisco. https://www.globalsyngas.org/uploads/eventLibrary/41SCHL.pdf (accessed 7 November 2018).
80. Air Liquide. Lurgi MPG™ – Multi-Purpose Gasifier. https://www.engineering-airliquide.com/lurgi-mpg-multi-purpose-gasifier (accessed 7 November 2018).
81. Frontline Bioenergy, LLC. Frontline Bioenergy, LLC. http://www.frontlinebioenergy.com (accessed 7 November 2018).
82. Smeenk, J. 2011. Operations Summary of a Commercial Biomass Gasifier Generating a Clean, Combustible Gas for Industrial Heat or Repowering an Existing Power Boiler. Presentation at 2011 TAPPI IBBC Conference. http://www.tappi.org/content/Events/11BIOPRO/15.2Smeenk.pdf (accessed 7 November 2018).
83. Smeenk, J. 2018. Interviewee: Personal Interview Regarding Frontline BioEnergy.
84. Outotec. Multifuel bubbling fluidized bed boilers and advanced staged gasifiers. Outotec, 2016. https://www.outotec.com/products/energy-production/waste-to-energy-plants (accessed 23 November 2018).
85. Enerkem. Enerkem Alberta Biofuels. Enerkem, 2017. https://enerkem.com/facilities/enerkem-alberta-biofuels (accessed 23 November 2018).
86. Babu, S.P. 2005. Biomass gasification for hydrogen production – process description and research needs. Executive Committee of IEA Bioenergy, Dublin, 2005.
87. Aries Clean Energy. Aries Clean Energy. https://ariescleanenergy.com (accessed 7 November 2018).
88. Rodden, G. (2009). Looking beyond the forest products industry. *Pulp & Paper International* **51** (9): 26–31.
89. Smeenk, J., Brown, R.C., and Yang, H. 2000. Development of a fluidized bed gasifier system for cooking gas in rural China, in Proceedings of the 9th Biennial Bioenergy Conference, Buffalo.
90. Community Power Corporation. Biomax® Specifications. http://www.gocpc.com/biomaxr-specifications.html (accessed 7 November 2018).
91. Babcock & Wilcox Vølund. Harboøre Varmeværk, Denmark. http://www.volund.dk/Biomass_energy/References/Harboore (accessed 7 November 2018).
92. Salo, K. and Horvath, A. 2009. Biomass Gasification in Skive: Opening Doors in Denmark. http://www.renewableenergyworld.com/rea/news/article/2009/01/biomass-gasification-in-skive-opening-doors-in-denmark-54341 (accessed 7 November 2018).
93. Tirone, J. 2007. 'Dead-end' Austrian town blossoms with green energy. http://www.nytimes.com/2007/08/28/business/worldbusiness/28iht-carbon.4.7290268.html (accessed 7 November 2018).
94. Larson, E.D., Williams, R.H., and Leal, M.R. (2001). A review of biomass integrated-gasifier/gas turbine combined cycle technology and its application in surgarcane industries, with an analysis for Cuba. *Energy for Sustainable Development* **V** (1): 54–76.
95. Klimantos, P., Koukouzas, N., Katsiadakis, A., and Kakaras, E. (2009). Air-blown biomass gasification combined cycles (BGCC): system analysis and economic assessment. *Energy* **34**: 708–714.
96. Bridgwater, A.V., Toft, A.J., and Brammer, J.G. (2002). A techno-economic comparison of power production by biomass fast pyrolysis with gasification and combustion. *Renewable and Sustainable Energy Reviews* **6**: 181–248.

97. Huber, G.W., Iborra, S., and Corma, A. (2006). Synthesis of transportation fuels from biomass: chemistry, catalysts, and engineering. *Chemical Reviews* **106** (9): 4044–4098.
98. Brown, R.C. (2005). Biomass refineries based on hybrid thermochemical/biological processing - an overview. In: *Biorefineries, Biobased Industrial Processes and Products* (ed. B. Kamm, P.R. Gruber and M. Kamm). Weinheim: Wiley-VCH Verlag.
99. LanzaTech. LanzaTech. http://www.lanzatech.com (accessed 7 November 2018).
100. Jupeng Bio. Fermentation Technology. http://www.jupengbio.com/fermentation-technology (accessed 7 November 2018).
101. Synata Bio, Inc.. Synata Bio, Inc.. https://www.globalsyngas.org/uploads/downloads/Synata-Bio_Company_Overview.pdf (accessed 7 November 2018).
102. Lane, J. 2015. LanzaTech: Biofuels Digest's 2015 5-Minute Guide. http://www.biofuelsdigest.com/bdigest/2015/01/13/lanzatech-biofuels-digests-2015-5-minute-guide (accessed 7 November 2018).

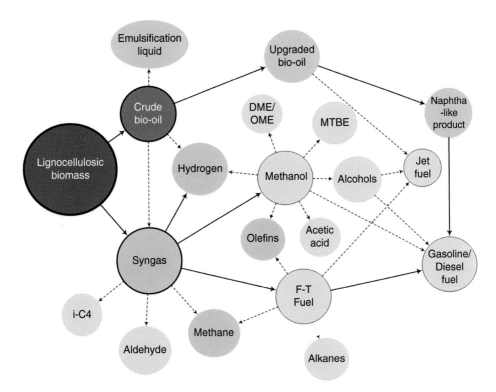

Figure 1.5 *Second-generation biofuels. F-T, Fischer-Tropsch; i-C4, isobutene and isobutane; DME, dimethyl ether; OME, oxymethylene ethers; MTBE, Methyl tert-butyl ether; HRJs, hydro-processed renewable jet fuels [20].*

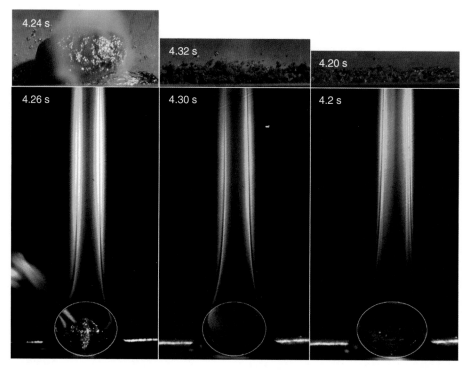

Figure 2.11 *Comparing pyrolysis of extracted lignin (left), extracted lignin mixed with silica matrix (center), and red oak (right) demonstrates how thermal ejection is an artifact of particle coalescence and non-uniform heating from below the sample. Source: figure adapted with permission from Tiarks [51].*

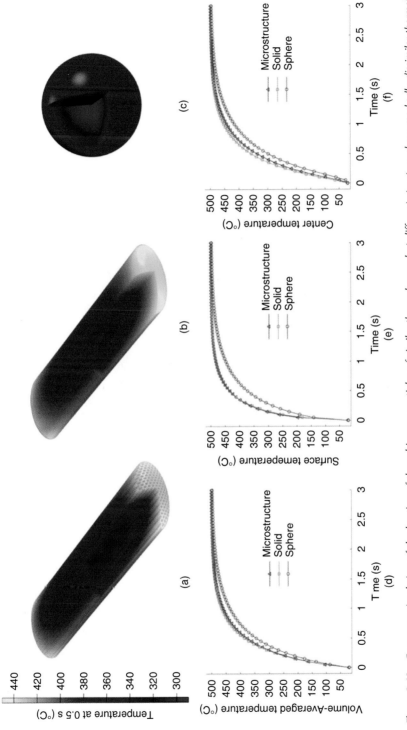

Figure 2.13 Computer simulation of the heating of three biomass particles of similar thermal mass but different structure show markedly dissimilar thermal gradients. Realistic morphology (a) develops larger temperature gradients than particles with the same proportions but without cell lumina (b). Spherical particles (c), often used for modeling simplicity, have dramatic thermal gradients that do not accurately represent biomass. Source: figure reproduced with permission from Reference [68].

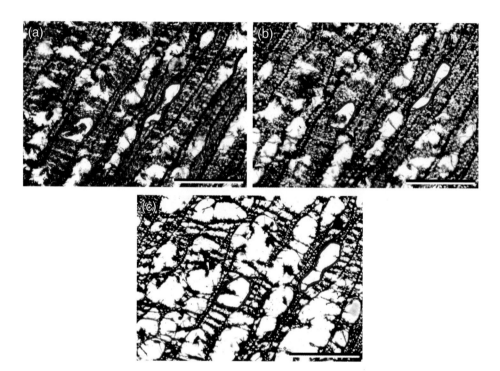

Figure 2.14 *Light microscope images of a cross section of poplar at 26 °C (a), 299 °C (b), and 501 °C (c) show how cell walls expand when heated before contracting as volatile products leave. The red outlines surround xylem. Scale bars equal 1 mm. Source: figure reproduced with permission from Reference [75].*

Figure 6.7 Photograph of the Ensyn plant. Source: with permission from Ensyn Technologies Inc. [27].

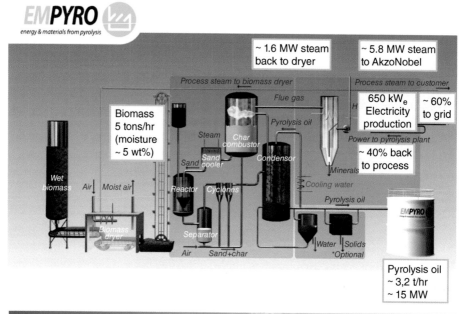

Figure 6.9 *The Empyro pyrolysis plant in The Netherlands.* Source: with permission from BTG Bioliquids B.V. [31].

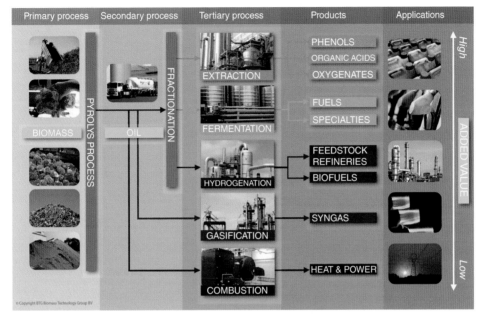

Figure 6.11 The "Bioliquid" refinery. Source: with permission from BTG Biomass Technology Group BV [25].

Figure 6.12 Combustion of Ensyn's liquids at the Memorial Hospital, North Conway, New Hampshire. Source: with permission from Ensyn Technologies Inc., Barry Freel, personal communications, January 2018.

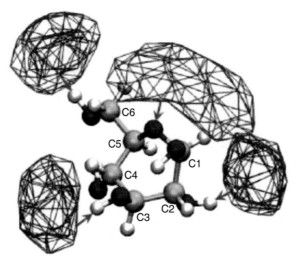

Figure 8.14 Volumetric spatial density maps (isosurfaces) of the time averaged distribution of THF (blue surfaces) and water (red surfaces) around glucose molecules in mixtures of 90/10 wt% THF and water. Source: taken from [222] with permission.

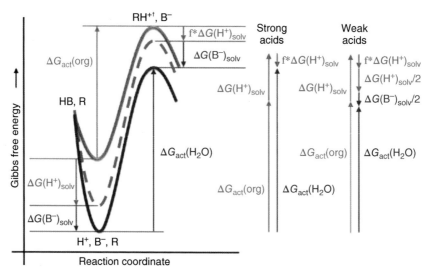

Figure 8.15 Energy diagram for destabilization of proton in water and organic solvent. Source: taken from [128] with permission.

5

Syngas Cleanup, Conditioning, and Utilization

David C. Dayton, Brian Turk and Raghubir Gupta

RTI International, Research Triangle Park, NC, USA

5.1 Introduction

Gasification is the thermal decomposition of solid carbonaceous material into predominately flammable gas known as producer gas or syngas. Gasification is attractive for its ability to convert very heterogeneous material into a uniform gaseous fuel suitable for heating, industrial process applications, electricity generation, and liquid fuels production. Thermal decomposition of biomass is an endothermic process requiring a source of energy to drive the process, which can be provided by heat transfer to the reactor (indirectly heated gasification) or by the addition of an oxidant to partially oxidize some of the products of thermal decomposition and release heat (directly heated gasification) [1]. Thermal decomposition of the solid fuel releases volatile matter, consisting of both condensable vapors and permanent gases, and produces a solid carbonaceous residue known as char. Secondary reactions include oxidation of the gases, vapors, and char (in the case of directly heated gasification) and reactions among the gaseous and vapor species that drive the products toward chemical equilibrium. In practice, the final product composition includes carbon monoxide (CO), carbon dioxide (CO_2), water (H_2O), hydrogen (H_2), methane (CH_4), other light hydrocarbons, tars, char, volatile inorganic constituents, and inorganic solid (ash). The volatile inorganic constituents include hydrogen cyanide (HCN), ammonia (NH_3), hydrogen chloride (HCl),

Thermochemical Processing of Biomass: Conversion into Fuels, Chemicals and Power, Second Edition.
Edited by Robert C. Brown.
© 2019 John Wiley & Sons Ltd. Published 2019 by John Wiley & Sons Ltd.

and hydrogen sulfide (H_2S). The generalized reaction of biomass gasification is given by:

$$\text{biomass} + O_2 \text{ (or } H_2O) \rightarrow CO + CO_2 + H_2O + H_2 + CH_4 + \text{other hydrocarbons}$$
$$+ \text{tar} + \text{char} + \text{ash}$$
$$+ HCN + NH_3 + HCl + H_2S + \text{other sulfur gases}$$

The actual amount of CO, CO_2, H_2O, H_2, tars, and hydrocarbons depends on the partial oxidation of the volatile products:

$$C_nH_m + \left(\frac{n}{2} + \frac{m}{4}\right)O_2 \rightarrow nCO + \frac{m}{2}H_2O$$

The char yield in a gasification process can be optimized to maximize carbon conversion, or the char can be thermally oxidized to provide heat for the process. Char is partially oxidized or gasified according to the following reactions:

$$C + \frac{1}{2}O_2 \rightarrow CO$$

$$C + H_2O \rightarrow CO + H_2$$

$$C + CO_2 \rightarrow 2CO \quad \text{(Boudouard reaction)}$$

The gasification product gas composition, particularly the H_2/CO ratio, can be further adjusted by reforming and shift chemistry. Additional hydrogen is formed when CO reacts with excess water vapor according to the water gas shift (WGS) reaction:

$$CO + H_2O \rightleftharpoons CO_2 + H_2$$

Reforming the light hydrocarbons and tars formed during biomass gasification also produces hydrogen. Steam reforming and so-called "dry" or CO_2 reforming occur according to the following reactions and are usually promoted by the use of catalysts:

$$C_nH_m + nH_2O \rightarrow nCO + \left(n + \frac{m}{2}\right)H_2$$

$$C_nH_m + nCO_2 \rightarrow (2n)CO + \frac{m}{2}H_2$$

The actual composition of the biomass gasification product gas depends heavily on the gasification process, the gasifying agent, and the feedstock composition [2, 3].

5.2 Syngas Cleanup and Conditioning

Syngas cleanup is a general term for removing the unwanted impurities from the gaseous product of gasification and generally involves an integrated, multistep approach that depends on the end use of the product gas [4–7]. Gas-phase impurities to be removed from syngas include NH_3, HCN, other nitrogen-containing gases, H_2S, other sulfur gases, HCl, alkali metals, organic hydrocarbons (including tar), and particulates. The concentration of these components strongly depends on the feedstock. Gasification of biomass containing high levels of nitrogen and sulfur yields high levels of NH_3 and H_2S in the syngas stream, and HCl concentration in biomass-derived syngas directly correlates with the chlorine content of the feedstock. Alkali metal (mostly potassium) in syngas is related to the alkali content in the biomass ash. Likewise, ash particles entrained in syngas affect the alkali metal content in syngas. The concentration of alkali vapors or aerosols in syngas depends on the ash chemistry of the selected biomass feedstock and the temperature of the gasification process [8–14].

The extremely heterogeneous nature of solid carbonaceous feedstocks makes comprehensive syngas cleaning and conditioning extremely challenging. These challenges include:

- effectively treating multiple contaminants present at significantly different concentrations;
- effectively treating syngas with varying contaminant concentrations associated with natural variations in feedstock composition and different gasification processes and process conditions;
- effectively treating syngas to meet different product requirements for various syngas utilization processes (catalytic fuel synthesis, chemical production, fuel cells, combustion turbine, etc.);
- developing treatment processes to simultaneously remove multiple contaminants, including trace elements; and
- designing treatment systems despite large variation in published or predicted concentrations of trace metals in syngas, resulting from inaccurate and imprecise measurement techniques.

Previous attempts to minimize cost and maximize efficiency in coal gasification processes relied on well-known commercial technologies, but with reduced thermal efficiency and increased capital and operating costs. The current commercial basis for syngas cleaning in integrated coal gasification processes involves cooling the syngas for treatment in a liquid scrubbing/absorption system based on either chemical (methyl diethanolamine, or MDEA) or physical (Selexol™ and Rectisol®) absorption.

All current syngas desulfurization systems cool the syngas significantly below the dew point, resulting in water condensation, which absorbs and removes most of the HCl and a majority of NH_3 from the syngas, depending upon the cooling temperature. The liquid scrubbing/absorption systems treat the water-free syngas stream to remove H_2S and COS (a COS hydrolysis unit may be necessary to effectively remove COS). The H_2S-rich streams from these absorption systems are sent to a Claus plant for final conversion into elemental sulfur.

The commercial status of these syngas cleaning systems also reflects a large degree of integration in coal gasification systems that have minimized the capital and operating costs and maximized plant efficiency. In spite of this integration, the syngas cooling requirements result in a significant loss of thermal efficiency. Because the regeneration temperature for these systems is significantly below 300 °F (149 °C), there is no potential to raise the operating temperature of these systems to achieve better thermal efficiency. Furthermore, these systems are primarily acid gas removal systems and are not designed to remove other contaminants, including trace metals. The Rectisol process removes HCN, NH_3, and trace metals, including mercury. However, a mercury carbon trap is typically installed upstream of the Rectisol process to eliminate the formation of metal amalgams with mercury in the low-temperature sections of the Rectisol process.

An obvious strategy is to integrate commercial quench systems with biomass gasifiers to produce clean syngas. As noted, these processes have been optimized for large-scale coal gasification to capitalize on economies of scale. Biomass is a widely distributed resource with relatively low bulk density, limiting the amount that can be transported to a central facility, and constraining its size compared to facilities that process energy-dense fossil fuels. Accordingly, lessons learned from coal gasification may have limited applicability to biomass gasification.

One of the advantages of biomass gasification compared to coal gasification is that biomass has a much higher volatile content than coal so less severe process conditions (temperature and pressure) are needed for high carbon conversion efficiency. A result of this, however, is significant amounts of tar contaminating the syngas, which is difficult to rectify in a thermally efficient and cost-effective manner. Tars are defined as the condensable organic fraction of gasification products consisting mostly of aromatic hydrocarbons larger than benzene. Tars are less of a problem in coal gasification because the temperature is high enough to decompose them into low molecular weight, noncondensable hydrocarbons. The energy content of biomass is lower than coal, resulting in a prohibitively low carbon efficiency if biomass gasification is operated at temperatures high enough to avoid tar formation.

The diversity in the operational definitions of tars reflects the diversity of product gas compositions required for particular end-use applications and how tar is collected and analyzed. Tar sampling protocols have been developed [15–21] to standardize the way tars are collected and new on-line tar analysis techniques are being developed [12–14]. Nevertheless, tar yields in biomass-derived syngas can range from 0.1% (downdraft) to 20% (updraft) in the product gases.

The chemical composition of these tars is also a function of process conditions: temperature, steam/biomass ratio, equivalence ratio, and pressure. The physical properties of tars, mainly their dew points, greatly impacts mitigation strategies, especially processes that rely on quenching the syngas. Hot gas cleanup strategies for tar removal or conversion can be implemented to avoid potentially problematic high-viscosity liquid hydrocarbon streams and improve overall energy efficiency of the integrated process. If tar mitigation involves catalysts, then their potential poisoning by inorganic impurities in the syngas must be addressed.

The contaminants in biomass-derived syngas are often managed separately. Ideally, the contaminants are collectively managed with an integrated approach to syngas cleaning, which has better prospects for minimizing capital and operating costs and maximizing thermal and carbon efficiencies. The following sections highlight specific technologies and strategies for each contaminant. Further details are available in the literature [22–28].

5.2.1 Particulates

Particulate matter in gasification product gas streams originates from several sources, depending on the reactor types, feedstock, and process conditions. Particle carryover from fluidized-bed reactors consists of char and attritted bed material that is entrained in the gas flow. Particle size distribution of is a function of the initial size of bed material. Char tends to be more friable and less dense than bed material and typically has a smaller particle size distribution. The smallest particles exiting a gasifier tend to be alkali metal vapor condensation aerosols. The concentration of these particles is a strong function of the ash content and ash chemistry of the feedstock.

Particulate removal requirements depend on the end use of the syngas. Gasification coupled with gas engines for stationary power applications requires particulate loadings below 50 mg/N-m^3 (N-m^3 is normal cubic meters). Particulate loadings less than 15 mg/N-m^3 with a maximum particle size of 5 µm are required to protect gas turbines in integrated gasification combined cycle processes. The most stringent requirements for particulate removal occur for fuel synthesis, which require particulate loadings of less than 0.02 mg/N-m^3 to protect syngas compressors and minimize catalyst poisoning by alkali fumes and ash mineral matter.

Several technologies are commercially available for particulate removal from high-temperature gas streams. Choosing the most appropriate technology for biomass gasification depends on the desired particle separation efficiency to achieve acceptable particulate loading in the cleaned syngas. Pressure drop through the particle removal unit operation and thermal integration are also key design parameters to be considered. Tars produced during biomass gasification also have significant impact on particulate removal strategies. Operating temperatures of most particulate removal devices should be above the tar dew point to avoid tar condensation and prevent particulate matter from becoming sticky and agglomerating.

Gas cyclone technology for particulate removal is well known and proven. Dirty gas enters a cyclone separator with high tangential velocity and angular momentum, which forces particles close to the wall into the boundary layer where they lose momentum and gravitational forces separate them from the gas flow. Separation efficiency is a function of particle size, gas flow, temperature, and pressure. Cyclones can be designed for optimum removal of particles of a specific size distribution, usually down to a lower size limit. Multiple cyclones of different design can be used in series to achieve near sub-micrometer particle removal at high efficiency.

Barrier filters are another technology option for high-temperature particulate removal. Filter housing design and filter media selection are keys for optimizing particle capture efficiency within a manageable window of pressure drop across the filter. The initial pore size of the filter medium determines initial pressure drop and particulate removal; however, the filtration efficiency and pressure drop increase as particles collect on the surface to produce a filter cake. Pulsing inert or clean product gas back through the filters dislodges the filter cake, allowing pressure drop across the filter to closely approach its original performance. Filter housings need to accommodate regular backpulsing to remove the filter cake. They need to be designed to remove particulate before the filter element is brought back online. Improper design can cause the material that was removed to immediately recoat the surface such that pressure drop is not restored.

Operating temperature and product gas composition are the primary process parameters that need to be considered in selecting appropriate filter medium. Ceramic candles are being developed for high-temperature gas filtration (~500 °C) applications. Thermal shock from repeated backpulsing can cause filters to break, and pore blinding over extended operation can reduce long-term filter performance.

Sulfur, chlorine, and alkali metal salts can be present in product gases generated from certain feedstocks. When these impurities contact ceramic filters or supports, high-temperature reactions can lead to morphological changes and embrittlement, which can also reduce long-term filter performance. Optimizing the seal between the ceramic candle and its metal support plate has been a key technical challenge to overcome for the success of this technology.

Sintered metal filter elements are an alternative to ceramic candles. Operating temperatures for sintered metal filters are typically lower than that for ceramic filters to guard against sintering. Operated at appropriate temperatures, sintered metal filters are more robust than ceramic candles, as the risk for rupture or cracking is much lower. Fabric filters, such as flexible ceramic bags, are another alternative filter medium that has been commercially proven at lower temperatures, but materials and mechanical compatibility have limited use of these materials for high-temperature syngas filtration.

Electrostatic precipitators (ESPs) have been used commercially for particulate removal in the electric power industry for many years and have found wide application in petroleum refineries for capturing catalyst dust from fluid catalytic cracking (FCC) units. Electric charge is induced on the surfaces of particles, which are then removed from the gas stream as they follow electric field lines to a grounded collector plate. ESPs can be applied to particulate removal in high-temperature and high-pressure gas streams. However, maintaining stability of the corona discharge for reliable, steady-state operation is a technical challenge. Another technical challenge is ensuring materials compatibility of the high-voltage discharge electrode and other metal internal components with syngas impurities. Overall size and capital cost of ESPs tend to make them best suited for large-scale operation.

Wet scrubbing systems use liquid sprays, either water or chilled condensate from the process, to remove particulates that collide with liquid droplets. The droplets are then removed from the gas stream in a demister. Venturi scrubbers are the most common wet scrubbers but often require relatively high pressure drop to circulate quench liquid. Heat loss from wet scrubbing systems can adversely affect the energy efficiency of the overall process. On the other hand, for indirect gasification systems, the excess steam that is used as the gasifying agent needs to be quenched and recovered. Wet scrubbing systems are inevitable in indirect gasification systems to remove excess water vapor prior to compression and downstream syngas utilization.

5.2.2 Sulfur

Oxides of many metals react with H_2S, as described by $MeO + H_2S \rightleftharpoons MeS + H_2O$, where MeO represents a metal oxide and MeS represents a metal sulfide, effectively reducing H_2S concentrations in syngas [8–11]. The minimum H_2S concentration in treated syngas is determined by the equilibrium concentration based on the relative amount of syngas contaminants and metal oxide. Reaction kinetics determine whether equilibrium is achieved in a sulfur sorption process. ZnO possesses one of the highest thermodynamic efficiencies for H_2S removal and the most favorable reaction kinetics of all the active oxide materials within a temperature range of 149–580 °C.

Although several ZnO-based guard bed materials are commercially available, the low sulfur capacity of these materials makes them very costly for a once-through disposable material, especially for high sulfur loadings in syngas. However, economics improve if ZnO-based materials can be regenerated and used for multiple cycles. A number of regenerable ZnO-based sorbents have been developed for this purpose, as shown in Table 5.1. RTI's transport reactor desulfurization system offers significant advantages over fixed-bed

Table 5.1 Various ZnO-based desulfurization sorbents available commercially.

Sorbent	Organization	Desulfurization system
RVS-1	DOE/NETL	Fixed bed
Z-Sorb	ConocoPhillips	Fixed and fluidized beds
EX-S03	RTI	Transport reactor
RTI-3	RTI	Transport reactor

and fluidized-bed desulfurization systems. These include higher throughput and smaller footprints, resulting in lower cost and excellent control of the highly exothermic regeneration reaction, which allows use of neat air for regeneration.

ZnO-based materials are typically regenerated using oxygen and nitrogen mixtures according to the reaction $ZnS + 1.5O_2 \rightarrow ZnO + SO_2$. The concentration of SO_2 in the regeneration off-gas depends on the sorbent and reactor configuration. SO_2 concentrations range from about 1% to 14% by volume, depending upon O_2 concentration in the regeneration gas. SO_2 off-gas must be further treated to produce sulfuric acid or elemental sulfur. Technologies include a modified Claus plant, RTI's direct sulfur recovery process, or conventional flue gas desulfurization based on reaction with gypsum.

RTI's experience with ZnO-based materials has demonstrated that these materials can effectively remove large amounts of H_2S, reducing inlet concentrations from as high as 500–30 000 ppmv to below 10 ppmv [11]. Actual effluent concentration appears to have a temperature dependence that is related to both thermodynamic equilibrium concentrations and reaction kinetics. At lower temperatures, thermodynamics become more favorable, but reaction kinetics, specifically diffusion of H_2S through the ZnS product layer, drop off rapidly below about 232 °C.

RTI has also demonstrated that ZnO materials, in addition to reacting with H_2S, also react with COS and CS_2. Thermodynamics and kinetics for reactions with COS and CS_2 are similar to those for H_2S. RTI has demonstrated greater than 99.9% removal of both H_2S and COS from syngas generated by gasification of coal and mixtures of coal and petroleum coke during more than 3000 hours of pilot plant testing at Eastman Chemical Company's coal-to-chemicals facility in Kingsport, TN, and more than 3500 hours of pre-commercial demonstration at Tampa Electric Company's integrated gasification combined cycle plant in Tampa. FL. RTI has also conducted laboratory testing of ZnO- and FeO-based sorbents demonstrating the potential to also remove mercaptans, disulfides, thiophenes, and dibenzothiophenes. The specialized sulfur removal technologies that RTI has developed are very different from more traditional acid gas removal technologies, which remove acidic chemical species in syngas, primarily CO_2 and H_2S. Traditional acid gas removal technologies can remove multiple contaminants from syngas because the fundamental removal mechanism is based on physical properties like acidity and solubility. Unfortunately, this presents an added challenge if there is a need to separate removed species for subsequent treatment and use. For example, separating acid gas components for H_2S disposal and CO_2 capture and sequestration or utilization requires increased process complexity and parasitic energy requirements.

Techno-economic analyses were conducted in which RTI's desulfurization technology was integrated with activated amine technologies, Selexol, and Rectisol, for dedicated CO_2 removal. The resulting simplification of the CO_2 removal process reduced capital and operating costs compared to a single acid gas removal process designed for removal of both contaminants to the same specifications.

5.2.3 Ammonia Decomposition and HCN Removal

Although NH_3 is a not a highly stable molecule ($\Delta G_f^\circ = 0$ kJ/mol at $\sim$175 °C), its dissociation requires a very high temperature because of high activation energy (92 kcal/mol). Krishnan et al. [29] studied the removal of fuel-bound nitrogen compounds in simulated coal gas streams using a laboratory-scale reactor and simulated coal gas compositions representative of several types of gasifiers. HTSR-1, a catalyst proprietary to Haldor-Topsøe A/S, Copenhagen, Denmark, exhibited excellent activity (even in the presence of 2000 ppm of H_2S) and high temperature stability. G-65, an SRI catalyst, demonstrated superior activity in the temperature range 550–650 °C at H_2S levels below 10 ppm The presence of impurities such as HCl and HCN did not affect catalyst performance in the temperature range studied.

A wide variety of metals and metal oxides/carbides/nitrides can catalyze decomposition of ammonia. Group 8 metals (Fe, Ru, and Os), Group 9 metals (Co, Rh, and Ir), and Group 10 metals (Ni, Pd, and Pt) seem to be active mainly in the metallic state [30, 31]. Even though oxides of these metals are reactive in a reducing atmosphere of H_2 or CO, these metal oxides are likely reduced, creating a metallic surface. Activity for ammonia decomposition on smooth metal surfaces has been reported in the following decreasing order: Co > Ni > Cu > Zr.

Although the Group 8, 9, and 10 metals tend to be more active than many other elements, carbides and nitrides of Groups 5 (V, Nb) and 6 (Cr, Mo, W) can be especially active for ammonia decomposition [11, 32–38]. For example, Mo_2C is about twice as active as vanadium carbides, which are two to three times more active than Pt/C [39]. Vanadium nitrides were found to be comparable or superior catalytically to Ni supported on silica–alumina [40]. LaNi alloys are also very active due to the formation of a nitride phase [41]. CaO [42], MgO [43], and dolomite (CaO–MgO) are all active. MgO will decompose ammonia to N_2 and H_2 at temperatures as low as 300 °C.

Gas-phase composition can significantly impact catalyst activity. For example, CaO is almost completely deactivated when CO, CO_2, and H_2 are present [42, 44], probably due to the reaction of CO_2 with CaO. Ni, on the other hand, does not seem to be affected by the presence of such gases [7]. Small quantities (<2000 ppm) of H_2S did not severely poison calcined dolomite, CaO, or, surprisingly, Fe when decomposing ammonia at concentrations of a few thousand parts per million [44, 45]. Because ammonia synthesis is highly structure-sensitive, the activity of Fe in decomposing ammonia is highly dependent on catalyst particle size (20–50 nm) [44].

5.2.4 Alkalis and Heavy Metals

Compared to sulfur, chlorine, ammonia, and particulate matter, technologies for removing mercury, arsenic, selenium, and cadmium from coal-derived syngas are poorly developed. Available commercial technologies for removing trace metals are limited to mercury and arsenic. Mercury control technologies employed in the natural gas industry were developed to limit metallurgical failures resulting from the formation of amalgams with aluminum and other metals during gas processing. Adsorbents for mercury control in natural gas processing developed by UOP LLC are based on type X and Y zeolites that have been coated with elemental silver. To regenerate, the material is heated and purged with a sweep gas to remove the mercury. Since the maximum regeneration temperature is below the target warm syngas temperature for trace metal removal, this technology cannot be adapted or modified for warm-gas cleaning of syngas. Various forms of activated carbon have also been used to remove mercury from natural gas. A mercury removal unit based on a fixed bed of sulfur-impregnated activated carbon has been designed to reduce the concentration of mercury in water-saturated natural gas [46] from $1000\,\mu g/m^3$ to less than $5\,\mu g/m^3$. In their product marketing brochures, Synetix describes a process for removing mercury from natural gas using metal sulfides on inorganic supports. In this application, a fixed-bed reactor maintained at 15 °C is used to reduce the concentration of mercury from 5.0 to $0.01\,\mu g/m^3$. Both processes operate at temperatures below target temperatures for warm-gas cleaning.

5.2.5 Chlorides

For bulk removal of HCl vapor, different sorbent materials and processes are under development. The two-stage "Ultra-Clean Process" has tested various sorbents, including synthetic dawsonite, nahcolite, and trona ($Na_2CO_3 \cdot NaHCO_3 \cdot 2H_2O$) [47]. HCl exit concentration from Stage I polishing at 449 °C was below 3 ppm Trona was found to be the best sorbent for HCl. Krishnan et al. [29] evaluated several alkali minerals for removal of HCl at 300 ppm from hot syngas at atmospheric pressure in the temperature range 550–650 °C. All sorbents reacted rapidly with HCl, reducing its concentration to about 1 ppm Nahcolite had superior absorption capacity; the spent sorbent contained up to 54% chloride by weight. In a subsequent study conducted by the team of SRI International (SRI), RTI, and General Electric (GE), HCl removal in syngas streams was demonstrated in bench- and pilot-scale reactors [48]. The results of bench-scale experiments in fixed- and fluidized-bed reactors demonstrated that nahcolite pellets and granules were capable of reducing HCl levels to less than 1 ppm in syngas streams in the temperature range 400–650 °C. Tests conducted with the product gas of a pilot-scale, fixed-bed gasifier confirmed that nahcolite effectively reduced HCl to below the 15 ppm detection limit of the analyzer with a high degree of sorbent utilization (>70%), when operated in a circulating fluidized-bed reactor.

The equilibrium concentration of HCl in the presence of sodium carbonate-based sorbents depends on process conditions, especially temperature and partial pressure of steam. Thermodynamic equilibrium calculations show that concentrations of HCl in equilibrium with Na_2CO_3–NaCl mixtures vary from 0.003 to 0.16 ppb when used on the gas stream from a non-quench gasifier gas operated at 4.14 MPa. These sub-parts per million concentrations indicate that extremely low HCl concentrations are thermodynamically achievable. Reaction kinetics considerations indicate that diffusion through the NaCl product layer will be the rate-limiting step for achieving these extremely low equilibrium HCl concentrations. Hence, high surface-area sorbents are necessary to achieve both higher reactivity and reasonable sorbent capacities. In fact, commercially available chloride guards achieve sub-parts per million levels of HCl in petroleum refinery streams, but are relatively expensive. SRI observed that HCl concentrations of 0.3 ppm could be achieved at 550 °C using Katalco Chloride Guard 59-3 [29].

HCl reacts with all sorbent materials to produce stable chloride materials that cannot be regenerated. Thus, HCl treatment materials are assumed to be once-through disposable materials. A number of alkali minerals have been shown to be very effective for HCl removal at temperatures between 550 and 650 °C in real coal-derived syngas. Because these alkali minerals are available as natural deposits, they can be processed into HCl sorbents at low cost. Their primary disadvantage is their low reactivity compared to commercially produced sorbents, particularly at lower temperatures. The minerals typically have lower surface areas that are rapidly covered with NaCl product layer, reducing sorbent reactivity. The commercial prepared sorbents are more reactive, they are also significantly more expensive.

5.2.6 Tars and Soot

Tar cleaning is one of the greatest technical challenges in the commercial development of advanced biomass gasification systems [49]. Tars can condense in exit pipes and on particulate filters, leading to blockages and clogged filters. Tars also have varied impacts on other downstream processes. Tars can clog fuel lines and injectors in internal combustion engines. Product gas from an atmospheric-pressure gasifier, when compressed before it is burned in a gas turbine, can cause tar to condense as the product gas is cooled ahead of the gas turbine. Soot formed in high temperature gasifiers also causes problems in gas turbines including luminous combustion and erosion of metal surfaces.

Tar mitigation methods can be classified as physical, thermal, or catalytic processes. If the end use of the gas requires cooling to near-ambient temperatures, it is possible to use many physical removal methods, including wet scrubbing and filtration, to remove tars; these processes are also effective in removing soot.

If the end use requires that the product gas remains at high temperature, at or slightly below the gasifier exit temperature, then hot gas cleaning will be needed for tar and soot elimination. Tars are typically cracked to light gases, which has the advantage of recovering the heating value of the gas and eliminating the waste stream arising from removing tar from syngas. Soot and other particulate must be removed by high temperature barrier filters.

5.2.6.1 Wet Scrubbing

Direct contact with a scrubbing liquid (water or solvent) can be used to condense tars and soot from product gas. This is an effective gas conditioning technology that is commercially available and can be optimized for removal of these contaminants. Wet scrubbing is also effective for removing soluble impurities (NH_3, HCN, alkalis, and chlorides) so a single unit operation can be used to manage multiple contaminants [22, 27].

Wet scrubbing reduces the overall thermal and carbon efficiency of an integrated biomass gasification process. Gas is cooled to near ambient temperatures, resulting in significant loss of sensible enthalpy of the gas. Also, the heating value of the tar is lost when it is removed from the product gas stream.

Another disadvantage of wet scrubbing is generation of wastewater. Scrubbing does not eliminate tars, but merely transfers the problem from the gas phase to the condensed phase. Wastewater minimization and treatment are important considerations when wet scrubbing is used for tar removal.

The wide range of boiling points for tar components means that temperature management becomes important. Lower molecular weight components such as benzene, toluene, and xylenes (BTX) require low temperatures (80–130 °C) whereas the viscosity of higher molecular weight polycyclic aromatic hydrocarbons (PAHs) may be quite high at these low temperatures. Most tertiary tars are hydrophobic so phase separation is relatively straight forward; however, oxygenated tars; such as phenols, are water soluble.

Using non-aqueous scrubbing liquids such as diesel, vegetable oils, and other hydrocarbons have proven effective for tar removal. The Energy Centre of the Netherlands (ECN) has developed a process called OLGA [50, 51], a multi-stage process for removing tars based on dew points of various components of tar. The first stage collects heavy tars via high temperature scrubbing of vapors. The next stage is an ESP to collect tar aerosols and fine particulate matter. The heavy tars and particulates from these two stages are recycled back to the gasifier. The third stage removes light tars. The spent scrubbing liquid is sent to a stripper to separate light tars, which are also recycled to the gasifier. The final stage is a water scrubber to remove any residual contaminants. The output is a tar-free, particulate-free dry syngas at ambient temperature.

The sequential temperature reduction steps based on dew points for tar components and water are optimized to improve thermal efficiency of the integrated process. Returning the recovered tars and particulates back to the gasifier maximizes carbon conversion efficiency. Recycling of scrubbing liquid reduces operating costs by avoiding costly treatment of the effluent. The added cost of using hydrocarbons as scrubbing liquids and increased process complexity offset some of these cost savings.

5.2.6.2 Thermal Cracking

Thermal cracking of tar requires temperatures (>1100 °C) that are higher than typical gasifier exit temperatures. Partial oxidation of the product gas through addition of oxygen can provide this thermal energy but at the cost of overall product gas yield. Thermal destruction of tars can generate soot, which is an unwanted impurity in product gas. Adding a catalyst for thermal cracking is a strategy to reduce the severity of this tar mitigation option.

Thermal cracking of hydrocarbons is a well-established technology using solid acid catalysts such as silica–alumina and zeolites. In fact, any catalyst with strong acid sites will crack hydrocarbons at temperatures above 200 °C. Aromatics present during hydrocarbon cracking usually lead to higher molecular weight hydrocarbons and coke [52].

Gil et al. [53] have studied the use of spent FCC catalyst in the gasifier as a way of reducing tar content of the effluent gas. They found that use of the spent FCC catalyst resulted in a reduction of tar from 20 to 8.5 g/N m^3 at 800 820 °C, compared to a reduction from 20 to 2 g/N-m^3 at a comparable temperature when using dolomite as cracking catalyst. Fresh or equilibrium FCC catalyst was significantly more active than spent FCC catalyst. Overall, the results were inconclusive because the spent FCC catalyst was so small (70 μm diameter) that it quickly elutriated from the gasifier while the large dolomite particles (400–1000 μm diameter) had much larger residence times. Spent and fresh FCC catalysts caused tar cracking, whereas dolomite appeared to promote steam reforming of tar. Because many tar components are polycyclic aromatics, cracking is expected to deposit coke on the catalysts, requiring catalyst regeneration.

5.2.6.3 Hydrogenation

Tars can be eliminated through hydrogenolysis, or ring opening, of the polyaromatics. Unfortunately, hydrogenolysis is thermodynamically limited, decreasing with increasing temperature. Unfortunately, even at temperatures as low as 200 °C, naphthalene showed little ring opening in the presence of catalysts such as Rh, Pt, Ir, and Ru, suggesting that hydrogenolysis is not technically viable for tar elimination [54, 55].

5.2.6.4 Steam Reforming

Catalytic steam reforming is an attractive approach to tar elimination at elevated temperatures [45, 56–62]. This technique offers several advantages: the endothermal energy of steam reforming can be provided through thermal integration with the gasifier, the composition of the product gas can be catalytically adjusted, and steam can be added to the catalyst reactor to ensure complete reforming of tars. Catalytic tar destruction has been studied for several decades, and many reviews have been written on the subject [63–66]. Numerous catalysts have been tested for tar destruction activity at a broad range of scales, and novel catalyst formulations have been sought to increase the activity and lifetime of tar-reforming catalysts. Different approaches have been investigated for integrating catalytic tar destruction into biomass gasification systems.

Steam reforming usually incorporates nickel supported catalysts on thermal-resistant, silicon-free supports such as α-alumina, MgAl spinel, or ZrO_2. Such nickel catalysts are poisoned by sulfur, thus requiring the level of sulfur to be kept at less than 10–20 ppmv. Nickel catalysts may contain promoters such as iron, ruthenium, manganese, potassium, or barium. Selective steam reforming can be done on aromatic hydrocarbons. Mainly alkyl groups are split off by steam; the process is thus also called steam dealkylation.

Use of dolomite (CaO–MgO), either in the biomass gasifier or in a reactor downstream from the gasifier, decreases tar concentration in the effluent stream. Gil et al. [53] found that adding dolomite to the gasifier reduced tar from 20 g/m^3 to ~2 g/m^3 at 800–820 °C. They hypothesized that dolomite acts as a base catalyst and catalyzes steam reforming of tars. CaO also appears to catalyze steam reforming of higher hydrocarbons [7].

Calcined dolomites are the most widely used nonmetallic catalysts for tar conversion in biomass gasification [67–73]. Dolomites are relatively inexpensive and are considered disposable; however, they are not very robust and quickly undergo attrition in fluidized-bed reactors. Consequently, dolomites find most use in fixed-bed catalytic reactors. Tar conversion efficiency is high when calcined dolomites are operated at high temperatures (900 °C) with steam. Olivine, another naturally occurring mineral, has also demonstrated tar conversion activity similar to that of calcined dolomite [70, 74]. Olivine is a much more robust material than calcined dolomite and has been applied as a primary catalyst to reduce the output tar levels from fluidized-bed biomass gasifiers.

Commercial nickel catalysts, designed for use in fixed-bed applications, are not robust enough for fluidized-bed applications. However, they have been extensively used as secondary catalysts in separate fixed-bed reactors operated independently to optimize performance [75–77]. They have high tar destruction activity, with the added advantages of completely reforming methane and WGS activity that allows the H_2:CO ratio of the product gas to be adjusted. Some studies have also shown that nickel catalyzes the reverse ammonia reaction, thus reducing the amount of NH_3 in gasification product gas.

A limitation of nickel catalyst for hot gas cleaning of product gases is rapid deactivation, which leads to limited catalyst lifetimes [78]. Causes of deactivation include sintering, coke formation, and poisoning. Sulfur, chlorine, and alkali metals that may be present in product gases can poison catalysts. Coke formation on catalyst surfaces can be substantial when tar levels in product gases are high. Coke can be removed by regenerating the catalyst; however, repeated high-temperature processing of nickel catalysts can lead to sintering, phase transformations, and volatilization of nickel.

Using fixed dolomite guard beds to lower input tar concentrations can extend nickel catalyst lifetimes. Adding various promoters and support modifiers has been demonstrated to improve catalyst lifetime by reducing catalyst deactivation resulting from coke formation, sulfur and chlorine poisoning, and sintering. Several novel, nickel-based catalyst formulations have been developed that show excellent tar reforming activity, improved mechanical properties for fluidized-bed applications, and enhanced lifetimes.

Hot gas conditioning using current or future commercially available catalysts offers a promising solution for mitigating biomass gasification tars. Tars are eliminated, methane can be reformed if desired, and the H_2:CO ratio can be adjusted in a single step. The best currently available tar reforming process consists of a calcined dolomite guard bed followed by a fixed-bed nickel catalyst reforming reactor operating at about 800 °C. Selection of the ideal nickel catalyst is somewhat premature. Commercially available steam-reforming catalysts have been demonstrated; however, several of the novel research catalysts appear to have the potential of longer lifetimes [79–81]. The dual-bed hot gas conditioning concept has been demonstrated and can be used to condition product gas from any gasification process. A proprietary nickel monolith catalyst has also shown considerable promise for biomass gasification tar destruction [45, 82, 83].

5.3 Syngas Utilization

In its simplest form, syngas is composed of two diatomic molecules, CO and H_2, that provide the building blocks upon which an entire field of fuel science and technology is based [26, 84–89]. The synthesis of hydrocarbons from CO hydrogenation was discovered in 1902 by Sabatier and Sanderens, who produced methane by passing CO and H_2 over nickel, iron, and cobalt catalysts. At about the same time, the first commercial production of hydrogen was achieved using syngas produced from steam reforming of methane. Haber and Bosch invented a process for synthesizing ammonia from H_2 and N_2 in 1910, and the first industrial ammonia synthesis plant was commissioned in 1913. The production of liquid hydrocarbons and oxygenates from syngas conversion over iron catalysts was discovered in 1923 by Franz Fischer and Hans Tropsch. Variations on this synthesis pathway were soon to follow for selective production of methanol, mixed alcohols, and isosynthesis products. Another outgrowth of Fischer–Tropsch synthesis (FTS) was the hydroformylation of olefins discovered in 1938.

Syngas composition, most importantly the H_2:CO ratio, varies as a function of feedstock and thermochemical process. Steam reforming of methane yields H_2:CO ratios of 3 : 1, while ratios closer to unity or even lower result from gasification of coal and biomass. Typically, most synthesis processes yield fewer moles of product compared to the reactant mixture of H_2 and CO. Consequently, syngas conversion processes are more thermodynamically favorable at elevated pressures with the optimum pressure dependent on the synthesis process.

Synthesis processes with syngas are exothermic, requiring careful process design to remove the resulting heat energy and control reaction temperature. Maximizing product yields, minimizing side or competing reactions, and maintaining catalyst integrity dictate optimum synthesis reaction conditions.

Since the genesis of syngas conversion to fuels and chemicals, a tremendous amount of research and development have been devoted to optimizing product yields and process efficiencies. This includes the discovery of catalysts with optimized formulations containing the most active metals in combination with appropriate additives to improve activity and selectivity in a given process. Mechanistic studies have been conducted to interpret the fundamentals of specific conversion processes and measure kinetic parameters of key chemical reactions. Reactor design and engineering is another active area of research and development in syngas upgrading. Detailed process engineering and integration (with respect to heat integration and syngas recycling to improve conversion efficiencies) are used to optimize commercial synthesis processes. Given the rich history of syngas conversion and the extensive research and development efforts devoted to this field of study, it is not surprising that a vast amount of literature is available that tracks the scientific and technological advancements in syngas chemistry. A summary of some of the relevant syngas conversion processes follows, with references to recent literature.

5.3.1 Syngas to Gaseous Fuels

5.3.1.1 Hydrogen

Hydrogen is primarily consumed in ammonia and methanol synthesis and petroleum refining. Presently, 77% of the worldwide hydrogen production comes from petrochemicals, 18% from coal, 4% from water electrolysis, and 1% from other sources [90]. Steam reforming of methane accounts for 95% of hydrogen produced in the USA [91]. Hydrogen can also be produced by reforming other hydrocarbon feedstocks, including naphtha, heavy residues from petrochemical processes, coke oven gas, and coal. Partial oxidation and autothermal reforming of hydrocarbons are alternative technologies to steam reforming of methane.

Renewable hydrogen technologies are increasingly attractive as greenhouse gas emissions are regulated to reduce the impact of energy use on global climate change [92]. Water electrolysis for hydrogen production can have a very small carbon footprint if renewable electricity (wind or solar power) or nuclear power is used to generate the electricity. Bio-hydrogen processes are also being developed and include biomass gasification, aqueous-phase reforming of biomass-derived sugars, and biological hydrogen production (fermentation and photolysis). Recent reviews summarize these fossil and renewable hydrogen production technologies and others [92–95].

Steam reforming of methane consists of four steps: feed pretreatment, steam reforming, CO shift conversion, and hydrogen purification. The only pretreatment required when natural gas is the feedstock is desulfurization, which usually consists of a hydrogenator

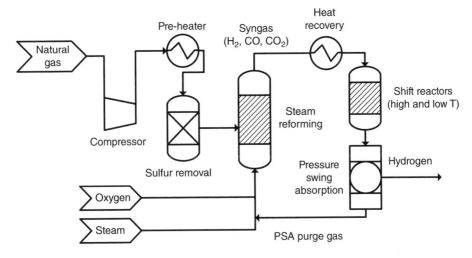

Figure 5.1 Block flow diagram of hydrogen via steam methane reforming.

followed by a zinc oxide bed. After desulfurization, natural gas is fed to a reformer reactor, where methane and other light hydrocarbons in natural gas react with steam to produce CO, CO_2, and H_2. The reformer reactor is comprised of catalyst-filled tubes, surrounded by a firebox that provides the thermal energy to support the endothermic reactions of steam reforming, which operates at about 850 °C and between 1.5 and 3 MPa. Gas exiting the reformer is cooled to about 350 °C to maximize the WGS reaction in a high temperature shift (HTS) converter. The gas can be further cooled to about 220 °C and introduced into a low-temperature shift (LTS) converter, followed by CO_2 scrubbing with monoethanolamine or hot potash. A methanation reactor removes trace amounts of CO and CO_2 ($CO + 3H_2 \rightarrow CH_4 + H_2O$). The final hydrogen product has a purity of 97–99%.

For higher purity hydrogen production, a pressure swing adsorption (PSA) unit is used downstream of the HTS converter. The PSA unit can easily remove CO and other components to produce high purity (99.99%) hydrogen. PSA off-gas, which contains unreacted CH_4, CO, and CO_2 and unrecovered hydrogen, is used to fuel the reformer. This stream usually supplies 80–90% of reformer heat duty, supplemented by natural gas to balance the remaining heat requirement. Figure 5.1 shows a block flow diagram of the process.

5.3.1.1.1 Chemistry

Steam reforming hydrocarbons involves catalytic conversion of hydrocarbons and steam to hydrogen and carbon oxides. Generally speaking, the chemical process of steam reforming of hydrocarbons is described by the following equation [90]:

$$C_nH_m + nH_2O \rightarrow \left(n + \frac{m}{2}\right)H_2 + nCO$$

WGS is another important reaction that occurs in the reformer. For steam reforming of methane, the following two reactions occur in the reformer:

$$CH_4 + H_2O \rightarrow 3H_2 + CO \quad \Delta H_r = 49.3 \text{ kcal/mol}$$

$$CO + H_2O \rightarrow H_2 + CO_2 \quad \Delta H_r = -9.8 \text{ kcal/mol}$$

Therefore, in this case, 50% of the hydrogen comes from steam. The reforming reaction is highly endothermic and is favored by high temperatures and low pressures. Higher pressures tend to lower methane conversion. In industrial reformers, reforming and shift reactions result in a product composition that closely approaches equilibrium.

The following side reactions produce carbon in the steam reformer:

$$2CO \rightarrow C(s) + CO_2 \quad \text{(Boudouard coking)}$$
$$CO + H_2 \rightarrow C(s) + H_2O \quad \text{(CO reduction)}$$
$$CH_4 \rightarrow C(s) + 2H_2 \quad \text{(methane cracking)}$$

The molar steam to carbon ratio is usually 2 : 6, depending on the feedstock and process conditions. Excess steam is used to prevent coking in the reformer tubes. The shift reaction is exothermic and favors low temperatures. Since it does not approach completion in the reformer (usually there is 10–15% by volume CO, dry basis, in reformer effluent), further conversion of CO is performed using shift conversion catalysts.

5.3.1.1.2 Catalysts

Conventional steam-reforming catalysts are 10–33% by weight NiO on a mineral support (alumina, cement, or magnesia). Reforming catalyst suppliers include BASF, Dycat International, Haldor Topsøe, Johnson Matthey Synetix (formerly ICI Katalco), and Clariant (formerly Süd Chemie) Heavy feedstocks tend to coke reforming catalyst, but promoters (potassium, lanthanum, ruthenium, and cerium) may be used to help mitigate this problem. Promoters increase steam gasification of solid carbon, thereby reducing coke formation but also reducing reforming activity of the catalyst. For feedstocks heavier than naphtha, nickel-free catalysts containing mostly strontium oxide, aluminum oxide, and calcium oxide have been successfully tested [90]. However, methane content in the exiting gas is high, requiring a secondary reformer.

HTS catalyst has an iron oxide–chromium oxide basis, while the major component in the LTS catalyst is copper oxide usually mixed with zinc oxide [90]. The HTS reactor operates in the temperature range 300–450 °C, while the LTS operates in the range 180–270 °C. Often, the LTS reactor operates near condensation conditions. Typical catalyst lifetimes for both HTS and LTS catalysts are three to five years.

Sulfur-tolerant or "sour shift" catalysts have also been developed that have high activity with larger concentrations of sulfur in the reactant gas. ICI Katalco makes dirty shift conversion catalysts that consist of cobalt and molybdenum oxides. They operate over a temperature range of 230–500 °C. The controlling factors are the ratio of steam to sulfur in the feed gas and catalyst temperature.

5.3.1.1.3 Pressure Swing Adsorption

About 65–75% of the CO and steam in the feed stream to the HTS reactor are converted to additional H_2 and CO_2. When an LTS reactor is used to convert additional CO to H_2, about 80–90% of the remaining CO is converted to H_2, increasing H_2 yield about 5%.

For the PSA unit, the minimum pressure ratio between feed and purge gas is about 4 : 1, and purge gas pressure is typically between 17 and 20 psi (~0.12–0.14 MPa) to obtain a high recovery of hydrogen. Hydrogen recovery is usually 85–90% at these conditions and drops to 60–80% at high purge gas pressures (55–95 psi, ~0.38–0.66 MPa). PSA efficiency is also affected by adsorption temperature. Fewer impurities are adsorbed at higher temperatures, because the equilibrium capacity of the molecular sieves decreases with increasing temperature. Additionally, nitrogen is weakly adsorbed onto the adsorbent bed in the PSA unit, reducing the hydrogen recovery rate for the same purity. Hydrogen recovery may be reduced by as much as 2.5% for a 10 ppm nitrogen concentration in the PSA feed stream.

5.3.1.1.4 Membrane Separation (Palladium and Polymers)

Research into in situ H_2 membranes has focused primarily on three types of membrane: microporous membranes [96], palladium-based membranes [97–101], and dense ceramic membranes [102, 103]. The primary challenges faced by these membrane materials include:

- maximizing H_2 selectivity;
- maximizing H_2 flux;
- minimizing membrane failure caused by thermal cycling or chemical interactions; and
- minimizing membrane cost.

Table 5.2 summarizes the advantages and disadvantages for these three types of membrane.

One generic disadvantage of all these in situ H_2 membrane systems is that the H_2-rich product is recovered at low pressure. Although this might be acceptable for fuel-cell applications, it is not particularly attractive for other applications where high-pressure H_2 product is necessary. This could be achieved simply by compressing the H_2-rich product to the desired pressure. However, because of its small molecular size, H_2 requires significantly more effort to compress. Thus, the advantages of in situ H_2 membranes must outweigh the associated energy and capital costs for compressing a highly H_2-rich product to useful operating pressures.

Because the primary function of the membrane is to remove H_2, allowing more complete shift of the CO into H_2 in a single reactor, this additional shift activity will also result in additional heat release. Thus, to avoid excessive temperature rise in the catalyst bed that would cause sintering and rapid deactivation of the catalyst, membrane reactor design will require effective heat removal. This represents a significant challenge in a fixed-bed reactor system, since typical fixed-bed systems tend to have hot spots that result from less-than-ideal mixing of the gases, subtle variations in the catalyst activity, and natural heat flow in the system.

5.3.1.2 Substitute (or Synthetic) Natural Gas

During the 1970s, ExxonMobil conducted an R&D program aimed at producing substitute natural gas (SNG) from coal. Equilibrium methane yields as a function of temperature and pressure are shown in Figure 5.2. Coal impregnated with an alkali metal salt catalyst (typically K_2CO_3) was introduced into a fluidized bed reactor [104] operating at temperatures between 593 and 815 °C. The coal was gasified in a mixture of steam and recycled H_2 and

Table 5.2 Comparison of membrane technologies for hydrogen separation.

	Microporous membrane	Metallic and metallic alloy membrane	Ceramic membrane	Dense ceramic membrane
Typical materials	Silica on ceramic, zeolites	Pd, Pd–Cu, Pd–Ag on stainless steel or ceramic	Pd mixed with dense ceramic	Perovskite, silica
Temperature (°C)	300–600	300–750	400–600	~900
H_2 purity (%)	>90	~100	~100	~100
H_2 flux[a]	100	36	36	2
Chemical stability	High, tolerant to sulfur, some form carbonates with CO_2 at high temperatures	Low, Pd easily poisoned by sulfur and CO, alloy causes defects and side reactions	Medium, tolerant to impurities	High
Thermal stability	High	Low, embrittled by H_2 at <300°C; phase change at high temperature reduces H_2 flux; Pd alloys preferred	Medium, phase change inhibited	High
Cost	Low	High (pure Pd) to medium (alloy)	Medium	Low

[a] H_2 flux is expressed as cubic feet of hydrogen per square foot of membrane area per 100 lbs (psi) of pressure drop.

CO to produce product gas composed primarily of CH_4, CO_2, H_2, and CO, which was separated in a cryogenic separation train into SNG product and recycle gas containing H_2 and CO. Char from the process was washed to recover alkali metal catalyst for reuse.

Catalytic conversion of coal into CH_4 and CO_2 at these low temperatures had several benefits. The first was that CH_4 was one of the most abundant components in the raw gasifier effluent. The second was that cheaper and more conventional materials of construction could be used instead of more costly and exotic alloys required by competing gasification technologies.

However, operating in this temperature range also has disadvantages. The alkali metal catalyst reacted with the mineral ash in the coal and reduced the amount of recoverable catalyst without a complex and costly recovery processes. Ultimately, the catalyst recovery and replacement costs could not be reduced enough to make the process economically competitive.

Although lower operating temperatures favor CH_4-rich product gas, lower temperatures also result in slower reactions. Increasing operating temperature would increase the carbon conversion rate, but this would also increase the rate of catalyst reacting with ash, resulting in greater catalyst losses. Similarly, adding more active catalyst would improve the coal conversion reaction rate, but this also increases the amount of catalyst available to react with

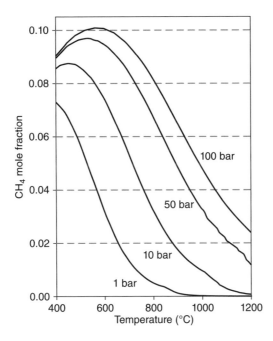

Figure 5.2 Equilibrium methane yields for the SNG process as a function of temperature and pressure.

coal ash, further hindering catalyst recovery. ExxonMobil determined that the optimum catalyst concentration was about 15–20% by weight. Furthermore, additional R&D demonstrated that coal type and origin significantly influenced coal conversion. Sub-bituminous western coals, generally with high moisture content, were identified as the most reactive for this process.

ExxonMobil was able to demonstrate this process successfully at the pilot-plant scale. However, unfavorable process economics prevented further development and commercialization was abandoned. Measured catalyst losses due to reaction with coal ash and the high cost of cryogenic separation of SNG and recycle gas made the process too costly compared with traditional natural gas recovery. A dramatic increase in price of natural gas in the late 1990s prompted a renewed interest in the production of SNG. A summary of the various coal-based technologies available is shown in Table 5.3. Many of these technologies were demonstrated at pilot scale in the 1970s but were abandoned as declining natural gas prices made these alternative technologies uneconomical.

Prior to about 2010, natural gas prices had been unpredictable although typically trended with petroleum fuels. Price volatility was a major risk factor that discouraged the use of natural gas for energy production. After 2010, increased production of shale gas in the United States drove down natural gas prices and reduced their volatility even as petroleum prices remained volatile. The Energy Information Administration (EIA) forecasts natural gas prices to stay low to moderate for several decades [111].

The impact of fossil CO_2 emissions on global climate change have recently renewed interest in producing SNG from renewable biomass resources [112–116], especially in Europe. Biomass gasification is being explored at the pilot, demonstration, and commercial

Table 5.3 Comparison of SNG production technologies.

Process	Gasifier operation	Key elements	Notes
Lurgi (Dakota Gasification Company) [91]	Moving bed	Produces approximately 12% methane in gasifier; CO shifted to H_2 before methanation; CO_2 removal by Rectisol	Gas cooling (with energy recovery, prior to Rectisol); extensive commercial experience
Exxon mobile catalytic gasification [105]	Low temperature (700 °C); 500 psig; catalytic fluidized bed; H_2 and CO recycled	Catalyst added to coal; must be recovered from ash; cryogenic separation of H_2 and CO required	Incomplete separation of the relatively expensive catalyst from the ash affected process economics
Nahas Process [106]	Low temperature (700 °C); 300 psig; catalytic gas; H_2 and CO recycled	Only suitable for no-ash materials (e.g. petroleum residues or coal pyrolysis products)	Conceptual design, but no experimental data to prove the concept.
ARCH hydrogasification [107]	900 °C; 1000 psig; entrained flow hydrogasification; H_2 recycled	"Dense phase" pulverized coal feed	Requires separate "partial combustion gasifier" to produce syngas, which is shifted to supply H_2 for the hydrogasifier
HYGAS (IGT) [108, 109]	>1000 psig; 800 °C; two-stage fluidized bed	Slurry feed using by-product oil	Requires imported hydrogen (e.g. from steam–iron process); caking coals require pretreatment.
CO_2 acceptor (Conoco) [110]	150 psig; 800 °C; fluidized bed; heat supplied by hot dolomite	Produces syngas for feed to methanation reactor; no WGS required	Applicable to lignite; requires separate char combustor to heat and regenerate dolomite
Synthane [110]	500–1000 psig; 1000 °C; dense-phase fluidized bed	Syngas must be shifted somewhat prior to methanation; char combusted to produce steam	Caking coals require (inline) fluidized bed pretreatment at gasifier pressure
Bi-Gas [110]	Two-stage reactor: entrained coal devolatilization in upper stage at 900 °C; slagging combustion in lower stage at 1500 °C	Requires shift and methanation	Rapid devolatilization in upper reactor produces methane; char separated in cyclone for return to lower stage

scales [117]. Hydrothermal treatment processes are also being developed to convert higher moisture content biomass feedstocks and waste materials into methane-rich product gases [116]. Nickel or ruthenium catalysts are co-mixed with biomass and water. The mixture is heated to around 300–400 °C at high pressure (sub- or super-critical water pressures) to produce a methane-rich gas.

5.3.2 Syngas to Liquid Fuels

5.3.2.1 Fischer–Tropsch Synthesis

In 1923, Fischer and Tropsch reported the use of alkalized iron catalysts, in what was termed the Synthol process, to produce liquid hydrocarbons rich in oxygenated compounds. Following these initial discoveries, considerable effort went into developing catalysts for this process. A precipitated cobalt catalyst with 100 parts by weight Co, 5 parts ThO_2, 8 parts MgO, and 200 parts kieselguhr (siliceous diatomaceous earth) became known as the "standard" atmospheric-pressure process catalyst. In 1936, Fischer and Pilcher developed the medium pressure (10–15 bar, ~0.07–0.10 MPa) FTS process. Following this development, alkalized iron catalysts were implemented into the medium-pressure FTS process. Collectively, the process of converting CO and H_2 mixtures to liquid hydrocarbons over a transition metal catalyst has become known as FTS.

Two main characteristics of FTS are the nonselective production of a wide range of hydrocarbon products and liberation of large amounts of heat from these highly exothermic synthesis reactions. Consequently, reactor design and process development have focused heavily on heat removal and temperature control. The focus of catalyst development is on improved catalyst lifetimes, activity, and selectivity. Single-pass FTS always produces a wide range of olefins, paraffins, and oxygenated products, such as alcohols, aldehydes, acids, and ketones, with water as by-product. Product distributions are influenced by temperature, feed gas composition (H_2/CO), pressure, catalyst type, and catalyst composition. Product selectivity can also be improved using multiple-step processes to upgrade FTS products.

A review by Frohning et al. [118] noted that by 1954 upwards of 4000 publications and a similar number of patents dealing with FTS could be found in the literature. Since then, FTS has attracted an enormous amount of research and development. A comprehensive bibliography of FTS literature, including journal and conference articles, books, government reports, and patents, can be found in the Fischer–Tropsch Archive at www.fischer-tropsch.org. This website, sponsored by Syntroleum Corporation in cooperation with Dr. Anthony Stranges, Professor of History at Texas A&M University, contains more than 7500 references and citations. This site has a bibliography of the large body of documents (from the 1920s through to the 1970s), which are important for researching the history and development of FTS and related processes, as well as an up-to-date listing of the latest publications in this field. Many excellent reviews of FTS are available [25, 26, 119–122], and this section attempts to summarize the chemistry [118], catalyst development [123, 124], commercial processes [120, 125–127], reactor development [121, 128–130], and economics [131, 132] of FTS.

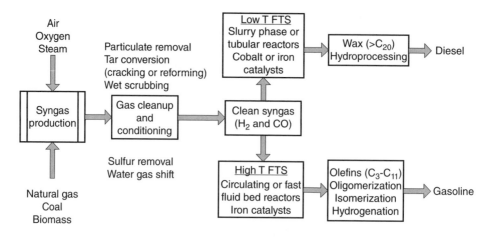

Figure 5.3 FTS general process flow diagram.

There are four main steps to producing Fischer–Tropsch products: syngas generation, gas purification, FTS, and product upgrading. See Figure 5.3 for a generic process flow diagram. When using natural gas as the feedstock, many authors [133–135] have recommended autothermal reforming or autothermal reforming in combination with steam reforming as the best option for syngas generation. This is primarily because of more favorable economies of scale for air separation units and H_2 to CO ratios compared to tubular steam-reforming reactors. If the feedstock is coal, syngas is produced via high-temperature gasification in the presence of oxygen and steam. Depending on the types and quantities of desired Fischer–Tropsch products, either low-temperature (200–240 °C) or high-temperature (300–350 °C) synthesis is used with either an iron or cobalt catalyst. FTS temperatures are usually kept below 400 °C to minimize CH_4 production. Generally, cobalt catalysts are used only at low temperatures, because at higher temperatures a significant amount of methane is produced. Low temperatures yield high molecular mass linear waxes, while high temperatures produce gasoline and low molecular weight olefins. If maximizing gasoline product fraction is desired, then it is best to use an iron catalyst at high temperature in a fixed fluid-bed reactor. If maximizing diesel product fraction is preferred, then a slurry reactor with cobalt catalyst is the best choice. Fischer–Tropsch reactors are operated at pressures ranging from 10 to 40 bar (145–580 psi, ~1–4 MPa). Upgrading usually involves a combination of hydrotreating, hydrocracking, and hydroisomerization in addition to product separation.

5.3.2.1.1 Chemistry

FTS is a polymerization process with basic steps of (1) reactant (CO) adsorption on the catalyst surface, (2) chain initiation by CO dissociation followed by hydrogenation, (3) chain growth by insertion of additional CO molecules followed by hydrogenation, (4) chain termination, and (5) product desorption from the catalyst surface. Chemisorbed methyl species are formed by dissociation of absorbed CO molecules and stepwise addition of hydrogen atoms. These methyl species can further hydrogenate to form methane or act as initiators for chain growth. Chain growth occurs via sequential addition of CH_2 groups, while the growing alkyl chain remains chemisorbed to the metal surface at the terminal methylene group. Chain termination can occur at any time during the chain growth process to yield either an α-olefin or an n-paraffin once the product desorbs.

The following is the FTS reaction [127]:

$$CO + 2H_2 \rightarrow -CH_2- + H_2O \quad \Delta H_r(227°C) = -165 \text{ kJ/mol}$$

The WGS reaction is a secondary reaction that readily occurs when iron catalysts are used. The required H_2 to CO ratio for the cobalt catalyst is 2 : 1, but since the iron catalyst performs WGS in addition to the Fischer–Tropsch reaction, the H_2 to CO ratio can be slightly lower (0.7 : 1) for the iron catalyst.

Specific FTS products are synthesized according to the following reactions:

$$CO + 3H_2 \rightarrow CH_4 + H_2O \quad \text{(methanation)}$$

$$nCO + (2n+1)H_2 \rightarrow C_nH_{2n+2} + nH_2O \quad \text{(paraffins)}$$

$$nCO + 2nH_2 \rightarrow C_nH_{2n} + nH_2O \quad \text{(olefins)}$$

$$nCO + 2nH_2 \rightarrow C_nH_{2n+1}OH + (n-1)H_2O \quad \text{(alcohols)}$$

The Boudouard reaction, also important in FTS, competes with fuel synthesis reactions for CO, producing coke that deposits on catalyst surfaces, causing deactivation:

$$2CO \rightarrow CO_2 + C$$

FTS is kinetically controlled, the intrinsic kinetics being stepwise chain growth; in effect, it is the polymerization of CH_2 groups on a catalyst surface. FTS product selectivities are determined by the ability to catalyze chain propagation versus chain termination reactions. Polymerization rates are independent of the products formed. The probability of chain growth and chain termination is independent of chain length. Therefore, selectivities of various hydrocarbons can be predicted based on simple statistical distributions calculated from chain growth probability and carbon number. The chain polymerization kinetics model known as the Anderson–Shulz–Flory (ASF) model is represented by the following equation:

$$W_n = n(1-\alpha)^2 \alpha^{n-1}$$

where W_n is the weight percent of a product containing n carbon atoms and α is the chain growth probability. This equation is represented graphically in Figure 5.4 and displays the predicted distributions for several products and product ranges of particular interest.

FTS produces a range of olefins, paraffins, and oxygenated compounds (alcohols, aldehydes, acids, and ketones) that are predominantly linear with a high percentage of olefinic hydrocarbons. In fact, the paraffin-to-olefin ratio is lower than thermodynamically predicted. The olefins that do form are predominantly terminal (alpha). A considerable amount of monomethyl chain branches form, and the degree of branching decreases as the chain length increases. Theoretically, only methane can be produced with 100% selectivity. The only other product that can be produced with high selectivity is heavy paraffin wax. The

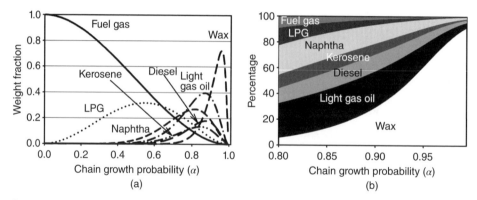

Figure 5.4 (a) Weight fraction of hydrocarbon products as a function of chain growth (propagation) probability during FTS. (b) Percentage of different hydrocarbon product cuts as a function of chain growth (propagation) probability, showing a range of operation for classical and developing Fischer–Tropsch catalysts and synthesis.

gasoline product fraction has a maximum selectivity of 48%, and the maximum diesel product fraction selectivity is closer to 40% and varies depending on the range of carbon numbers in the product cut. The variables that influence the distribution of these products are reactor temperature, pressure, feed gas composition, catalyst type, and promoters.

5.3.2.1.2 Catalysts

Transition metal oxides are generally regarded as good CO hydrogenation catalysts. The relative activity of these metals for FTS, in decreasing order of activity, is Ru > Fe > Ni > Co > Rh > Pd > Pt [129]. Nickel is basically a methanation catalyst and does not have the broad selectivity of other Fischer–Tropsch catalysts. Ruthenium has very high activity and quite high selectivity for producing high molecular weight products at low temperatures. Iron is also very active and has WGS activity. Iron readily forms carbides, nitrides, and carbonitrides with metallic character that also have FTS activity. Iron also has a stronger tendency than nickel or cobalt to produce carbon that deposits on the surface and deactivates the catalyst. Cobalt tends to have a longer lifetime than iron catalysts and does not have WGS activity, which leads to improved carbon conversion to products because CO_2 is not formed. Cobalt catalysts in FTS yield mainly straight-chain hydrocarbons (no oxygenates like iron). Although ruthenium is the most active FTS catalyst, it needs to be noted that it is 3×10^5 times more expensive than iron. Iron is by far the cheapest FTS catalyst of all of these metals. Cobalt catalysts are 230 times more expensive than iron but are still an alternative to iron catalysts for FTS because they demonstrate activity at lower synthesis pressures – so higher catalyst costs can be offset by lower operating costs.

The three key properties of Fischer–Tropsch catalysts are lifetime, activity, and product selectivity. Optimizing these properties for desired commercial application has been the focus of Fischer–Tropsch catalyst research and development since the processes were first discovered. Each one of these properties can be affected by a variety of strategies, including the use of promoters (chemical and structural), catalyst preparation and formulation, pretreatment and reduction, selective poisoning, and shape selectivity with zeolites.

The performance of cobalt catalysts is not very sensitive to the addition of promoters. Early work demonstrated that addition of ThO_2 improved wax production at atmospheric pressure but had little effect at higher pressures. With iron catalysts, however, promoters and supports are essential catalyst components. Since the discovery of FTS, potassium has been used as a promoter for iron catalysts to effectively increase the basicity of the catalyst surface. The objective is to increase adsorption of CO to metal surfaces, which tends to withdraw electrons from the metal by providing an electron donor. Potassium oxide addition to iron catalysts also tends to decrease hydrogenation of adsorbed carbon species, so chain growth is enhanced, resulting in a higher molecular weight product distribution that is more olefinic. Potassium promotion also tends to increase WGS activity and lead to a faster rate of catalyst deactivation because of the increased rate of carbon deposition on the surface of the catalyst.

Copper has also been successfully used as a promoter in iron FTS catalysts. Although it increases the rate of FTS more effectively than potassium, it decreases the rate of the WGS reaction. Copper has also been shown to facilitate iron reduction. Average molecular weight of products increases in the presence of copper, but not as much as when potassium is used. (Potassium promotion is not effective for cobalt catalysts.)

Catalyst preparation impacts performance of iron and cobalt catalysts. Iron catalysts can be prepared by precipitation onto catalyst supports, such as SiO_2 or Al_2O_3, or prepared as fused iron, where formulations are prepared in molten iron, then cooled and crushed. The role of supports in cobalt catalysts is also important. Since cobalt is more expensive than iron, precipitating the ideal concentration of metal onto a support can help reduce catalyst costs while maximizing activity and durability.

The combination of light transition metal oxides such as MnO with iron increases selectivity of light olefins in FTS. Iron/manganese/potassium catalysts have shown selectivity for C_2–C_4 olefins as high as 85–90%. Noble metal addition to cobalt catalysts increases FTS activity but not selectivity.

5.3.2.2 Methanol

Research and development efforts at the beginning of the twentieth century involving conversion of syngas to liquid fuels and chemicals led to the discovery of a methanol synthesis process concurrently with development of FTS. In fact, methanol is a by-product of FTS when alkali-metal-promoted catalysts are used. Methanol synthesis is now a well-developed commercial catalytic process with high activity and very high selectivity (>99%). The lowest cost process includes natural gas reforming to produce syngas followed by methanol synthesis (90% of worldwide methanol production [136]). However, a variety of feedstocks other than natural gas can be used.

The long-time interest in methanol is due to its potential fuel and chemical uses. In particular, methanol can be used directly or blended with other petroleum products as a clean-burning transportation fuel. Methanol is also an important chemical intermediate used to produce formaldehyde, dimethyl ether (DME), methyl *tert*-butyl ether, acetic acid, olefins, methyl amines, and methyl halides, to name a few.

Currently, the majority of methanol is synthesized from syngas that is produced via steam reforming of natural gas. It can also be reformed using autothermal reforming or a combination of steam methane reforming and autothermal reforming. Once natural gas is reformed,

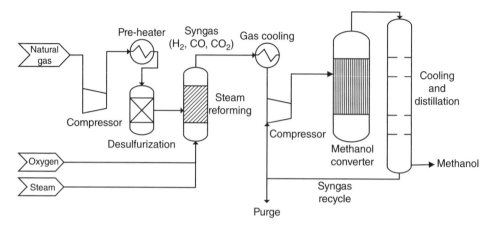

Figure 5.5 Simplified methanol synthesis process flow diagram.

the resulting synthesis gas is fed to a reactor vessel in the presence of a catalyst to produce methanol and water vapor. This crude methanol – which usually contains up to 18% water, plus ethanol, higher alcohols, ketones, and ethers – is fed to a distillation plant consisting of a unit that removes volatiles and a unit that removes water and higher alcohols. Unreacted syngas is recirculated back to the methanol converter, resulting in overall conversion efficiency of over 99%. A generic methanol synthesis process flow diagram is shown in Figure 5.5.

One of the challenges associated with commercial methanol synthesis is removing the large excess heat of reaction. Methanol synthesis catalyst activity increases at higher temperatures, but so does the chance for competing side reactions [137]. By-products of methanol formation are CH_4, DME, methyl formate, higher alcohols, and acetone. Catalyst lifetimes are also reduced by continuous high-temperature operation, and process temperatures are typically maintained below 300 °C to minimize catalyst sintering.

Overcoming the thermodynamic constraint is another challenge in commercial methanol synthesis. Maximum per-pass conversion efficiency of syngas to methanol is limited to 25% [88]. Higher conversion efficiencies per pass can be realized at lower temperatures where methanol equilibrium is shifted toward products; however, catalyst activities generally decrease as temperature is lowered. Removing methanol as it is produced is another strategy used for overcoming thermodynamic limitations and improving per-pass conversion efficiencies. Methanol is either physically removed (condensed out or physisorbed onto a solid) or converted to another product such as DME, methyl formate, or acetic acid.

Controlling and dissipating the heat of reaction and overcoming the equilibrium constraint to maximize per-pass conversion efficiency are two main process features that are considered when designing a methanol synthesis reactor, commonly referred to as a methanol converter. Numerous methanol converter designs have been commercialized over the years, and these can be roughly separated into two categories: adiabatic reactors or isothermal reactors. Adiabatic reactors often contain multiple catalyst beds separated by gas-cooling devices, either direct heat exchange or injection of cooled, fresh, or recycled syngas. Axial temperature profiles often have a sawtooth pattern that is low at the point of heat removal and increases linearly between the heat exchange sections. The isothermal reactors are designed to remove heat of reaction continuously, so they operate essentially like a heat exchanger with an isothermal axial temperature profile. A description of some of the many methanol converter designs can be found in the literature [89].

5.3.2.2.1 Chemistry

Catalytic methanol synthesis from syngas is a classic high-temperature, high-pressure, exothermic, equilibrium-limited synthesis reaction. The chemistry of methanol synthesis is as follows [138]:

$$CO + 2H_2 \rightarrow CH_3OH \quad \Delta H_r = -90.64 \text{ kJ/mol}$$
$$CO_2 + 3H_2 \rightarrow CH_3OH + H_2O \quad \Delta H_r = -49.67 \text{ kJ/mol}$$
$$CO + H_2O \rightarrow CO_2 + H_2 \quad \Delta H_r = -41.47 \text{ kJ/mol}$$

For methanol synthesis, a stoichiometric ratio, defined as $(H_2 - CO_2)/(CO + CO_2)$, of about 2 is preferred. This means that just the stoichiometric amount of hydrogen needed for methanol synthesis will be provided. For kinetic reasons, and in order to control by-products, a value slightly above 2 is normally preferred [139].

Although methanol is made from mixtures of H_2 and CO, the reaction is about 100 times faster when CO_2 is present [88]. Until as recently as the 1990s, the role of CO_2 in methanol synthesis was not clear. The WGS activity of copper catalysts is so high that it was difficult to deconvolute the role of CO and CO_2 in methanol synthesis. Isotopic labeling studies have unequivocally proved that CO_2 is the source of C in methanol [88, 140]. CO is involved in the reverse WGS reaction to make H_2 and CO_2. CO_2 is also thought to keep the catalyst in an intermediate oxidation state Cu^0/Cu^+, preventing ZnO reduction followed by brass formation [140]. The proposed mechanism for catalytic methanol synthesis is believed to proceed through a long-lived formate intermediate. CO_2 is adsorbed on a partially oxidized metal surface as a carbonate and hydrogenated. This intermediate is then hydrogenated in the rate-limiting step. The copper catalyst sites have high activity for splitting the first C—O bond in CO_2 that helps maintain the oxidation state of the active copper sites. At high concentrations, however, CO_2 actually reduces catalyst activity by inhibiting methanol synthesis. The feed gas composition for methanol synthesis is typically adjusted to contain 4–8% CO_2 for maximum activity and selectivity. Even though copper has WGS activity, excessive amounts of H_2O also lead to active site blocking, which is poor for activity but improves selectivity by reducing by-product formation by 50% [141].

5.3.2.2.2 Catalysts

The first high-temperature, high-pressure methanol synthesis catalysts were ZnO/Cr_2O_3 and were operated at 350 °C and 250–350 bar (25–35 MPa). Catalyst compositions contained 20–75% (atomic) zinc. These catalysts demonstrated high activity and selectivity for methanol synthesis and proved robust enough to resist sulfur poisoning inherent when generating syngas from coal gasification. Over the years, as gas purification technologies improved (i.e. removal of impurities such as sulfur, chlorine, and metals), interest in the easily poisoned copper catalysts for methanol synthesis was renewed. In 1966, ICI introduced a new, more active $Cu/ZnO/Al_2O_3$ catalyst that began a new generation of methanol production by a low-temperature (220–275 °C), low-pressure (50–100 bar, 5–10 MPa) process. The last high-temperature methanol synthesis plant closed in the mid-1980s [142]; presently, low-temperature, low-pressure processes based on copper catalysts are used for all commercial production of methanol from syngas. The synthesis process has been optimized to the point that modern methanol plants yield 1 kg of MeOH per liter of catalyst per hour with >99.5% selectivity for methanol. Commercial methanol synthesis catalysts have lifetimes on the order of three to five years under normal operating conditions.

The copper crystallites in methanol synthesis catalysts have been identified as the active catalytic sites, although the actual state (oxide, metallic, etc.) of the active copper site is still being debated. The most active catalysts all have high copper content; optimum ~60% copper is limited by the need to have enough refractory oxide to prevent sintering of the copper crystallites. ZnO in the catalyst formulation creates a high copper metal surface area; it is suitably refractory at methanol synthesis temperatures and hinders the agglomeration of copper particles. ZnO also interacts with Al_2O_3 to form a spinel that provides a robust catalyst support. Acidic materials such as alumina are known to catalyze methanol dehydration reactions to produce DME. By interacting with the Al_2O_3 support material, ZnO effectively improves methanol selectivity by reducing the potential for DME formation. Table 5.4 shows catalyst formulations provided by several commercial manufacturers.

Additional catalyst formulations have been presented in the literature for the purpose of improving per-pass methanol yields [143]. Addition of cesium to Cu/ZnO mixtures has shown improved methanol synthesis yields. This holds true only for the heavier alkali metals, as addition of potassium to methanol synthesis catalysts tends to enhance higher alcohol yields. Cu/ThO_2 intermetallic catalysts have also been investigated for methanol synthesis [143]. These catalysts have demonstrated high activity for forming methanol from CO_2-free syngas. Thoria-based methanol catalysts deactivate very rapidly in the presence of CO_2. Cu/Zr catalysts have proven active for methanol synthesis in CO-free syngas at 5 atm (~0.5 MPa) and 160–300 °C [144]. Supported palladium catalysts have also demonstrated methanol synthesis activity in CO_2-free syngas at 5–110 atm (~0.5–111 MPa) at 260–350 °C.

Table 5.4 Commercial methanol synthesis catalyst formulations.

Manufacturer	Cu (atom%)	Zn (atom%)	Al (atom%)	Other	Patent date
IFP	45–70	15–35	4–20	Zr, 2–18 atom%	1987
ICI	20–35	15–50	4–20	Mg	1965
BASF	38.5	48.8	12.9		1978
Shell	71	24		Rare earth oxide, 5 atom%	1973
Süd Chemie	65	22	12		1987
Dupont	50	19	31		None found
United catalysts	62	21	17		None found
Haldor Topsøe MK-121	>55	21–25	8–10		None found

Source: Reproduced from Spath and Dayton [89].

5.3.2.3 Methanol to Gasoline

The methanol to gasoline (MTG) process developed by ExxonMobil involves conversion of methanol to hydrocarbons over zeolite catalysts. The MTG process, although considered the first major new synthetic fuel development since FTS, was discovered by accident in the 1970s by two independent groups of ExxonMobil scientists trying to convert methanol to ethylene oxide and attempting to methylate isobutene with methanol over a ZSM-5 zeolite catalyst [145].

The MTG process occurs in two steps. First, crude methanol (17% water) is superheated to 300 °C and partially dehydrated over an alumina catalyst at 27 atm (~2.7 MPa) to yield an equilibrium mixture of methanol, DME, and water (75% of the methanol is converted). This effluent is then mixed with heated recycled syngas and introduced into a reactor containing ZSM-5 zeolite catalyst at 350–366 °C and 19–23 atm (~1.9–2.3 MPa) to produce hydrocarbons (44%) and water (56%) [146]. The overall MTG process usually contains multiple gasoline conversion reactors in parallel because the zeolites have to be regenerated frequently to burn off the coke formed during the reaction. The reactors are then cycled so that individual reactors can be regenerated without stopping the process, usually every two to six weeks [147].

MTG reactions may be summarized as follows [88]:

$$2CH_3OH \rightarrow CH_3OCH_3 + H_2O \quad \Delta H = -5.6\,\text{kcal/mol}$$

$$CH_3OCH_3 \rightarrow C_2 - C_5 \text{ olefins}$$

$$C_2 - C_5 \text{ olefins} \rightarrow \text{paraffins, cycloparaffins, aromatics}$$

Selectivity to gasoline-range hydrocarbons is greater than 85%, with the remainder of the product being primarily liquefied petroleum gas [88]. Nearly 40% of the gasoline produced from the MTG process is aromatic hydrocarbons with the following distribution: 4% benzene, 26% toluene, 2% ethylbenzene, 43% xylenes, 14% trimethyl-substituted benzenes, plus 12% other aromatics [88]. The shape selectivity of the zeolite catalyst results in a relatively high durene (1,2,4,5-tetramethylbenzene) concentration, 3–5% of the gasoline produced [148]. Therefore, MTG gasoline is usually distilled, and the heavy fraction is processed in the heavy gasoline treating unit to reduce durene concentration to below 2%. This results in a high-quality gasoline with a high octane number.

The first commercial MTG plant came on stream in 1985 in New Zealand (ExxonMobil's Motunui plant) producing both methanol and high-octane gasoline from natural gas. The plant produced 14 500 BPD of gasoline before gasoline manufacture was abandoned in favor of methanol production. Today, this plant, along with a nearby methanol plant at Waitara, produces 2.43×10^6 t/year of chemical-grade methanol for export (http://www.teara.govt.nz/en/oil-and-gas/5).

A fluid-bed MTG plant was jointly designed and operated near Cologne, Germany, by ExxonMobil Research and Development Corp., Union Rheinische Braunkohlen Kraftstoff AG, and Uhde GmbH [145]. A demonstration plant (15.9 m^3/day) operated from 1982 to 1985. Although, the fluid-bed technology is ready for commercialization, no commercial plants have been built.

5.3.2.4 TIGAS (Topsøe Integrated Gasoline Synthesis)

Topsøe integrated gasoline synthesis (TIGAS) was developed by Haldor Topsøe with the intent of minimizing capital and energy costs by integrating methanol synthesis with the MTG step into a single loop without isolation of methanol as an intermediate [145, 149, 150]. This process was developed with the idea that future plants would be constructed in remote areas for recovery of low-cost natural gas. In ExxonMobil's MTG process, different pressures are preferred for syngas production, methanol synthesis, and the fixed-bed MTG step. These pressures are 15–20 atm (221–294 psig, ~1.5–2.0 MPa), 50–100 atm (735–1470 psig, ~5–10 MPa), and 15–25 atm (221–368 psig, ~1.5–2.5 MPa) respectively [88]. TIGAS involves modified catalysts and conditions so that system pressure levels out and separate compression steps are not required. To do this, a mixture of methanol and DME is made prior to gasoline synthesis, and then there is only one recycle loop, which goes from the gasoline synthesis step back to the MeOH/DME synthesis step. A 1000 ton/day demonstration plant was built in Houston, Texas, in 1984 and operated for three years [150]. The gasoline yield for the process, defined as the amount of gasoline produced divided by the amount of natural gas feed and fuel, was 56.5% by weight [150].

Recent interest in advanced hydrocarbon fuels from biomass and the potential for capturing and monetizing stranded natural gas has brought renewed interest in the TIGAS process [151, 152]. This includes a project to demonstrate the integration of woody biomass gasification (25 ton/day), gas cleanup and conditioning, and gasoline synthesis using the TIGAS process with a GSK-10 gasoline catalyst [151]. The GSK-10 catalyst has been used extensively for industrial applications in China but the first commercial-scale production of synthetic gasoline from stranded natural gas is planned for operation in 2018 in a 600 000 tonne/year plant in Turkmenistan. [153].

5.3.2.5 Methanol to Olefins and Methanol to Gasoline and Diesel

Along with the MTG process, ExxonMobil developed several other processes for converting methanol to hydrocarbons based on zeolite catalysts. Since light olefins are intermediates in the MTG process, it is possible to optimize the methanol-to-olefins (MTO) synthesis. Higher reaction temperatures (~500 °C), lower pressures, and lower catalyst (acidity) activity favor light olefin production [145]. The rate of olefin production could be modified so that 80% of the product consists of C_2–C_5 olefins rich in propylene (32%) and butenes (20%) with an aromatic-rich C_5^+ gasoline fraction (36%) [88, 148]. The process can also be modified for high ethylene and propylene yield (>60%).

ExxonMobil also developed a methanol-to-gasoline-and-diesel (MOGD) process. Oligomerization, disproportionation, and aromatization of the olefins produced in the MTO synthesis are the basis for the MOGD process. In the MOGD process, the selectivity of gasoline and distillate from olefins is greater than 95% [145]. One source gives the gasoline product from MOGD to be 3% (by weight) paraffins, 94% olefins, 1% naphthenes, and 2% aromatics [154].

Neither the MTO nor the MOGD process are currently in commercial practice [88]; however, UOP and HYDRO of Norway license their own MTO process where the primary products are ethylene and propylene [145]. They use a fluidized-bed reactor at 400–450 °C and achieve roughly 80% carbon selectivity to olefins at nearly complete methanol conversion [155]. The operating parameters can be adjusted so that either more ethylene is produced (48% by weight ethylene, 31% propylene, 9% butenes, and 1.5% other olefins) or else more propylene (45% by weight propylene, 34% ethylene, 12% butenes, and 0.75% other olefins).

5.3.2.6 Dimethyl Ether

DME is industrially important as the starting material in the production of the methylating agent dimethyl sulfate and is also used as an aerosol propellant. DME has potential to be used as diesel fuel or cooking fuel, refrigerant, or chemical feedstock [156–158]. Commercial production of DME originated as a by-product of high-pressure methanol production.

DME is formed in a two-step process where methanol is first synthesized and then dehydrated over an acid catalyst such as γ-alumina at methanol synthesis conditions. The DME reaction scheme is as follows [159, 160]:

$CO + 2H_2 \rightarrow CH_3OH$ (methanol synthesis) $\Delta H = -21.6\,\text{kcal/mol}$

$2CH_3OH \rightarrow CH_3OCH_3 + H_2O$ (methanol dehydration) $\Delta H = -5.6\,\text{kcal/mol}$

$H_2O + CO \rightarrow H_2 + CO_2$ (WGS) $\Delta H = -9.8\,\text{kcal/mol}$

Net reaction:

$3H_2 + 3CO \rightarrow CH_3OCH_3 + CO_2$ $\Delta H = -58.6\,\text{kcal/mol}$

Note that one product in each reaction is consumed by another reaction. Because of the synergy between these reactions, syngas conversion to DME gives higher conversions than syngas conversion to methanol [157]. Table 5.5 gives the per-pass and total conversion for the synthesis of methanol, methanol/DME, and DME.

The optimum H_2 to CO ratio for DME synthesis is lower than for methanol synthesis and ideally should be around 1 : 1 [156, 157, 160]. Recent improvements to DME synthesis involve development of bifunctional catalysts to produce DME in a single gas-phase step (i.e. one reactor) [156, 161] and the use of a slurry reactor for liquid-phase DME synthesis [162, 163].

5.3.2.7 Ethanol and Mixed Alcohols

The production of higher alcohols from syngas has been known since the beginning of the last century. There are several processes that can be used to make mixed alcohols

Table 5.5 Conversions for methanol, methanol/DME, and DME.

Conversion	MeOH	MeOH/DME	DME
Per pass (%)	14	18	50
Total (%)	77	85	95

Source: Reproduced from Spath and Dayton [89].

from CO and H_2, including isosynthesis, variants of FTS, oxosynthesis involving hydroformylation of olefins, and homologation of methanol and lower molecular weight alcohols to make higher alcohols. With the development of various gas-to-liquid processes such as Fischer–Tropsch and methanol synthesis, it was recognized that higher alcohols were by-products of these processes when catalysts or conditions were not optimized. Modified Fischer–Tropsch or methanol synthesis catalysts can be promoted with alkali metals to shift the products toward higher alcohols. Higher alcohol synthesis (HAS) is also optimized at higher temperatures and lower space velocities than methanol synthesis is and with an H_2 to CO ratio of around 1 : 1 instead of 2 : 1 or greater.

While other syngas-to-liquids processes were being commercialized, the commercial success of HAS is limited by poor selectivity and low product yields. Single-pass yields of HAS are on the order of 10% syngas conversion to alcohols, with methanol typically being the most abundant alcohol produced [88, 164]. Methanol can be recycled to produce more higher alcohols or removed and sold separately. Despite these shortcomings, in 1913 BASF patented a process to synthesize a mixture of alcohols, aldehydes, ketones, and other organic compounds from CO and H_2 over an alkalized cobalt oxide catalyst at 10–20 MPa and 300–400 °C [165].

Fisher and Tropsch developed the "Synthol" process for alcohol production in 1923. They used an alkalized iron catalyst to convert syngas to alcohols at >10 MPa and 400–450 °C. Between 1935 and 1945, commercial mixed alcohol synthesis was performed with alkalized ZnO/Cr_2O_3 catalysts. The demand for mixed alcohol production from syngas decreased after 1945 with increasing availability of petroleum and the desire for neat alcohols for manufacturing chemicals [166]. Much of this early work on HAS is detailed in the review by Natta et al. [167].

The conversion of syngas to ethanol via direct synthesis and methanol homologation pathways has been performed using a wide range of homogeneous and heterogeneous catalysts. For the direct synthesis of ethanol from syngas, two types of catalyst currently hold promise: rhodium-based and copper-based catalysts. For rhodium-based catalysts, there are reports in the literature of selectivities as high as 50% at higher pressure [168]. However, most often, this high selectivity is obtained at the expense of conversion; that is, high selectivities are seen only at very low conversions. Depending on the type of catalyst used, both the direct synthesis and indirect synthesis via methanol homologation are accompanied by a host of side reactions leading to methane, C_2–C_5 alkanes and olefins, ketones, aldehydes, esters, and acetic acid. Methanation can be particularly significant via hydrogenation of CO. To increase ethanol selectivity, the catalyst and the reaction conditions need to be better designed to suppress methanation activity.

A review of the literature on the conversion of syngas into ethanol and higher alcohols [169] indicates that higher selectivity may be achieved with homogeneous catalysts, but commercial processes based on these catalysts require extremely high operating pressures. Rhodium-based heterogeneous catalysts preferentially produce ethanol over other alcohols. However, the high cost of rhodium and low ethanol yield make such catalysts less attractive for commercial application, especially if high metal loadings are required.

Modified methanol synthesis catalysts based on CuZn, CuCo, and Mo have been developed and demonstrated in pilot plant testing. Alcohol production rates are significantly less compared with methanol synthesis, requiring significant improvements of at least two-to three-fold in alcohol production rate for commercial viability.

Reactor designs employed in the HAS catalyst R&D have typically adapted standard fixed-bed reactor technology with specialized cooling designs used for methanol synthesis or FTS of hydrocarbons. Improved product yield and selectivity could be achieved by performing the reactions in slurry reactors with efficient heat removal and temperature control.

A review of more than 220 recent publications and patents on syngas-to-ethanol conversion can be found in the literature [169]. The review looked at various routes and chemistries of converting syngas to ethanol. Thermodynamic calculations were also presented to understand the limits on various reactions as a function of process parameters. Past research efforts in developing catalysts and reactor designs were extensively discussed to finally summarize the R&D needs in commercializing syngas conversion to ethanol.

Increasing interest in second-generation lignocellulosic ethanol processes spurred on by the U.S. Energy Independence and Security Act of 2007 led to renewed efforts to demonstrate thermochemical ethanol production [170]. A multi-year effort culminated in a pilot demonstration of an integrated woody biomass gasification, gas cleanup and conditioning, and mixed alcohol synthesis process [171]. Pilot-scale results provided the basis for an updated techno-economic analysis of the process that demonstrated cost-competitive biofuel production from a scalable and sustainable process [172, 173].

5.3.2.7.1 Chemistry

The mechanism for HAS involves a complex set of numerous reactions with multiple pathways leading to a variety of products that are impacted by kinetic and thermodynamic constraints. No kinetic analysis of HAS has been published that is capable of globally predicting product compositions over ranges of operating conditions [174]. Depending on the process conditions and catalysts used, the most abundant products are typically methanol and CO_2. The first step in HAS is the formation of a C—C bond by CO insertion into CH_3OH. Linear alcohols are produced in a stepwise fashion involving the synthesis of methanol followed by its successive homologation to ethanol, propanol, butanol, etc. [175]. Therefore, the HAS catalyst should have methanol synthesis activity because methanol can be considered a recurrent C_1 reactant. Branched higher alcohols are typically formed from modified methanol synthesis and modified FTS catalysts, and straight-chain alcohols are formed when alkalized MoS_2 catalysts are used.

The mechanism for HAS over modified high-temperature methanol synthesis catalysts has been described as a unique carbon-chain growth mechanism that is referred to as oxygen retention reversal (ORR) aldol condensation with β-carbon (adjacent to the alcohol oxygen) addition [164]. Individual reactions in HAS can be grouped into several distinct reaction types [176]:

- linear chain growth by C_1 addition at the end of the chain to yield primary linear alcohols;
- beta addition between the C_1 and C_n ($n \geq 2$) to yield, for example, 1-propanol and branched primary alcohols such as 2-methyl-1-propanol (isobutanol) for $n = 2$;
- beta addition between C_m ($m = 2$ or 3) and C_n ($n \geq 2$);
- methyl ester formation via carboxylic acids formed from synthesized alcohols; and
- carbonylation of methanol to yield methyl formate.

Linear alcohols can proceed along the reaction path, but branched alcohols are terminal products of the aldol condensation pathways because they lack the two α-hydrogen atoms required for chain growth [177].

The general HAS reaction mechanism has the following overall stoichiometry [178, 179]:

$$nCO + 2nH_2 \rightarrow C_nH_{2n+1}OH + (n-1)H_2O \quad \Delta H_r = -61.2 \text{ kcal/mol}$$

with n typically ranging from 1 to 8 [166]. The reaction stoichiometry suggests that the optimum H_2 to CO ratio is 2 : 1; however, the simultaneous occurrence of the WGS reaction means that the optimum ratio is closer to 1 : 1. The major reactions in HAS are methanol synthesis, Fischer–Tropsch reactions, HAS reactions, and the WGS reaction [180]. The following is a list of some of these more important reactions described above that are associated with HAS:

$CO + 2H_2 \rightleftharpoons CH_3OH$	methanol synthesis
$CO + H_2O \rightleftharpoons CO_2 + H_2$	WGS
$CH_3OH + CO \rightleftharpoons CH_3CHO + H_2OCO$	beta addition – aldehydes
$CH_3OH + CO + 2H_2 \rightleftharpoons CH_3CH_2OH + H_2O$	ethanol homologation
$C_nH_{2n-1}OH + CO + 2H_2 \rightleftharpoons CH_3(CH_2)_nOH + H_2O$	HAS homologation
$2CH_3OH \rightleftharpoons CH_3CH_2OH + H_2O$	condensation/dehydration
$2CH_3OH \rightleftharpoons (CH_3)_2CO + H_2O$	DME formation
$(CH_3)_2CO + H_2 \rightleftharpoons (CH_3)_2CHOH$	branched iso-alcohols
$2CH_3CHO \rightleftharpoons CH_3COOCH_2CH_3$	methyl ester synthesis

Competing reactions:

$nCO + 2nH_2 \rightleftharpoons C_nH_{2n} + nH_2O$	olefins
$nCO + (2n+1)H_2 \rightleftharpoons C_nH_{2n+2} + nH_2O$	paraffins

Methanol formation is favored at low temperatures and high pressures [181]. At high pressures, HAS increases as the temperature is increased at the expense of methanol formation and minimizing hydrocarbon formation. To maximize higher alcohols, H_2 to CO ratio should be close to the usage ratio, which is about 1. Lower H_2 to CO ratios favor CO insertion and C–C chain growth. In general, the reaction conditions for HAS are more severe than those for methanol production. To increase yield of higher alcohols, methanol can be recycled for subsequent homologation, provided the catalyst shows good hydrocarbonylation activity [175, 181]. Unavoidably, the main reactions stated above produce H_2O and CO_2 as by-products. WGS plays a major role, and, depending on the catalyst's shift activity, some chemical dehydration of alcohols can be undertaken in situ to produce higher alcohols, esters, and ethers [181]. Secondary reactions also produce hydrocarbons, including aldehydes and ketones [181, 182]. Also, frequently, substantial quantities of methane are formed [183]. Thermodynamic constraints limit the theoretical yield of HAS and, as in other syngas-to-liquids processes, one of the most important limitations to HAS is removing the considerable heat of reaction to maintain control of process temperatures [182]. Compared with methanol, less alcohol product is made per mole of CO, more by-product is made per mole of alcohol product, and the heat release is greater.

5.3.2.7.2 Catalysts

HAS catalysts are essentially bifunctional base–hydrogenation catalysts and are typically categorized into several groups based on their composition. Common to all HAS catalysts is the addition of alkali metals to the formulation. The activating character of alkali metal promoters is a function of their basicity. Alkali metals provide a basic site to catalyze the aldol condensation reaction by activating surface-adsorbed CO and enhancing the formation of the formate intermediate. Information pertaining to the four primary groups of catalysts used for HAS is summarized in Tables 5.6 and 5.7. The catalyst groups include [165]:

- modified high-pressure methanol synthesis catalysts – alkali-doped ZnO/Cr_2O_3;
- modified low-pressure methanol synthesis catalysts – alkali-doped Cu/ZnO and $Cu/ZnO/Al_2O_3$;
- modified Fischer–Tropsch catalysts – alkali-doped $CuO/CoO/Al_2O_3$; and
- alkali-doped sulfides, mainly MoS_2.

Table 5.6 summarizes the typical range of process conditions for each catalyst and a measure of catalyst performance in terms of CO conversion and product selectivity. Table 5.7 details the compositions of typical catalyst materials in each group and highlights key research findings reported in the literature.

One of the major hurdles to overcome before HAS becomes an economically feasible commercial process is the development of improved catalysts that increase productivity and selectivity to higher alcohols [204]. To date, modified methanol and modified Fischer–Tropsch catalysts have been more effective in the production of mixed alcohols; sulfide-based catalysts tend to be less active than the oxide-based catalysts [164]. Rhodium-based catalysts are another group of catalysts that are not specifically used for HAS but have been developed for selective ethanol synthesis. Other C_2 oxygenates (i.e. acetaldehyde and acetic acid), as well as increased levels of methane production, are also synthesized over rhodium-based catalysts [205]. The high cost and limited availability of rhodium for ethanol synthesis catalysts will impact any commercialization of these synthetic processes for converting syngas to ethanol [180].

5.4 Summary and Conclusions

Increasing global energy demand within the constraint of reducing the impact of CO_2 emissions from fossil fuel consumption on global climate change keeps the focus on renewable energy technologies for the future. Sustainable biofuels from first-generation technologies and developing advanced conversion technologies offer an option for reducing the carbon intensity of transportation. Integrated biomass gasification and catalytic fuel synthesis is a thermochemical conversion process that can utilize a wide variety of biomass feedstocks to produce syngas. Once purified, that syngas can be converted to transportation fuels through

Table 5.6 Process and performance summary for HAS catalysts.

Catalyst	Operating conditions		Products	Conversion	Selectivity	Company
	Temperature (°C)	Pressure (MPa)				
Modified high-temperature/ high-pressure methanol synthesis	300–425	12.5–30	Branched primary alcohols	5–20% CO		
Modified low-temperature/ low-pressure methanol synthesis	275–310	5–10	Primary alcohols	21–29% CO	29–45% for $\geq C_2$ 17–25% CO_2 [180]	Lurgi/Süd Chemie
Modified Fischer–Tropsch	260–340	6–20	Linear alcohols	5–30% CO and CO_2	30–50% for higher alcohols	Institut Francais du Petrol
Alkali-doped sulfides	260–350	3–17.5	Linear alcohols	~10% CO [144]	75–90% for higher alcohols [144]	Dow Chemical, Union Carbide

Table 5.7 HAS catalyst composition and key research findings.

Catalyst	Composition	Comments
Modified high-temperature/high-pressure methanol synthesis catalyst	Alkalized Cu–Zn–Cr oxides	(1) 15–21 wt% Cr optimized HAS yields for non-alkalized Cu–Zn–Cr oxides (2) Chromia acts as a structural promoter increasing surface area and inhibiting Cu sintering [184] (3) Threefold decrease in ~C_2 alcohol production observed with 6% CO_2 at 400 °C [185] (4) 0.3–0.5 mol% Cs addition maximized HAS product yields [176] (5) Cs addition increases ethanol synthesis rate [176] (6) Systematic description of properties and effectiveness of Zn/Cr HAS catalysts by Hoflund and his group [186–192]
Modified low-temperature/low-pressure methanol synthesis catalyst	25–40 wt% CuO 10–18 wt% Al_2O_3 30–45 wt% ZnO 1.7–2.5 wt% K_2O 0.4–1.9 Cu/Zn ratio 3–18 wt% promoter (Cr, Ce, La, Mn or Th) [152]	(1) Methanol is most abundant product (~80%) (2) Average carbon number of oxygenated products is lower than products from modified high-temperature methanol catalysts [166] (3) Additional literature on catalyst effectiveness [176, 193–195]
Modified Fischer–Tropsch catalyst	10–50% Cu 5–25% Co 5–30% Al 10–70% Zn (on elemental basis) 0–0.2 alkali/Zn ratio 0.4–2.0 Zn/Al ratio 0.2–0.75 Co/Al ratio 1–3.0 Cu/Al ratio [152]	(1) CO_2 is reactant [180] (2) Anderson–Schultz–Flory (ASF) distribution for chain growth observed in alcohol and hydrocarbon products (3) Good catalyst activity correlates with catalyst homogeneity (4) Cu and Co are active components which have been modified with Zn and alkali (5) Low activity and lack of long-term stability hinders commercial application (6) Little deactivation observed during an 8000 h pilot plant test. Observed deactivation was caused by coke formation and sintering that decreased homogeneity of catalyst [180] (7) Additional available catalyst literature [181, 182, 196–198]

(continued)

Table 5.7 Continued

Catalyst	Composition	Comments
Alkali-doped sulfides	Alkali-doped MoS_2 or $CoMoS_2$	(1) Alkali additions suppress hydrogenation activity of Mo and provide additional sites for alcohol synthesis (2) Cs is most effective alkali promoter [164] (3) Higher alcohols and hydrocarbon products have ASF molecular weight distribution (4) >30% CO_2 retards catalyst activity, whereas moderate to low concentrations do not significantly impact catalyst activity [144] (5) Higher alcohol selectivity is reduced even in the presence of low CO_2 concentrations [144] (6) Co promotes the homologation of methanol to ethanol resulting in higher production of ethanol and other higher alcohols [166] (7) Activity of sulfide catalysts depends on catalyst support materials [199–203] (8) 50–100 ppm sulfur in the feed gas is required to maintain sulfidity of catalyst (9) H_2S in the feed gas moderates hydrogenation and improves selectivity for higher alcohols by reducing methanol production

a number of catalytic fuel synthesis processes. Removing contaminants from syngas efficiently and cost effectively, however, remains the most challenging technical hurdles to the commercial deployment of biomass gasification technologies.

This chapter provides a summary of the R&D efforts focused on syngas cleanup, conditioning, and conversion that will enable the effective use of syngas generated from biomass gasification. Throughout the discussion, the primary objective has been to describe the technical challenges that need to be overcome to enable the commercial deployment of biomass gasification technologies for power, fuels, and chemicals production.

Since the publication of the first edition of this book, one could argue that few significant technical breakthroughs have occurred in syngas cleanup and conditioning, but technologies are being scaled up and demonstrated at the pilot and demonstration scale. These studies have provided data for updating techno-economic analyses for selected technology options but the technical challenges remain and the economic competitiveness of integrated biomass gasification is still a significant challenge [172, 173, 206–210].

The high capital cost of integrated biomass gasification technologies presents a substantial financial risk for commercial deployment. Syngas cleanup, conditioning, and conversion represent a large fraction ($\sim$50%) of the total capital costs for a biomass gasification to liquid fuels process. Consequently, implicit in efforts to overcome technical barriers is the need reduce capital costs by developing processes with higher conversion efficiencies at less severe conditions (temperature and pressure) to improve product yields.

Process intensification strategies in biomass gasification, syngas cleanup and conditioning, and syngas conversion are also needed to reduce capital costs of large-scale plants and enable the development of small-scale to medium-scale technology options that can be deployed more widely with lower financial risk. One option is to reduce the number of unit operations by combining several syngas cleanup steps. Catalytic barrier filters have been tested for simultaneous particulate removal and tar cracking/reforming [211–213]. In another novel biomass gasification concept, candle filters are installed in the gasifier vessel itself to mitigate particulate matter and tars [214]. Other options look to prevent the formation of syngas impurities by adding catalytic materials into the fluidized bed gasifier for in situ tar cracking and sulfur and CO_2 capture [24, 215].

The lower energy content and bulk density of biomass feedstocks – compared to coal, for example – limits the amount of biomass that can be economically collected, prepared, and transported to a large centralized location. This limits the capital cost savings that can be gained from economies of scale. Nevertheless, as the scale of biomass conversion facilities increases, biomass must be transported over longer distances to meet the feedstock needs of the plant and the infrastructure for transporting large quantities of biomass over long distances is not widely available. Consequently, feedstock costs are a major contribution to the operating cost of a biomass conversion facility. In fact, techno-economic assessments of thermochemical biomass conversion technologies estimate that feedstock costs represent nearly 50% of the cost of production of biofuels in a biomass-to-liquids process [216].

Clearly, the entire value chain – from biomass production, collection, and delivery; through biomass conversion; through power, fuels, or chemicals production; through to product end use – needs to be integrated and optimized to commercialize biomass thermochemical conversion technologies successfully. The extensive research and development efforts devoted to syngas cleanup, conditioning, and conversion to fuels and chemicals are documented in a vast amount of literature that tracks the scientific and technological advancements in syngas chemistry. In many cases, multiple, integrated approaches are being actively pursued to find the most technically robust and economically feasible solutions. This wealth of information and technical experience can be leveraged to accelerate commercial deployment of biomass thermochemical conversion technologies, but the ultimate test will be market acceptance on a cost competitive basis.

References

1. Tabatabaieraissi, A. and Trezek, G.J. (1987). Parameters governing biomass gasification. *Industrial & Engineering Chemistry Research* **26** (2): 221–228.
2. Beenackers, A.A.C.M. and Van Swaaij, W.P.M. (1984). Gasification of biomass, a state of the art review. In: *Thermochemical Processing of Biomass* (ed. A.V. Bridgwater), 91–136. London: Butterworths.
3. Hos, J.J. and Groeneveld, M.J. (1987). Biomass gasification. In: *Biomass* (ed. D.O. Hall and R.P. Overend), 237–255. Chichester: Wiley.
4. Caballero, M.A., Corella, J., Aznar, M.P., and Gil, J. (2000). Biomass gasification with air in fluidized bed. Hot gas cleanup with selected commercial and full-size nickel-based catalysts. *Industrial & Engineering Chemistry Research* **39** (5): 1143–1154.
5. Cummer, K.R. and Brown, R.C. (2002). Ancillary equipment for biomass gasification. *Biomass & Bioenergy* **23** (2): 113–128.
6. Devi, L., Ptasinski, K.J., and Janssen, F. (2003). A review of the primary measures for tar elimination in biomass gasification processes. *Biomass & Bioenergy* **24** (2): 125–140.
7. Simell, P.A., Hepola, J.O., and Krause, A.O.I. (1997). Effects of gasification gas components on tar and ammonia decomposition over hot gas cleanup catalysts. *Fuel* **76** (12): 1117–1127.
8. Gangwal, S.K., Harkins, S.M., Woods, M.C. et al. (1989). Bench-scale testing of high-temperature desulfurization sorbents. *Environmental Progress* **8** (4): 265–269.
9. Woods, M.C., Gangwal, S.K., Jothimurugesan, K., and Harrison, D.P. (1990). Reaction between H_2S and zinc-oxide titanium-oxide Sorbents.1. Single-pellet kinetic-studies. *Industrial & Engineering Chemistry Research* **29** (7): 1160–1167.
10. Gasper-Galvin, L.D., Atimtay, A.T., and Gupta, R.P. (1998). Zeolite-supported metal oxide sorbents for hot-gas desulfurization. *Industrial & Engineering Chemistry Research* **37** (10): 4157–4166.
11. Gupta, R.P., Turk, B.S., Portzer, J.W., and Cicero, D.C. (2001). Desulfurization of syngas in a transport reactor. *Environmental Progress* **20** (3): 187–195.
12. Baumhakl, C. and Karellas, S. (2011). Tar analysis from biomass gasification by means of online fluorescence spectroscopy. *Optics and Lasers in Engineering* **49** (7): 885–891.
13. Defoort, F., Thiery, S., and Ravel, S. (2014). A promising new on-line method of tar quantification by mass spectrometry during steam gasification of biomass. *Biomass & Bioenergy* **65**: 64–71.
14. Gredinger, A., Schweitzer, D., Dieter, H., and Scheffknecht, G. 2014. Online Tar Monitoring Via Fid – Laboratory And Pilot Plant Experiments Of An Advanced Online Tar Analyzer Prototype. Papers of the 22nd European Biomass Conference: Setting the Course for a Biobased Economy, pp. 569–572.
15. Simell, P., Stahlberg, P., Kurkela, E. et al. (2000). Provisional protocol for the sampling and analysis of tar and particulates in the gas from large-scale biomass gasifiers. Version 1998. *Biomass & Bioenergy* **18** (1): 19–38.
16. Knoef, H.A.M. and Koele, H.J. (2000). Survey of tar measurement protocols. *Biomass & Bioenergy* **18** (1): 55–59.
17. Maniatis, K. and Beenackers, A. (2000). Tar protocols. IEA bioenergy gasification task. *Biomass & Bioenergy* **18** (1): 1–4.
18. Xu, M., Brown, R.C., and Norton, G. (2006). Effect of sample aging on the accuracy of the international energy agency's tar measurement protocol. *Energy & Fuels* **20** (1): 262–264.

19. Xu, M., Brown, R.C., Norton, G., and Smeenk, J. (2005). Comparison of a solvent-free tar quantification method to the international energy agency's tar measurement protocol. *Energy & Fuels* **19** (6): 2509–2513.
20. Horvat, A., Kwapinska, M., Xue, G. et al. (2016). Detailed measurement uncertainty analysis of solid-phase adsorption-total gas chromatography (GC)-detectable tar from biomass gasification. *Energy & Fuels* **30** (3): 2187–2197.
21. Neubert, M., Reil, S., Wolff, M. et al. (2017). Experimental comparison of solid phase adsorption (SPA), activated carbon test tubes and tar protocol (DIN CEN/TS 15439) for tar analysis of biomass derived syngas. *Biomass & Bioenergy* **105**: 443–452.
22. Abdoulmoumine, N., Adhikari, S., Kulkarni, A., and Chattanathan, S. (2015). A review on biomass gasification syngas cleanup. *Applied Energy* **155**: 294–307.
23. Mondal, P., Dang, G.S., and Garg, M.O. (2011). Syngas production through gasification and cleanup for downstream applications – recent developments. *Fuel Processing Technology* **92** (8): 1395–1410.
24. Richardson, Y., Blin, J., and Julbe, A. (2012). A short overview on purification and conditioning of syngas produced by biomass gasification: catalytic strategies, process intensification and new concepts. *Progress in Energy and Combustion Science* **38** (6): 765–781.
25. Siedlecki, M., de Jong, W., and Verkooijen, A.H.M. (2011). Fluidized bed gasification as a mature and reliable technology for the production of bio-syngas and applied in the production of liquid transportation fuels-a review. *Energies* **4** (3): 389–434.
26. Sikarwar, V.S., Zhao, M., Fennell, P.S. et al. (2017). Progress in biofuel production from gasification. *Progress in Energy and Combustion Science* **61**: 189–248.
27. Woolcock, P.J. and Brown, R.C. (2013). A review of cleaning technologies for biomass-derived syngas. *Biomass & Bioenergy* **52**: 54–84.
28. Xu, C.B., Donald, J., Byambajav, E., and Ohtsuka, Y. (2010). Recent advances in catalysts for hot-gas removal of tar and NH_3 from biomass gasification. *Fuel* **89** (8): 1784–1795.
29. Krishnan, G., Wood, B.J., Tong, G.T., and Kothari, V.P. 1988. Removal of hydrogen chloride vapor from high-temperature coal gases. *Abstracts of Papers of the American Chemical Society* 195: 32-FUEL.
30. Friedlander, A.G., Courty, R., and Montarnal, R.E. (1977). Ammonia decomposition in presence of water-vapor:1. Nickel, ruthenium and palladium catalysts. *Journal of Catalysis* **48** (1–3): 312–321.
31. Friedlander, A.G., Courty, R., and Montarnal, R.E. (1977). Ammonia decomposition in presence of water-vapor:2. Kinetics of reaction on nickel-catalyst. *Journal of Catalysis* **48** (1–3): 322–332.
32. Bera, P. and Hedge, M.S. (2002). Oxidation and decomposition of NH_3 over combustion synthesized Al_2O_3 and CeO_2 supported Pt, Pd and Ag catalysts. *Indian Journal of Chemistry Section a-Inorganic Bio-Inorganic Physical Theoretical & Analytical Chemistry* **41** (8): 1554–1561.
33. Cholach, A.R., Sobyanin, V.A., and Gorodetskii, V.V. (1981). Decomposition of ammonia on rhenium:3. Interaction of ammonia with rhenium. *Reaction Kinetics and Catalysis Letters* **18** (3–4): 391–396.
34. Grosman, M. and Loffler, D.G. (1983). Kinetics of ammonia decomposition on polycrystalline tungsten. *Journal of Catalysis* **80** (1): 188–193.
35. Papapolymerou, G. and Bontozoglou, V. (1997). Decomposition of NH_3 on Pd and Ir – comparison with Pt and Rh. *Journal of Molecular Catalysis a-Chemical* **120** (1–3): 165–171.
36. Sano, K., Sugishima, N., Ikeda, M. et al. (1999). The new technology for selective catalytic oxidation of ammonia to nitrogen. In: *Science and Technology in Catalysis 1998* (ed. H. Hattori and K. Otsuka), 399–402. Amsterdam: Elsevier Science Publ B V.
37. Sheu, S.P., Karge, H.G., and Schlogl, R. (1997). Characterization of activated states of ruthenium-containing zeolite NaHY. *Journal of Catalysis* **168** (2): 278–291.
38. Sobyanin, V.A., Gorodetskii, V.V., and Cholach, A.R. (1982). Decomposition of ammonia on tungsten at low-pressure. *Kinetics and Catalysis* **23** (1): 89–93.

39. Choi, J.G. (1999). Ammonia decomposition over vanadium carbide catalysts. *Journal of Catalysis* **182** (1): 104–116.
40. Choi, J.G., Jung, M.K., Choi, S. et al. (1997). Synthesis and catalytic properties of vanadium nitrides. *Bulletin of the Chemical Society of Japan* **70** (5): 993–996.
41. Vladov, D., Dyakovit, V., and Dinkov, S. (1966). Catalytic decomposition of ammonia – electric field effect on catalytic activity. *Journal of Catalysis* **5** (3): 412–418.
42. Chambers, A., Yoshii, Y., Inada, T., and Miyamoto, T. (1996). Ammonia decomposition in coal gasification a atmospheres. *Canadian Journal of Chemical Engineering* **74** (6): 929–934.
43. Kagami, S., Onishi, T., and Tamaru, K. (1984). Fourier-transform infrared spectroscopic study of adsorption and decomposition of ammonia over magnesium-oxide. *Journal of the Chemical Society-Faraday Transactions I* **80**: 29–35.
44. Ohtsuka, Y., Xu, C.B., Kong, D.P., and Tsubouchi, N. (2004). Decomposition of Ammonia with Iron and Calcium Catalysts Supported on Coal Chars. *Fuel*
45. Simell, P., Kurkela, E., Stahlberg, P., and Hepola, J. (1996). Catalytic hot gas cleaning of gasification gas. *Catalysis Today* **27** (1–2): 55–62.
46. Horvat, M., Jeran, Z., Spiric, Z. et al. (2000). Mercury and other elements in lichens near the INA Naftaplin gas treatment plant, Molve, Croatia. *Journal of Environmental Monitoring* **2** (2): 139–144.
47. Lau, F.S., Slimane, R.B., Roberts, M.J., et al. 2004. Flex-Fuel Testing of the Ultra-Clean Process for the Control of Sulfur, Halide, and Mercury Compounds in Coal Gasification Gases, in 21st Annual Pittsburgh Coal Conference. Osaka, Japan.
48. Krishnan, G.N., Gupta, R.P., Canizales, A. et al. (1996). Removal of hydrogen chloride from hot coal gas streams. In: *High Temperature Gas Cleaning* (ed. E. Schmidt, P. Gang, T. Pilz and A. Dittler), 405–414. Karlsruhe: Institut fur Mechanische Verfahrenstechnik und Mechanik.
49. Milne, T.A., Abatzoglou, N., and Evans, R.J. (1998). *Biomass Gasifier "Tars": Their Nature, Formation, and Conversion*, 202. Golden, CO: National Renewable Energy Laboratory.
50. Rabou, L., Zwart, R.W.R., Vreugdenhil, B.J., and Bos, L. (2009). Tar in biomass producer gas, the Energy research Centre of The Netherlands (ECN) experience: an enduring challenge. *Energy & Fuels* **23**: 6189–6198.
51. Zwart, R.W.R., Van der Drift, A., Bos, A. et al. (2009). Oil-based gas washing-flexible tar removal for high-efficient production of clean heat and power as well as sustainable fuels and chemicals. *Environmental Progress & Sustainable Energy* **28** (3): 324–335.
52. Jentoft, F.C. and Gates, B.C. (1997). Solid-acid-catalyzed alkane cracking mechanisms: evidence from reactions of small probe molecules. *Topics in Catalysis* **4** (1–2): 1–13.
53. Gil, J., Caballero, M.A., Martin, J.A. et al. (1999). Biomass gasification with air in a fluidized bed: effect of the in-bed use of dolomite under different operation conditions. *Industrial & Engineering Chemistry Research* **38** (11): 4226–4235.
54. Jacquin, M., Jones, D.J., Roziere, J. et al. (2003). Novel supported Rh, Pt, Ir and Ru mesoporous alumino silicates as catalysts for the hydrogenation of naphthalene. *Applied Catalysis a-General* **251** (1): 131–141.
55. Mandreoli, M., Vaccari, A., Veggetti, E. et al. (2002). Vapour phase hydrogenation of naphthalene on a novel Ni-containing mesoporous aluminosilicate catalyst. *Applied Catalysis a-General* **231** (1–2): 263–268.
56. Abu El-Rub, Z., Bramer, E.A., and Brem, G. (2004). Review of catalysts for tar elimination in Biomass gasification processes. *Industrial & Engineering Chemistry Research* **43** (22): 6911–6919.
57. Gerhard, S.C., Wang, D.N., Overend, R.P. et al. (1994). Catalytic conditioning of synthesis gas produced by biomass gasification. *Biomass & Bioenergy* **7** (1–6): 307–313.
58. Kinoshita, C.M., Wang, Y., and Zhou, J. (1994). Tar formation under different biomass gasification conditions. *Journal Of Analytical And Applied Pyrolysis* **29** (2): 169–181.
59. Marsak, J. and Skoblja, S. (2002). Role of catalysts in tar removal from biomass gasification. *Chemicke Listy* **96** (10): 813–820.

60. Sutton, D., Kelleher, B., and Ross, J.R.H. (2001). Review of literature on catalysts for biomass gasification. *Fuel Processing Technology* **73** (3): 155–173.
61. Yung, M.A., Jablonski, W.S., and Magrini-Bair, K.A. (2009). Review of catalytic conditioning of biomass-derived syngas. *Energy & Fuels* **23**: 1874–1887.
62. Zhang, R.Q., Brown, R.C., Suby, A., and Cummer, K. (2004). Catalytic destruction of tar in biomass derived producer gas. *Energy Conversion and Management* **45** (7–8): 995–1014.
63. Baker, E.G. and Mudge, L.K. (1987). Catalytic tar conversion in coal-gasification systems. *Industrial & Engineering Chemistry Research* **26** (7): 1390–1395.
64. Baker, E.G., Mudge, L.K., and Brown, M.D. (1987). Steam gasification of biomass with nickel secondary catalysts. *Industrial & Engineering Chemistry Research* **26** (7): 1335–1339.
65. Mudge, L.K., Baker, E.G., Mitchell, D.H. et al. (1985). Catalytic steam gasification of biomass for methanol and methane production. *Journal of Solar Energy Engineering-Transactions of the Asme* **107** (1): 88–92.
66. Mudge, L.K., Sealock, L.J., and Weber, S.L. (1979). Catalyzed steam gasification of biomass. *Journal Of Analytical And Applied Pyrolysis* **1** (2): 165–175.
67. Corella, J., Aznar, M.P., Gil, J., and Caballero, M.A. (1999). Biomass gasification in fluidized bed: where to locate the dolomite to improve gasification? *Energy & Fuels* **13** (6): 1122–1127.
68. Corella, J., Toledo, J.M., and Padilla, R. (2004). Olivine or dolomite as in-bed additive in biomass gasification with air in a fluidized bed: which is better? *Energy & Fuels* **18** (3): 713–720.
69. Delgado, J., Aznar, M.P., and Corella, J. (1997). Biomass gasification with steam in fluidized bed: effectiveness of CaO, MgO, and CaO-MgO for hot raw gas cleaning. *Industrial & Engineering Chemistry Research* **36** (5): 1535–1543.
70. Devi, L., Ptasinski, K.J., Janssen, F. et al. (2005). Catalytic decomposition of biomass tars: use of dolomite and untreated olivine. *Renewable Energy* **30** (4): 565–587.
71. Orio, A., Corella, J., and Narvaez, I. (1997). Performance of different dolomites on hot raw gas cleaning from biomass gasification with air. *Industrial & Engineering Chemistry Research* **36** (9): 3800–3808.
72. Simell, P.A., Leppalahti, J.K., and Kurkela, E.A. (1995). Tar-decomposing activity of carbonate rocks under high CO_2 partial-pressure. *Fuel* **74** (6): 938–945.
73. Gusta, E., Dalai, A.K., Uddin, A., and Sasaoka, E. (2009). Catalytic decomposition of biomass tars with Dolomites. *Energy & Fuels* **23**: 2264–2272.
74. Devi, L., Ptasinski, K.J., and Janssen, F. (2005). Pretreated olivine as tar removal catalyst for biomass gasifiers: investigation using naphthalene as model biomass tar. *Fuel Processing Technology* **86** (6): 707–730.
75. Aznar, M.P., Caballero, M.A., Gil, J. et al. (1998). Commercial steam reforming catalysts to improve biomass gasification with steam-oxygen mixtures:2. Catalytic tar removal. *Industrial & Engineering Chemistry Research* **37** (7): 2668–2680.
76. Caballero, M.A., Aznar, M.P., Gil, J. et al. (1997). Commercial steam reforming catalysts to improve biomass gasification with steam-oxygen mixtures. 1. Hot gas upgrading by the catalytic reactor. *Industrial & Engineering Chemistry Research* **36** (12): 5227–5239.
77. Courson, C., Makaga, E., Petit, C. et al. (2000). Development of Ni catalysts for gas production from biomass gasification. Reactivity in steam- and dry-reforming. *Catalysis Today* **63** (2–4): 427–437.
78. Bain, R.L., Dayton, D.C., Carpenter, D.L. et al. (2005). Evaluation of catalyst deactivation during catalytic steam reforming of biomass-derived syngas. *Industrial & Engineering Chemistry Research* **44** (21): 7945–7956.
79. Bangala, D.N., Abatzoglou, N., and Chornet, E. (1998). Steam reforming of naphthalene on Ni-Cr/Al_2O_3 catalysts doped with MgO, TiO_2, and La_2O_3. *AICHE Journal* **44** (4): 927–936.
80. Magrini-Bair, K.A., Czernik, S., French, R. et al. (2007). Fluidizable reforming catalyst development for conditioning biomass-derived syngas. *Applied Catalysis A: General* **318**: 199–206.

81. Magrini-Bair, K.A., Jablonski, W.S., Parent, Y.O. et al. (2012). Bench- and pilot-scale studies of reaction and regeneration of Ni-mg-K/Al_2O_3 for catalytic conditioning of biomass-derived syngas. *Topics in Catalysis* **55** (3–4): 209–217.
82. Corella, J., Toledo, J.M., and Padilla, R. (2004). Catalytic hot gas cleaning with monoliths in biomass gasification in fluidized beds. 2. Modeling of the monolithic reactor. *Industrial & Engineering Chemistry Research* **43** (26): 8207–8216.
83. Corella, J., Toledo, M., and Padilla, R. (2004). Catalytic hot gas cleaning with monoliths in biomass gasification in fluidized beds. 1. Their effectiveness for tar elimination. *Industrial & Engineering Chemistry Research* **43** (10): 2433–2445.
84. Gesser, H.D. and Hunter, N.R. (1998). A review of C-1 conversion chemistry. *Catalysis Today* **42** (3): 183–189.
85. Green, A.E.S. (1991). Overview of fuel conversion. In: *Solid Fuel Conversion for the Transportation Sector*, vol. **12**, 3–15. FACT (American Society of Mechanical Engineers).
86. Keim, W. 1984 C1 chemistry: present status and aspects for the future. *Chemistry Future*, Procceedings of the 29th IUPAC Congress, pp. 53–62.
87. Rostrup-Nielsen, J.R. (2002). Syngas in perspective. *Catalysis Today* **71** (3–4): 243–247.
88. Wender, I. (1996). Reactions of synthesis gas. *Fuel Processing Technology* **48** (3): 189–297.
89. Spath, P.L. and Dayton, D.C. (2003). *Preliminary Screening -- Technical and Economic Assessment of Synthesis Gas to Fuels and Chemicals with Emphasis on the Potential for Biomass-Derived Syngas*, 160. Golden, CO: National Renewable Energy Laboratory.
90. Häussinger, P., Lohmüller, R., and Watson, A.M. (2000). Hydrogen. In: *Ullmann's Encyclopedia of Industrial Chemistry*. https://doi.org/10.1002/14356007.a13_297.
91. Spivey, J.J. and Egbebi, A. (2007). Heterogeneous catalytic synthesis of ethanol from biomass-derived syngas. *Chemical Society Reviews* **36** (9): 1514–1528.
92. Dincer, I. and Acar, C. (2015). Review and evaluation of hydrogen production methods for better sustainability. *International Journal of Hydrogen Energy* **40** (34): 11094–11111.
93. Holladay, J.D., Hu, J., King, D.L., and Wang, Y. (2009). An overview of hydrogen production technologies. *Catalysis Today* **139** (4): 244–260.
94. Nikolaidis, P. and Poullikkas, A. (2017). A comparative overview of hydrogen production processes. *Renewable and Sustainable Energy Reviews* **67**: 597–611.
95. Sharma, S. and Ghoshal, S.K. (2015). Hydrogen the future transportation fuel: from production to applications. *Renewable and Sustainable Energy Reviews* **43**: 1151–1158.
96. Rothenberger, K.S., Howard, B.H., Killineyer, R.P. et al. (2003). Evaluation of tantalum-based materials for hydrogen separation at elevated temperatures and pressures. *Journal of Membrane Science* **218** (1–2): 19–37.
97. Kamakoti, P., Morreale, B.D., Ciocco, M.V. et al. (2005). Prediction of hydrogen flux through sulfur-tolerant binary alloy membranes. *Science* **307** (5709): 569–573.
98. Kulprathipanja, A., Alptekin, G.O., Falconer, J.L. et al. (2004). Effects of water gas shift gases on Pd-Cu alloy membrane surface morphology and separation properties. *Industrial & Engineering Chemistry Research* **43** (15): 4188–4198.
99. Roa, F. and Way, J.D. 2003. Alloy composition effect on the N-value for hydrogen-selective palladium-copper membranes. *Abstracts of Papers of the American Chemical Society* **225**: 155-FUEL.
100. Roa, F. and Way, J.D. (2003). Influence of alloy composition and membrane fabrication on the pressure dependence of the hydrogen flux of palladium-copper membranes. *Industrial & Engineering Chemistry Research* **42** (23): 5827–5835.
101. Roa, F., Way, J.D., McCormick, R.L. et al. (2003). Preparation and characterization of Pd-Cu composite membranes for hydrogen separation. *Chemical Engineering Journal* **93** (1): 11–22.

102. Tong, H.D., Gielens, F.C., Gardeniers, J.G.E. et al. (2005). Microsieve supporting palladium-silver alloy membrane and application to hydrogen separation. *Journal of Microelectromechanical Systems* **14** (1): 113–124.
103. Tong, J.H., Matsumura, Y., Suda, H. et al. (2005). Thin and dense Pd/CeO$_2$/MPSS composite membrane for hydrogen separation and steam reforming of methane. *Separation and Purification Technology* **46** (1–2): 1–10.
104. Vadovic, C.J. and Eakman, J.M. (1978). Kinetics of potassium catalyzed gasification. *Abstracts of Papers of the American Chemical Society* **176** (SEP): 9–9.
105. Gallagher, J.E. and Euker, C.A. (1980). Catalytic coal-gasification for Sng manufacture. *International Journal of Energy Research* **4** (2): 137–147.
106. Nahas, N.C. (2004). Catalytic methane synthesis can extend hydrocarbon supply. *Oil & Gas Journal* **102** (37): 18.
107. Zhang, A., Kaiho, M., Yasuda, H. et al. Fundamental studies on hydrogasification of Taiheiyo coal. *Energy* **30** (11–12): 2243–2250.
108. Lee, B.S. (1972). Development of HYGAS process for converting coal to synthetic pipeline gas. *Journal of Petroleum Technology* **24** (12): 1407–1410.
109. Vorres, K.S. (1977). Hygas process update. *Energy Communications* **3** (6): 613–624.
110. Linden, H.R., Bodle, W.W., Lee, B.S., and Vyas, K.C. (1976). Production of high-BTU gas from coal. *Annual Review of Energy* **1** (1): 65–86.
111. Annual Energy (2015). *Outlook 2015 with Projections to 2040* (ed. U.S.E.I. Administration). U.S. Energy Information Administration.
112. Luterbacher, J.S., Froling, M., Vogel, F. et al. (2009). Hydrothermal gasification of waste biomass: process design and life cycle assessment. *Environmental Science & Technology* **43** (5): 1578–1583.
113. Vogel, F., Waldner, M.H., Rouff, A.A., and Rabe, S. (2007). Synthetic natural gas from biomass by catalytic conversion in supercritical water. *Green Chemistry* **9** (6): 616–619.
114. Waldner, M.H., Krumeich, F., and Vogel, F. (2007). Synthetic natural gas by hydrothermal gasification of biomass selection procedure towards a stable catalyst and its sodium sulfate tolerance. *Journal of Supercritical Fluids* **43** (1): 91–105.
115. Waldner, M.H. and Vogel, F. (2005). Renewable production of methane from woody biomass by catalytic hydrothermal gasification. *Industrial & Engineering Chemistry Research* **44** (13): 4543–4551.
116. Zwart, R.W.R. and Boerrigter, H. (2005). High efficiency co-production of synthetic natural gas (SNG) and Fischer-Tropsch (FT) transportation fuels from biomass. *Energy & Fuels* **19** (2): 591–597.
117. Ronsch, S., Schneider, J., Matthischke, S. et al. (2016). Review on methanation – from fundamentals to current projects. *Fuel* **166**: 276–296.
118. Frohning, C., Kolbel, H., Ralek, M. et al. (1982). Fischer-Tropsch process. In: *Chemical Feedstocks from Coal* (ed. J. Falbe), 309–432. New York: John Wiley and Sons.
119. Mills, G.A. (1982). Catalytic concepts in coal conversion. *Chemtech* **12** (5): 294–303.
120. Dry, M.E. (2002). The Fischer-Tropsch process: 1950–2000. *Catalysis Today* **71** (3–4): 227–241.
121. Dry, M.E. and Hoogendoorn, J.C. (1981). Technology of the Fischer-Tropsch process. *Catalysis Reviews-Science and Engineering* **23** (2): 265–278.
122. Ail, S.S. and Dasappa, S. (2016). Biomass to liquid transportation fuel via Fischer Tropsch synthesis – technology review and current scenario. *Renewable & Sustainable Energy Reviews* **58**: 267–286.
123. Bartholomew, C.H. (1991). Recent technological developments in Fischer-Tropsch catalysis. *Catalysis Letters* **7** (1–4): 303–315.
124. Bartholomew, C.H. (1991). Recent developments in Fischer-Tropsch catalysis. In: *New Trends in CO Activation* (ed. L. Guczi), 158–224. Elsevier.
125. Dry, M.E. (1982). Catalytic aspects of industrial Fischer-Tropsch synthesis. *Journal of Molecular Catalysis* **17** (2–3): 133–144.
126. Senden, M.M.G. (1998). The Shell middle distillate synthesis process: commercial plant experience and outlook into the future. *Petrole et Techniques* **415**: 94–97.

127. Haid, M.O., Schubert, P.F., and Bayens, C.A. (2000). Synthetic fuel and lubricants production using gas-to-liquids technology. *DGMK Tagungsbericht* **2000** (3): 205–212.
128. Dry, M.E. (1988). The Sasol route to chemicals and fuels. *Studies in Surface Science and Catalysis* **36** (Methane Convers): 447–456.
129. Adesina, A.A. (1996). Hydrocarbon synthesis via Fischer-Tropsch reaction: travails and triumphs. *Applied Catalysis a-General* **138** (2): 345–367.
130. Cooper, C.G., Nguyen, T.H., Lee, Y.J. et al. (2008). Alumina-supported cobalt-molybdenum catalyst for slurry phase Fischer-Tropsch synthesis. *Catalysis Today* **131** (1–4): 255–261.
131. Larson, E.D., Jin, H.M., and Celik, F.E. (2009). Large-scale gasification-based coproduction of fuels and electricity from switchgrass. *Biofuels Bioproducts & Biorefining* **3** (2): 174–194.
132. Steynberg, A.P., Espinoza, R.L., Jager, B., and Vosloo, A.C. (1999). High temperature Fischer-Tropsch synthesis in commercial practice. *Applied Catalysis a-General* **186** (1–2): 41–54.
133. Rostrup-Nielsen, J., Dybkjaer, I., and Aasberg-Petersen, K. (2000). Synthesis gas for large scale Fischer-Tropsch synthesis. *Preprints – American Chemical Society, Division of Petroleum Chemistry* **45** (2): 186–189.
134. Vosloo, A.C. (2001). Fischer-Tropsch: a futuristic view. *Fuel Processing Technology* **71** (1–3): 149–155.
135. Wilhelm, D.J., Simbeck, D.R., Karp, A.D. et al. (2001). Syngas production for gas-to-liquids applications: technologies, issues and outlook. *Fuel Processing Technology* **71** (1–3): 139–148.
136. Davenport, B. (2002). *Methanol*. Menlo Park, CA: SRI International.
137. Supp, E. and Quinkler, R.F. (1984). The Lurgi low-pressure methanol process. In: *Handbook of Synfuels Technology* (ed. R.A. Meyers), 113–131. McGraw-Hill.
138. Rostrup-Nielsen, J.R. (2000). New aspects of syngas production and use. *Catalysis Today* **63** (2–4): 159–164.
139. Dybkjaer, I. and Christensen, T.S. (2001). Syngas for large scale conversion of natural gas to liquid fuels. *Studies in Surface Science and Catalysis* **136** (Natural Gas Conversion VI): 435–440.
140. Ladebeck, J. (1993). Improve methanol synthesis. *Hydrocarbon Processing, International Edition* **72** (3): 89–91.
141. Chinchen, G.C., Mansfield, K., and Spencer, M.S. (1990). The methanol synthesis – how does it work. *Chemtech* **20** (11): 692–699.
142. Fiedler, E., Grossmann, G., Kersebohm, D.B. et al. (2003). Methanol. In: *Ullmann's Encyclopedia of Industrial Chemistry Release 2003*, 6e. Wiley-VCH Verlag GmbH & Co.KGaA.
143. Klier, K. (1982). Methanol synthesis. *Advances in Catalysis* **31**: 243–313.
144. Herman, R.G. (1991). Classical and Non-Classical Route for Alcohol Synthesis, Chapter 7. In: *New Trends in CO Activation* (ed. L. Guczi), 281–285. Amsterdam: Elsevier.
145. Keil, F.J. (1999). Methanol-to-hydrocarbons: process technology. *Microporous and Mesoporous Materials* **29**: 49–66.
146. Hancock, E.G. (1985). *The Manufacture of Gasoline and the Chemistry of its Components*. Critical Reports on Applied Chemistry, vol. **10** (Technology of Gasoline), 20–56. Blackwell Scientific Publications.
147. Kam, A.Y., Schreiner, M., and Yurchak, S. (1984). Mobil methanol-to-gasoline (MTG) process, Chapter 2-3. In: *Handbook of Synfuels Technology* (ed. R.A. Meyers), 2-75–2-111. New York: McGraw-Hill Book Company.
148. MacDougall, L.V. (1991). Methanol to fuels routes – the achievements and remaining problems. *Catalysis Today* **8**: 337–369.
149. Topp-Jorgensen, J. (1987). The Topsoe integrated gasoline synthesis. *Petrole et Techniques* **333**: 11–17.

150. Topp-Jorgensen, J. (1988). *Topsoe Integrated Gasoline Synthesis – The TIGAS Process*, in Methane Conversion : Proceedings of a Symposium on the Production of Fuels and Chemicals from Natural Gas, Auckland, April 27–30, 1987 (ed. D.M. Bibby), 293–305. Amsterdam, New York: Elsevier.
151. Bellussi, G., Millini, R., and Pollesel, P. (2015). An industrial perspective on the impact of Haldor Topsoe on research and development in catalysis by zeolites. *Journal of Catalysis* **328**: 11–18.
152. Dybkjaer, I. and Aasberg-Petersen, K. (2016). Synthesis gas technology large-scale applications. *Canadian Journal of Chemical Engineering* **94** (4): 607–612.
153. Haldor Topsoe 2014. Groundbreaking in Turkmenistan to build major plant for producing synthetic gasoline. http://blog.topsoe.com/2014/08/groundbreaking-turkmenistan-build-major-plant-producing-synthetic-gasoline (accessed 8 November 2018).
154. Quann, R.J., Green, L.A., Tabak, S.A., and Krambeck, F.J. (1988). Chemistry of olefin oligomerization over Zsm-5 catalyst. *Industrial & Engineering Chemistry Research* **27** (4): 565–570.
155. Wurzel, T. (2007). Lurgi MegaMethananol technology – delivering the building blocks for the future fuel and monomer demand. *Oil Gas-European Magazine* **33** (2): 92–96.
156. Peng, X.D., Wang, A.W., Toseland, B.A., and Tijm, P.J.A. (1999). Single-step syngas-to-dimethyl ether processes for optimal productivity, minimal emissions, and natural gas-derived syngas. *Industrial & Engineering Chemistry Research* **38** (11): 4381–4388.
157. Shikada, T., Ohno, Y., Ogawa, T. et al. (1999). Synthesis of dimethyl ether from natural gas via synthesis gas. *Kinetics and Catalysis (Translation of Kinetika i Kataliz)* **40** (3): 395–400.
158. Gunda, A., Tartamella, T., Gogate, M., and Lee, S. (1995). Dimethyl ether synthesis from CO_2-rich syngas in the LPDME process. In: *Proceedings – Annual International Pittsburgh Coal Conference*, vol. **12**, 710–715.
159. Hansen, J.B. and Joensen, F. (1991). High conversion of synthesis gas into oxygenates. *Studies in Surface Science and Catalysis* **61** (Nat. Gas Convers.): 457–467.
160. Peng, X.D., Toseland, B.A., and Tijm, P.J.A. (1999). Kinetic understanding of the chemical synergy under LPDME conditions-once-through applications. *Chemical Engineering Science* **54** (13–14): 2787–2792.
161. Ge, Q., Huang, Y., Qiu, F., and Li, S. (1998). Bifunctional catalysts for conversion of synthesis gas to dimethyl ether. *Applied Catalysis, A: General* **167** (1): 23–30.
162. Sardesai, A. and Lee, S. (1998). Liquid phase dimethyl ether (DME) synthesis: a review. *Reviews in Process Chemistry and Engineering* **1** (2): 141–178.
163. Brown, D.M., Bhatt, B.L., Hsiung, T.H. et al. (1991). Novel technology for the synthesis of dimethyl ether from syngas. *Catalysis Today* **8** (3): 279–304.
164. Herman, R.G. (2000). Advances in catalytic synthesis and utilization of higher alcohols. *Catalysis Today* **55** (3): 233–245.
165. Herman, R.G. (1991). Classical and non-classical routes for alcohol synthesis, Chapter 7. In: *New Trends in CO Activation* (ed. L. Guczi), 265–349. New York: Elsevier.
166. Forzatti, P., Tronconi, E., and Pasquon, I. (1991). Higher alcohol synthesis. *Catalysis Reviews-Science and Engineering* **33** (1–2): 109–168.
167. Natta, G., Colombo, U., and Pasquon, I. (1957). Direct catalytic synthesis of higher alcohols from carbon monoxide and hydrogen. In: *Catalysis* (ed. P.H. Emmett), 131–174. New York: Reinhold.
168. Hu, J.L., Wang, Y., Cao, C.S. et al. (2007). Conversion of biomass-derived syngas to alcohols and C_2 oxygenates using supported Rh catalysts in a microchannel reactor. *Catalysis Today* **120** (1): 90–95.
169. Subramani, V. and Gangwal, S.K. (2008). A review of recent literature to search for an efficient catalytic process for the conversion of syngas to ethanol. *Energy & Fuels* **22** (2): 814–839.
170. Luk, H.T., Mondelli, C., Ferre, D.C. et al. (2017). Status and prospects in higher alcohols synthesis from syngas. *Chemical Society Reviews* **46** (5): 1358–1426.
171. Bain, R.L., Magrini-Bair, K.A., Hensley, J.E. et al. (2014). Pilot scale production of mixed alcohols from wood. *Industrial & Engineering Chemistry Research* **53** (6): 2204–2218.

172. Dutta, A., Hensley, J., Bain, R. et al. (2014). Technoeconomic analysis for the production of mixed alcohols via indirect gasification of biomass based on demonstration experiments. *Industrial & Engineering Chemistry Research* **53** (30): 12149–12159.
173. Dutta, A., Talmadge, M., Hensley, J. et al. (2012). Techno-economics for conversion of lignocellulosic biomass to ethanol by indirect gasification and mixed alcohol synthesis. *Environmental Progress & Sustainable Energy* **31** (2): 182–190.
174. Beretta, A., Tronconi, E., Forzatti, P. et al. (1996). Development of a mechanistic kinetic model of the higher alcohol synthesis over a Cs-doped Zn/Cr/O catalyst .1. Model derivation and data fitting. *Industrial & Engineering Chemistry Research* **35** (7): 2144–2153.
175. Quarderer, G.J. 1986. Mixed alcohols from synthesis gas. in in *78th American Institute of Chemical Engineers, National Meeting*. New Orleans, LA.
176. Nunan, J.G., Bogdan, C.E., Klier, K. et al. (1989). Higher alcohol and oxygenate synthesis over cesium-doped copper/zinc oxide catalysts. *Journal of Catalysis* **116** (1): 195–221.
177. Hilmen, A.-M., Xu, M., Gines, M.J.L., and Iglesia, E. (1998). Synthesis of higher alcohols on copper catalysts supported on alkali-promoted basic oxides. *Applied Catalysis A: General* **169** (2): 355–372.
178. Hutchings, G.J., Copperthwaite, R.G., and Coville, N.J. (1988). Catalysis for hydrocarbon formation and transformations. *South African Journal of Science* **84** (1): 12–16.
179. Wong, S.F., Patel, M.S., and Storm, D.A. 1986. Retrofitting methanol plants for higher alcohols. in *78th American Institute of Chemical Engineers, National Meeting*. New Orleans, LA.
180. Xiaoding, X., Doesburg, E.B.M., and Scholten, J.J.F. (1987). Synthesis of higher alcohols from syngas – recently patented catalysts and tentative ideas on the mechanism. *Catalysis Today* **2** (1): 125–170.
181. Courty, P., Chaumette, P., Raimbault, C. et al. (1990). Production of methanol-higher alcohol mixtures from natural gas via syngas chemistry. *Revue de l'Institut Francais du Petrole* **45** (4): 561–578.
182. Courty, P., Arlie, J.P., Convers, A. et al. (1984). C1-C6 alcohols from syngas. *Hydrocarbon Processing* **63** (11): 105–108.
183. Roberts, G.W., Lim, P.K., McCutchen, M.S. et al. (1992). The thermodynamics of higher alcohol synthesis. *Preprints – American Chemical Society, Division of Petroleum Chemistry* **37** (1): 225–233.
184. Campos-Martin, J.M., Guerreroruiz, A., and Fierro, J.L.G. (1995). Structural and surface properties of CuO-ZnO-Cr_2O_3 catalysts and their relationship with selectivity to higher alcohol synthesis. *Journal of Catalysis* **156** (2): 208–218.
185. Tronconi, E., Lietti, L., Forzatti, P. et al. (1989). Synthesis of alcohols from carbon oxides and hydrogen .17. Higher alcohol synthesis over alkali metal-promoted high- temperature methanol catalysts. *Applied Catalysis* **47** (2): 317–333.
186. Epling, W.S., Hoflund, G.B., Hart, W.M., and Minahan, D.M. (1997). Reaction and surface characterization study of higher alcohol synthesis catalysts, II. Cs-promoted commercial Zn/Cr spinel. *Journal of Catalysis* **172** (1): 13–23.
187. Epling, W.S., Hoflund, G.B., and Minahan, D.M. (1998). Reaction and surface characterization study of higher alcohol synthesis catalysts. VII. Cs- and Pd-promoted 1: 1 Zn/Cr spinel. *Journal of Catalysis* **175** (2): 175–184.
188. Epling, W.S., Hoflund, G.B., and Minahan, D.M. (1999). Higher alcohol synthesis reaction study. VI: effect of Cr replacement by Mn on the performance of Cs- and Cs, Pd-promoted Zn/Cr spinel catalysts. *Applied Catalysis, A: General* **183** (2): 335–343.
189. Hoflund, G.B., Epling, W.S., and Minahan, D.M. (1997). Higher alcohol synthesis reaction study using K-promoted ZnO catalysts. III. *Catalysis Letters* **45** (1,2): 135–138.
190. Hoflund, G.B., Epling, W.S., and Minahan, D.M. (1999). An efficient catalyst for the production of isobutanol and methanol from syngas. XI. K- and Pd-promoted Zn/Cr/Mn spinel (excess ZnO). *Catalysis Letters* **62** (2–4): 169–173.

191. Minahan, D.M., Epling, W.S., and Hoflund, G.B. (1998). An efficient catalyst for the production of isobutanol and methanol from syngas. VIII: Cs- and Pd-promoted Zn/Cr spinel (excess ZnO). *Catalysis Letters* **50** (3, 4): 199–203.
192. Minahan, D.M., Epling, W.S., and Hoflund, G.B. (1998). Higher-alcohol synthesis reaction study. V. Effect of excess ZnO on catalyst performance. *Applied Catalysis, A: General* **166** (2): 375–385.
193. Elliott, D.J. and Pennella, F. (1988). Mechanism of ethanol formation from synthesis gas over copper oxide/zinc oxide/alumina. *Journal of Catalysis* **114** (1): 90–99.
194. Smith, K.J. and Anderson, R.B. (1984). A chain growth scheme for the higher alcohols synthesis. *Journal of Catalysis* **85** (2): 428–436.
195. Smith, K.J. and Klier, K. 1992. An overview of the higher alcohol synthesis. *Abstracts of Papers of the American Chemical Society* 203: 82-PETR.
196. Courty, P., Durand, D., Freund, E., and Sugier, A. (1982). C1-C6 alcohols from synthesis gas on copper-cobalt catalysts. *Journal of Molecular Catalysis* **17** (2–3): 241–254.
197. Dai, L., Chen, Z., Li, G. et al. (1989). *Proceedings – Annual International Pittsburgh Coal Conference* **6** (2): 739–746.
198. Dalmon, J.A., Chaumette, P., and Mirodatos, C. (1992). Higher alcohols synthesis on cobalt based model catalysts. *Catalysis Today* **15** (1): 101–127.
199. Avila, Y., Kappenstein, C., Pronier, S., and Barrault, J. (1995). Alcohol synthesis from syngas over supported molybdenum catalysts. *Applied Catalysis, A: General* **132** (1): 97–109.
200. Bian, G.-z., Fan, L., Fu, Y.-l., and Fujimoto, K. (1998). Mixed alcohol synthesis from syngas on sulfided K-Mo-based catalysts: influence of support acidity. *Industrial & Engineering Chemistry Research* **37** (5): 1736–1743.
201. Iranmahboob, J. and Hill, D.O. (2002). Alcohol synthesis from syngas over K_2CO_3/CoS/MoS_2 on activated carbon. *Catalysis Letters* **78** (1–4): 49–55.
202. Iranmahboob, J., Toghiani, H., Hill, D.O., and Nadim, F. (2002). The influence of clay on K_2CO_3/CO-MoS_2 catalyst in the production of higher alcohol fuel. *Fuel Processing Technology* **79** (1): 71–75.
203. Li, X., Feng, L., Liu, Z. et al. (1998). Higher alcohols from synthesis gas using carbon-supported doped molybdenum-based catalysts. *Industrial & Engineering Chemistry Research* **37** (10): 3853–3863.
204. Fierro, J.L.G. (1993). Catalysis in C1 chemistry: future and prospect. *Catalysis Letters* **22** (1–2): 67–91.
205. Nirula, S.C. (1994). *Dow/Union Carbide Process for Mixed Alcohols from Syngas*. Menlo Park, CA: SRI International.
206. Brown, T.R. (2015). A techno-economic review of thermochemical cellulosic biofuel pathways. *Bioresource Technology* **178**: 166–176.
207. Dutta, A., Bain, R.L., and Biddy, M.J. (2010). Techno-economics of the production of mixed alcohols from lignocellulosic biomass via high-temperature gasification. *Environmental Progress & Sustainable Energy* **29** (2): 163–174.
208. Phillips, S.D., Tarud, J.K., Biddy, M.J., and Dutta, A. (2011). Gasoline from woody biomass via thermochemical gasification, methanol synthesis, and methanol-to-gasoline technologies: a technoeconomic analysis. *Industrial & Engineering Chemistry Research* **50** (20): 11734–11745.
209. Tan, E.C.D., Snowden-Swan, L.J., Talmadge, M. et al. (2017). Comparative techno-economic analysis and process design for indirect liquefaction pathways to distillate-range fuels via biomass-derived oxygenated intermediates upgrading. *Biofuels Bioproducts & Biorefining* **11** (1): 41–66.
210. Snehesh, A.S., Mukunda, H.S., Mahapatra, S., and Dasappa, S. (2017). Fischer-Tropsch route for the conversion of biomass to liquid fuels – technical and economic analysis. *Energy* **130**: 182–191.
211. Nacken, M., Ma, L., Heidenreich, S., and Baron, G.V. (2009). Performance of a catalytically activated ceramic hot gas filter for catalytic tar removal from biomass gasification gas. *Applied Catalysis B: Environmental* **88** (3–4): 292–298.

212. Nacken, M., Ma, L., Heidenreich, S., and Baron, G.V. (2010). Catalytic activity in naphthalene reforming of two types of catalytic filters for hot gas cleaning of biomass-derived syngas. *Industrial & Engineering Chemistry Research* **49** (12): 5536–5542.
213. Nacken, M., Ma, L., Heidenreich, S. et al. (2012). Development of a catalytic ceramic foam for efficient tar reforming of a catalytic filter for hot gas cleaning of biomass-derived syngas. *Applied Catalysis B: Environmental* **125**: 111–119.
214. Heidenreich, S. and Foscolo, P.U. (2015). New concepts in biomass gasification. *Progress in Energy and Combustion Science* **46**: 72–95.
215. Meng, X., de Jong, W., Pal, R., and Verkooijen, A.H.M. (2010). In bed and downstream hot gas desulphurization during solid fuel gasification: a review. *Fuel Processing Technology* **91** (8): 964–981.
216. Phillips, S.D. (2007). Technoeconomic analysis of a lignocellulosic biomass indirect gasification process to make ethanol via mixed alcohols synthesis. *Industrial & Engineering Chemistry Research* **46** (26): 8887–8897.

6

Fast Pyrolysis

Robbie H. Venderbosch

Biomass Technology Group B.V., AV Enschede, the Netherlands

6.1 Introduction

Fast pyrolysis is the thermal decomposition of biomass in the absence oxygen at temperatures in the range of 400–600 °C to produce mostly liquid product but also smaller amounts of char and non-condensable gases. The liquid can be burned as boiler fuel, can be upgraded to transportation fuels, or can be used as a source of organic chemicals. The char and gas are often used to provide thermal energy to the pyrolysis plant, although other applications have been proposed for the char – including activated carbon and soil amendment. Fast pyrolysis has gained attention for its potential as a cost-effective method to decouple liquid fuel production from its utilization, in scale, time, and location. It has the promise to connect (conventional) agricultural business to (petro)chemical processes. In this approach, the existing assets (and related existing infrastructure) in existing refineries can be deployed as a very cost-effective tool to create "new" hybrid biorefineries.

6.2 Fundamentals of Pyrolysis

Many reviews on pyrolysis can be found in literature. A broad overview is presented by Venderbosch and Prins [1]. Some reviews focus on fundamental phenomena associated with thermal deconstruction of biomass [2, 3], although these suffer from insufficient information on chemical kinetic mechanisms and poor understanding of physical transport processes (momentum, mass, and energy) at the molecular, cellular, and particle scale.

Pyrolysis is very complex, and involves thousands of elementary reactions across three phases which generate hundreds of volatile species. Thermal decomposition of biomass produces condensable vapors (the liquid product aimed at in pyrolysis) and by-products – char and non-condensable gas. The liquid fraction can be increased by heating the biomass faster and quenching the vapors more rapidly. Realizing that wood is a very good insulation material, the heat penetration depths of the woody particles should be typically less than a few mm. In this way, liquid yields from woody feedstock in continuously operated laboratory reactors and pilot plants are usually in the range of 60–70 wt% (dry-feed basis), while exceptionally yields over 70% are reported.

6.2.1 Effects of the Chemical and Physical Structure of Biomass and Intermediate Products

The intention of *fast* pyrolysis is (i) to prevent over-cracking of vapors (over the char formed already) to small non-condensable gas molecules and (ii) to prevent recombination/polymerization of molecules into larger molecules (such as char or its precursors). For this reason it is reportedly crucial to obtain a short residence time for the primary products at the surface of the decomposing particle and to avoid a high residence time in any equipment before vapor condensation takes place. Early process developers therefore suggested small particles (<1 mm), relatively high temperatures, and use of inert gases to limit the gas-phase residence time of the produced gases. New insights, however, show that the overall liquid yield seems much less dependent on biomass particle size and vapor residence time than originally assumed if no catalysts surface (for example char, and particularly the ash in the char) is present [4, 5]. Some other new insights and views from the present author that have been gained over the last few years are presented below.

High external heat transfer to the biomass particles might indeed be realized by mixing the biomass intensively with an excess of a hot heat carrier (e.g. hot sand), but efficient heat transfer *through* the biomass particle is also required. For this it is not the particle size itself that is important, conventionally limiting the size of biomass particles to 1 mm, but the effective heat penetration depth and its level of anisotropy [6]. Usually the particle size is determined as mesh size, but this does not really reflect the preferred size. For example, the Empyro commercial plant uses larger particles while their demonstration unit in Malaysia deployed fibers with diameters below 1 mm but lengths of 10 mm or even larger.

Both heat transfer into the biomass and the composition of the biomass influences the products of pyrolysis. Lignocellulosic biomass is a highly diverse in its chemical compositino, with major components being cellulose, hemicellulose, and lignin, along with smaller amounts of inorganic compounds and a wide range of extractive compounds including alkaloids, oils, fats, glycosides, gums, mucilages, pectins, proteins, resins, saponins, terpenes, and waxes. [7]. All of these influence to different degrees the overall rate of pyrolysis and product composition with synergetic effects likely to occur. For the main components, the following general observations can be made.

- Hemicellulose, mostly xylans with an average composition of $(C_5H_8O_4)_n$ with $n = 50$–200, is the first to decompose, starting at about 200 °C and completed around 400 °C. Hemicellulose produces mainly CO_2; it is likely the smaller components (including the acids) are derived from the hemicellulose.

- Cellulose, with a rough composition of $(C_6H_{10}O_5)_n$ with $n = 500$–4000, appears rather stable up to around 300 °C, after which almost all cellulose is converted to non-condensable gas and condensable organic vapors in the range from 320 to 420 °C. Cellulose yields more CO than CO_2, while overall dehydrated carbohydrates (monomeric and oligomeric) are the main products.
- Lignin, which consists of highly branched, substituted, mononuclear aromatic polymers, has a very broad decomposition temperature range from as low as 160 °C extending up to 800–900 °C. Its cracking and deformation releases the largest fraction of the CH_4. Part of the hemicellulose that is closely linked in the biomass structure with the lignin ends up in the liquid product, associated and linked with fragments from lignin.

Interestingly, below temperatures of around 500 °C, conversion of lignin is limited (less than 40%), from which it can be concluded that, contrary to the overall perception, it is reasonable to state that the generally accepted view that short reaction times are critical is only partially correct: the first 90% of the conversion takes place in 2 seconds (probably the holocellulose and part of the lignin), but the final 10% in 20 seconds for the remainder of the lignin. While the elemental composition of char is often close to that of the lignin, it seems justified to state that at atmospheric pressure the solid residues (char) is mainly derived from the lignin (and partially from the hemicellulose), while most of the cellulose can be converted to a liquid.

Extensive research has been carried out over the years to measure and describe the kinetics of pyrolysis. A general concern here is always which kinetics are actually determined among the biomass components. Even when this can be discerned, the question arises whether real kinetics are being measured, which is only assured in the absence of heat transport limitations (both internally in the biomass as well as those from an external body to the biomass). That is, heat transfer to and within the particles must be much faster than chemical reaction. Complicating the case even further, significant endothermic or exothermic reaction enthalpies of one or more of the reactions can cause the actual reaction temperature to be different from the reactor temperature. For this reason, kinetic expressions reported in the literature are to be considered with caution [6].

Not surprisingly, the various kinetic expressions (reaction rate, heat rate, product distribution, and so on) are substantially different for different types of feedstock. High concentrations of potassium, for example, increase the rate of biomass decomposition during the first stages of reaction but slows the second stage. The general rate expressions are also hugely over-simplified: in most cases the kinetic parameters in these equations assume decomposition of a single reactant representing a composite feedstock such as wood that decomposes by a number of parallel first-order reactions to form classes of products (permanent gases, char, and liquids) [7–10].

The composition of biomass varies widely among softwoods, hardwoods, and grasses. Compared to wood, grasses are usually richer in hemicellulose, with typical values of 35–50%. Hemicellulose consists of various monosaccharide units but the exact composition is different for each type of biomass, and differences are expected in how these units pyrolyze. Cellulose structure can be classified into crystalline and amorphous forms, that is to say a uniform structure versus a disordered structure, again both requiring different fragmentation temperatures, yielding different products. Lignin also varies with the parent biomass. For example, softwood lignin predominantly contains guaiacyl groups, a polymer with a higher proportion of coniferyl phenylpropane units. A guaiacyl-syringyl lignin is typically found in hardwoods and is a copolymer of both coniferyl and sinapyl phenylpropane units, leaving hardwoods richer in the methoxyl content, which will have different conversion rates. In the authors view, it is unrealistic to expect that the thermal degradation of biomass can be described accurately with a simple set of rate/selectivity equations, and distinct kinetics should be developed for the three main components of biomass.

For the reasons given above, measuring reaction kinetic rates to obtain generalized kinetic rate expressions for the conversion of biomass is not only far too ambitious, but of limited use in practical applications of pyrolysis. Kinetics, if they are determined at all, should address a specific feedstock, and cover the full decomposition range of both the carbohydrate and lignin fractions. Specific considerations are listed here.

(i) The decomposition at reaction times greater than 2 seconds is relevant for practical applications, as this allows a substantial increase in liquid yield (probably from the lignin).
(ii) The kinetic expressions merely serve to describe and partially understand the overall reaction kinetics and mechanisms but have little practical application in the design of pyrolysis reactors.
(iii) Fast pyrolysis occurs in at least a minimum of three distinct component levels that must be addressed accordingly (cellulose, hemicellulose, and lignin), each having its own reaction mechanisms, kinetics, and product yields.
(iv) Interaction among these components potentially affects liquid composition but is poorly understood.

An approach covering the components of biomass (including ash) prevails over models that just focus on the physical internal structure of the biomass (such as its anisotropic nature to explain the role of transport phenomena). Pyrolysis of the individual components alone is already very complicated, not only because many sequential and parallel reactions take place, but also because the chemical structure of the biomass most likely has a much larger effect on yields and quality than the physical structure (apart from particle size).

Pyrolysis of cellulose can yield very high yields of liquids at more extreme conditions of very small particles ($<50\,\mu m$), very high heating rates ($\geq 5000\,°C/s$), and rapid quenching of the products. More interestingly, however, reducing the absolute pressure down to about 5 mbar while very quickly heating the biomass (in a so-called "screen heater") yields condensed products, and no non-condensable gases [11]. This was attributed to the absence of secondary reactions of these products outside the reacting particle due to a very short residence of the vapors at high temperatures (estimated <20 ms, compared with seconds in a fluid bed). The low pressure and absence of secondary reactions has a significant impact on the product distribution as well, which is shown in Figure 6.1. The condensed products from atmospheric liquid (both from the fluid bed and from the screen heater) contain large amounts of monomeric levoglucosan (here referred to as DP_1) and minor amounts of di-, tri-, and other oligomers, while the liquids from the vacuum tests also had high concentrations of oligomeric compounds measured (up to cellopentosan). It is hypothesized that the sugars and anhydrosugars volatilize faster at lower pressures, which is supported by tests with levoglucosan and cellobiosan. An interesting conclusion here is that lower operating pressures may lead to substantially higher liquid yields but, moreover, may also lead to a

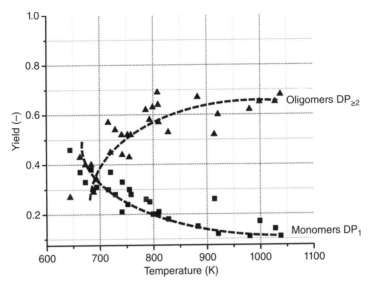

Figure 6.1 *Yields ($kg_{product}/kg_{cellulose}$) of monomeric anhydrosugars (DP1) and oligomeric (DP≥2) as a function of the final pyrolysis temperature, measured at the screen's surface. Indicated are trend lines. Source: adapted from [11].*

better quality of the liquid (in terms of water content), and may, in hindsight, provide further evidence for the relatively high yields and reasonable quality of liquid from Pyrovac's slow pyrolysis unit [12].

An interesting visual observation is the appearance of an intermediate liquid state, recently confirmed [13], as shown in Figure 6.2. Also visible in these photographs are aerosol particles ejected from the molten cellulose, which suggests both vapors and aerosols released from pyrolyzing biomass contribute to liquid products. On-going studies with in situ laser diagnostic tools are expected to provide further insights on transport processes during pyrolysis [15, 16]. Technical lignin appears to decompose through an intermediate liquid with volatiles released as bursting bubbles, although it is not clear that technical lignin pyrolyzes in the same manner as lignin imbedded in the structure of lignocellulosic biomass.

Although screen heater experiments differ from actual heating of biomass in the presence of solid particles such as sand, an important conclusion is that pyrolysis is initiated with a network of largely unidentified solid-phase reactions which fractionate and depolymerize the solid to a short-lived intermediate liquid [13, 14]. Within this liquid phase a complexity of depolymerization, rearrangement, and dehydration reactions occur until products volatilize or are ejected in the form of aerosols. The mechanisms of fast pyrolysis are much more complex and models must involve, along with primary liquefaction reaction kinetics with heat and mass transfer characteristics, mechanisms explaining the evaporation and other physical behavior (including bubble bursting, particle ejection, and the like).

6.2.2 Effects of Ash

Ash has an important effect in the mechanisms of biomass decomposition, affecting both the liquid yield and the liquid's composition [17]. Ash significantly complicates understanding

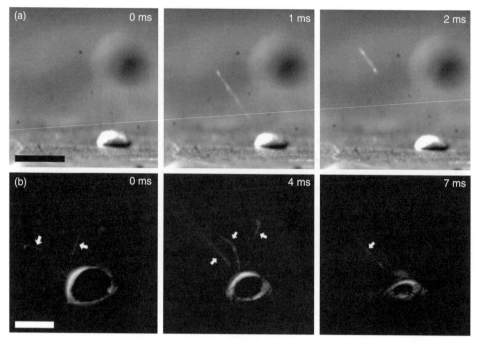

Figure 6.2 *Aerosol ejection from molten intermediate cellulose. (a) A particle of microcrystalline cellulose thermally decomposes to molten intermediate cellulose on 700 °C α-alumina and ejects liquid visible as a white streak. Subsequent evaporation completely vaporizes the molten droplet resulting in a clean surface. (b) A droplet of molten intermediate cellulose on 700 °C α-alumina exhibits multiple aerosol ejections highlighted with white arrows. Source: adapted from [13, 14].*

of pyrolysis in terms of yield and quality. Specifically, biomass usually contains significant inorganic components, in amounts which – next to the biomass variety itself – also depend on environmental conditions (soil quality, water quality), harvesting methods, and time. Common elements are, in decreasing order of mass abundance, Ca, K, Si, Mg, Al, S, Fe, P, Cl, Na, Mn, and Ti [17–19], with trace elements such as As, Ba, Co, Cr, Cu, Mo, Ni, Se, Sn, U, V, and Zn.

Ash can be present as a free ions, dissolved in fluid matter inside the biomass material, structured mineral forms, or covalently boned with organic biomass structures. The main alkali metals K and Na are present mainly as salts, usually in free ion form (Na^+, K^+) with counter-ions such as chloride (Cl^-), carbonate (CO_3^{2-}), or malate ($C_4H_4O_5^{2-}$), dissolved in the fluid matter inside the biomass cell structure. Alternatively, the alkali metals can exist as solid salt structures fixed on the cell wall. In comparison, alkaline earth metals such as Ca and Mg are predominately bonded with the organic parts of the biomass and are less commonly present in free ionic form [19]. Sulfur and phosphorus are both found in organic and inorganic components, as is chlorine (but due to the complexity of the analysis of chlorine, not much is known on how it is present in the biomass).

Demineralization of woody feedstock prior to pyrolysis results in a significant change in the composition of the resulting liquid [20]. Recently, the effect of potassium (1–10 000 ppm) has been systematically investigated while pyrolyzing cellulose [21]. A higher concentration of potassium resulted in a lower yield of condensable products from atmospheric and vacuum pyrolysis, as shown in Figure 6.3.

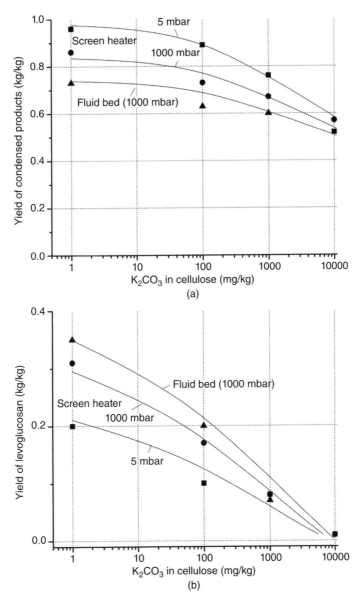

Figure 6.3 (a) The product yield of liquids from cellulose and (b) the levoglucosan yield, both in kg/kg$_{dry\ biomass}$, versus the potassium concentration, in the screen heater (both at 5 mbar and atmospheric) and in the fluid bed (atmospheric). Source: adapted from [20].

A clear difference in condensed product yield can be observed between vacuum and atmospheric pressure experiments at low concentrations of potassium, especially in the range up to 1000 ppm. In this range, the condensed product yield under vacuum was significantly higher compared to at 1000 mbar experiments, suggesting that pyrolysis products produced from the hot reacting particles are much less or perhaps not contacted with the potassium [21]. Earlier investigation reported that the vapor residence time alone already caused a drop in liquid yield, ascribed to cracking and polymerization reactions of vapors to gases and solids, respectively. However, alkalis are crucial in such secondary vapor phase cracking and thus the amount of char present as catalysts also has a significant effect on the extent of this cracking.

The product composition of the liquid changes dramatically with increasing potassium content. Whereas levoglucosan is the main product in pyrolysis of pure cellulose, even very low concentrations of alkali and alkaline earth metals are destructive to sugars and anhydrosugars [21].

The moisture content of pyrolysis liquids is observed to increase with increasing ash content of biomass. However, this does not appear to be the result of increasing yield of water as observed in (previously unpublished) experiments at BTG laboratories using a semi-continuous small-scale pyrolysis unit [17]. As illustrated in Figure 6.4, pyrolysis of pine wood, softwood, and mixtures of pine wood and grass found the yield of water ($kg/kg_{dry\ feed}$) was almost independent of ash content while the yield of organic compounds decreased with increasing ash content, thus explaining the increased concentration of water in the pyrolysis liquid product. Overall, it is reasonable to state that maximizing the overall liquid yield also improves the quality of the liquids as it goes hand in hand with a lower water content (in wt%) in the liquids. A systematic approach on a variety of feedstocks is required to confirm this statement.

While ash has an effect on the organic yield and thus the quality of the liquids with respect to water content and phase stability, the actual fate of ash is not fully clear, whether they are retained in the char or, less favorably, report to the pyrolysis liquids. Overall, monovalent elements Na and K are released to a larger extent in the free ion form rather than as hydroxides or carbonates [17]. In contrast, the divalent elements Ca and Mg are retained in the solid phase [23, 24].

There are several main routes available for the ash to end up in liquids and solids. For example, ash can leave the reactor as char, entrained with the vapor phase, or it can be evaporated and trapped in the condensing pyrolysis liquid [17]. The fraction of entrained/evaporated ash constituents in the pyrolysis liquid is seen to depend largely on the amount of alkalis present in the feedstock. As a typical example, Figure 6.5 shows the content of Na/K in the pyrolysis liquid versus the content in the original feed for a dedicated set of data from various researchers (adapted from [17]).

For the non-metal S a similar trend is also observed, but – despite the wide variation in data – absolute values for S are substantially higher, and the concentration in the pyrolysis liquid is usually close to the concentration in the original biomass. This high sulfur concentration is believed to be due to organic sulfur, while data for manure, containing a large amount of inorganic sulfur, in turn shows a low degree of transfer to the pyrolysis liquid.

Transfer of phosphorus is much lower than for sulfur, usually less than 5% of the biomass sulfur is transferred into the liquid and the remainder into the char. Information on the

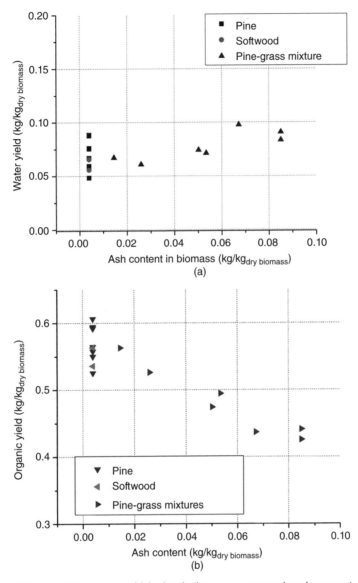

Figure 6.4 (a) Water and (b) organic yield, both in kg/kg$_{drybiomass}$, versus the ash content in the biomass. Source: with permission from BTG Biomass Technology Group BV [22].

transfer of phosphorus, on the other hand, is scarce, but it is believed that it is mainly transformed into inorganic phosphorus in the char upon some reactions during pyrolysis. As a consequence, phosphorus in the liquids can be minimized by filtrating the particles, while sulfur requires varying of the operating conditions (such as lower heating rates and/or dilution of the vapors with a carrier gas).

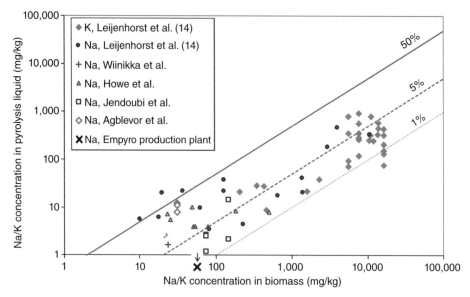

Figure 6.5 Concentration of sodium and potassium in the pyrolysis liquids as a function of the biomass concentration: comparison of the experimental data with results from literature. The percentage lines are defined as the concentration in the pyrolysis liquid divided by the concentration in the biomass. Source: adapted from [17].

6.3 Properties of Pyrolysis Liquids

The composition of the liquid product of fast pyrolysis is very complex and apt to change with time as many components are highly reactive. The dark red-brown to almost black liquid comprises a vast range of oxygenated compounds. It can be characterized as an emulsion of lignin-derived compounds in an aqueous phase of carbohydrate-derived compounds. In contrast to many reports, the emulsion can be relatively stable if the water content is limited to around 23 wt%. In some cases pyrolysis liquids can be stored for periods longer than 12 months without any visual change. However, phase separation readily occurs when water content reaches 25 wt% water, with the rate depending on the water content. The naturally occurring acid content of pyrolysis liquids can help stabilize the emulsion, an advantage not acknowledged by all researchers. The most abundant acid in the liquid is acetic acid (usually 3–4 wt%).

Water is present in amounts between 15 and 35 wt%. This water is derived from the biomass as such, but – as a rule of thumb – roughly 10 wt% points is pure reaction water. While water reduces the heating value of pyrolysis liquid, it helps reduce viscosity. In combustion applications, water lowers flame temperature, which is thought to favor lower NO_x emissions.

Water can be evaporated from pyrolysis liquid to increase heating value and increase viscosity, but this should be done under vacuum to avoid high temperatures that accelerate polymerization of the liquid.

Gas chromatography (GC) analysis (including two-dimensional GC, GC–mass spectrometry [MS], etc.) is widely used in the measurement of liquid quality. However, the

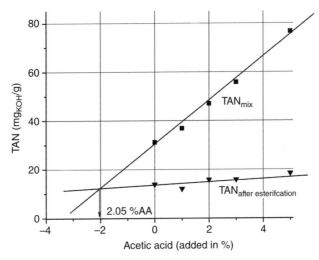

Figure 6.6 Total acid number (in mg_{KOH}/g to reach pH = 7) for a mixture of pyrolysis liquids, butanol, and acetic acid versus the acetic acid addition (in wt%), before and after esterification. Source: with permission from BTG Biomass Technology Group BV [25].

usefulness of GC data is limited because it cannot measure the nonvolatile constituents of pyrolysis liquid. GC injection includes vaporization of the feed, which is known to be very difficult for the pyrolysis liquid and results in coke forming in the injection part of the system. Moreover, chemical reactions occurring in the GC column cannot be excluded, and, as a consequence, it is questionable whether the components actually detected are present in the liquid feed or result from further cracking reactions in the injection system. Even the content of simple molecules such as acetic acid is difficult to measure as GC will provides levels that are too high if cracking occurs in the injection.

An alternative measurement technique for acids is provided in Figure 6.6 [25]. The plot is constructed by measuring the total acid number (TAN) (to reach pH = 7) for a mixture of pyrolysis liquid in butanol (50:50%), both as received and after esterification using conventional Amberlyst-15 at a temperature of 70 °C. Further, the TAN is increased by adding acetic acid (in equivalent wt%) to the mixture of pyrolysis liquid in butanol, and repeating the procedure. TAN increases linearly with increasing acetic acid, as expected. However, after esterification of the liquid, TAN is almost constant with acetic acid addition, suggesting that practically all acetic acid is transformed by esterification to form butyl acetate. The intersect of both lines now points at the acetic acid equivalent of around 4.1% in the pure bio-liquid, corresponding to 2.05% for the diluted 50:50% liquid. Interestingly, a "residual TAN number" can be noted, which is caused by components that do not form esters with the butanol present in excess (likely due to steric hindering), but that have an acidity similar to that of acetic acid (pK_a = 4.76). These can be acids derived from the holo-cellulose or lignin fraction of course.

In conventional GC, a substantial amount of the liquid's components – up to 50% in some cases – cannot be detected. Non-disruptive techniques are required to analyze and quantify the composition of the highly reactive liquids. Development of new techniques to identify the constituents of pyrolysis liquid is ongoing: next to HPLC, solvent fractionation before analysis has potential, and can be important to understand the liquid's oxygen functionalities. The most interesting non-disruptive technique in this context is 2D-NMR. In a manner identical to one-dimensional FT-NMR, the two dimensions of a 2D-NMR analysis are used to identify the functionalities of the carbon and hydrogen atoms and bonds, with the intensity of the peaks providing information on the quantity of these functionalities. From the observation that the liquid is an emulsion at acidic conditions, a particular phenomenon observed in the liquid is the apparent shift in the continuous – dispersed phases if the acidity is changed: above pH > 11 or so, the physical and visual properties of the liquid changes considerably. However, not much work has been done in this specific field yet.

6.4 Fast Pyrolysis Process Technologies

The aim of fast pyrolysis is to convert biomass into a maximum quantity of liquid. In this way, the energy in the biomass feedstock is concentrated in a smaller volume and as a result transport costs and storage space can be reduced. This liquid, though still a very complicated mixture of oxygenated compounds, is more uniform than solid biomass, rather stable, ash-free, and pumpable, Moreover, it is a much better feed when used for any further chemistry, forming either (final or intermediate) energy carriers or chemicals. As a bonus, the ash from the original biomass is largely concentrated in the char (see section above), allowing recycling of the minerals as a natural fertilizer that can be returned to biomass production. The excess energy from gaseous and solid by-products must preferably be used for heat or electricity production.

The essential characteristics of fast pyrolysis for maximal liquid production include very rapid heating of the biomass and rapid quenching of the vapors produced. Of these two characteristics, the most crucial one is the heating rate of the biomass particles.

Commercially applied systems are the circulating-bed process (Ensyn, Valmet) and BTL's rotating cone, which are described in the following sections. Further details are found elsewhere [1].

6.4.1 Ensyn (CFB)

Ensyn's technology is referred to as rapid thermal processing or RTP™ [26]. RTP comprises mixing of biomass and hot carrier materials such as sand in a circulating fluid bed system with a separate section for the char combustion. A photograph of an Ensyn unit is shown in Figure 6.7.

The company was established in 1984, based on research performed at the University of Western Ontario, Canada. Since 1989, Red Arrow Products Company LLC applied Ensyn's technology for commercial production of liquid smoke for use in the food industry. Several RTP facilities have been delivered to Red Arrow since then, and six are still operating. From 1990 onwards the technology was scaled up to a facility of 45 tonnes per day (dry) biomass input. A 20 kg/h process development unit (PDU) was delivered to VTT in 1995 and, after some modifications recently, is still operated by VTT on a regular basis for the production of liquids.

In 1995, Ensyn shipped a larger 625 kg/h unit to Bastardo (Perugia, Italy), to be operated by ENEL. Over the last 15 years the plant is stated to be "running on demand," but

Figure 6.7 Photograph of the Ensyn plant. Source: with permission from Ensyn Technologies Inc. [27]. See the plate section for a color representation of this figure.

in practice this appears hardly ever or never. In 1998 Ensyn adapted the 45 tonnes per day unit to upgrade heavy oil (approx. 20 barrels per day). In the period 2000–2005 focus was on this petroleum application, resulting in a 1000 barrels per day heavy oil RTP facility in California in 2004. From 2005 onwards, renewable fuels were aimed at, and a 75 tonnes per day unit was commissioned in Renfrew Ontario in 2007 to produce renewable fuel oil or RFO™, a petroleum-replacement to be used for heating purposes. In 2008 Ensyn established a joint venture with UOP (the latter a Honeywell company) to create "Envergent Technologies LLC". In 2011 UOP (with Envergent and Ensyn) started construction of a small pyrolysis unit (20 kg/h) in Hawaii, to combine it with a hydrotreater to produce transportation fuels. In 2014 it is reported that the hydrotreater was skipped, as interest was redirected toward another application for the liquids (co-FCC [fluid catalytic cracking], see Section 6.4.3) [28]. The Hawaii project was terminated in 2016.

In 2014, Ensyn's alliance with UOP expanded to explore integration of RFO into petroleum refineries. In that year 4 four million Canadian Dollars (CAD) was invested in the facility in Renfrew by Ensyn "to transform the forest products industry in Northern Ontario" with the purpose of converting the facility to a dedicated biofuels facility with enhanced production capacity of 3 million gallons/year. In 2016 the construction of a 200 tonnes per day facility was started by a consortium of Ensyn, Arbec Forest Products, and Groupe Rémabec, at Arbec's sawmill in Port Cartier, Quebec (approx. 70 million €). Forest slash feed is to be used as input, with liquids to be sold to heating and refining customers in the northeast US. The liquids from the facility are used to produce heat, while some is co-processed in petroleum refineries.

6.4.2 Valmet/UPM (CFB)

In 2010, the experience VTT gained over the years, together with incremental improvements, culminated in a patent application by VTT and subsequent licensing by Valmet. The technology is based on integration of a conventional biomass-based fluidized-bed boiler with a pyrolysis reactor based on a circulating fluid bed concept. The pyrolysis unit utilizes the circulating hot sand from the boiler as a heat source. The first production was accomplished at Valmet's test unit in Tampere (Finland) in 2010.

In 2007 VTT applied for a patent in which the fast pyrolysis process is integrated into a fluidized bed boiler [29]. A schematic diagram of the integrated fast pyrolysis concept is shown in Figure 6.8. The concept combines fast pyrolysis with existing CHP plant technology, where the heat needed for the pyrolysis is provided by the CHP plant by feeding hot sand to the circulating fluidized-bed (CFB) pyrolysis unit (800 °C), and cold sand together with the char (at around 500 °C) is sent back to the fluidized bed boiler, which is also where the non-condensables are fired. The concept is a symbiosis of two processes, with efficient use of back-pressure steam/condensate and low temperature heat for drying, combustion of char and non-condensable gases in the boiler, and utilization of existing infrastructure. An overall plant energy efficiency around 90% is mentioned. The combination allows lower investment costs by integration into an existing unit, but of course limits a commercial out roll to a certain number of existing boilers.

The concept has been demonstrated by Metso. A smaller 7 t/d pyrolysis liquids production unit was constructed at Metso's premises in Tampere (Finland) in 2008 and since 2009 more than 140 tons of liquid have been produced. A larger commercial-scale demonstration unit has been delivered by Metso to Fortum Power and Heat. The plant is located in Joensuu (Finland) with a capacity to produce 50 000 tonnes per year pyrolysis liquids to be mainly used instead of heavy fuel oil. The plant started in the autumn of 2013. No update has been presented on the production figures (runtimes, yields, and the like) nor quality of the produced liquids.

6.4.3 BTG-BtL (Rotating-cone)

Developments by BTG led to a high intensity pyrolyzer that does not require inert gases. This considerably simplifies the reactor concept and makes scale-up easier – particularly the peripheral equipment such as condenser, gas cleaning, absence of gas recirculation, and the like. The pyrolysis reactions take place upon mechanical mixing of biomass and hot sand, and the sand and char are further transported to a separate fluid bed where the combustion of char takes place. Direct heat transfer from fresh hot sand and biomass takes

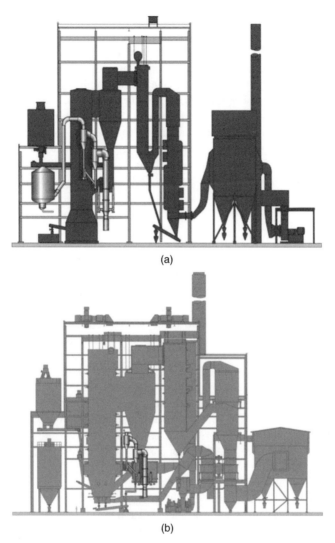

Figure 6.8 *(a) Stand alone pyrolysis liquid production plant. Green = pyrolysis equipment; Blue = char combustion and heat recovery. (b) Integration of pyrolysis in existing boiler. Green = pyrolysis equipment; Gray = existing boiler plant. Source: with permission from Valmet [30].*

place, avoiding the limitations reported for fluid bed operation and expanding the size of the biomass particles that can be used up to a few mm.

A first 50 kg/h prototype was shipped to China in 1994. The so-called RCR technology has been further developed at BTG in the period 1997–2006, integrating a 250 kg/h reactor in a circulating sand system further composed of a riser, a fluid bed char combustor, and a down comer. In 2004, BTG sold a first commercial unit (50 t/d) using so-called empty fruit bunch (EFB) in Malaysia to Genting Bio-Oil Sdn Bhd (GBO). The feedstock EFB is left over from palm oil mills. In the Malaysian plant, EFB is taken directly from a nearby palm mill, pressed, shredded, dried, and converted to a pyrolysis liquid. GBO commissioned its bio-liquid pilot plant in Ayer Itam, Johor (Malaysia) in 2006. From mid-2005 the plant was running on a daily basis, showing the potentials but also the local shortcomings.

In 2007, BTG established BTG-Bioliquids with the objective to commercialize the technology. Start-up of the first commercial unit, Empyro, took place at the end of 2015 (Empyro), illustrated in Figure 6.9. The 5 tonne per hour unit is operated on a continuous basis, with a crew of around seven operators (two during the day, one at night). Unmanned operation is foreseen in the next year. More than 15 million liter was produced by July 2017, which has been used to replace natural gas in a boiler . The plant is operating as expected: the total running time increased over the last three years to the design value. The capacity varies between 2.7 and 3.3 tonnes per hour and liquid yields (65 wt%) are around design values. As shown in Figure 6.10, around 90% of the energy input is converted into liquid product, steam, or electricity [31].

6.4.4 Dynamotive Technologies Corp

Dynamotive Technologies Corporation (DMTF) was established in 1991, and acquired its current pyrolysis technology from Resource Transforms Ltd. (RTI) in 2000. The basis is an indirectly heated fluidized bed for fast pyrolysis of biomass, from which the char is elutriated. In 2001 DMTF commissioned a first 10 t/d lab unit, and from 2003 a 100 t/d commercial unit in West Lorne, Ontario (Canada) was operated. This unit encountered major challenges, for instance in the biomass feeding system and in the reactor heating, but also related to the fines in the feedstock. The pyrolysis liquids contained 2–3% of char, an order of magnitude higher than in the liquids from the lab units. This high solid content dramatically affected the foreseen combustion in an Orenda 2.5 MW_e gas turbine (Magellan Aerospace). Despite these problems, a 200 t/d "commercial unit" was launched in 2005 (located in Guelph, also in Ontario), with a total investment of about $18.5 million.

Several other initiatives have been reported, amongst others two 200 t/d plants in Canada and a master license granted to an Australian partner. At the same time, DMTF established R&D offices & a laboratory on the premises of the University of Waterloo in 2006 (the "Waterloo Research Laboratory"), and offices in the US and Argentina in 2007. Research was initiated on the gasification of bio-liquids to syngas for synthetic fuels, and a first fully commercial industrial bio-fuel plant was announced in the US in Willow Springs, MO.

In May 2007, DMTF announced the initial production runs in Guelph, producing a so-called intermediate grade "bio-oil" or BioOil Plus, actually a pyrolysis liquid containing high quantities of char. Soon after that the West Lorne plant went into receivership, and the assets were sold. The Guelph plant was shut-down and all larger projects disappeared from sight. As of the date of writing, the Guelph plant has been dismantled while the West Lorne plant still seems to be there, but not operational. DMTF refocused to work on the use of the liquids for biofuels production. For this proposal a memorandum of understanding was signed with IFP Energies Nouvelles and Axens (both from France) for the development of the liquid-upgrading process (see also below). Gasoline and diesel were to be produced from pyrolysis liquids through a two-stage high-pressure catalytic upgrading process – the Biomass IN to GasOil or BINGO. The process involves a hydro-reforming of the liquid to a fuel that can either be directly utilized in blends with hydrocarbon fuel for power and heating, or be further upgraded to transportation grade liquid hydrocarbon fuels in a further hydrotreating process. A patent application related to this was published in 2012 [32].

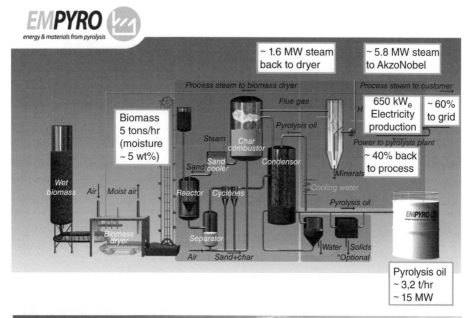

Figure 6.9 The Empyro pyrolysis plant in The Netherlands. Source: with permission from BTG Bioliquids B.V. [31]. *See the plate section for a color representation of this figure.*

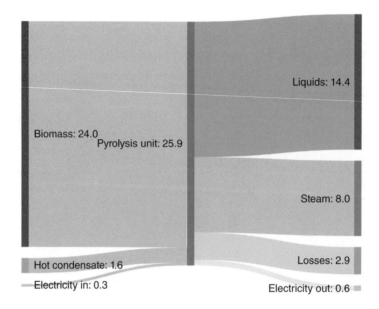

Figure 6.10 Energy balance Empyro. Source: with permission from BTG Bioliquids B.V. [31].

It is probable that, technical problems hindered these indirectly heated fluid bed pyrolyzers, which eventually forced Dynamotive to cease its operation at West Lorne and Guelph. Gas-to-coil wall heat transfer coefficients are limited to an estimated 100–200 W/m² K, with a driving force of maximal 300 °C (800 down to 600 °C in coils versus 500–550 °C in the fluid bed). Optimistically at least 10–20 m² surface area is required per ton/h of biomass fed, which is difficult to achieve in practice. In addition, problems with cyclonic removal of char led to a high solid content in the liquids produced.

6.5 Applications of Pyrolysis Liquids

The properties of fast pyrolysis liquids are very different from those transportation fuels derived from crude oil. Modification/upgrading is required to improve the compatibility of fast pyrolysis liquids with such fuels. As illustrated in Figure 6.11, various approaches are possible – ranging from simple physical treatment to severe thermochemical chemical treatment. The concept behind this bio-liquid refinery is the decentralized production of liquids from biomass ("solve logistics first"), followed by large-scale centralized use for low added valued products as in combustion, syngas production, and co-refining, or more local uses to produce specialties and/or chemicals.

6.5.1 Combustion

The simplest use of pyrolysis liquids still is its combustion to produce heat. Compared to diesel fuel, the flames derived from pyrolysis liquids in the same combustion conditions are shorter, wider, and brighter. Compared to natural gas, the flame is more yellow and

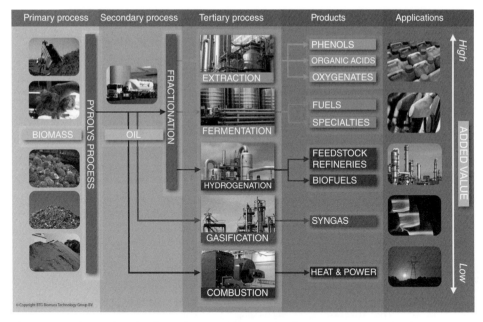

Figure 6.11 The "Bioliquid" refinery. Source: with permission from BTG Biomass Technology Group BV [25]. *See the plate section for a color representation of this figure.*

longer. Both Empyro and Ensyn deploy their pyrolysis liquid primarily for combustion systems – Empyro at FrieslandCampina in Borculo (the Netherlands) and Ensyn in three relatively small-sized boilers.

Ensyn have announced three contracts. A first one aims to replace up to 50% of natural gas in a boiler with deliveries from the fourth quarter of 2015 onwards (9500 t/y pyrolysis liquids). The boiler operated by Youngstown Thermal provides steam for heat and hot water for the central business district of Youngstown, including the Youngstown State University. Youngstown Thermal owns and operates four boilers at the site. Two other contracts are aiming at heating purposes for two hospitals, the Memorial Hospital (see Figure 6.12) and Valley Regional Hospital, both in New Hampshire, and both roughly deploying 1000 t/y of those liquids. The total amount of liquids combusted by Ensyn is not known.

FrieslandCampina is located around 35 km from Empyro's liquid production site, and uses the liquids to replace natural gas in the production of steam. The overall energy balance for Borculo is provided in Figure 6.13 [31].

6.5.2 Diesel Engines

Use of the liquids in diesel engines to produce heat *and* power seems an even more interesting approach; even at small scales <1 MW_e high efficiencies can be achieved. Use of pyrolysis liquids in diesel is very challenging, as, for example, standard injectors and pumps are not corrosion resistant, cannot withstand abrasive wear and high impact, and the atomization of pyrolysis liquids is quite difficult, while the Cetane number for pyrolysis liquids – although data are scarce – is in any case much lower than for diesel [33, 34].

194 *Thermochemical Processing of Biomass*

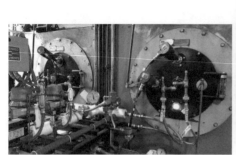

Figure 6.12 Combustion of Ensyn's liquids at the Memorial Hospital, North Conway, New Hampshire. Source: with permission from Ensyn Technologies Inc., Barry Freel, personal communications, January 2018. See the plate section for a color representation of this figure.

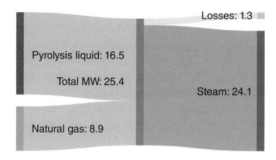

Figure 6.13 Energy balances for Borculo. Source: with permission from BTG Bioliquids B.V. [31].

Nevertheless, some significant progress has been made in the last few years, and a small one-cylinder engine (max output 23 kW$_e$) has been demonstrated using pyrolysis liquids mixed with 10–20% ethanol for periods over 100 hours, with the longest continuous run of over 50 hours [35]. In these runs, both the compression ratio and the air inlet temperature were increased. A four-cylinder 50 kW$_e$ diesel engine has been modified and has been successfully started on pyrolysis liquids as well, running for over 10 hours.

6.5.3 Co-refining Options

A high-impact perspective for pyrolysis liquids is to convert them in existing refining units. This possibly represents a boost to the fast pyrolysis technology, to become "the (*ligno-*)cellulosic game changer" [36]. This huge potential impact justifies a more detailed description of the status of co-FCC of pyrolysis liquids.

Opportunities to convert pyrolysis liquids into new generation bio-fuels for the existing transport sector include various routes, schematically outlined in Figure 6.14. A straightforward option is biomass fast pyrolysis to liquids (PL), followed by (co-)hydrotreating of the oxygenated components and possibly (co-)processing with FCC catalysts. The co-refining (FCC or hydrotreating) can occur by mixing (emulsification) the liquid with crude oil

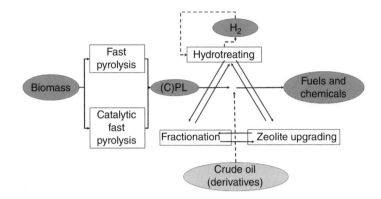

Figure 6.14 Routes from biomass to fuels and chemicals through (catalytic) pyrolysis.

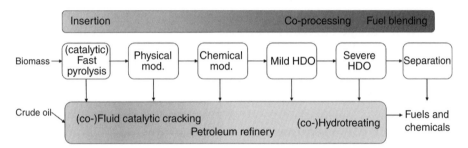

Figure 6.15 Pathways for the introduction of biomass in a petroleum refineries. Source: adapted from [38, 39].

derivatives. Alternatives include the use of an intermediate fraction of the pyrolysis liquid obtained by a fractionation step (for example, separation of the lignitic fraction from the carbohydrate fraction) and separate upgrading (FCC or hydrotreating) of one or more of these fractions.

One approach indicated in Figure 6.14 is to reduce the oxygen content upfront of the FCC, for example by catalytic pyrolysis of biomass (catalytic pyrolysis liquids, CPL). It is suggested that the liquids from catalytic pyrolysis have higher heating values, low acid and water content, and even may yield so-called drop-in biofuels miscible with crude oil derivatives. It is likely though that these liquids require further upgrading. Furthermore, extensive coking of catalysts contributes to relatively low carbon yields for liquids from catalytic pyrolysis. A close-to-commercial and relatively low-risk option instead is to co-feed pyrolysis liquids (or upgraded liquids) in an existing refinery as a replacement of fossil fuels. This facilitates the introduction of renewable carbon into the existing fuels infrastructure, re-purposing the units. Feeding of pyrolysis liquid in such complexes, however, introduces significant uncertainties, and related to this, some potential problems and pitfalls, as summarized elsewhere [37].

Possibilities for introducing pyrolysis liquids into petroleum refineries are presented in Figure 6.15. A key conversion technology for pyrolysis liquids though is FCC.

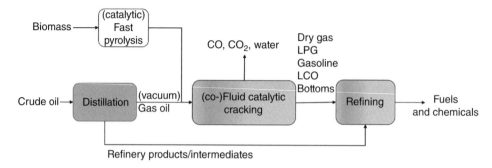

Figure 6.16 Pyrolysis liquid co-refining in a FCC. Source: adapted from [40].

The FCC, located in the heart of the refinery complex, usually is operated at very large scales (ranging up to 200 t petroleum derivative/h), but still is a rather flexible process unit as it can readily be tuned to accommodate different feedstocks. Relevant for the present work, the main valuable outputs from the FCC are gasoline (roughly 40–50%), light cycle oil or LCO (10–20%), and the dry gas and LPG (5–15%). Less preferred products are the bottoms (merely the unconverted vacuum gas oil [VGO] or "slurry oil," 10–15%) and coke (5–10%). Co-feeding of pyrolysis liquids can – in principle – be rapidly integrated. A schematic diagram is shown in Figure 6.16.

6.5.3.1 Co-FCC of Pyrolysis Liquids and Crude Oil Derivatives

Co-treating pyrolysis liquids in existing FCC units dates back to the end of the 1990s, but at that time no incentives were in place to promote biofuels, neither were refineries willing to take any risk deploying any biomass-derived liquids [41–43]. Moreover, various experiments showed that conversion of pure pyrolysis liquids over conventional FCC catalysts resulted in hydrocarbons, but at very low overall yields well below 30 wt% (from pyrolysis liquids to final fuels). These values correspond to maximum carbon yield to liquids far below 40%, the remaining carbon being lost in char, coke, and gas. The final product liquids could still contain oxygen, and additional treatments of the liquids seemed required to produce transport fuels, while the deployed catalysts (in FCC as well as in the upgrading) rapidly deactivated [40, 44, 45].

It took over 10 years before blending of pyrolysis liquids with conventional crude-derived oils (vacuum gas oil or VGO) became of renewed interest, initially deploying standardized techniques for the simulation of FCC units. The results were ambiguous.

Quite positive results were reported by UOP [46]; although high yields of coke were noted, significant amounts of carbon were transferred to gas (fuel gas and LPG) but, surprisingly, also to gasoline. Indicatively, replacement of 20% of the crude oil effectively reduces the total amount of carbon fed to the FCC by approx. 13% (due to the oxygen in the pyrolysis liquid), but the apparent yield of naphtha was reduced by only less than 5%. As a further positive aspect, the pyrolysis liquid appeared to increase the crackability of the VGO, being a very positive synergy.

Negative conclusions were drawn by Grace, processing a blend of 3 wt% pure pyrolysis liquids and VGO in a continuously operated system [47]. The co-feeding of pyrolysis

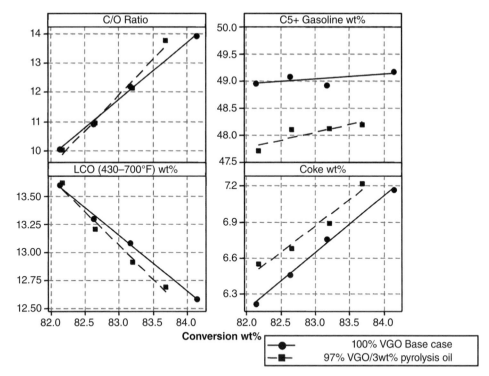

Figure 6.17 Yields (wt%) versus conversion for the co-processing of PL in FCC. (C/O = catalyst-to-oil ratio, LCO = light cycle oil, or diesel fraction) [47].

liquids into FCC units with VGO is not beneficial, with a pessimistic estimate of 10% of the carbon from the bio-liquids ending up in useable products (LPG and liquids) [47]. Small amounts of pyrolysis liquids already significantly affect the yields, causing higher yields of coke (see also Figure 6.17). More bottoms were formed, at the expense of gasoline and LCO yields. Less LPG is formed, and yields of useable products (LPG and liquids) are around 40%.

These data did not provide a convincing reason to co-feed pyrolysis liquids in an FCC. Nevertheless, recognizing that small-scale laboratory reactors substantially deviate from continuously operated systems, the Brazilian company Petrobras deployed a near-commercial FCC vacuum gasoil unit ("Six", located in São Mateus do Sul in Brazil) to co-process pure pyrolysis liquids [40, 44, 45]. "Six" is one of the few large FCC demonstration units left to carry out demonstration-scale testing. It has a vertical riser reactor (50 mm ID and 18 m height), a fluid bed regenerator reactor, a stripper, and a lift line, with a total catalyst inventory of 450 kg. VGO and pure pyrolysis liquids were injected at two different heights into the riser reactor, at a combined feed rate of 150 kg/h at conditions close to that used in commercial FCC units.

From 2012 onwards, Petrobras tested pyrolysis liquids in this unit, and provided a large dataset – varying catalyst circulation rates, substitution levels, and cracking severities. These larger scale experiments initially revealed technical problems while co-feeding pure PL, specifically due to nozzle plugging. But this could well be circumvented by modifications to the liquid injection system. The main conclusion from the Petrobras work supports the general comments from the work by UOP and Grace; however, the reduction in useable products were shown to be far less. For a 10% substitution of VGO by pyrolysis liquids similar yields in gasoline and coke are found; a significant reduction was seen if the pyrolysis liquid was increased up to 20%.

From the Petrobras data, carbon yields (from biomass to fuels) in the range of 25–30% can be estimated. As suggested earlier, this is partially due to the presence of the reactive carbohydrate fraction in the pyrolysis liquids. At elevated temperatures these components are rapidly converted into coke. By converting the reactive oxygenates into alcoholic groups by hydrogenation ("stabilized pyrolysis oils"), such molecules become far less reactive and this approach opened the way to increase the overall carbon yield in the co-FCC option (see below).

6.5.3.2 Co-Fluid Catalytic Cracking of Hydrotreated Pyrolysis Liquids

The approach to deploy stabilized pyrolysis oils in co-FCC is a natural choice while converting pyrolysis liquids into oxygen-free transportation fuels. To produce transportation fuels (diesel, kerosene, gasoline) from fast pyrolysis liquids, a severe thermochemical treatment of the fast pyrolysis is needed (upgrading) to obtain full deoxygenation of the bio-liquid. Research on this topic dates back to the 1990s, from which it was concluded that a two-step approach must be applied: a first step is carried out at 200–250 °C to stabilize the (very) reactive components in the pyrolysis liquids, and the second step is carried out at 350–400 °C to deoxygenate the stabilized oil product in a further hydrotreating step [48]. Carbohydrate chemistry is the main mechanism in the low temperature step, while lignin chemistry prevails at higher temperatures [49]. Obviously, the reactive cellulose-derived fraction of the pyrolysis liquid must be transformed first (possibly to alcohols) before further dehydration and hydrogenation can take place at more severe conditions. This latter high temperature step may be more similar to petrochemical hydrotreatment, enabling conventional (sulfided) NiMo or CoMo catalysts. As an alternative to full deoxygenation of the pyrolysis liquid in the catalytic hydrotreatment process, mixtures of transportation fuels can also be obtained by co-feeding partially upgraded fast pyrolysis liquids into the FCC This is a hybrid form of co-FCC, where a combination of dehydration (by hydrogenation) and decarboxylation (co-FCC) is chosen to optimize the overall carbon yield, hydrogen consumption, and process costs (see Figure 6.18).

The first successful results using hydrotreated liquid date back to the end of the 1990s [50], but a more detailed study is from 2010 [51]. Liquids catalytically upgraded with hydrogen showed less coke and char than pure untreated bio-liquids. Interestingly, substantial deoxygenation to low oxygen levels in the hydrotreated product, requiring ~800 liters hydrogen per kilogram of pyrolysis liquid, is probably not needed to obtain the best performance. Liquids with a relatively high remaining oxygen content in between 10 and 30 wt%, and associated low hydrogen use required for preparation in the order of 100–300 NL/kg, could already be successfully co-processed with VGO.

Important objectives for the hydrotreatment step in this integrated hydroprocessing – FCC approach (HT/FCC) is (i) to reduce the overall hydrogen consumption, (ii) to maintain a high overall carbon yield, and (iii) to meet specific product properties targets for further FCC processing (such as low water content, low acid number, volatility, miscibility

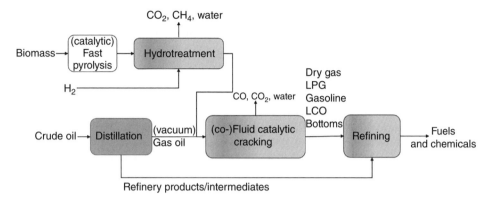

Figure 6.18 Co-refining hydrotreated pyrolysis liquid and VGO in a FCC.

with a polar component, low charring tendency and high yields of useable products). Pyrolysis liquids treated over Ru/C could be co-processed well, even at small laboratory scale [52, 53]. The severity of the prior hydrotreatment of pyrolysis liquids has a major effect on the product distribution. Interestingly, high and constant gasoline yields are obtained at low treatment temperatures. As in case of pure pyrolysis liquids, synergetic effects between high molecular weight compounds in the oxygenated bio-liquids and VGO are apparent. For severely treated pyrolysis liquid (330 °C over Ru/C) no or very limited differences in product distributions are noted between VGO and mixtures of VGO and the treated bio-liquids.

As a main conclusion, there is good potential for co-feeding of such post-treated liquids in a FCC unit. A mild prior treatment, yielding a 20+ wt% residual oxygen in the products, appears suitable for co-FCC purposes, while it also allows one to get rid of most of the water in the liquids. Co-FCC is technically possible, and no unexpected deviation in the product spectrum derived from standard VGO is observed. There are no substantial differences between the products derived from pure VGO and the co-feeding, while small deviations can be attributed to the significantly lower amounts of carbon fed as the upgraded liquid contains water and oxygen.

6.5.4 Gasification

A direct and straightforward application of pyrolysis liquids is in gasification. Here, the liquids are converted into syngas with steam and/or oxygen, and then this syngas can be converted catalytically to a desired product (e.g. methanol). Pilot-scale test work has been performed by several consortia.

Back in 2002 gasification of pyrolysis liquids was demonstrated by Shell and BTG in a 1 MW_{th} oxygen-blown atmospheric pressure entrained flow gasifier from Future Energy GmbH (formerly Choren) in Freiberg [54]. Feeding the pyrolysis liquid at 140 kg/hour, a run was performed lasting 10 hours, consuming all available pyrolysis liquid. No significant tar concentrations (aromatic or phenolic) were found in the final product gas.

More recently, Chemrec in Sweden successfully demonstrated the co-gasification of pyrolysis liquids with black liquor. Advantages claimed compared to solid biomass gasification included the ease of pressurization of a liquid and the higher energy density of the liquid. In 2012 and 2013, tests were performed in a pressurized pilot-scale entrained flow gasifier from ETC in Sweden [55]. Pyrolysis liquids derived from clean wood as well as from wheat straw produced by BTG were gasified, varying the equivalence ratios. Typical dry gas compositions of 46% CO, 30% H_2, and 23% CO_2 were obtained. Gas-phase contaminants included about 2% CH_4 and ppm levels of H_2S, COS, and benzene.

An alternative approach to produce syngas from biomass via pyrolysis and entrained flow gasification is explored by KIT under the name Bioliq [56]. Bioliq operates a 500 kg/h pilot plant for the production of a slurry of pyrolysis liquid and char, which is further gasified in a high pressure 5 MW_{th} entrained flow gasifier. In 2013 high-pressure gas cleaning and dimethyl ether (DME)/synfuel synthesis plants were constructed and in July 2013 the DME synthesis plant was successfully tested using artificial syngas.

The disadvantage is that, unlike pure pyrolysis liquid, the slurry contains most of the ash from the original biomass. These ash components cause slag formation inside the gasifier, requiring the use of a so called "slagging gasifier," which is considerably more expensive compared to the "non-slagging" variant. Recently a full test campaign was finalized to covert approx. 40 tonnes of straw and produced around 1 m^3 of synthetic fuel. Tests deploying liquids from Empyro, as well as from a slow pyrolysis unit (Profagus) have been carried out. Some experimental results for the gasifier have been recently published [57].

Other gasification systems have been tested as well, but these are mainly only on a laboratory scale [58–61].

– *CPO*: in catalytic partial oxidation (CPO) a catalyst is added to the system, which allows higher conversion at lower temperatures. Operational difficulties associated with particulate formation in the atomization of pyrolysis liquids has led to the investigation of alternative catalyst systems, such as the use of monolithic catalysts [58].
– *Steam reforming*: steam reforming produces hydrogen, and sometimes produces synthesis gas as intermediate product for further processing. Research has been carried out regarding the steam reforming of model compounds of pyrolysis liquids. An example is acetic acid to mimic steam reforming. The main challenge, however, is to prevent catalyst contamination, as carbon deposition in the form of whiskers is troublesome even for pure methane [59]. Recently, results were published for the catalytic and non-catalytic steam reforming of only the aqueous phase [62–64]. High hydrogen yields of approx. 70–80% of the theoretically achievable amount were shown. However, catalyst stability is the main bottleneck here also.

As a main conclusion, gasification of pyrolysis liquids is technically possible and existing entrained flow gasifiers enable the direct use of these liquids. The remaining challenges are mainly economic, because such gasifiers are primarily used for low-value feedstocks. Higher efficiency gasifiers or steam reformers require catalysts, and these systems are less far developed. In all these systems, catalyst activity and stability remain important for future process development.

6.6 Chemicals

Hundreds of compounds have been identified in GC analysis as fragments from lignin (phenols, eugenols, and guaiacols amongst others) and from holocellulose (sugars, acetaldehyde, and formic acid). Methanol (wood spirit) and acetic acid are so far the only products that could be isolated from these liquids on a near-commercial economic scale. Indeed, large fractions of acetic acid, acetol, and hydroxyacetaldehyde are identified but still, until now, only 40–50% of the liquid's identity (excluding the water) has been revealed. The large, less severely cracked or depolymerized molecules in the liquids (derived from the cellulose and the lignin) are not identified.

All types of (oxygen) functionalities are present: acids, sugars, alcohols, ketones, aldehydes, phenols and their derivatives, furans, and other mixed oxygenates. Also (poly)phenols are present, sometimes in rather high concentrations. These phenolic fractions include phenol, eugenol, guaiacols and their derivatives, and the so-called pyrolytic lignin (poly-phenols) representing the water-insoluble components derived mostly from lignin. Nevertheless, for all these compounds, the rule of thumb is simply that they are present in amounts too low to be isolated cost effectively.

From this perspective, fast pyrolysis is merely an effective tool to pretreat biomass, to further facilitate its subsequent fractionation. Upon pyrolysis the (ash-free) liquid can easily be fractionated into three product streams, namely the pyrolytic lignin (~35% of the original carbon); the pyrolytic sugars (~45% of the original carbon), and a watery phase containing smaller organic components, e.g. acetic acid (~20% of the original carbon). A good option is to separate the pyrolysis liquids on basis of functionality rather than molecular (monomeric) structure. Subsequently, the obtained fractions can be further processed to produce chemicals and/or green products.

- Use of pyrolytic lignin as a renewable substituent for fossil phenol in phenol/formaldehyde resins and derivatives. These types of resins are widely used in wood products like particle boards, plywood etc. Recently it has been demonstrated that the phenol in phenol/formaldehyde resins can be substituted up to 75 wt% by pyrolytic lignin standards for this type of resin. Another application of the pyrolytic lignin is in the replacement of fossil bitumen in various bitumen-based materials, e.g. in asphalt and roofing materials, the latter recently demonstrated as a bitumen replacement [65]. The pyrolytic lignin could be used in the production of green phenolic (mono-) derivatives, and as a possible raw material for various coatings, composites, and preservatives.
- The sugar phase could be a renewable source for the production (via glucose) of bio-ethanol, levulinic acid, and polyols, etc., for example. It has a high potential as a renewable sugar source due to the fact that it is not extracted from food sources such as corn, sugarcane, etc. Exploratory experiments by different researchers have indicated that a substantial amount of the pyrolytic sugars can be converted to fermentable sugar monomers (especially the C_6 sugars). Subsequently, fermentation of this phase by means of a yeast culture resulted in the production of ethanol.
- The water phase contains carboxylic acids, which can be converted to carboxylate salts by reaction with hydroxides or carbonates.

A European project, Bio4Products, started to demonstrate a two-step conversion method to transform four feedstocks (straw, bark, forest residues, and sunflower husks) into four types of intermediates for renewable products. A 5 ton fractionation unit will be built and operated to obtain phenolic resins, roofing materials, sand molding resins, engineered wood, and natural fiber-reinforced products [66].

6.7 Catalytic Pyrolysis

Another option is to produce hydrocarbonaceous oxygen-free liquids from biomass, using catalysts to deoxygenate the liquids in situ (in-bed or ex-bed). An extensive critical view on this concept has been published by the author elsewhere [67]. The most relevant conclusion from this work is that, despite much effort on a large variety of catalysts and process conditions over more than 30 years, no significant improvement in yield and quality is reported compared to those in the early days.

An extensive, very detailed, inside story on KiOR was published by Lane in 2016, and merely told the story of "too fast with too little knowledge" [68]. High yields of 67 gal oxygen-free fuel per ton biomass that was claimed was never justified by experiments, the low minimum selling costs (around $1.80 per gallon) presented could never be justified and was not based on realistic assumptions, and the proposed catalysts were never shown to remain active for sufficient long times, and the basis that this could be the case was not well justified.

Two other companies are involved in the commercialization of catalytic pyrolysis technology, viz. the Dutch company BioBTX and the US-based company Anellotech. They use different feedstocks, and focus on chemicals and not fuels. These are still in relatively early development and will not be discussed here.

6.8 Concluding Remarks

Fast pyrolysis is a cost-effective technique to depolymerize plant materials into an energy-rich liquid. The process development is maturing, having been developed steadily during the last 30 years. With two or three large-scale failures in the past 30 years (Dynamotive, Pyrovac, and KiOR), current developments are promising, considering the success of BTG Bioliquids (Malaysia/The Netherlands), Ensyn Technologies Inc. (USA and Canada), and Valmet (Finland). Although transportation fuels and high-value chemicals are attractive long-term goals, more immediate targets for the main liquid product are process heat and electricity. Large amounts of those liquids are available for the development and commercial-scale demonstration of a wide range of applications, and high-quality liquids can be purchased on the market.

Immediate challenges for the biomass pyrolysis sector are to improve the reliability of plant operations and to use cheaper residual feedstocks. Although pyrolysis processes may accept a variable feed, the varying nature of the feedstock leads to inconsistency in feed as well as product behavior. In addition, each feedstock will require a dedicated feed section. Establishment of standards for feedstock quality but also further detailing of product quality can help in avoiding inconsistencies.

Despite almost 30 years of experience, analysis of pyrolysis liquids is incomplete, it cannot be easily replicated in other laboratories, and repeatable/reproducible analysis techniques to do so are limited. GC analysis does not provide a complete picture of the liquid composition, as, by definition, the low injection temperature affects the actual composition of the products effectively being analyzed. New non-destructive techniques, such as 2D-NMR, are required.

The general lack of understanding of pyrolysis chemistry is caused by a number of aspects, including, but not limited to, the following: (i) the vast amount of functionalities in biomass, the liquids, and the intermediate product, (ii) the temperature sensitivity of all these materials and the short lifetime (less than 0.1 s) of most of the intermediates, (iii) the relatively slow heat transfer that complicates isothermal testing, and (iv) the dependence of product yields on the residence time of volatiles within the liquid and gas phases. A clue to better understand pyrolysis is that it is initiated with a network of largely unidentified solid-phase reactions which fractionate and depolymerize the solid to a short-lived intermediate liquid. Within this liquid phase a complexity of depolymerization, rearrangement, and dehydration reactions occur until products volatilize or are ejected in the form of aerosols. Understanding the condensed-phase chemistry is critical to gaining more insight in pyrolysis, since the former determines the quality of the bio-liquid intermediate and therefore the overall economics of the process.

Acknowledgement

This chapter is an update of *Fast Pyrolysis* prepared by Robbie H. Venderbosch and Wolter Prins for the first edition of this book (*Thermochemical Processing of Biomass: Conversion into Fuels, Chemicals and Power, First Edition*, edited by Robert C. Brown, © 2011 John Wiley & Sons, Ltd).

References

1. Venderbosch, R.H. and Prins, W. (2011). Fast pyrolysis. In: *Thermochemical Processing of Biomass*, 124–156. Wiley.
2. Mettler, M.S., Vlachos, D.G., and Dauenhauer, P.J. (2012). *Energy & Environmental Science* **5**: 7797–7809.
3. Kersten, S. and Garcia-Perez, M. (2013). *Current Opinion Biotechnology* **24**: 414–420.
4. Wang, X., Kersten, S.R.A., Prins, W., and van Swaaij, W.P.M. (2005). Biomass pyrolysis in a fluidized bed reactor. Part 2: experimental validation of model results. *Industrial & Engineering Chemistry Research* **44** (23): 8786–8795.
5. Wang, X. (2006). Biomass fast pyrolysis in a fluidized bed, Ph.D. Thesis, University of Twente.
6. Kersten, S.R.A., Wang, X., Prins, W., and Van Swaaij, W.P.M. (2005). Biomass pyrolysis in a fluidized bed reactor. Part 1: literature review and model simulations 2005. *Industrial & Engineering Chemistry Research* **44** (23): 8773–8785.
7. Wagenaar, B.M., Prins, W., and van Swaaij, W.P.M. (1993). Flash pyrolysis kinetics of pine wood. *Fuel Processing Technology* **36**: 291–298.
8. Chan, W.-C.R., Kelbon, M., and Krieger, B.B. (1985). Modelling and experimental verification of physical and chemical processes during pyrolysis of a large biomass particle. *Fuel* **64**: 1505–1513.
9. Thurner, F. and Mann, U. (1981). Kinetic investigation of wood pyrolysis. *Industrial & Engineering. Chemistry Process Design and Development* **20**: 482–488.
10. Di Blasi, C. and Branca, C. (2001). Kinetics of primary product formation from wood pyrolysis. *Industrial and Engineering Chemistry Research* **40**: 5547–5556; Di Blasi, C. (2008). Modelling chemical and physical processes of wood and biomass pyrolysis. *Progress in Energy and Combustion Science*, **34**, 47–90.
11. Westerhof, R.J.M., Oudenhoven, S.R.G., Marathe, P.S. et al. (2016). The interplay between chemistry and heat/mass transfer during the fast pyrolysis of cellulose. *Reaction Chemistry & Engineering* **1** (5): 555–566.

12. Roy, C., Morin, D., and Dubé, F. (1997). The biomass Pyrocycling™ process. In: *Biomass Gasification and Pyrolysis: State of the Art and Future Prospects* (ed. M. Kaltschmidt and A.V. Bridgwater), 307–315. Stuttgart, Germany: CPL Press.
13. Teixeira, A.R., Mooney, K.G., Kruger, J.S. et al. (2011). Aerosol generation by reactive boiling ejection of molten cellulose. *Energy & Environmental Science* **4** (10): 4306–4321.
14. Dauenhauer, P. (2018). Personal communication.
15. Dedic, C., Tiarks, J., Sanderson, P.D., et al. (2015). Optical diagnostic techniques for investigation of biomass pyrolysis, Fourth International Conference on Thermochemical Biomass Conversion Science, Chicago, IL, November 2–5.
16. Tiarks, J.A., Dedic, C.E., Brown, R.C., et al. (2016). Laboratory optical diagnostics for the investigation of physical transport mechanisms for biomass fast pyrolysis, in: 2016 Spring Technical Meeting, Central States Section of the Combustion Institute, May 15–17, 2016.
17. Leijenhorst, E.J., Wolters, W., Van De Beld, L. et al. (2016). Inorganic element transfer from biomass to fast pyrolysis oil: review and experiments. *Fuel Processing Technology* **149**: 96–111.
18. Vassilev, S.V., Baxter, D., Andersen, L.K. et al. (2012). An overview of the organic and inorganic phase composition of biomass. *Fuel* **94** (1): 1–33.
19. Vassilev, S.V., Vassileva, C.G., and Baxter, D. (2014). Trace element concentrations and associations in some biomass ashes. *Fuel* **129**: 292–313.
20. Keown, D.M., Hayashi, J.-i., and Li, C.-Z. (2008). Effects of volatile–char interactions on the volatilisation of alkali and alkaline earth metallic species during the pyrolysis of biomass. *Fuel* **87** (7): 1187–1194.
21. Marathe, P.S., Oudenhoven, S.R.G., Heerspink, P.W. et al. (2017). Fast pyrolysis of cellulose in vacuum: the effect of potassium salts on the primary reactions. *Chemical Engineering Journal* **329** (Suppl. C): 187–197.
22. Leijenhorst, E. (2018). BTG Biomass Technology Group BV. Personal communication.
23. Li, C.Z., Sathe, C., Kershaw, J.R., and Pang, Y. (2000). Fates and Roles of Alkali and Alkaline Earth Metals During the Pyrolysis of a Victorian Brown Coal. *Fuel* **79** (3–4): 427–438.
24. Quyn, D.M., Wu, H., Bhattacharya, S.P., and Li, C.-Z. (2002). Volatilisation and catalytic effects of alkali and alkaline earth metallic species during the pyrolysis and gasification of Victorian brown coal. Part II. Effects of chemical form and valence. *Fuel* **81** (2): 151–158.
25. BTG Biomass Technology Group BV (2018). Personal communication.
26. Piskorz, J., Radlein, D., Scott, D.S., and Czernik, S. (1988). Liquid products from the fast pyrolysis of wood and cellulose. In: *Research in Thermochemical Biomass Conversion* (ed. A.V. Bridgwater and J.L. Kuester), 557. London: Elsevier Applied Science.
27. Freel, B. (2018). Ensyn Technologies Inc. Personal communication.
28. UOP (2016). Final project report pilot-scale biorefinery: Sustainable transport fuels from biomass and algal residues via integrated pyrolysis, catalytic hydroconversion and co-processing with vacuum gas oil , award No. EE0002879.
29. Sipila, K., Solantausta, Y., Jokela, P., and Raiko, M. (2008). Method for carrying out pyrolysis. World Patent WO2009047387, filed 8 October 2008 and issued 11 August 2009.
30. Autio, J. (2018). Valmet. Personal communication.
31. Muggen, G. (2018). BTG Bioliquids. Personal communication.
32. Radlein, D. and Quignard, A. (2012). Methods of upgrading biooil to transportation grade hydrocarbon fuel, CA2812974 (A1).
33. Chiaramonti, D., Oasmaa, A., and Solantausta, Y. (2007). Power generation using fast pyrolysis liquids from biomass. *Renewable and Sustainable Energy Reviews* **11** (6): 1056–1086.
34. Hossain, A.K. and Davies, P.A. (2013). Pyrolysis liquids and gases as alternative fuels in internal combustion engines – A review. *Renewable and Sustainable Energy Reviews* **21** (Suppl. C): 165–189.
35. Van de Beld, B., Holle, E., and Florijn, J. (2018). The use of a fast pyrolysis oil – ethanol blend in diesel engines for CHP applications. *Biomass and Bioenergy* **110**: 114–122.

36. Ensyn Technologies Inc. (2014). Cellulosic game changer, Bioenergy Insight, December 1, 5(6), p. 58–59.
37. Venderbosch, R.H. and Heeres, H.J. (2015). Chapter 16: Coprocessing of (Upgraded) Pyrolysis Liquids in Conventional Oil Refineries. In: *Biomass Power for the World* (ed. W.P.M. Van Swaaij, S.R.A. Kersten and W. Palz), 760. Pan Stanford. ISBN: 9789814613880.
38. Zacher, A.H., Olarte, M.V., Santosa, D.M. et al. (2014). A review and perspective of recent bio-oil hydrotreating research. *Green Chemistry* **16** (2): 491.
39. Talmadge, M.S., Baldwin, R.M., Biddy, M.J. et al. (2014). A perspective on oxygenated species in the refinery integration of pyrolysis oil. *Green Chemistry* **16** (2): 407–453.
40. Pinho, A. de Rezende, de Almeida, M.B.B., Mendes, F.L. et al. (2017). Fast pyrolysis oil from pinewood chips co-processing with vacuum gas oil in an FCC unit for second generation fuel production. *Fuel* **188** (Suppl. C): 462–473.
41. Adjaye, J.D. and Bakhshi, N.N. (1995). Production of hydrocarbons by catalytic upgrading of a fast pyrolysis bio-oil. Part I: Conversion over various catalysts. *Fuel Processing Technology* **45**: 161–183.
42. Adjaye, J.D. and Bakhshi, N.N. (1995). Catalytic conversion of a biomass-derived oil to fuel and chemicals I: model compound studies and reaction pathways. *Biomass and Bioenergy* **8** (3): 131–149.
43. Adjaye, J.D. and Bakhshi, N.N. (1995). Production of hydrocarbons by catalytic upgrading of a fast pyrolysis bio-oil. Part II: comparative catalyst performance and reaction pathways. *Fuel Processing Technology* **45**: 185–202.
44. Pinho, A. de Rezende, de Almeida, M.B.B., Mendes, F.L. et al. (2015). Co-processing raw bio-oil and gasoil in an FCC Unit. *Fuel Processing Technology* **131** (0): 159–166.
45. Pinho, A. de Rezende, de Almeida, M.B.B., Mendes, F.L. et al. (2014). Production of lignocellulosic gasoline using fast pyrolysis of biomass and a conventional refining scheme. *Pure and Applied Chemistry* **86** (5).
46. Marinangeli, R., Marker, T., and Petri, J. (2006). Opportunities for biorenewables in oil refineries, Report No. DE-FG36-05GO15085, UOP.
47. Bryden, K., Weatherbee, G., and Habib, T.E. (2013). *FCC Pilot Plant Results with Vegetable Oil and Pyrolysis Oil Feeds, Biomass 2013*, July 31 – August 1. Washington: US Department of Energy. Also available at http://www1.eere.energy.gov/bioenergy/pdfs/biomass13_habib_2-d.pdf (last accessed Match 2018).
48. Elliott, D.C. (2007). Historical developments in hydroprocessing bio-oils. *Energy and Fuels* **21**: 1792–1815.
49. Venderbosch, R.H., Ardiyanti, A.R., Wildschut, J. et al. (2010). Stabilization of biomass-derived pyrolysis oils. *Journal of Chemical Technology & Biotechnology* **85** (5): 674–686.
50. Samolada, M.C., Baldauf, W., and Vasalos, I.A. (1998). Production of a bio-gasoline by upgrading biomass flash pyrolysis liquids via hydrogen processing and catalytic cracking. *Fuel* **77** (14): 1667–1675.
51. de Miguel Mercader, F., Groeneveld, M.J., Kersten, S.R.A. et al. (2010). Production of advanced biofuels: co-processing of upgraded pyrolysis oil in standard refinery units. *Applied Catalysis B: Environmental* **96** (1–2): 57–66.
52. Fogassy, G., Thegarid, N., Schuurman, Y. et al. (2011). From biomass to bio-gasoline by FCC co-processing: effect of feed composition and catalyst structure on product quality. *Energy & Environmental Science* **4** (12): 5068–5076.
53. Thegarid, N., Fogassy, G., Schuurman, Y. et al. (2014). Second-generation biofuels by co-processing catalytic pyrolysis oil in FCC units. *Applied Catalysis B: Environmental* **145**: 161–166.
54. Knoef, H.A.M. (ed.) (2012), Chapter 8). *Handbook Biomass Gasification*, 2e, 219–250. ISBN: 978-90-819385-0-1.
55. Leijenhorst, E.J., Assink, D., van de Beld, L. et al. (2015). Entrained flow gasification of straw- and wood-derived pyrolysis oil in a pressurized oxygen blown gasifier. *Biomass and Bioenergy* **79**: 166–176.

56. Dahmen, N., Henrich, E., Dinjus, E., and Weirich, F. (2012). *Energy, Sustainability and Society* **2**: 3.
57. Eberhard, M., Santo, U., Böning, D. et al. (2018). Der bioliq®-Flugstromvergaser – ein Baustein der Energiewende. The bioliq® entrained-flow gasifier – a module for the German Energiewende. *Chemie Ingenieur Technik* **90** (1–2): 85–98.
58. Leijenhorst, E.J., Wolters, W., Van De Beld, B. et al. (2014). Autothermal catalytic reforming of pine-wood-derived fast pyrolysis oil in a 1.5 kg/h pilot installation: performance of monolithic catalysts. *Energy and Fuels* **28**: 5212–5221.
59. van Rossum, G. (2009). Steam reforming and gasification of pyrolysis oil. PhD thesis. University of Twente.
60. Rennard, D., French, R., Czernik, S. et al. (2010). Production of synthesis gas by partial oxidation and steam reforming of biomass pyrolysis oils. *International Journal of Hydrogen Energy* **35**: 4048–4059.
61. Trane, R., Dahl, S., Skjoth-Rasmussen, M.S., and Jensen, A.D. (2012). Catalytic steam reforming of bio-oil. *International Journal of Hydrogen Energy* **37** (8): 6447–6472.
62. Remón, J., Broust, F., Valette, J. et al. (2014). Production of a hydrogen-rich gas from fast pyrolysis bio-oils: comparison between homogeneous and catalytic steam reforming routes. *International Journal of Hydrogen Energy* **39** (1): 171–182.
63. Garcia, L., French, R., Czernik, S., and Chornet, E. (2000). Catalytic steam reforming of bio-oils for the production of hydrogen: effects of catalyst composition. *Applied Catalysis A: General* **201** (2): 225–239.
64. Czernik, S., Evans, R., and French, R. (2007). Hydrogen from biomass-production by steam reforming of biomass pyrolysis oil. *Catalysis Today* **129** (3–4): 265–268.
65. http://www.innovatievematerialen.nl/index.php/biotumen?id=95 (accessed 16 November 2018).
66. https://bio4products.eu (accessed 16 November 2018).
67. Venderbosch, R.H. (2015). A critical view on catalytic pyrolysis of biomass. *ChemSusChem* **8**: 1306–1316. https://doi.org/10.1002/cssc.201500115.
68. Lane, J. (2016). KiOR: The inside true story of a company gone wrong, in: *Biofuels Digest*. http://www.biofuelsdigest.com/bdigest/2016/05/17/kior-the-inside-true-story-of-a-company-gone-wrong (accessed 16 November 2018).

7
Upgrading Fast Pyrolysis Liquids

Karl O. Albrecht, Mariefel V. Olarte and Huamin Wang

Chemical and Biological Process Development Group, Pacific Northwest National Laboratory, Richland, WA, USA

7.1 Introduction

The liquid formed from fast pyrolysis (FP) of biomass is commonly termed bio-oil. Biomass is converted to bio-oil through FP at 400–550 °C typically at 1 atm. During FP, the biomass is held at temperature for a residence time on the order of 1–2 s. A large portion of the total mass and carbon contained in the biomass is converted and retained in the liquid fraction of pyrolysis products. However, bio-oil is acidic, viscous, and thermally unstable. Bio-oil must be upgraded to be useful as a liquid transportation fuel or separated and further treated to recover value-added chemicals. Bio-oil is generally regarded as difficult to upgrade due to its thermal instability and chemical complexity. This results in part from the presence of highly reactive species such as aldehydes and terminal olefins as well as the presence of sugars and anhydrosugar species, which undergo secondary polymerization reactions when heated. Upgrading strategies considered in this chapter include:

- stabilization of bio-oil through low temperature hydroprocessing followed by higher temperature "deep hydrotreating" to remove the most of the remaining oxygen,
- converting bio-oil catalytically (often with zeolites such as HZSM-5) prior to condensation with catalysts inside the FP reactor or in a separate fixed or fluidized bed after the FP reactor,
- catalytic cracking with acid catalysts such as zeolites after condensation, and
- physical or chemical separations to improve bio-oil stability, extract valuable chemicals, or produce H_2.

Thermochemical Processing of Biomass: Conversion into Fuels, Chemicals and Power, Second Edition.
Edited by Robert C. Brown.
© 2019 John Wiley & Sons Ltd. Published 2019 by John Wiley & Sons Ltd.

Bio-oil upgrading by co-processing with refinery streams such as vacuum gas oil (VGO) in a fluid catalytic cracking (FCC) unit in a refinery has also been investigated. Numerous excellent reviews, perspectives, and other chapters have been published since 2010 [1–10].

This chapter is an update of the original article prepared by Prof. Anthony V. Bridgwater for the first edition of this book (*Thermochemical Processing of Biomass: Conversion into Fuels, Chemicals and Power*, First Edition, Edited by Robert C. Brown, © 2011 John Wiley & Sons, Ltd). We gratefully acknowledge the original contribution by Prof. Bridgwater, much of which is retained with updated references in this second edition. The scope of this updated chapter includes references published since about 2010. In the interest of conciseness, many references included from the original chapter are not included in this edition. Thus, Prof. Bridgwater's original chapter should be viewed as a companion to the present chapter to provide a holistic set of references to the numerous bio-oil upgrading processes discussed herein.

7.2 Bio-oil Characteristics and Quality

Bio-oil, also known as pyrolysis liquid or pyrolysis oil, is a dark brown liquid produced from the rapid quenching of condensable vapors produced during FP of biomass. This process occurs under inert conditions, typically at 400–550 °C and very short residence times (less than 2 s) to optimize liquid product yield [1]. A solid product, called char, and non-condensable gases, mainly carbon dioxide, carbon monoxide, and methane, are also produced. The characteristics of the bio-oil produced are highly dependent on feedstock [11–13], pyrolysis conditions [13–16] and reactor configuration [14, 17, 18]. Simulations of FP are able to provide semi-quantitative predictions of bio-oil composition [13, 19–22]. The reader is referred to Chapter 5 *Fast Pyrolysis* in this book for a more detailed review of the recent advances in the FP of biomass.

Bio-oil was initially explored for use in boiler combustion applications as a fuel oil substitute and was found to burn at almost similar rates as other commercial fuels [23]. However, its low heating value, high water content, acidity, chemical and thermal instability, and presence of solids were challenging for this application. Heating to 80 °C was sufficient to cause phase change and increase viscosity [24]. The general properties of wood-derived bio-oil are summarized in Table 7.1.

During FP, cellulose, hemicellulose, and lignin are decomposed to vapor and aerosol products, which may undergo secondary reactions before products are quenched (condensed) [13]. The myriad of complex bio-oil products give rise to its multi-phasic nature [25, 26]. The stability of the bio-oil emulsions is important during processing on its own as well as in co-processing with other fuels such as diesel and other bio-based oils [27, 28]

Hemicellulose degradation to low molecular weight acids is mainly responsible for the acidity of bio-oil [29]. Bio-oil contains a wide variety of other oxygen-containing functional groups such as aldehydes and ketones, phenols, ethers, and alcohols [30]. These functional groups have been identified as the main cause of bio-oil chemical and thermal instability due to their susceptibility to condensation reactions [31–33]

Bio-oil is denser than water, at around 1.2 kg/l. Bridgwater [34] estimated that the liquid has about 42% by weight and 61% by volume of the energy content of a light fuel with a density of 0.85 kg/l. In one study, the effect of temperature and pressure on the liquid

Table 7.1 Typical properties of wood-derived crude bio-oil.

Physical property	Typical value
Moisture content (%)	25
pH	2.5
Specific gravity	1.20
Elemental analysis	
C (%)	56
H (%)	6
O (%)	38
N (%)	0–0.1
HHV[a] as produced (MJ/kg)	17
Miscibility with hydrocarbons	Very low
Viscosity (40 °C and 25% water) (cP)	40–100
Solids (char) (%)	0.1
Stability	Poor
Vacuum distillation residue (%)	up to 50

[a] HHV: higher heating value.

density of Netherlands-based Biomass Technology Group BV (BTG) produced bio-oil sample was studied [35]. Density was shown to have an inverse linear dependency to temperature – based on measurements from ambient temperature to 343 K. On the other hand, bio-oil density increased while pressure was increased from 0.14 to 10 MPa.

Another physical property, bio-oil viscosity, can affect bio-oil transportation and processing [36]. The viscosity of bio-oils from various feedstocks were measured and reported [37]. The effect of temperature [35, 37], water content [37], pressure [35] and acidity [37] on viscosity was determined, with temperature and water content showing more significant effect. Bio-oil viscosity is also an important indicator of bio-oil aging. Aging occurs when the physical and/or chemical composition of the freshly made bio-oil changes due to exposure to the atmosphere and/or elevated temperatures (e.g. for a certain amount of time during storage [31, 38, 39]

7.2.1 Feedstock Factors Affecting Bio-oil Characteristics

Characteristics of the starting feed material can impact yield and quality of the bio-oil, which are described in the following sections.

7.2.1.1 Inorganic Content and Composition

Minerals and other inorganic compounds in biomass occur as plant nutrients or contaminants picked up from the soil during harvesting. Typically, these inorganic compounds become sequestered in the char. Sulfur is expelled to the gas phase depending on the pyrolysis temperature [40, 41]. About 1% of the incoming inorganic content can be retained in the condensed liquid [41]. The presence of ash in the starting biomass has been correlated inversely with the bio-oil yield [42, 43] and bio-oil short-chain acids [44]. Ash can also contribute to reactor degradation. As such, ways to remove the inorganic content from biomass have been implemented [45, 46]. Lastly, a dependency of bio-oil phase separation on the feedstock ash content has been reported, with more ash resulting in phase separation [26]

7.2.1.2 Water Content of Biomass

The water content of the biomass feed to the FP unit is usually limited to 10 wt%, resulting in bio-oil water content between 25 and 30 wt% [1]. Torrefaction, a thermal pretreatment of biomass around 200–280 °C, produces a lower initial water and oxygen content of the feed. Pyrolysis of torrefied biomass resulted in lower acid and water content of the bio-oil product, with the lower water content contributing to a higher heating value of the liquid product [47, 48]

7.2.1.3 Composition of Biomass

The complex nature of lignocellulosic biomass translates to a very complex composition of the bio-oil product. Primary reactions from carbohydrates are mainly characterized by depolymerization and dehydration to levoglucosan followed by further dehydration, C—O bond scission and C—C bond scission [49]. For hemicellulose, the source determines its reactivity, with the xylan from hardwoods being more reactive than softwood's glucomannan [40]. Similar to cellulose, the xylan depolymerizes and rearranges to form 1,4-anhydro-D-xylopyranose to further decompose into smaller compounds such as acids and furans [50]. For lignin, free-radical catalyzed depolymerization is the main pyrolysis route [51]. Phenols, syringols, and guaiacols in the bio-oil are mainly attributed to lignin pyrolysis [51]. As such, various feedstocks with differing percentages of polymeric content are expected to have different bio-oil composition as well. In comparing four types of feedstocks consisting of woody biomass, straw, shells (peanut shell, rice husk, corn cob, and rape pod), and algae, the authors reported that woody biomass produced more phenols, straw produced more ketones, shells produced more furans, and algae produced more fatty acids [44]

7.2.2 Effect of Pyrolysis Operating Conditions on Bio-oil Composition

The effects of pyrolysis operating temperature and residence time are discussed in this section.

7.2.2.1 Effect of Operating Temperature

The actual pyrolysis temperature, the rate of heat up, and rate of cool down affect the yield distribution between solid char, non-condensable gas, and condensed bio-oil. Pyrolysis typically occurs between 400 and 550 °C. Higher temperatures (>650 °C) tend to produce more gases and is characterized as gasification. A slow heat-up rate tends to produce more char such that the liquid yield goes down. Slow pyrolysis favors char formation and is thus used in charcoal production. On the other hand, slower cooling rates produce more gas, due to the preponderance of secondary reactions, which in turn reduce liquid yield. Recently, the production of pyrolysis liquid fractions with unique composition through staged condensation was demonstrated by researchers at Iowa State University [52, 53]

7.2.2.2 Effect of Vapor Residence Time

The vapor residence time for FP is typically 2 s or less. Rapid heating and quenching results in the highest liquid yields by minimizing secondary reactions. Higher yields contribute to a more favorable techno-economic assessment of the technology [54]

7.2.3 Need for Upgrading Bio-oil

Bio-oil aging is characterized by phase separation, increased viscosity, increased molecular weight, acidity changes, and increases in water content [55, 56]. These changes are attributed to the presence of more than 400 oxygenated species in bio-oils, classified into functional groups such as acids, esters, alcohols, ketones, aldehydes, sugars, furans, phenolics, and high molecular weight species [57]. The changes that occur at relatively low temperatures over extended periods of time (e.g. 37 °C for 56 days) are similar to changes that occur at higher temperatures over shorter times (e.g. 90 °C for 6 hours) [58]. Accelerated aging of bio-oil is performed on sealed samples exposed to 80 °C in an oven for 24 hours, which was used to conduct an IEA round-robin evaluation of bio-oil aging [38, 39, 55, 59, 60]. Important aging reactions were reported to include sugar decomposition and condensation to form humins, aldol condensation of furfural and ketone, acid-catalyzed lignin condensation, radical initiated lignin condensation, and phenol-glycoaldehyde coupling [55]. Through NMR studies, consistent observations were found: the amount of aliphatic C—O bonds and aromatic C—H bonds decreased while aliphatic C—C bonds, aromatic C—C bonds, and C—O bonds increased [59]. Condensation reactions also resulted in increased molecular weights, with accompanying increases in viscosity and even phase separation [39]. In another study using birch wood bio-oil [60], aging at 80 °C resulted in the reduction of the carbon content associated with olefins, aldehydes, and hydroxyl groups, suggesting condensation of aldehydes and ketones, phenol-aldehyde reaction, conversion of sugars to humins, alcohol etherification, and carboxylic acid and alcohol esterification. These reactions need to be slowed in order to stabilize bio-oil for storage and processing. Physical means to stabilize bio-oil are discussed in Section 7.4 while the catalytic treatment is reported in Section 7.5.

Efforts to upgrade this renewable oxygenated product into hydrocarbons through hydrotreating has gained traction over the years. The catalytic upgrading of bio-oil is proposed as a pathway to produce liquid transportation fuels [2, 7–10, 54, 61]. As shown in Figure 7.1, the van Krevelen plot is a convenient way to compare biomass, bio-oils, and crude oil. These plots show that bio-oils have lower O/C ratio than the starting biomass but similar H/C levels. In order to approximate petroleum hydrocarbons, the oxygen content needs to be decreased while the hydrogen content increased. This is typically accomplished through catalytic hydrotreating.

Traditional hydrotreating of petroleum to remove heteroatoms involves catalytic treatment of the feedstock at elevated temperatures (350–450 °C) and high H_2 pressures. This treatment is also needed to remove oxygen from bio-oil but the presence of oxygenates such as sugars, aldehydes, and ketones, make the bio-oil susceptible to condensation reactions. In a continuous, trickle-bed reactor, these polymerized structures can deactivate the catalyst bed through surface deposition that can eventually bridge catalysts, forming solid

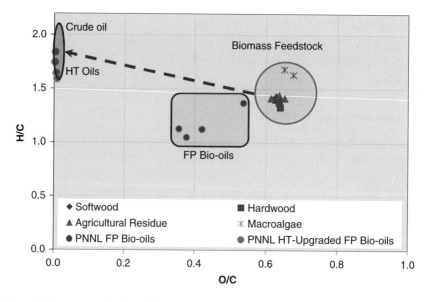

Figure 7.1 van Krevelen plot of biomass, raw bio-oil, hydrotreated (HT) bio-oil, and crude oil.

carbon deposits (plugs) that will eventually lead to process shutdown to avoid reactor overpressure [61, 62]. A three-step process consisting of an intermediate stabilization process at lower temperature under reducing conditions (around 140 °C) followed by two-step sulfided hydrotreating enabled long-term operation [7, 32, 33]. The extent of carbonyl content after the reductive treatment was found to be crucial. A metric of 1.5 mmol carbonyl per gram of bio-oil fed became a requirement to produce bio-oil sufficiently stabilized to be fed to high temperature reactors without plugging [63]. The process of converting bio-oil into hydrocarbons will be discussed in detail in Section 7.5.

7.3 Norms and Standards

Successful conversion of bio-oil into a commodity product requires norms and standards to establish an acceptable metric of quality across industries [64]. Development of standard analytical methods is very important in establishing quality for marketing purposes. Initial efforts to specify bio-oil quality was performed in 1995 [65]. This led to certification of standards for use of bio-oil in biopower applications by CEN in Europe and ASTM in North America [66, 67]. Characteristics such as gross heat of combustion, water content, pyrolysis solids content, kinematic viscosity, density, sulfur content, ash content, pH, flash point, and pour point were controlled to assign quality to the bio-oil [68]. The initial certifications have since been updated and new standards published, including a standard method for measuring pyrolysis solids content (Table 7.2) [68–71]. The very first ASTM standard on bio-oil functional group quantification, ASTM E3146, was recently approved [72].

Inter-laboratory validation is an important component in determining the reliability of proposed characterization methods. Since the first round-robin organized by the International Energy Agency in 1988 [73] several international round-robins on bio-oil analysis

Table 7.2 List of standards related to bio-oil.

Standard designation	Organization	Title
ASTM D7544–12	ASTM International	Standard Specification for Pyrolysis Liquid Biofuel
ASTM D7579–09	ASTM International	Standard Test Method for Pyrolysis Solids Content in Pyrolysis Liquids by Filtration of Solids in Methanol
ASTM E3146–18	ASTM International	Standard Test Method for the Determination of Carbonyls in Pyrolysis Bio-Oils by Potentiometric Titration
EN 16900:2017	CEN	FP bio-oils for industrial boilers - Requirement and test methods
CEN/TR 17103:2017	CEN	FP bio-oil for stationary internal combustion engines - Quality determination

methods as well as bio-oil production have been reported [14, 38, 39, 74, 75] including a round-robin on functional group determination of bio-oils [75].

7.4 Physical Pre-treatment of Bio-oil

This section covers the various physical methods that have been studied and reported to improve the quality and stability of bio-oil for storage and processing.

7.4.1 Physical Filtration

The char and inorganic content of bio-oils are reported to catalyze bio-oil polymerization during storage, contributing to bio-oil aging [76]. Filtration to remove char has been proposed to mitigate bio-oil aging [39, 56, 77–79]. Condensed-phase filtration was evaluated through an IEA round-robin [39]. The participants reported an overall lower viscosity of the filtered bio-oil compared to an unfiltered sample (about 30% lower) after undergoing an accelerated aging test [39]. On the other hand, hot gas filtration during FP was found to reduce the alkali and alkaline earth metals and total solids content of the bio-oil, resulting in a 10-fold reduction in viscosity increase during an accelerated aging test [77]. A similar trend was reported for bio-oils derived from rice husk and rice straw that were hot-vapor filtered [78]. In conclusion, hot gas filtration was able to decrease the viscosity increase rate during accelerated aging more successfully than condensed-phase filtration.

7.4.2 Solvent Addition

Addition of solvents, especially alcohols, is another method to mitigate bio-oil aging [80]. Among several alcohols tested (methanol, ethanol, and isopropanol), methanol was found to be the most effective. The action of alcohol was proposed to include (i) dilution of active species, especially the hydrophobic high molecular mass lignin-derived fractions and (ii) capping reactions with aldehydes, ketones, and sugars. Addition of a mixture of solvents consisting of 1 wt% methanol, 5.064 wt% N,N-dimethylformamide, and 1.940 wt% acetone showed some improvement compared to the effect of a single solvent but was ineffective in completely preventing bio-oil aging [81]

7.4.3 Fractionation

Fractionation was also considered as a physical means of stabilizing bio-oil [52, 53, 82]. By controlling condensation temperature after pyrolysis, specific fractions of condensable vapors can be selectively recovered. By lowering the temperature in the first stage from 345 to 102 °C and controlling the last stage to 18 °C, water and acids were condensed in the last stage, while most of the lignin-derived oil and sugar-based compounds were recovered in the first and second stages [53]. Comparison between fractionation during liquid recovery by adjusting scrubber/condenser temperature from 36 to 66 °C and fractionation via spontaneous phase separation by using feedstock with 25 wt% moisture content instead of 10 wt% have been reported [82]. Water content in the bio-oil decreased from 24 to 7 wt% by controlling scrubber temperature, with a concomitant loss of light volatile compounds, decrease in acidity, and increase in viscosity. Better fractionation between lignin- and sugar-derived fractions was achieved through the spontaneous phase separation mode induced by the greater water content present initially in the feed [82].

7.5 Catalytic Hydrotreating

Catalytic hydrotreating rejects oxygen from bio-oil to substantially increase the H/C ratio. Petroleum hydrotreating is typically done between 350 and 450 °C and high H_2 pressure (20.3 MPa) in the presence of sulfided Mo-based catalysts on Al_2O_3. Adjustments to the process developed for petroleum feeds are necessary to enable bio-oil processing due to the thermal and chemical instability of bio-oil.

7.5.1 Stabilization Through Low Temperature Hydrotreating

Early studies into bio-oil hydrotreating revealed that direct heating caused bio-oil to form a heavy plug requiring termination of operations within 24 hours [61]. The plug was attributed to condensation and polymerization reactions as temperature increased, despite the presence of hydrogen and active catalysts, which soon became encased in carbonaceous deposits. By using a surrogate mixture consisting of sugars, acids, aldehydes, ketones, furans, and phenols in a batch reactor, the contribution of these compounds to bio-oil condensation and repolymerization reactions was studied [83]. In experiments done at 190 °C, glucose was found to decompose into reactive compounds containing hydroxyl, carbonyl, or π-bond conjugations, which in turn are reactive towards polymerization. Carboxylic acids were shown to catalyze polymerization while phenols containing carbonyl groups, such as vanillin, were susceptible to polymerization. Phenols also participated in the acid-catalyzed condensation reactions with carbonyl-containing molecules [83]. Since these compounds are typically found in bio-oils, catalytic means to stabilize their propensity to polymerize is proposed through mild hydrotreating in order to reduce the more active carbonyl and conjugated olefin species, thus stabilizing the bio-oil.

Unlike petroleum oil fractions that are thermally stable at hydrotreating temperatures (350 °C or higher), mild temperature hydrogenation pretreatment is needed for bio-oils to convert the reactive species, such as aldehydes, into less reactive species. Several reviews have identified and discussed this requirement when dealing with bio-oil hydrotreating [6, 7, 61, 84]. A two-step hydrotreating process utilizing a low temperature sulfided catalyst in the first step (170 °C Ru/C or 250 °C CoMo/Al_2O_3) followed by a high temperature sulfided hydrotreating catalyst (400 °C CoMo/Al_2O_3) was found to be sufficient to effect almost complete deoxygenation for about 90 hours [85]. This two-step process was originally patented in the 1980s [86]. Beyond 90 hours, excessive carbon fouling formed a plug at the reactor interface where the bed temperature transitioned from 200 to 350 °C. This suggested rapid thermal polymerization of reactive bio-oil species that the low temperature catalyst bed was unable to convert to more stable species. The less-than-100 hour processing time is not sufficient to make this process technically or economically feasible. Reaction studies in a batch system suggested that at temperatures exceeding 250 °C, thermal polymerizations become rapid, resulting in decreased hydrogen consumption in the presence of Ru/C catalyst with the formation of species recalcitrant to further hydrotreating [87, 88]. Batch experiments in the presence of Pd/C and at 10 MPa showed increasing coke yields from 150 to 300 °C [89].

The use of reduced noble metal catalysts was considered for catalyst stabilization. Model compound batch reactor studies between 150 and 300 °C using Ru/C and Pd/C catalysts showed Ru/C to be the more active catalyst between the two [90]. Both catalysts were able to convert furfural and guaiacol at low temperatures (less than 200 °C). However, the Pd catalyst required a temperature of 300 °C to convert acetic acid. Another study corroborated the conversion of acetic acid to ethanol with Ru at high selectivity at 160 °C [91]. As such, an additional hydrogenation step using reduced Ru/C to allow further stabilization of the bio-oil constituents at 140 °C was implemented before the two-step process of sulfided Ru/C and promoted sulfided Mo catalyst [32]. Treatment of bio-oil with Ru metal catalyst between 70 and 100 °C and between 4 and 14 MPa of H_2 was claimed to improve bio-oil stability in aging tests [92]. At 140 °C, limited bio-oil deoxygenation with a slight increase in water content was observed in the presence of non-sulfided Ru/C. The additional stabilization step (i.e. a three-bed system) enabled the cumulative continuous operation of the sulfided Mo-based deep hydrotreating catalyst bed for more than 1440 hours, a significant improvement over the previously reported 100 hours [32]. Functional group analysis using bio-oil in a batch reactor with a 4 hour reaction time in the presence of the non-sulfided Ru/C catalyst showed that carbonyl groups were significantly reduced through hydrogenation at lower temperatures (125–150 °C) but increased at higher temperature where condensation followed by subsequent dehydration occurred [93]. Another study utilizing a Ru/TiO_2 catalyst between 120 and 160 °C demonstrated the catalyst's ability to hydrogenate bio-oil at different degrees at various temperatures [33]. A hydrogenation reaction rate sequence was proposed: sugar conversion to sugar alcohols, followed by ketone and aldehyde conversion to alcohols, followed by alkene and aromatic hydrogenation, and, lastly, carboxylic acid conversion to alcohols [33]. About 50% of hydrogen consumption was attributed to the hydrogenation of carbonyls in sugar, aldehydes, and ketones [33]. The facile hydrogenation of aldehydes was also observed in batch studies [89, 93]. Figure 7.2 shows the hydrogen selectivity for each functional group as a function of hydrogen addition. The batch reactor experiments were in good agreement with the reappearance of carbonyl groups after catalyst deactivation was suspected to have occurred in continuous bio-oil stabilization tests [32].

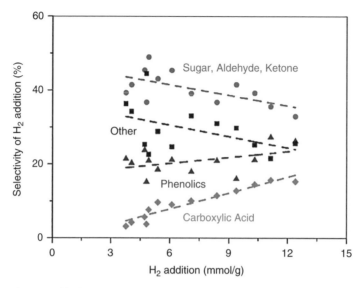

Figure 7.2 Selectivity of hydrogen additions to several components of bio-oil at various hydrogen additions. Source: adapted with permission from [33] Copyright 2016 American Chemical Society.

Both catalyst and process development are still necessary to address the techno-economic cost considerations of stabilizing bio-oil [54]. Ru is a relatively expensive noble metal catalyst. An active catalyst robust enough to hydrogenate the reactive bio-oil species at low temperatures that do not favor thermal polymerization while being available at a low cost or able to be utilized for a substantially longer time on-stream than Ru/C is needed. Additionally, the three-bed configuration entails significant capital cost incompatible with economic operation [54]. Setting a quality metric on the carbonyl content of the low temperature Ru-stabilized bio-oil that can be directly processed at high temperature allowed the second Ru/C bed to be eliminated, improving process economics [63]. However, this requires the low temperature Ru/TiO$_2$ catalyst to be robust enough to maintain activity for an extended period of time. Evidence of catalyst deactivation was observed in the re-appearance of carbonyls at longer time on-stream, demonstrating that significant improvement in catalyst robustness is needed [32]. During stabilization reactions, different modes of catalyst deactivation are possible, such as: (i) carbon fouling and catalyst deposition due to competing bio-oil component thermal polymerization; and, (ii) catalyst poisoning by bio-oil contaminants such as sulfur and other inorganic compounds. Nitrogen physisorption analysis of spent stabilization catalysts showed complete loss of microporosity and an over-all 93% decrease in catalyst pore volume, in addition to increased S and Ca concentrations [32]. Analysis of sulfided Ru catalysts showed the presence of carbonaceous deposition due to bio-oil species condensation [94]. Sulfur poisoning is believed to be the main mode of reduced catalyst activity [33, 95]. It is thought that as the active catalyst becomes poisoned by S, the rate of hydrogenation of the bio-oil reactive species becomes slower compared to the competing thermal polymerization reactions, thus increasing the carbon fouling of the catalyst.

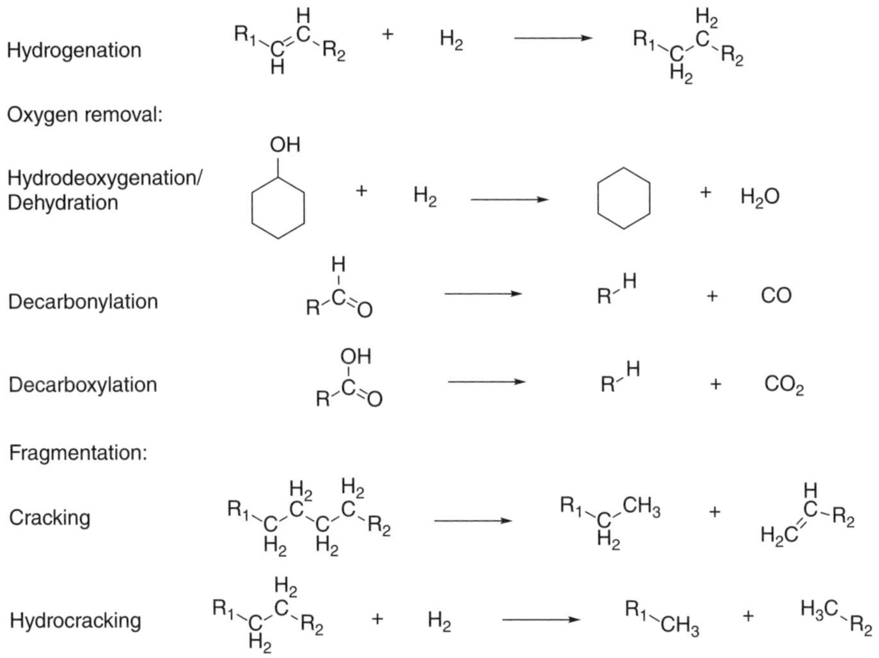

Figure 7.3 Major reactions during deep hydrotreating of bio-oils. Source: reproduced from [7] with permission from The Royal Society of Chemistry.

7.5.2 Deep Hydrotreating

After reaching the target quality of bio-oil through stabilization or catalytic FP (Section 7.6), bio-oil is expected to be stable enough to be processed continuously by high temperature, deep hydrotreating. Deep hydrotreating aims to remove oxygen from bio-oils for hydrocarbon production to generate fuels or chemicals. This process typically occurs around 400 °C using various catalysts and requires a hydrogen supply or source. This would be expected to take place very similarly to the hydrotreating processes in a conventional refinery to take advantage of know-how and existing process and infrastructure.

Various reactions occur simultaneously during this upgrading step, as shown in Figure 7.3. The reactions are loosely classified as hydrogenation to saturate C—C bonds, oxygen removal via hydrodeoxygenation (HDO), decarbonylation and decarboxylation, and fragmentation through hydrocracking, cracking, and isomerization [6, 7]. Oxygen can be expelled from the bio-oil through HDO to produce water, decarboxylation to produce carbon dioxide, or decarbonylation to produce carbon monoxide. HDO is preferred as it conserves carbon yield. For this reason, HDO has been extensively studied in recent years and the large portion of available reports are on catalyst evaluation or mechanistic and kinetic studies using model compounds.

Conventional Mo-based sulfide catalysts, which are commonly used in petroleum refineries, have been widely used for bio-oil deep hydrotreating and demonstrated to have good performance for oxygen removal. Non-sulfide catalysts, including noble metal catalysts and base metals in reduced, carbide, nitride, or phosphide form have also been tested, primarily for model compound studies [6] with limited examples for actual bio-oil hydrotreating [96–98]. Secondary catalyst functions, such as acidic sites on zeolites, have also been added to obtain bifunctional catalysts for dehydration or alkylation reactions, which were primarily evaluated by model compound studies [6].

Using various hydrotreating catalysts and operating conditions, several research groups from different parts of the world performed the continuous deep hydrotreating of actual bio-oils. These are summarized in Table 7.3 Hydrotreating of bio-oils using other types of reactor was also accomplished [114, 115]. Analysis of the fuel properties of the hydrocarbon products are discussed in Section 7.8.1.

7.6 Vapor Phase Upgrading via Catalytic Fast Pyrolysis

The significant challenge of stabilizing and upgrading bio-oil is increasing interest in ways to improve the quality of bio-oil. Catalysis integrated or combined with FP to directly upgrade the pyrolysis vapor, a process called catalytic FP (CFP), is a promising approach to producing better quality bio-oil. CFP was the first explored in the 1980s [116], with proliferating interest since 2010 as evidenced by extensive publications and commercial developments [117–119]

The goal of CFP is to lower pyrolysis temperature and/or improve selectivity towards compounds that improve quality of bio-oil. Several research efforts have demonstrated the strong impact of chemical composition and variation of bio-oil quality through the use of different catalysts. Catalysis can be integrated/combined with FP in two ways: *in situ* catalytic fast pyrolysis (*in situ* CFP) where the catalyst is mixed directly with the feedstock in the pyrolysis reactor, or *ex situ* catalytic fast pyrolysis (*ex situ* CFP) where the catalyst is contacted with the pyrolysis vapors after they exit the pyrolysis reactor. These two schemes are illustrated in Figure 7.4. CFP generally produces liquid in two phases: an aqueous phase (rich in water and light oxygenates such acids) and an oil phase (CFP bio-oil). CFP bio-oil typically has lower oxygen and water content and smaller amounts of reactive and corrosive species compared to conventional FP bio-oil. The improved physical and chemical properties of the resulting bio-oil makes possible hydroprocessing without a difficult stabilization process.

Although CFP can also completely deoxygenate bio-oil, producing hydrocarbon fuel blendstocks, yields are generally low. Accordingly, most work in CPF focuses on improving bio-oil quality while retaining high carbon yield for subsequent upgrading to hydrocarbons. In this aspect, in order to maximize overall biomass to fuel carbon efficiency and lower the overall conversion cost, it is critical to optimize the CFP process with regard to not only bio-oil yield and oxygen content but also CFP bio-oil processability and fuel yield after upgrading.

Since 2010, a wide range of studies have investigated CFP over a range of process configurations (*in situ* and *ex situ*), process parameters (temperature, residence time, etc.), catalysts, feedstocks, and scales. Much of this work involves catalyst evaluation in laboratory-scale batch reactors while only limited work has been performed in larger continuous reactors appropriate to commercial development. Several papers provide comprehensive reviews of the literature [117–124]. Here the focus is on work performed since 2010 on chemistry, catalyst, process conditions, reactor design, and practical examples.

Table 7.3 Representative bio-oil hydrotreating research since 2010.

Organization	Catalyst	Conditions	Feedstocks	Ref.
Continuous flow systems				
PNNL, USA	$CoMoS_x/C$	400°C, 14 MPa, LHSV 0.19 h^{-1}	Pine bio-oil, after stabilization on RuS_x catalyst at 170°C	[85]
	$CoMoS_x/Al_2O_3$	400°C, 10 MPa, LHSV 0.20 h^{-1}	Oak and switchgrass bio-oil, after stabilization on RuS_x catalyst at 220°C	[99] [100]
	$CoMoS_x/C$	405°C, 14 MPa, LHSV 0.18 h^{-1}	Pine bio-oil and lumber mill residue bio-oil, after stabilization on RuS_x catalyst at 170°C	[101]
	$CoMoS_x/Al_2O_3$	320–400°C, 12 MPa, LHSV 0.20 h^{-1}	Native and dehydrated bioCRACK bio-oil (from spruce and liquid phase pyrolysis in VGO), without stabilization	[102]
	Commercial Mo based sulfide	400°C, 14 MPa, LHSV 0.20 h^{-1}	Pine bio-oil, after stabilization on Ru/C catalyst at 120–140°C and then on RuS_x/C catalyst at 170°C	[32]
	$CoMoS_x/ZrO_2$	400°C, 12 MPa, LHSV 0.13–0.15 h^{-1}	Catalytic FP bio-oil from pinyon juniper, without stabilization	[103]
	Pd/C	370°C, 12 MPa, LHSV 0.10 h^{-1}	Phenolic fraction of red oak bio-oil or corn stover bio-oil, after stabilization on Ru catalyst at 140°C	[96]
	Ni-, Cu- and Ca-doped Mo_2C	400°C, 12 MPa, LHSV 0.24–0.30 h^{-1}	Softwood residue bio-oil, after stabilization of same type of catalyst at 180°C	[97]
	Pd- and Ru-supported on oxide + zeolite acid catalyst	320°C, 12 MPa, LHSV 0.2 h^{-1}	Pine bio-oil, after stabilization with Ni-, Pd-, and Ru/TiO_2 at 160°C	[98]
	Commercial Mo based sulfide	400°C, 14 MPa, LHSV 0.22 h^{-1}	Oak bio-oil, after stabilization on Ru/C catalyst at 140°C and then on RuS_x/C catalyst at 170-190°C	[104]
	Pd/C	395–405°C, 14 MPa, LHSV 0.27 h^{-1}		
U. of New Brunswick, Canada	$NiMoS_x/Al_2O_3$	280–350°C, 10 MPa, LHSV 1 h^{-1}	Diluted pine bio-oil at bio-oil to 1-methylnaphthalene ratio of 1:9, without stabilization	[105]

(Continued)

Table 7.3 Continued

Organization	Catalyst	Conditions	Feedstocks	Ref.
RTI, USA	NiMoSx	290–350 °C, 10–14 MPa, LHSV 0.25–0.50 h^{-1}	Catalytic pyrolysis bio-oil from pine, without stabilization	[106]
UCRE, Czech	CoMoS$_x$/Al$_2$O$_3$, NiMoS$_x$/Al$_2$O$_3$	320–450 °C, 13 MPa, WHSV 0.2–0.4 h^{-1}	Woody bio-oil with 0–5 wt% methanol, after stabilization on the same sulfide catalyst at 170–230 °C	[107]
Curtin University of Technology, Australia	NiMoS or supported NiMoS	375–450 °C, 7 MPa LHSV 1–5 h^{-1}	Mallee wood bio-oil, after passing through a Pd/C bed (< 250 °C) above the sandbath	[108–110]
Batch systems				
U. of Groningen, Netherlands	Pd/ZrO$_2$, Rh/ZrO$_2$, Pt/ZrO$_2$, RhPd/ZrO$_2$, PdPt/ZrO$_2$, RhPt/ZrO$_2$, CoMoS/Al$_2$O$_3$	350 °C, 20 MPa total pressure, 4 h reaction time, 1300 rpm	Pine bio-oil	[111]
	NiCu/Al$_2$O$_3$	350 °C, 10 MPa initial pressure, 3 h reaction time, 1300 rpm	Pine bio-oil exposed to 150 °C for 1 h before increasing temperature	[112]
	NiCu on several supports (CeO$_2$-ZrO$_2$, ZrO$_2$, SiO$_2$, TiO$_2$, rice husk carbon, and Sibunite)	350 °C, 20 MPa initial pressure, 3 h reaction time, 1300 rpm	Pine bio-oil exposed to 150 °C for 1 h before increasing temperature	[113]

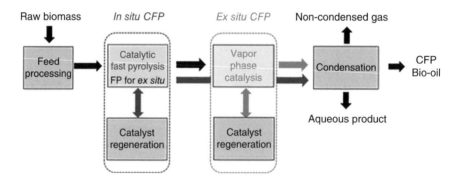

Figure 7.4 In situ and ex situ configuration of catalytic fast pyrolysis.

7.6.1 CFP Chemistry

CFP begins with the thermal decomposition of biomass, as described previously for non-catalytic pyrolysis, producing a wide range of oxygenate intermediates containing numerous functional groups such as acids, aldehydes, ketones, esters, phenolics, and alcohols. The role of a catalyst in CFP is to promote selective removal of oxygen from the compounds and convert reactive species such as aldehydes and ketones into more stable alcohols, ethers, and/or aromatics. Recent research on model compounds or biomass fractions as well as detailed comparison of composition of CFP bio-oil with regular bio-oil provided useful information regarding CFP chemistry. In general, the following reaction classes occur simultaneously in CFP [117, 124]:

- decarbonylation and decarboxylation to eliminate oxygen from aldehydes and acids as CO or CO_2
- ketonization and aldol condensation of acids, ketones, and aldehydes to reject oxygen as CO_2 and H_2O and form C—C bonds
- aromatization of small oxygenates and olefins to aromatics with oxygen rejected as CO, CO_2, and H_2O and larger polyaromatics converted to coke
- hydrodeoxygenation to remove oxygen and hydrogenation to saturate C=C bonds using externally supplied H_2 or H_2 from stream reforming
- cracking, isomerization, and alkylation to rearrange carbon backbones
- other reactions such as dehydration and demethoxylation.

The result is oxygen removal and carbon backbone rearrangement. The distribution of functional groups are widely changed. Ideally, oxygen is removed as CO_2 and H_2O while retaining most of the hydrogen and carbon in the bio-oil. Deoxygenation via CFP decreased the polarity of many of the compounds and leads to phase-separated liquid products consisting of bio-oil with reduced oxygen content and aqueous product containing light oxygenated compounds such as carboxylic acids.

7.6.2 Key Factors Impacting Catalytic Fast Pyrolysis

7.6.2.1 Catalyst

One of the critical challenges for CFP is selection of efficient and economical catalysts that are selective for the desired products. The catalyst must be resistant to mineral deposition and coke formation and spent catalyst should be readily regenerated. Table 7.4 categorizes the major catalysts investigated for use in CFP. Acid/base catalysts, especially zeolites such as ZSM-5, have been the central focus. Metal catalysts and metal/acid bifunctional catalysts have recently garnered increased attention. Low-cost waste slag materials, such as red mud, have also been used as a CFP catalyst and shown promising performance.

7.6.2.2 Biomass Feedstock

Various biomass feedstocks can be converted to bio-oil through CFP, including forest residue of softwood or hard woods such as saw dust [186, 187] or forest thinnings [188], agriculture residues such as rice/wheat straw and husk [189], palm shell [190] jatropha wastes [191] energy crops such as switch grass [128] and algae (including both micro- and

Table 7.4 Summary of catalyst used for catalytic fast pyrolysis.

Category	Catalyst
Porous oxides (zolite)	*Microporous zeolite*: ZSM-5 [125–130], HY [131–134], H-Beta [134–137], Ferrierite [138], SAPO-11 [139], FCC catalyst [140] *Mesoporous zeolite*: (Al)SBA-15 [141], (Al)MCM-41 [142–144], MCM-22 [145], Meso-ZSM-5 [146, 147], Meso-Y [148] *Modified zeolites*: Zn-ZSM-5 [149], Ga-ZSM-5 [150], Mg-ZSM-5 [151, 152], Fe-ZSM-5 [153–155], Fe-Beta [155], M/Al-MCM-41 [156–158], La-SBA-15 [159]
Metal oxide	Al_2O_3 [160] SiO_2-Al_2O_3 [160] MgO [161, 162]ZrO_2 [123, 163] TiO_2 [123], CeO_2 [164]Fe_2O_3 [165] CuO [166] WO_x [165] MoO_x [165, 167] CaO [168] and their binary or trinary mixtures [160, 166] Modified oxides: sulfated ZrO_2 [169] sulfated TiO_2 [170] Complex oxide mixtures: Red mud [103, 171–173] kaolin [174]
Metal, metal carbide and phosphide	Supported Ni [175–177] Pd [178–181] Pt [177, 182] Fe [175] Cu [183] Metal phosphide such as Ni_2P and Rh_2P [177] Metal carbide such Mo_2C [184]
Soluble inorganics	$MgCl_2$ [117] $ZnCl_2$ [168] $CaCl_2$ [185], $FeCl_3$ [185]

macro-algae [192]. Lignin is also an important feedstock for recent CFP research [120]. Co-processing biomass feedstocks with other feedstocks such as plastic [158], tires [193], and food wastes [194] has been evaluated as well. The choice of feedstock impacts product properties and can dictate the requirement of catalyst functions and process parameters significantly [128].

7.6.2.3 Process Configuration

As discussed throughout Section 7.6, *in situ* and *ex situ* CFP configurations have been widely applied. In general, *in situ* CFP using a fluidized reactor is simple and straightforward with low capital cost but requires robust catalysts with resistance to attrition, mineral deposition, and coke formation, which limit flexibility towards operation parameters (e.g. temperature). FCC-like circulating fluidized bed reactors with catalyst regenerators are normally used [188]. Auger reactors are also used. although with some difficulty in transporting fine catalyst [195]. Bubbling fluidized bed reactors with off-line catalyst regeneration are used mainly for bench-scale research [128]. Microwave-assisted CFP, normally as *in situ* configuration, has been studied [196–198]. *Ex situ* CFP requires a more complicated multi-reactor system with higher capital cost but which may operate different catalytic reactors with more diverse catalysts and chemistries combined with hot gas filtration for removing minerals and chars from pyrolysis vapor prior to contact with the catalyst. A catalytic reactor after the FP reactor can be a fluidized bed [199, 200] moving bed [195] entrained flow, or a traditional fixed bed reactor [201].

7.6.2.4 Process Parameters

The process parameters, including temperature, residence time, heating rate, and catalyst-to-biomass ratio, are important factors impacting bio-oil yield and quality.

In situ CFP requires carefully tuning of temperature, residence time, heating rate, and catalyst-to-biomass ratio to ensure optimized pyrolysis and catalytic reaction. For *ex situ* CFP, the catalytic reactor operates similar to a conventional catalytic reactor in which temperature and residence time are critical. Inert gas such as N_2 is generally used as a diluent gas or to aid in consistent feedstock flow through the reactor. H_2 at low pressure is also applied with certain catalysts to facilitate HDO reactions, as a variation of CFP called catalytic fast hydropyrolysis [202].

7.6.2.5 Catalyst Deactivation

Catalyst deactivation is a major technical and economic challenge in CFP. Deactivation is mainly caused by coke formation which covers the active surface of the catalyst or blocks its pores. Coke deposition can be mitigated by catalyst design, process modification, and especially catalyst regeneration through oxidative treated (i.e. burning coke). Minerals such as alkali and alkaline earth metals present in the biomass feed serve to deactivate catalysts by poisoning acid sites [133, 186, 203]. Mineral deposition is often irreversible and therefore catalysts resistant to mineral deposition are required for *in situ* CFP or hot gas filtration is required to clean the pyrolysis vapor before the catalytic reactor in *ex situ* designs. Certain catalysts, such red mud, showed the promising characteristic of resistance to mineral deposition during *in situ* CFP [103].

7.6.3 Practical Catalytic Fast Pyrolysis of Lignocellulosic Biomass

Table 7.5 summarizes recent research reporting bio-oil yield and quality for *in situ* and *ex situ* CFP of biomass feedstocks using continuous reactors. *In situ* CFP has been widely studied because it can be conducted with unmodified FP reactors. In contrast, *ex situ* CFP has been primarily studied in small-scale reactors because of their relative complexity. Many studies report improved CFP bio-oil quality as a result of lower content of oxygen, acids, and reactive species such as carbonyls. Fewer studies have investigated the subsequent upgrading of CFP bio-oil to determine ease of processing and overall carbon efficiency. PNNL has recently reported success in hydrotreating USU CFP bio-oil from pinyon juniper using red mud CFP catalyst [103] and RTI has reported success in hydrotreating CFP bio-oil from pine using an alumina-based CFP catalyst [106, 213]. A single-stage catalytic reactor was used for the two studies, with results similar to the deep hydrotreating of stabilized FP bio-oil, as previously described in Section 7.5.2.

7.7 Other Upgrading Strategies

This section includes non-physical methods and catalytic processes not previously described.

7.7.1 Liquid Bio-oil Zeolite Upgrading and Co-processing in FCC

In addition to use in CFP, zeolite catalysts have been used to crack bio-oil liquids from non-catalytic pyrolysis. Similar to the chemistry of CFP, zeolite cracking rejects oxygen as CO_2, CO, and H_2O. Catalyst deactivation and carbon loss to coke and tar formation is

Table 7.5 Representative CFP research since 2010.

Organization	Catalyst	Biomass	Reactor	Products	Ref.
In situ catalytic fast pyrolysis					
U. of Massachusetts, USA	ZSM-5	Pine sawdust, cellulose	Bubbling fluidized bed	Aromatics	[187, 204–206]
USDA-ARS, USA	H-ZSM-5, Ca-Y, Beta	Oak	Bubbling fluidized bed	Bio-oil with reduced O content	[131] [207]
VTT, Finland	ZSM-5	Forest Thinning	Bubbling fluidized bed	Bio-oil with reduced O content	[188]
		Pine sawdust	Circulating fluidized-bed, 20 kg/h PDU	Bio-oil with reduced O content	[186]
Southeast U., China	ZSM-5, spent FCC, Al_2O_3, CaO, MCM-41	Rice stalk	Internally interconnected fluidized bed	Aromatics and olefins	[189] [208]
SK Innovat, S. Korea	ZSM-5, Equilibrium FCC	Palm kernel shell	Bubbling fluidized bed	Bio-oil with reduced O content	[190]
Utah State U., USA	ZSM-5	Poplar wood, pine, Pinyon-juniper, Switch grass, Corn Stover, Pine bark	Bubbling fluidized bed	Bio-oil with reduced O content	[128]
	ZSM-5 + FCC catalyst Red mud	Poplar Pinyon-Juniper			[132] [103, 172]
Iowa State U., USA	MgO, SiO_2-Al_2O_3	Red oak	Bubbling fluidized bed	Bio-oil with varied properties	[209]
NREL, USA	ZSM-5	Pine	Bubbling fluidized bed	Bio-oil with reduced O content	[129]
Petrobras, Brazil	ZSM-5	Pine, sugarcane bagasse	Circulating fluidized bed	Bio-oil with reduced O content	[210]
CERTH, Greece	ZSM-5	Beech wood	Bubbling fluidized bed	Bio-oil with reduced O content	[203] [211]

Institution	Catalyst	Feedstock	Reactor	Product	Ref.
RTI, USA	Alumina-based catalyst	Wood, grasses, crop residue	Bubbling fluidized bed	Bio-oil with reduced O content	[212]
		Pine	Circulating fluidized-bed, 45 kg/h		[213]
	W, Fe, or Mo based oxides, hydrotreating catalyst	Pine	Bubbling fluidized bed, with H_2	Bio-oil with reduced O content	[165]
ICB-CSIC, Spain	CaO, CaO-MgO	Pine	Auger reactor	Bio-oil with reduced O content	[162]
Ghent U., Belgium	ZSM-5	Pine	Auger reactor	Bio-oil with reduced O content	[195]
Ex situ catalytic fast pyrolysis					
Ghent U., Belgium	ZSM-5	Pine	Moving bed catalytic reactor	Bio-oil with reduced O content	[195]
NREL, USA	ZSM-5	Pine	Bubbling fluidized bed catalytic reactor	Bio-oil with reduced O content	[214] [199] [129]
Virginia Polytechnic Institute and Utah State U., USA	ZSM-5	Hybrid poplar	Fluidized bed catalytic reactor	Bio-oil with reduced O content	[200]
Integrated in situ catalytic hydropyrolysis and ex situ vapor catalytic upgrading					
GTI, USA	Undisclosed for two catalytic reactors	Maple and pine	Fluidized bed for catalytic hydropyrolysis; Fixed bed for vapor upgrading	Hydrocarbon fuel blendstock	[201]

the major drawback [215, 216]. Recent studies examined the upgrading of organic phase of Napier grass pyrolysis bio-oil over micro- or mesoporous ZSM-5 zeolite at 400 °C in a batch reactor to produce deoxygenated cyclic olefin, cycloalkane, monoaromatic, and polyaromatic hydrocarbons [216]. The phenolic-rich fraction of bio-oil produced by methanol extraction was treated over ZSM-5 catalyst at 400 °C to produce aromatic hydrocarbons [217]

Co-processing of bio-oil with petroleum refinery intermediates such as VGO in FCC units is another promising method of upgrading bio-oil with zeolite catalysts. The major advantage is that most refineries are equipped with FCC units that could be available for co-processing without additional hydrogen or energy inputs. The biogenic carbon from bio-oil could end up as fuels, coke, gases, CO_2, or CO. Compared to cracking VGO alone, co-processing with bio-oil tends to form more coke, CO, CO_2, gas, and water because of the instability and high oxygen content (HOC) of bio-oil [218, 219]. Co-processing CFP bio-oil or hydrotreated FP bio-oil with VGO improved processing stability and carbon efficiency. Most work to date employed 5–20% blending levels of bio-oil. Table 7.6 summarizes recent reported work on co-processing of bio-oil in a FCC unit.

Table 7.6 Recent reported work on co-processing of bio-oil in FCC unit.

Organization	Feedstock	Process	Ref.
U. of Twente, Netherland	20% of oil fraction of hydrotreated forest residue bio-oil and 80% long residue	FCC equilibrium catalyst; 520 °C; Lab-scale fluidized bed catalytic cracking unit	[220]
Université Lyon 1, France	10, 20% oil fraction of hydrotreated forest residue bio-oil and balanced VGO	FCC equilibrium catalyst, HY, or HZSM-5; 500 °C; Lab-scale fixed-bed reactor	[221–223]
VTT, Finland	Up to 20% raw low-water forest thinning bio-oil, CFP pine bio-oil, or hydrotreated forest thinning bio-oil and balanced VGO	FCC equilibrium catalyst; 482 °C; MAT fixed bed reactor.	[219]
Petrobras, Brazil	10 and 20% raw pine bio-oil and balanced VGO	FCC equilibrium catalyst; 540 and 560 °C; Demonstration-scale FCC unit with pseudoadiabatic riser reactor	[218]
UPV/EHU, Spain	20% raw black poplar bio-oil and 80% VGO	FCC equilibrium catalyst; 540 °C; Demonstration-scale FCC unit with pseudoadiabatic riser reactor	[224]
NREL, USA, Petrobras, Brazil	5, 10% raw pine bio-oil and balanced VGO	FCC equilibrium catalyst; 500 °C; Riser simulator reactor	[225]
Beijing U. of Chem., Tech., China	10% CFP beech wood bio-oil and 90% VGO	Fe/ZSM-5; 525 °C; pilot scale FCC riser reactor	[226]

7.7.2 Reactions with Alcohols

Significant recent work has explored catalyzed reactions of bio-oil with various alcohols. Table 7.7 lists numerous organizations investigating bio-oil upgrading in the presence of alcohols. Two reviews were recently published [253, 254]. Alcohols are often mixed

Table 7.7 Organizations involved in investigating upgrading bio-oil in the presence of alcohols.

Organization	Alcohol	Catalyst	Latest known activity	Example Ref.
Zhejiang University, Hangzhou, China	Supercritical ethanol; Supercritical methanol	Ru/HZSM-5; Pt/C; Pd/C/ Ru/C; Pt/(SO$_4$)$_2$-ZrO$_2$/SBA-15; PtNi/MgO; Pt/MgO; Pt/Al$_2$O$_3$-SiO$_2$	2017	[227–231]
University of Georgia, Athens, GA	Ethanol vapor	Non-catalytic; Ethanol vapor contacted with uncondensed bio-oil	2010	[232]
Curtin University of Technology, Australia	Methanol	Amberlyst® 70	2017	[233–238]
University of Limerick, Ireland	Ethanol	Sulfated ZrO$_2$-TiO$_2$	2015	[239]
South China University of Technology, Guangzhou, China	n-butanol; Methanol	Amberlyst® 36; NKC-9 cation-exchange resin; Silica sulfuric acid	2017	[240–242]
Aston University, UK	Methanol	Propyl sulfonic-modified SBA-15	2017	[243]
ETH Zurich, Switzerland	Model compounds: Acetic acid and o-cresol	Zeolites: Ferrierite; ZSM-5; Mordenite; Beta; Faujasite	2014	[244]
University of Aveiro, Portugal	Ethanol	Al-TUD-1-type mesoporous aluminosilicates	2013	[245]
Shangdong University of Technology, China	Butanol	Sulfuric acid	2014	[246]
Western University, London, ON, Canada	Glycerol	Amberlyst® 35	2014	[247]
VTT, Finland	Methanol or n-butanol	Para-toluenesulfonic acid; sulfuric acid; Amberlyst® 15; Amberlyst® 36; Smopex-101	2015	[248]
Mississippi State University, Starkville, MS	Butanol; 1-octene	Ru/γ-Al$_2$O$_3$ with KOH; Dowex50WX2; Amberlyst® 70; Amberlyst® 15; Nafion NR50.	2014	[249, 250]
Zhengzhou University, Zhengzhou, China	Supercritical n-butanol	Ru/C	2014	[251]
Guangzhou Institute of Energy Conversion, Chinese Academy of Sciences	Supercritical ethanol	Ni/SiO$_2$–ZrO$_2$	2015	[252]

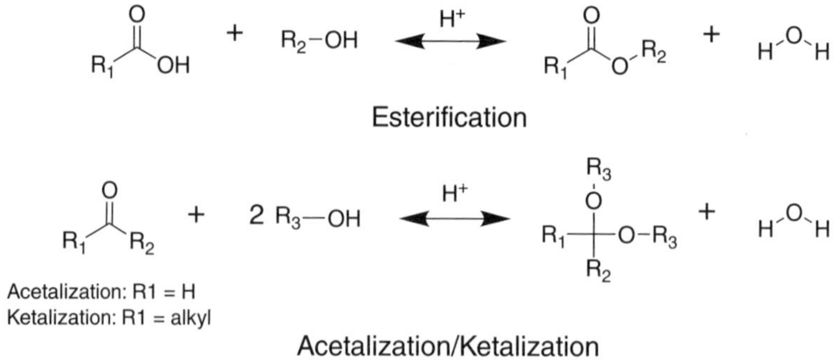

Figure 7.5 The acid-catalyzed reactions of alcohols with carboxylic acids, aldehydes (acetalization), or ketones (ketalization) to form esters, acetals, or ketals, respectively.

directly with bio-oil to serve as solvents in addition to reagents. Methanol, ethanol, and n-butanol have received the bulk of the attention. While numerous reactions take place in the presence of an acid catalyst with an alcohol reagent, the primary reactions are with carboxylic acids to produce esters and reactions with the carbonyls of aldehydes (acetalization) or ketones (ketalization) to produce ethers (see Figure 7.5).

Esterification significantly diminishes the TAN of bio-oil. Bio-oil stability is also improved through the conversion of highly reactive aldehydes and ketones to acetals or ketals, respectively. Other properties such as viscosity are also improved. Acid catalysts are most often employed. Cation ion-exchange resins such as Amberlyst®, sulfated ZrO_2 or SiO_2, and zeolites have been utilized as acid catalysts. Bi-functional catalysts possessing both acid and metal sites for hydrogenation have been employed.

An important challenge in the esterification of bio-oil is recovery of alcohols. Because esterification, acetalization, and ketalization are equilibrium limited, an excess of alcohol is often employed. Recovery of the alcohols in a continuous process will most likely be necessary for economic feasibility. Catalyst deactivation due to site inhibition from polymerization and adsorption from heavy products is also a challenge. Some catalysts including zeolites can be regenerated via oxidative treatment to remove coke but this method is not feasible for catalysts with lower maximum operating temperatures and poor oxidative stability such as ion-exchange resins. Catalysts with acid site density similar to those of strong acid cation-exchange resins suitable for higher temperature operation are needed. Even high temperature cation-exchange resins such as Amberlyst 70 are limited to temperatures of about 190 °C. Hydrothermally stable catalysts are also needed for higher temperature applications. Zeolites may feasibly be used and regenerated via oxidative treatment, but lack the site density of solid cation-exchange resins and suffer from generally poor hydrothermal stability.

7.8 Products

7.8.1 Liquid Transportation Fuel Properties

A major goal of bio-oil upgrading is production of liquid transportation fuels such as gasoline or diesel. Fuels are often considered in terms of the atmospheric boiling range of the fuel mixtures, such as gasoline (55–225 °C), and diesel (160–340 °C) [255]. Raw bio-oil cannot be distilled directly as it is unstable and forms a significant portion of solid residue when heated. Bio-oil upgraded via hydrotreating may be fractionated, although bio-oils containing significant residual oxygen (4–9%) produce about 10 wt% solid char residuals [256]. Whether processed as a neat material, upgraded as a blendstock with petroleum through a refinery unit operation such as an FCC unit, or physically blended with other products, finished fuels must meet ASTM specifications for the individual fuel types. In the USA, these specifications include ASTM D4814 for spark ignition fuels (i.e. gasoline) and ASTM D975 for diesel fuel oils.

7.8.1.1 Gasoline

7.8.1.1.1 Hydrotreated Low Oxygen Content Bio-Oil (<1.5 wt%)

Elliott and Schiefelbein [257] reported that hydroprocessed bio-oil with a boiling range between 23 and 225 °C with 1.3 wt% O exhibited a Research Octane Number (RON) of 78 and a Motor Octane Number (MON) of 72. Anti-knock index (AKI), defined as (RON + MON)/2, gives a value of 75 for this material. More recent work by Christensen et al. found gasoline-range light and naphtha distillate cuts with 0.3 wt% O exhibited estimated RON/MON values of 64/61 and 71/68, respectively [256]. The naphtha distillate fraction exhibited 2 mg KOH/g due to the presence of phenolics. It should be noted that a TAN of greater than 0.5 mg KOH/g may be potentially corrosive in refinery pipelines [258]. More recently, Olarte et al. [104] found the TAN of a naphtha-boiling LOC material to be below detection limits, with an estimated RON and MON of 65 and 60, respectively, based on an estimate from PIONA analysis. Gasoline in the USA is often produced via blending of a petroleum blendstock for oxygenate blending (BOB), which is mixed with 10 vol% ethanol to create finished fuel meeting the specifications of ASTM D4814 [259]. BOBs generally exhibit RON of 86 and MON of 81 prior to ethanol blending [260]. Thus, hydroprocessing bio-oil to relatively low oxygen content creates a material with lower RON and MON than not only finished gasoline but also with lower octane of the BOB destined for mixing with ethanol. The relatively low RON and MON values of deeply hydroprocessed bio-oil is likely due to the fact that a high concentration of the mixture consists of naphthenes such as alkyl-substituted cyclohexanes. Cyclohexanes have intrinsically low RON and MON values; e.g. the RON/MON of methylcyclohexane is 75/71 and ethylcyclohexane is 46/41 [261]

7.8.1.1.2 Hydrotreated Higher Oxygen Content Bio-oil

Christensen et al. [256] investigated hydrotreated bio-oils with varying residual oxygen content: high oxygen content (HOC – 8.2 wt% O) and medium oxygen content (MOC – 4.9 wt% O). When fractionated, the HOC sample had about 14 wt% O present in the lights and naphtha cuts, consisting largely of carboxylic acids. Acids were not detected in the MOC gasoline fractions but these fractions also contained a higher concentration of phenolics versus the HOC fractions. For example, the TAN for the HOC gasoline fractions was 102–123 mg KOH/g with all of the TAN contribution attributed to carboxylic acids. In contrast, the MOC fraction exhibited 100 TAN but the titration curve suggested the acidity was due to the presence of phenolics with no contribution to the TAN from carboxylic acids. The

PIONA-estimated RON/MON of the naphtha fractions was 73/72 for the MOC and 88/87 for the HOC, with the greater RON and MON values of the HOC attributed to the greater aromatic content. Similarly, more recent work by Olarte et al. [104] estimated a RON and MON of 78 and 59 for a naphtha fraction with about 10 wt% O. However, the PIONA estimate was applied with caution for these higher oxygenate fractions as the relationship was developed for petroleum mixtures without oxygenates.

7.8.1.1.3 Gasoline Properties of Oxygenated Model Compounds

McCormick et al. considered the properties of specific oxygenates found in partially hydroprocessed bio-oil with about 7 wt% O remaining [255]. 2,5-dimethylfuran (DMF) blended at 13 vol% and 4-methylanisole (4MA) blended at 10 vol% significantly increase the RON and MON values when blended with a BOB. However, DMF produced a substantial amount of gum when blended at 13% into the BOB. Several of the compounds investigated such as p-cresol or 2,4-xylenol have good blending octane numbers, but boil at the upper end of what is acceptable for gasoline, which limits the concentration of aryl ethers or phenolics that could be added to gasoline but remain within the distillation specification. 2,4-Xylenol was later found to significantly increase the T_{90} distillation temperature of a BOB when blended at 10% [262]. Anisole, 4MA, and 2,4-xylenol caused significant increases in particulates when studied as blends in gasoline, likely due to their high boiling points and unsaturated structures [262]. p-Cresol blended at 2% into an 10% ethanol gasoline absorbed about three times as much water as the mixture without p-cresol present [255]. Thus, while many components of the oxygenates present in bio-oil have potential as high octane blendstocks, challenges exist due to their effects in altering the boiling range and emissions of the gasoline when blended above a few volume percent.

7.8.1.1.4 Gasoline from FCC Upgraded Bio-oil

A partnership between Petrobras and the US Department of Energy investigated the co-processing of bio-oil with VGO [218, 225]. Processing of up to 10 wt% bio-oil with 50 wt% oxygen was found to be technically feasible when co-fed with VGO [225]. The co-processed material made gasoline and diesel range products. Oxygenated products, primarily alkyl phenolics, were observed with or without bio-oil co-processing. Earlier work from the same group determined that oxygenate concentrations increased about a factor of 6 to approximately 1.9 wt% when co-processing VGO with 20 wt% bio-oil [218]. The RON and MON of the FCC product processed with 20 wt% bio-oil increased slightly to 96.5 and 84.4 from 95.8 and 83.3 when processing without bio-oil. When the co-processed gasoline mixture was sent to the USA for final hydrotreating, the resulting gasoline blendstock met specifications under Title 40 CFR Part 79 of the Clean Air Act required to sell renewable fuels in the USA [225]

7.8.1.2 Jet Fuel

Jet fuel must meet stringent requirements, including the approval of an annex to ASTM D7566-17a, which is the Standard Specification for Aviation Turbine Fuel Containing

Synthesized Hydrocarbons. The synthesized paraffinic kerosene (SPK) products which are currently approved for blending as a portion of finished jet fuel include:

- Fischer–Tropsch hydroprocessed SPK (SPK-FT),
- hydroprocessed esters and fatty acids (SPK-HEFA),
- hydroprocessed fermented sugars – specifically, the compound faransane
- synthesized kerosene with aromatics derived by alkylation of light aromatics from non-petroleum sources (SPK/A)
- alcohol-to-jet materials made from dehydration, oligomerization, hydrogenation, and fractionation (ATJ-SPK).

While no FP-upgraded bio-oil process is currently approved for use as a jet fuel blendstock, it is instructive to compare the properties of upgraded FP bio-oil with those of other approved processes. Christensen et al. [256] found that the fraction of a low oxygen content bio-oil (0.7 wt% O) upgraded by hydroprocessing and distilled to simulate jet fuel exhibited a TAN of 14 mg KOH/g, exceeding the 0.015 mg KOH/g maximum for all SPK blendstocks laid out in ASTM D7566. Ajam et al. [263] reported that a 400 °C hydrotreated bio-oil obtained from BTG distilled about 45 wt% of the total upgraded bio-oil in a range between 150 and 280 °C. The fractionated bio-oil contained between 100 and 1000 ppm phenolics after hydrotreating and distillation. Additionally, the fractionated material was denser than the specification laid out for Jet A-1 grade fuel. The thermal oxidative stability of phenolics was shown to be poor through the use of a model compounds. However, the thermal oxidative stability of aryl ethers (e.g. anisole) was found to be significantly better than phenolics and proposed as a potential route to diminish concerns with phenolics, including elevated TAN. A study by Wang et al. [264] found that upgrading via zeolite cracking, alkylation, and hydrotreating produced a material with generally similar properties compared to Jet-A and JP-8 fuels. Hydrogenation of the aromatic-rich zeolite-processed material increased the heat of combustion from 42.8 to 46.5 MJ/kg. Generally, jet fuel blendstocks from bio-oil will need to be low in oxygen to meet the TAN requirement of less 0.015 mg KOH/g and the energy density requirement of the finished jet fuel blend of 42.8 MJ/kg. Regarding the combustion of various hydrocarbon types in a turbine, ASTM D7566 reports that paraffins are the most desirable, followed by cycloparaffins. Olefins have good combustion properties but typically have poor gum stability. Aromatics are less desirable because they tend to produce a smoky flame and generate more thermal radiation when compared to other hydrocarbon types. Thus, a hypothetical upgraded FP jet fuel may be limited in aromatic content to very low levels based on these considerations coupled with the history of other approved hydroprocessed mixtures such as SPK-HEFA or SPK-FT which allow a maximum of 0.5 wt% aromatics.

7.8.1.3 Diesel

7.8.1.3.1 Raw Bio-oil

The use of raw bio-oil as diesel fuel blendstock was recently reviewed [265]. Generally, the ignition quality of raw bio-oil was found to be poor, with most cetane estimates between 0 and 14. For reference, the minimum cetane number for #2 diesel fuel is 40 per ASTM D975. No engine tests utilizing raw bio-oil were successful for longer than twelve consecutive hours. Several challenges, such as coking at fuel injectors and within cylinders, corrosion to common engine materials, plugging, and the need for ignition quality improvers such as diglyme or ethylhexyl nitrate, were noted. Work at the USDA investigated emulsions of raw bio-oil into diesel fuel [266]. Small bubbles would slowly settle out of the emulsion, but this could be overcome with slight mechanical agitation. Levoglucosan was identified as a likely cause for the immiscibility. Bio-oil produced by CFP of was found to be miscible with diesel, although the aromatic nature of the material would likely have low cetane and higher sooting potential.

7.8.1.3.2 Hydrotreated Bio-oil

The oxygen content of upgraded bio-oil will significantly impact the potential for distilling a diesel fraction blendstock. The diesel fraction of bio-oil initially hydrotreated to 8.2 wt% O prior to distillation was found to have a TAN of 20 and an oxygen content of 7.5 wt% after distillation, suggesting deeper hydrotreating would be required prior to distillation to diminish the oxygen content and by extension the TAN [256]. Upgraded bio-oils with lower oxygen concentration had TAN numbers for the diesel fraction of 0.3 and 0.1 for bio-oils with 4.9 and 1.3 wt% O, respectively. At relatively low blend levels (2 vol%) several monolignol model compounds representative of those in upgraded bio-oil were shown to slightly reduce the derived cetane number when mixed with a certification diesel fuel [267]. At 2 vol%, NO_x and particulate matter emissions decreased while increasing CO, CH_2O, and total hydrocarbon emissions. At 6 vol%, these compounds produced deleterious effects, including increased NO_x and reduced thermal efficiency, resulting in a recommendation to keep the blending of these oxygenates to 2 vol% or less.

7.8.2 Chemicals

Utilizing bio-oil as a source for renewable chemicals is often cited as a required portion of a biorefinery to improve process economics due to their greater value [268, 269] As with the production of fuels, bio-oil yield and the selection of upgrading process has a significant impact on overall potential process economics [270]. Aromatics are often targeted as CFP often employs zeolite catalysts such as H-ZSM-5 [271]. Since 2010, several reviews have been written on elucidating reaction mechanisms towards improving the yields of value-added chemicals from biomass pyrolysis [272–274]

Bio-oil is a complex mixture which generally cannot be distilled. Thus, solvent separations are frequently utilized in the pursuit of chemical recovery and purification. Liquid–liquid extraction with organic solvents [275–277] or water [278, 279] or both [280] may be performed. Super critical CO_2 may also be utilized [281]. Amines such as trioctyl amines [282] or ionic liquids [283] may be utilized to selectively extract organic acids. Salt may also be added to facilitate phase separation [284]. Fractionation of bio-oil by controlled condensation is a promising method to separate out bio-oil components into fractions which may be more readily extracted [53]. Biomass fractionation prior to pyrolysis may also facilitate more directed chemical targets such as levoglucosan or phenolics [285]. CFP may generate a spontaneous aqueous phase containing mono-lignols, which may be converted to bio-polymer precursors [286]. Table 7.8 lists the activities

Table 7.8 Organizations investigating the recovery of chemicals from bio-oil.

Organization	Chemical	Latest known activities	Refs
Acetic acid			
University of Science and Technology, South Korea		2016	[287]
University of Seoul, South Korea		2017	[288, 289]
Xiamen University, Xiamen, China		2013	[282]
University of Science and Technology of China, Hefei, China		2013	[290]
Indian Institute of Technology Guwahati, India		2015	[283]
Aromatics			
Iowa State University, Ames, IA		2017	[270, 291]
ETH Zurich, Switzerland		2015	[272, 292]
University of Malaya, Kuala Lumpur, Malaysia		2014	[271]
Annellotech, Inc.		2017	[293]
Guangzhou Institute of Energy Conversion, Chinese Academy of Sciences		2017	[294]
Furfural & 5-hydroxymethyl furfural (HMF)			
Indian Institute of Technology, Guwahati, Inida		2015	[283]
University of Seoul, South Korea		2013	[289]
University of Science and Technology of China, Hefei, China		2013	[290]
North China Electric Power University, Beijing, China		2015	[295]
Mississippi State University, Starkville, MS		2014	[278]
NREL		2015	[296]
Institute of Chemical Industry of Forestry Products, Nanjing, China		2011	[297]
Nanjing Forestry University, Nanjing, China		2017	[298]
Hydrogen			
See Section 7.8.3			
Levoglucosan			
Southeast University, Nanjing, China		2016	[299]
Mississippi State University, Starkville, MS		2013	[300]

(*Continued*)

Table 7.8 Continued

Organization	Chemical	Latest known activities	Refs
Institute of Chemical Industry of Forestry Products, Nanjing, China		2011	[297]
Guangzhou Institute of Energy Conversion, Chinese Academy of Sciences		2017	[285, 301, 302]
Levoglucosenone			
North China Electric Power University, Beijing, China		2015	[170, 295]
Institute of Chemical Industry of Forestry Products, Nanjing, China		2011	[297]
Southeast University, Nanjing, China		2016	[299]
Phenol and phenolics			
University of Science and Technology, South Korea		2016	[287]
National Renewable Energy Laboratory, Golden, CO		2014	[296]
North China Electric Power University, Beijing, China		2016	[303–306]
Guangzhou Institute of Energy Conversion, Chinese Academy of Sciences		2017	[285]
Synthesis gas			
Lulea University of Technology, Sweden		2016	[307]
Northwest A&F University, China		2016	[308]
Karlsruhe Institute of Technology bioliq® Process		2017	[309]
Mixed and miscellaneous			
University of Seoul, South Korea	Magnesium acetate deicer	2017	[288]
Iowa State University, Ames, IA	Biocement	2017	[269, 310]
National Renewable Energy Laboratory, Golden, CO	Resol biopolymer	2017	[286]
Department of Agriculture and Agri-Food, Canada	Insecticide	2015	[311]
Curtin University of Technology, Australia	Levulinic acid and esters	2012	[236]
University of Notre Dame, South Bend, IN	Furans, ketones, aldehydes	2012	[127]
Hokkaido University, Japan	Iron (bio-oil used as carbon-based reducing agent)	2012	[312]
University of Oklahoma	Gluconic acid	2014	[313]

Table 7.8 Continued

Organization	Chemical	Latest known activities	Refs
USDA-ARS Southern Regional Research Center, New Orleans, LA	Phosphorus fertilizer	2015	[314]
Key Laboratory of Biomass Energy and Material, Nanjing, China	Methyl-α-D-glucopyranoside	2017	[315]
Chang'an University, Xi'an, China	Asphalt	2017	[316]
Pacific Northwest National Laboratory, Richland, WA	Coke for metal refining	2013	[317]

of numerous groups working to recover a wide variety of value-added compounds from bio-oil.

7.8.3 Hydrogen Production

Hydrogen production from bio-oil continues to receive significant interest as evidenced by the publication of several dedicated reviews [318–321] as well as coverage in the review on oxygenate reforming by Li et al. [322] Bio-oil is difficult to steam reform as traditionally practiced. Metal catalysts such as Ni tend to form carbonaceous deposits (i.e. coke) requiring frequent regeneration or higher steam-to-carbon ratios, but high steam requirements are detrimental to overall economics. Precious metals such as Pt or Ru are more resistant to coking but significantly increase cost. The use of supports with dopants with increased basicity or redox properties such as CeO_2, MgO, or La_2O_3 diminish the rate of coking. Table 7.9 represents numerous accounts of research with the goal of producing H_2 from FP bio-oil.

7.9 Summary

The corrosiveness, chemical and thermal instability, viscosity, and high water content of bio-oil require it to be upgraded before it can be used in many applications including transportation fuels. Bio-oil yield and quality are influenced by biomass properties and FP reaction conditions. Significant progress has been made in the past several years towards a series of standards to describe and produce consistent bio-oil that regulating bodies around the world will recognize. Commoditization of bio-oil towards a common product as a feedstock for commercial fuel and chemical processes will be greatly facilitated by the development of these common standards and advanced characterization techniques.

Continuous hydrotreating of FP bio-oil to reduce oxygen will require either an initial mild stabilization step to reduce the most reactive species or catalytic processing of the bio-oil vapors prior to condensation to generate a thermally stable material. Significant progress in both reductive stabilization as well as CFP has occurred, but the need for further catalyst and process improvements remain. For reductive stabilization, economic catalysts that are active at low temperature (~140 °C) and resistant to coking are needed. In CFP, economic

Table 7.9 Organizations investigating the production of hydrogen from bio-oil since about 2010.

Organization	Catalyst	Feed or model compounds	Latest known activity	Example ref.
Chemical looping redox				
Southeast University, Nanjing China	Fe-oxide (oxygen carrier)	Heavy bio-oil	2017	[323–325]
Chulalongkorn University, Thailand	NiO; CaO as CO_2 sorbent	Bio-oil – modeled study	2016	[326]
Catalytic steam reforming				
University of Nevada Las Vegas, Las Vegas, NV	Nickel; activated C	Nebulized bio-oil	2015	[327]
University of the Basque Country, Vizcaya, Spain	Ni/Al_2O_3; $Ni/CeO_2-Al_2O_3$; $Ni/La_2O_3-Al_2O_3$	1-butanol; m-xylene; furfural; bio-oil/ethanol	2016	[328–330]
Anhui University of Science and Technology, Ahhui, China	$Ni-Fe/Ca_xLa_yCe_zO_m$	Bio-oil	2016	[331]
Institute of Chemical Technology, Mumbai, India	Ni/Al_2O_3; Ru/Al_2O_3; Ru/C; Pt/C; Pd/C;	Propylene glycol	2016	[332]
Zhejiang University, Hangzhou, China	Ni/Al_2O_3; Ni/coal ash	Bio-oil; acetic acid; phenol	2014	[333, 334]
Northeastern University, Liaoning China	Ni-Mg/Co; Ce-Ni/Co	Ethanol; acetone; phenol	2015	[335]
University of Science and Technology of China, Hefei, China	Ni/carbon nanofibers; Cu-Mg-Ce-Ni-Al; Ni/HZSM-5	Bio-oil; acetic acid; ethanol	2013	[336–338]
Pacific Northwest National Laboratory, Richland, WA	Co; Ni; Rh; Co/CeO_2; $Co/MgAl_2O_4$; Co/ZnO; Co/Carbon	Acetic acid; bio-oil aqueous phase	2017	[339, 340]
Xi'an Jiaotong University, Shanxi, China	Fe/olivine	Bio-oil	2017	[341]
Tianjin University, Tianjin, China	$La_{1-x}K_xMnO_3$ perovskite-type catalysts	Bio-oil	2016	[342]
Technical University of Denmark, Lyngby, Denmark	$Ni/MgAl_2O_4$; $Ni/CeO_2-K/MgAl_2O_4$	2-methylfuran; furfural; guaiacol	2015	[343, 344]
Central University of Ecuador, Quito, Ecuador	Thermodynamic evaluation	Simulated bio-oil	2015	[345]
Chengdu University of Technology, Chengdu, China	Ni-Co/MgO	Acetic acid	2014	[346]
Shangdong University of Technology, Zibo, China	Ni/Al_2O_3; $Ni/Ce/Al_2O_3$	Bio-oil	2014	[347]
East China University of Science and Technology, Shanghai, China	$Ni-Co/\gamma-Al_2O_3$; $Co/\gamma-Al_2O_3$; $Ni/\gamma-Al_2O_3\|Co/\gamma-Al_2O_3$; $Ni/\gamma-Al_2O_3$	Bio-oil volatiles	2011	[348]

Table 7.9 Continued

Organization	Catalyst	Feed or model compounds	Latest known activity	Example ref.
University of Zaragoza, Spain	Co-Cu-Ni/MgAlO	Bio-oil	2013	[349, 350]
University of Calgary, Canada	Ru-Ni/Al$_2$O$_3$; Mg-Ni/Al$_2$O$_3$	Bio-oil	2011	[351]
Steam reforming with CO$_2$ capture				
Northeastern University, Liaoning, China	Ce-Ni/Co/Al$_2$O$_3$; CaO as CO$_2$ sorbent	Bio-oil	2016	[352]
Instituto Nacional del Carbon, Oviedo, Spain	Pd/Ni-Co; CaO as CO$_2$ sorbent	Bio-oil	2014	[353]
Combined pyrolysis and gasification with CO$_2$ capture				
Southeast University, Nanjing, China	Zr/CaO/H-ZSM-5	Bio-oil	2017	[354]
Combined partial oxidation and steam reforming processes				
Lanzhou Institute of Chemical Physics, Chinese Academy of Sciences	Ni/Al$_2$O$_3$; Ni/La$_2$O$_3$	Bio-oil; CO$_2$	2010	[355]
Chulalongkorn University, Thailand	Thermodynamic model; no physical catalyst	Bio-oil aqueous fraction	2016	[356]
Cracking and steam reforming				
University of Science and Technology of China, Hefei, China	H-ZSM-5; Ce/H-ZSM5	Bio-oil	2016	[357]
Aqueous phase reforming and hydrodeoxygenation				
Zhejiang University, Hangzhou, China	Pt/Al$_2$O$_3$	Low boiling bio-oil fraction	2012	[358]
Korea Research Institute of Chemical Technology, Daejon, South Korea	Pt/mesoporous C	Ethylene glycol	2011	[359]
Electrochemical catalytic reforming				
University of Science and Technology of China, Hefei, China	CoZnAl; Cu-Mg-Ce-Ni-Al	Bio-oil; Acetic acid; ethanol	2013	[338, 360]

catalysts with balanced active sites to remove oxygen but minimize liquid yield loss are needed. Catalysts which are impervious to poisoning by sulfur and dissolved inorganics are needed for both reductive and CFP applications. Opportunities for significant and impactful new research exists to correlate studies of model compounds to studies with real bio-oil. Once thermally stabilized, sulfided base metal catalysts such as those currently used in petroleum refining are prime candidates to generate hydrocarbon fuel blendstocks. The fuel properties of the gasoline and diesel fractions of upgraded bio-oil are generally lower in RON or cetane number, respectively, of their finished fuel counterparts. Further research into improving the fuel properties of upgraded FP bio-oil is needed.

Significant progress continues to be made in producing value-added chemicals by maximizing the yield and recovering individual compounds from the complex chemical mixture of FP bio-oil. Hydrogen produced from bio-oil may be used for other green applications such as fuel cells or applied to upgrading bio-oil through hydrotreatment. Commercialization of FP bio-oil upgrading to produce liquid transportation fuels will be facilitated by the ability to produce value-added chemicals and generate hydrogen economically in order to maximize profits and diminish greenhouse gas footprints.

References

1. Bridgwater, A.V. (2012). Review of fast pyrolysis of biomass and product upgrading. *Biomass and Bioenergy* **38**: 68–94.
2. Mortensen, P.M., Grunwaldt, J.D., Jensen, P.A. et al. (2011). A review of catalytic upgrading of bio-oil to engine fuels. *Applied Catalysis A: General* **407**: 1–19.
3. Pham, T.N., Shi, D., and Resasco, D.E. (2014). Evaluating strategies for catalytic upgrading of pyrolysis oil in liquid phase. *Applied Catalysis B: Environmental* **145**: 10–23.
4. Resasco, D.E. and Crossley, S.P. (2015). Implementation of concepts derived from model compound studies in the separation and conversion of bio-oil to fuel. *Catalysis Today* **257** (Part 2): 185–199.
5. Talmadge, M.S., Baldwin, R.M., Biddy, M.J. et al. (2014). A perspective on oxygenated species in the refinery integration of pyrolysis oil. *Green Chemistry* **16**: 407–453.
6. Wang, H., Male, J., and Wang, Y. (2013). Recent advances in hydrotreating of pyrolysis bio-oil and is oxygen-containing model compounds. *ACS Catalysis* **3**: 1047–1070.
7. Zacher, A.H., Olarte, M.V., Santosa, D.M. et al. (2014). A review and perspective of recent bio-oil hydrotreating research. *Green Chemistry* **16**: 491–515.
8. Elliott, D.C. (2015). Biofuel from fast pyrolysis and catalytic hydrodeoxygenation. *Current Opinion in Chemical Engineering* **9**: 59–65.
9. Elliott, D.C. (2016). 19 - Production of biofuels via bio-oil upgrading and refining. In: *Handbook of Biofuels Production*, 2e (ed. R. Luque, C. Sze Ki Lin, K. Wilson and J. Clark), 595–613. Woodhead Publishing. https://www.sciencedirect.com/science/article/pii/B9780081004555000199 (accessed 28 November 2018).
10. Elliott, D.C. (2013). Transportation fuels from biomass via fast pyrolysis and hydroprocessing. *Wiley Interdisciplinary Reviews: Energy and Environment* **2**: 525–533.
11. Carpenter, D., Westover, T., Howe, D. et al. (2017). Catalytic hydroprocessing of fast pyrolysis oils: impact of biomass feedstock on process efficiency. *Biomass and Bioenergy* **96**: 142–151.
12. Howe, D., Westover, T., Carpenter, D. et al. (2015). Field-to-fuel performance testing of lignocellulosic feedstocks: an integrated study of the fast pyrolysis-hydrotreating pathway. *Energy & Fuels* **29**: 3188–3197.
13. Neves, D., Thunman, H., Matos, A. et al. (2011). Characterization and prediction of biomass pyrolysis products. *Progress in Energy and Combustion Science* **37**: 611–630.
14. Elliott, D.C., Meier, D., Oasmaa, A. et al. (2017). Results of the International Energy Agency Round Robin on fast pyrolysis bio-oil production. *Energy & Fuels* **31**: 5111–5119.
15. Carrier, M., Hugo, T., Gorgens, J., and Knoetze, H. (2011). Comparison of slow and vacuum pyrolysis of sugar cane bagasse. *Journal of Analytical and Applied Pyrolysis* **90**: 18–26.
16. Pighinelli, A.L.M.T., Boateng, A.A., Mullen, C.A., and Elkasabi, Y. (2014). Evaluation of Brazilian biomasses as feedstocks for fuel production via fast pyrolysis. *Energy for Sustainable Development* **21**: 42–50.
17. Funke, A., Morgano, M.T., Dahmen, N., and Leibold, H. (2017). Experimental comparison of two bench scale units for fast and intermediate pyrolysis. *Journal of Analytical and Applied Pyrolysis* **124**: 504–514.

18. Carvalho, W.S., Santana, J.A., de Oliveira, T.J.P., and Ataide, C.H. (2017). Fast pyrolysis of sweet sorghum bagasse in a fluidized bed reactor: product characterization and comparison with vapors generated in analytical pyrolysis. *Energy* **131**: 186–197.
19. Mellin, P., Zhang, Q.L., Kantarelis, E., and Yang, W.H. (2013). An Euler-Euler approach to modeling biomass fast pyrolysis in fluidized-bed reactors - Focusing on the gas phase. *Applied Thermal Engineering* **58**: 344–353.
20. Xue, Q., Dalluge, D., Heindel, T.J. et al. (2012). Experimental validation and CFD modeling study of biomass fast pyrolysis in fluidized-bed reactors. *Fuel* **97**: 757–769.
21. Mellin, P., Kantarelis, E., and Yang, W.H. (2014). Computational fluid dynamics modeling of biomass fast pyrolysis in a fluidized bed reactor, using a comprehensive chemistry scheme. *Fuel* **117**: 704–715.
22. Ranzi, E., Debiagi, P.E.A., and Frassoldati, A. (2017). Mathematical modeling of fast biomass pyrolysis and bio-oil formation. Note I: kinetic mechanism of biomass pyrolysis. *ACS Sustainable Chemistry & Engineering* **5**: 2867–2881.
23. Oasmaa, A. and Czernik, S. (1999). Fuel oil quality of biomass pyrolysis oils state of the art for the end users. *Energy & Fuels* **13**: 914–921.
24. Boucher, M.E., Chaala, A., Pakdel, H., and Roy, C. (2000). Bio-oils obtained by vacuum pyrolysis of softwood bark as a liquid fuel for gas turbines. Part II: Stability and ageing of bio-oil and its blends with methanol and a pyrolytic aqueous phase. *Biomass and Bioenergy* **19**: 351–361.
25. Oasmaa, A., Fonts, I., Pelaez-Samaniego, M.R. et al. (2016). Pyrolysis oil multiphase behavior and phase stability: a review. *Energy & Fuels* **30**: 6179–6200.
26. Oasmaa, A., Sundqvist, T., Kuoppala, E. et al. (2015). Controlling the phase stability of biomass fast pyrolysis bio-oils. *Energy & Fuels* **29**: 4373–4381.
27. Lin, B.J., Chen, W.H., Budzianowski, W.M. et al. (2016). Emulsification analysis of bio-oil and diesel under various combinations of emulsifiers. *Applied Energy* **178**: 746–757.
28. Krutof, A. and Hawboldt, K. (2016). Blends of pyrolysis oil, petroleum, and other bio-based fuels: a review. *Renewable and Sustainable Energy Reviews* **59**: 406–419.
29. Oasmaa, A., Elliott, D.C., and Korhonen, J. (2010). Acidity of biomass fast pyrolysis bio-oils. *Energy & Fuels* **24**: 6548–6554.
30. Strahan, G.D., Mullen, C.A., and Boateng, A.A. (2016). Prediction of properties and elemental composition of biomass pyrolysis oils by NMR and partial least squares analysis. *Energy & Fuels* **30**: 423–433.
31. JP, D. (2000). *A Review of the Chemical and Physical Mechanisms of the Storage Stability of Fast Pyrolysis Bio-oils*. Lakewood, CO: Thermalchemie, Inc.
32. Olarte, M.V., Zacher, A.H., Padmaperuma, A.B. et al. (2016). Stabilization of softwood-derived pyrolysis oils for continuous bio-oil hydroprocessing. *Topics in Catalysis* **59**: 55–64.
33. Wang, H., Lee, S.-J., Olarte, M.V., and Zacher, A.H. (2016). Bio-oil stabilization by hydrogenation over reduced metal catalysts at low temperatures. *ACS Sustainable Chemistry & Engineering* **4**: 5533–5545.
34. Bridgwater, A.V. (2011). Upgrading fast pyrolysis liquids. In: *Thermochemical Processing of Biomass*, 157–199. Wiley.
35. Abedi, J., Nourozieh, H., Kariznovi, M., and Seyedeyn-Azad, F. (2015). Thermo-physical properties of bio-oil and its fractions: measurement and analysis. *Canadian Journal of Chemical Engineering* **93**: 500–509.
36. Ghezelchi, M.H., Garcia-Perez, M., and Wu, H.W. (2015). Bioslurry as a fuel. 7: Spray characteristics of bio-oil and bioslurry via impact and twin-fluid atomizers. *Energy & Fuels* **29**: 8058–8065.
37. Nolte, M.W. and Liberatore, M.W. (2010). Viscosity of biomass pyrolysis oils from various feedstocks. *Energy & Fuels* **24**: 6601–6608.
38. Elliott, D.C., Oasmaa, A., Meier, D. et al. (2012). Results of the IEA round robin on viscosity and aging of fast pyrolysis bio-oils: long-term tests and repeatability. *Energy & Fuels* **26**: 7362–7366.

39. Elliott, D.C., Oasmaa, A., Preto, F. et al. (2012). Results of the IEA Round Robin on viscosity and stability of fast pyrolysis bio-oils. *Energy & Fuels* **26**: 3769–3776.
40. Liu, W.J., Li, W.W., Jiang, H., and Yu, H.Q. (2017). Fates of chemical elements in biomass during its pyrolysis. *Chemical Reviews* **117**: 6367–6398.
41. Wiinikka, H., Johansson, A.C., Sandstrom, L., and Ohrman, O.G.W. (2017). Fate of inorganic elements during fast pyrolysis of biomass in a cyclone reactor. *Fuel* **203**: 537–547.
42. Li, W.Q., Dang, Q., Brown, R.C. et al. (2017). The impacts of biomass properties on pyrolysis yields, economic and environmental performance of the pyrolysis-bioenergy-biochar platform to carbon negative energy. *Bioresource Technology* **241**: 959–968.
43. Reichel, D., Klinger, M., Krzack, S., and Meyer, B. (2013). Effect of ash components on devolatilization behavior of coal in comparison with biomass - Product yields, composition, and heating values. *Fuel* **114**: 64–70.
44. Li, J., Chen, Y.Q., Yang, H.P. et al. (2017). Correlation of feedstock and bio-oil compound distribution. *Energy & Fuels* **31**: 7093–7100.
45. Edmunds, C.W., Hamilton, C., Kim, K. et al. (2017). Using a chelating agent to generate low ash bioenergy feedstock. *Biomass and Bioenergy* **96**: 12–18.
46. Oudenhoven, S.R.G., Westerhof, R.J.M., and Kersten, S.R.A. (2015). Fast pyrolysis of organic acid leached wood, straw, hay and bagasse: improved oil and sugar yields. *Journal of Analytical and Applied Pyrolysis* **116**: 253–262.
47. Chen, D.Y., Zheng, Z.C., Fu, K.X. et al. (2015). Torrefaction of biomass stalk and its effect on the yield and quality of pyrolysis products. *Fuel* **159**: 27–32.
48. Zheng, A.Q., Zhao, Z.L., Chang, S. et al. (2012). Effect of torrefaction temperature on product distribution from two-staged pyrolysis of biomass. *Energy & Fuels* **26**: 2968–2974.
49. Zhang, X.L., Yang, W.H., and Blasiak, W. (2013). Kinetics study on thermal dissociation of levoglucosan during cellulose pyrolysis. *Fuel* **109**: 476–483.
50. Shen, D.K., Gu, S., and Bridgwater, A.V. (2010). Study on the pyrolytic behaviour of xylan-based hemicellulose using TG-FTIR and Py-GC-FTIR. *Journal of Analytical and Applied Pyrolysis* **87**: 199–206.
51. Kosa, M., Ben, H.X., Theliander, H., and Ragauskas, A.J. (2011). Pyrolysis oils from CO_2 precipitated Kraft lignin. *Green Chemistry* **13**: 3196–3202.
52. Rover, M.R., Johnston, P.A., Whitmer, L.E. et al. (2014). The effect of pyrolysis temperature on recovery of bio-oil as distinctive stage fractions. *Journal of Analytical and Applied Pyrolysis* **105**: 262–268.
53. Pollard, A.S., Rover, M.R., and Brown, R.C. (2012). Characterization of bio-oil recovered as stage fractions with unique chemical and physical properties. *Journal of Analytical and Applied Pyrolysis* **93**: 129–138.
54. S. Jones, P. Meyer, L. Snowden-Swan, et al. 2013 Process Design and Economics for the Conversion of Lignocellulosic Biomass to Hydrocarbon Fuels: Fast Pyrolysis and Hydrotreating Bio-oil Pathway, Pacific Northwest National Laboratory. PNNL-23053.
55. Meng, J.J., Moore, A., Tilotta, D. et al. (2014). Toward understanding of bio-oil aging: accelerated aging of bio-oil fractions. *ACS Sustainable Chemistry & Engineering* **2**: 2011–2018.
56. Yang, Z.X., Kumar, A., and Huhnke, R.L. (2015). Review of recent developments to improve storage and transportation stability of bio-oil. *Renewable and Sustainable Energy Reviews* **50**: 859–870.
57. Oasmaa, A., Kuoppala, E., and Elliott, D.C. (2012). Development of the basis for an analytical protocol for feeds and products of bio-oil hydrotreatment. *Energy & Fuels* **26**: 2454–2460.
58. Czernik, S., Johnson, D.K., and Black, S. (1994). Stability of wood fast pyrolysis oil. *Biomass and Bioenergy* **7**: 187–192.
59. Ben, H.X. and Ragauskas, A.J. (2012). In Situ NMR characterization of pyrolysis oil during accelerated aging. *ChemSusChem* **5**: 1687–1693.

60. Alsbou, E. and Helleur, B. (2014). Accelerated aging of bio-oil from fast pyrolysis of hardwood. *Energy & Fuels* **28**: 3224–3235.
61. Elliott, D.C. (2007). Historical developments in hydroprocessing bio-oils. *Energy & Fuels* **21**: 1792–1815.
62. Weber, R.S., Olarte, M.V., and Wang, H.M. (2015). Modeling the kinetics of deactivation of catalysts during the upgrading of bio-oil. *Energy & Fuels* **29**: 273–277.
63. S. Jones, L.J. Snowden-Swan, P. Meyer, et al. 2016 Fast Pyrolysis and Hydrotreating: 2015 State of Technology R&D and Projections to 2017, Pacific Northwest National Laboratory. PNNL-25312.
64. Oasmaa, A., van de Beld, B., Saari, P. et al. (2015). Norms, standards, and legislation for fast pyrolysis bio-oils from lignocellulosic biomass. *Energy & Fuels* **29**: 2471–2484.
65. Diebold, J.P., Milne, T.A., Czernik, S. et al. (1997). Proposed specifications for various grades of pyrolysis oils. In: *Developments in Thermochemical Biomass Conversion: Volume 1 / Volume 2* (ed. A.V. Bridgwater and D.G.B. Boocock), 433–447. Dordrecht: Springer Netherlands.
66. ASTM International 2009 Standard Specification for Pyrolysis Liquid Biofuel
67. Oasmaa, A., Elliott, D.C., and Muller, S. (2009). Quality control in fast pyrolysis bio-oil production and use. *Environmental Progress & Sustainable Energy* **28**: 404–409.
68. ASTM International(2017). ASTM D7544-12, Standard Specification for Pyrolysis Liquid Biofuel.
69. ASTM International (2013) ASTM D7579-09, Standard Test Method for Pyrolysis Solids Content in Pyrolysis Liquids by Filtration of Solids in Methanol.
70. CEN European Committee for Standardization 2017 EN 16900:2017, Fast Pyrolysis Bio-oils for Industrial Boilers – Requirement and Test Methods.
71. CEN European Committee for Standardization 2017 CEN/TR 17103:2017, Fast Pyrolysis Bio-oil for Stationary International Combustion Engines – Quality determination.
72. ASTM International 2018 E3641-18: Standard Test Method for Determination of Carbonyls in Pyrolysis Bio-oils by Potentiometric Titration.
73. J. McKinley, R. Overend, and D. Elliott 1994. The Ultimate Analysis of Biomass Liquefaction Products: The Results of the IEA Round Robin #1, Biomass Pyrolysis Oil Properties and Combustion Meeting, Estes Park, CO.
74. Oasmaa, A. and Meier, D. (2005). Norms and standards for fast pyrolysis liquids: 1. Round robin test. *Journal of Analytical and Applied Pyrolysis* **73**: 323–334.
75. Ferrell, J.R., Olarte, M.V., Christensen, E.D. et al. (2016). Standardization of chemical analytical techniques for pyrolysis bio-oil: history, challenges, and current status of methods. *Biofuels, Bioproducts and Biorefining* **10**: 496–507.
76. F. Agblevor, S. Besler, and R. Evans 1994. Inorganic compounds in biomass feedstocks: Their role in char formation and effect on the quality of fast pyrolysis oils, Biomass Pyrolysis Oil Properties and Combustion Meeting, Estes Park, CO.
77. Baldwin, R.M. and Feik, C.J. (2013). Bio-oil stabilization and upgrading by hot gas filtration. *Energy & Fuels* **27**: 3224–3238.
78. Pattiya, A. and Suttibak, S. (2012). Influence of a glass wool hot vapour filter on yields and properties of bio-oil derived from rapid pyrolysis of paddy residues. *Bioresource Technology* **116**: 107–113.
79. Naske, C.D., Polk, P., Wynne, P.Z. et al. (2012). Postcondensation filtration of pine and cottonwood pyrolysis oil and impacts on accelerated aging reactions. *Energy & Fuels* **26**: 1284–1297.
80. Oasmaa, A., Kuoppala, E., Selin, J.F. et al. (2004). Fast pyrolysis of forestry residue and pine. 4. Improvement of the product quality by solvent addition. *Energy & Fuels* **18**: 1578–1583.
81. Zhu, L., Li, K., Zhang, Y., and Zhu, X. (2017). Upgrading the storage properties of bio-oil by adding a compound additive. *Energy & Fuels* **31**: 6221–6227.
82. Lindfors, C., Kuoppala, E., Oasmaa, A. et al. (2014). Fractionation of bio-oil. *Energy & Fuels* **28**: 5785–5791.
83. Hu, X., Wang, Y., Mourant, D. et al. (2013). Polymerization on heating up of bio-oil: a model compound study. *AICHE Journal* **59**: 888–900.

84. Furimsky, E. (2013). Hydroprocessing challenges in biofuels production. *Catalysis Today* **217**: 13–56.
85. Elliott, D.C., Hart, T.R., Neuenschwander, G.G. et al. (2012). Catalytic hydroprocessing of fast pyrolysis bio-oil from Pine Sawdust. *Energy & Fuels* **26**: 3891–3896.
86. D.C. Elliott E.G. Baker 1989. Process for Upgrading Biomass Pyrolyzates. US Patent #4,795,841, issued January 3, 1989.
87. Mercader, F.D., Koehorst, P.J.J., Heeres, H.J. et al. (2011). Competition between hydrotreating and polymerization reactions during pyrolysis oil hydrodeoxygenation. *AICHE Journal* **57**: 3160–3170.
88. Mercader, F.D., Groeneveld, M.J., Kersten, S.R.A. et al. (2010). Pyrolysis oil upgrading by high pressure thermal treatment. *Fuel* **89**: 2829–2837.
89. Gunawan, R., Li, X., Lievens, C. et al. (2013). Upgrading of bio-oil into advanced biofuels and chemicals. Part I. Transformation of GC-detectable light species during the hydrotreatment of bio-oil using Pd/C catalyst. *Fuel* **111**: 709–717.
90. Elliott, D.C. and Hart, T.R. (2009). Catalytic hydroprocessing of chemical models for bio-oil. *Energy & Fuels* **23**: 631–637.
91. Wan, H.J., Chaudhari, R.V., and Subramaniam, B. (2013). Aqueous phase hydrogenation of acetic acid and its promotional effect on p-cresol hydrodeoxygenation. *Energy & Fuels* **27**: 487–493.
92. D.C. Elliott A. Oasmaa2012. Process for stabilizing fast pyrolysis oil, and stabilized fast pyrolysis oil, WO2012153001 A1.
93. Stankovikj, F., Tran, C.-C., Kaliaguine, S. et al. (2017). Evolution of functional groups during pyrolysis oil upgrading. *Energy & Fuels* **31**: 8300–8316.
94. Wang, H.M. and Wang, Y. (2016). Characterization of deactivated bio-oil hydrotreating catalysts. *Topics in Catalysis* **59**: 65–72.
95. Zhao, C., He, J.Y., Lemonidou, A.A. et al. (2011). Aqueous-phase hydrodeoxygenation of bio-derived phenols to cycloalkanes. *Journal of Catalysis* **280**: 8–16.
96. Elliott, D.C. and Wang, H. (2015). Hydrocarbon liquid production via catalytic hydroprocessing of phenolic oils fractionated from fast pyrolysis of Red Oak and Corn Stover. *ACS Sustainable Chemistry & Engineering* **3**: 892–902.
97. Choi, J.S., Zacher, A.H., Wang, H.M. et al. (2016). Molybdenum carbides, active and in situ regenerable catalysts in hydroprocessing of fast pyrolysis bio-oil. *Energy & Fuels* **30**: 5016–5026.
98. Wang, H. and Drennan, C. (2015). *Novel and robust catalyst for bio-oil hydrotreating Pacific Northwest National Laboratory, the 2015 Bioenergy Technologies Office Peer Review*. Washington, DC.
99. Elliott, D.C., Wang, H., French, R. et al. (2014). Hydrocarbon liquid production from biomass via hot-vapor-filtered fast pyrolysis and catalytic hydroprocessing of the bio-oil. *Energy & Fuels* **28**: 5909–5917.
100. Wang, H.M., Elliott, D.C., French, R.J. et al. (2016). Biomass conversion to produce hydrocarbon liquid fuel via hot-vapor filtered fast pyrolysis and catalytic hydrotreating. *Jove-Journal of Visualized Experiments*.
101. Zacher, A.H., Elliott, D.C., Olarte, M.V. et al. (2014). Pyrolysis of woody residue feedstocks: upgrading of bio-oils from mountain-pine-beetle-killed trees and hog fuel. *Energy & Fuels* **28**: 7510–7516.
102. Schwaiger, N., Elliott, D.C., Ritzberger, J. et al. (2015). Hydrocarbon liquid production via the bioCRACK process and catalytic hydroprocessing of the product oil. *Green Chemistry* **17**: 2487.
103. Agblevor, F.A., Elliott, D.C., Santosa, D.M. et al. (2016). Red Mud catalytic pyrolysis of pinyon juniper and single-stage hydrotreatment of oils. *Energy & Fuels* **30**: 7947–7958.
104. Olarte, M.V., Padmaperuma, A.B., Ferrell, J.R. et al. (2017). Characterization of upgraded fast pyrolysis oak oil distillate fractions from sulfided and non-sulfided catalytic hydrotreating. *Fuel* **202**: 620–630.
105. Wang, Y., Lin, H., and Zheng, Y. (2014). Hydrotreatment of lignocellulosic biomass derived oil using a sulfided NiMo/[gamma]-Al_2O_3 catalyst. *Catalysis Science & Technology* **4**: 109–119.

106. Mante, O.D., Dayton, D.C., Gabrielsen, J. et al. (2016). Integration of catalytic fast pyrolysis and hydroprocessing: a pathway to refinery intermediates and "drop-in" fuels from biomass. *Green Chemistry* **18**: 6123–6135.
107. Horacek, J. and Kubicka, D. (2017). Bio-oil hydrotreating over conventional CoMo & NiMo catalysts: the role of reaction conditions and additives. *Fuel* **198**: 49–57.
108. Gholizadeh, M., Gunawan, R., Hu, X. et al. (2016). Different reaction behaviours of the light and heavy components of bio-oil during the hydrotreatment in a continuous pack-bed reactor. *Fuel Processing Technology* **146**: 76–84.
109. Gholizadeh, M., Gunawan, R., Hu, X. et al. (2016). Effects of temperature on the hydrotreatment behaviour of pyrolysis bio-oil and coke formation in a continuous hydrotreatment reactor. *Fuel Processing Technology* **148**: 175–183.
110. Gholizadeh, M., Gunawan, R., Hu, X. et al. (2016). Importance of hydrogen and bio-oil inlet temperature during the hydrotreatment of bio-oil. *Fuel Processing Technology* **150**: 132–140.
111. Ardiyanti, A.R., Gutierrez, A., Honkela, M.L. et al. (2011). Hydrotreatment of wood-based pyrolysis oil using zirconia-supported mono- and bimetallic (Pt, Pd, Rh) catalysts. *Applied Catalysis A: General* **407**: 56–66.
112. Ardiyanti, A.R., Khromova, S.A., Venderbosch, R.H. et al. (2012). Catalytic hydrotreatment of fast-pyrolysis oil using non-sulfided bimetallic Ni-Cu catalysts on a δ-Al_2O_3 support. *Applied Catalysis B: Environmental* **117-118**: 105–117.
113. Ardiyanti, A.R., Khromova, S.A., Venderbosch, R.H. et al. (2012). Catalytic hydrotreatment of fast pyrolysis oil using bimetallic Ni–Cu catalysts on various supports. *Applied Catalysis A: General* **449**: 121–130.
114. French, R.J., Stunkel, J., and Baldwin, R.M. (2011). Mild hydrotreating of bio-oil: effect of reaction severity and fate of oxygenated species. *Energy & Fuels* .
115. French, R.J., Black, S.K., Myers, M. et al. (2015). Hydrotreating the organic fraction of biomass pyrolysis oil to a refinery intermediate. *Energy & Fuels* **29**: 7985–7992.
116. Frankiewicz, T.C. (1980). *The Conversion of Biomass Derived Pyrolytic Vapors to Hydrocarbons, Specialist's workshop on fast pyrolysis of biomass proceedings*. Colorado: Copper Mountain.
117. Liu, C.J., Wang, H.M., Karim, A.M. et al. (2014). Catalytic fast pyrolysis of lignocellulosic biomass. *Chemical Society Reviews* **43**: 7594–7623.
118. Venderbosch, R.H. (2015). A critical view on catalytic pyrolysis of biomass. *ChemSusChem* **8**: 1306–1316.
119. Kabir, G. and Hameed, B.H. (2017). Recent progress on catalytic pyrolysis of lignocellulosic biomass to high-grade bio-oil and bio-chemicals. *Renewable and Sustainable Energy Reviews* **70**: 945–967.
120. Xu, L.J., Zhang, Y., and Fu, Y. (2017). Advances in upgrading lignin pyrolysis vapors by ex situ catalytic fast pyrolysis. *Energy Technology* **5**: 30–51.
121. Yildiz, G., Ronsse, F., van Duren, R., and Prins, W. (2016). Challenges in the design and operation of processes for catalytic fast pyrolysis of woody biomass. *Renewable and Sustainable Energy Reviews* **57**: 1596–1610.
122. Ma, Z.Y., Wei, L., Zhou, W. et al. (2015). Overview of catalyst application in petroleum refinery for biomass catalytic pyrolysis and bio-oil upgrading. *RSC Advances* **5**: 88287–88297.
123. Lu, Q., Zhang, Z.B., Wang, X.Q. et al. (2014). Catalytic upgrading of biomass fast pyrolysis vapors using ordered mesoporous ZrO_2, TiO_2 and SiO_2. In: *International Conference on Applied Energy, Icae2014* (ed. J. Yan, D.J. Lee, S.K. Chou, et al.), 1937–1941.
124. Ruddy, D.A., Schaidle, J.A., Ferrell, J.R. et al. (2014). Recent advances in heterogeneous catalysts for bio-oil upgrading via "ex situ catalytic fast pyrolysis": catalyst development through the study of model compounds. *Green Chemistry* **16**: 454–490.
125. Carlson, T.R., Jae, J., Lin, Y.C. et al. (2010). Catalytic fast pyrolysis of glucose with HZSM-5: the combined homogeneous and heterogeneous reactions. *Journal of Catalysis* **270**: 110–124.

126. Foster, A.J., Jae, J., Cheng, Y.T. et al. (2012). Optimizing the aromatic yield and distribution from catalytic fast pyrolysis of biomass over ZSM-5. *Applied Catalysis A: General* **423**: 154–161.
127. Neumann, G.T. and Hicks, J.C. (2012). Novel hierarchical cerium-incorporated MFI zeolite catalysts for the catalytic fast pyrolysis of lignocellulosic biomass. *ACS Catalysis* **2**: 642–646.
128. Mante, O.D. and Agblevor, F.A. (2014). Catalytic pyrolysis for the production of refinery-ready biocrude oils from six different biomass sources. *Green Chemistry* **16**: 3364–3377.
129. Iisa, K., French, R.J., Orton, K.A. et al. (2016). In situ and ex situ catalytic pyrolysis of pine in a bench-scale fluidized bed reactor system. *Energy & Fuels* **30**: 2144–2157.
130. Zhou, G.Q., Li, J., Yu, Y.Q. et al. (2014). Optimizing the distribution of aromatic products from catalytic fast pyrolysis of cellulose by ZSM-5 modification with boron and co-feeding of low-density polyethylene. *Applied Catalysis A: General* **487**: 45–53.
131. Mullen, C.A., Boateng, A.A., Mihalcik, D.J., and Goldberg, N.M. (2011). Catalytic fast pyrolysis of white oak wood in a bubbling fluidized bed. *Energy & Fuels* **25**: 5444–5451.
132. Mante, O.D., Agblevor, F.A., Oyama, S.T., and McClung, R. (2014). Catalytic pyrolysis with ZSM-5 based additive as co-catalyst to Y-zeolite in two reactor configurations. *Fuel* **117**: 649–659.
133. Ma, Z.Q. and van Bokhoven, J.A. (2012). Deactivation and regeneration of H-USY zeolite during lignin catalytic fast pyrolysis. *ChemCatChem* **4**: 2036–2044.
134. Compton, D.L., Jackson, M.A., Mihalcik, D.J. et al. (2011). Catalytic pyrolysis of oak via pyroprobe and bench scale, packed bed pyrolysis reactors. *Journal of Analytical and Applied Pyrolysis* **90**: 174–181.
135. Kim, Y.M., Park, R.S., Lee, E.H. et al. (2017). In-situ catalytic pyrolysis of dealkaline lignin over MMZ-H beta. *Journal of Nanoscience and Nanotechnology* **17**: 2760–2763.
136. Du, Z.Y., Ma, X.C., Li, Y. et al. (2013). Production of aromatic hydrocarbons by catalytic pyrolysis of microalgae with zeolites: catalyst screening in a pyroprobe. *Bioresource Technology* **139**: 397–401.
137. Xu, L.J., Yao, O., Zhang, Y., and Fu, Y. (2017). Integrated production of aromatic amines and N-doped carbon from lignin via ex situ catalytic fast pyrolysis in the presence of ammonia over zeolites. *ACS Sustainable Chemistry & Engineering* **5**: 2960–2969.
138. Chagas, B.M.E., Dorado, C., Serapiglia, M.J. et al. (2016). Catalytic pyrolysis-GC/MS of Spirulina: evaluation of a highly proteinaceous biomass source for production of fuels and chemicals. *Fuel* **179**: 124–134.
139. Kim, Y.M., Lee, H.W., Jeon, J.K. et al. (2017). In-situ catalytic pyrolysis of xylan and dealkaline lignin over SAPO-11. *Topics in Catalysis* **60**: 644–650.
140. Mante, O.D., Agblevor, F.A., and McClung, R. (2013). A study on catalytic pyrolysis of biomass with Y-zeolite based FCC catalyst using response surface methodology. *Fuel* **108**: 451–464.
141. Park, Y.K., Yoo, M.L., and Park, S.H. (2016). Catalytic fast pyrolysis of wild reed over nanoporous SBA-15 catalysts. *Journal of Nanoscience and Nanotechnology* **16**: 4561–4564.
142. Casoni, A.I., Nievas, M.L., Moyano, E.L. et al. (2016). Catalytic pyrolysis of cellulose using MCM-41 type catalysts. *Applied Catalysis A: General* **514**: 235–240.
143. Custodis, V.B.F., Karakoulia, S.A., Triantafyllidis, K.S., and van Bokhoven, J.A. (2016). Catalytic fast pyrolysis of lignin over high-surface-area mesoporous aluminosilicates: effect of porosity and acidity. *ChemSusChem* **9**: 1134–1145.
144. Kim, Y.M., Jae, J.H., Lee, H.W. et al. (2016). Ex-situ catalytic pyrolysis of citrus fruit peels over mesoporous MFI and Al-MCM-41. *Energy Conversion and Management* **125**: 277–289.
145. Naqvi, S.R., Uemura, Y., Yusup, S. et al. (2015). In situ catalytic fast pyrolysis of paddy husk pyrolysis vapors over MCM-22 and ITQ-2 zeolites. *Journal of Analytical and Applied Pyrolysis* **114**: 32–39.
146. Gou, J.S., Wang, Z.P., Li, C. et al. (2017). The effects of ZSM-5 mesoporosity and morphology on the catalytic fast pyrolysis of furan. *Green Chemistry* **19**: 3549–3557.
147. Gamliel, D.P., Cho, H.J., Fan, W., and Valla, J.A. (2016). On the effectiveness of tailored mesoporous MFI zeolites for biomass catalytic fast pyrolysis. *Applied Catalysis A: General* **522**: 109–119.

148. Lee, H.W., Kim, T.H., Park, S.H. et al. (2013). Catalytic fast pyrolysis of lignin over mesoporous Y zeolite using Py-GC/MS. *Journal of Nanoscience and Nanotechnology* **13**: 2640–2646.
149. Fanchiang, W.L. and Lin, Y.C. (2012). Catalytic fast pyrolysis of furfural over H-ZSM-5 and Zn/H-ZSM-5 catalysts. *Applied Catalysis A: General* **419**: 102–110.
150. Zheng, Y.W., Wang, F., Yang, X.Q. et al. (2017). Study on aromatics production via the catalytic pyrolysis vapor upgrading of biomass using metal-loaded modified H-ZSM-5. *Journal of Analytical and Applied Pyrolysis* **126**: 169–179.
151. Zhang, H.Y., Zheng, J., and Xiao, R. (2013). Catalytic pyrolysis of willow wood with Me/ZSM-5 (Me = Mg, K, Fe, Ga, Ni) to produce aromatics and olefins. *BioResources* **8**: 5612–5621.
152. Gao, L.J., Sun, J.H., Xu, W., and Xiao, G.M. (2017). Catalytic pyrolysis of natural algae over Mg-Al layered double oxides/ZSM-5 (MgAl-LDO/ZSM-5) for producing bio-oil with low nitrogen content. *Bioresource Technology* **225**: 293–298.
153. Mullen, C.A. and Boateng, A.A. (2015). Production of aromatic hydrocarbons via catalytic pyrolysis of biomass over Fe-modified HZSM-5 zeolites. *ACS Sustainable Chemistry & Engineering* **3**: 1623–1631.
154. Li, P., Li, D., Yang, H.P. et al. (2016). Effects of Fe-, Zr-, and co-modified zeolites and pretreatments on catalytic upgrading of biomass fast pyrolysis vapors. *Energy & Fuels* **30**: 3004–3013.
155. Rezaei, P.S., Shafaghat, H., and Daud, W. (2016). Aromatic hydrocarbon production by catalytic pyrolysis of palm kernel shell waste using a bifunctional Fe/HBeta catalyst: effect of lignin-derived phenolics on zeolite deactivation. *Green Chemistry* **18**: 1684–1693.
156. Karnjanakom, S., Suriya-Umporn, T., Bayu, A. et al. (2017). High selectivity and stability of Mg-doped Al-MCM-41 for in-situ catalytic upgrading fast pyrolysis bio-oil. *Energy Conversion and Management* **142**: 272–285.
157. Yu, F.W., Gao, L.C., Wang, W.J. et al. (2013). Bio-fuel production from the catalytic pyrolysis of soybean oil over Me-Al-MCM-41 (Me = La, Ni or Fe) mesoporous materials. *Journal of Analytical and Applied Pyrolysis* **104**: 325–329.
158. Lee, H.W., Kim, Y.M., Jeong, C.S. et al. (2017). Catalytic pyrolysis of waste wood plastic composite over H-V-MCM-41 catalysts. *Science of Advanced Materials* **9**: 934–937.
159. Zhang, Y.L., Xiao, R., Gu, X.L. et al. (2014). Catalytic pyrolysis of biomass with Fe/La/SBA-15 catalyst using TGA-FTIR analysis. *BioResources* **9**: 5234–5245.
160. Kulyk, K., Palianytsia, B., Alexander, J.D. et al. (2017). Kinetics of valeric acid ketonization and ketenization in catalytic pyrolysis on nanosized SiO_2, gamma-Al_2O_3, CeO_2/SiO_2, Al_2O_3/SiO_2 and TiO_2/SiO_2. *ChemPhysChem* **18**: 1943–1955.
161. Fan, L.L., Chen, P., Zhang, Y.N. et al. (2017). Fast microwave-assisted catalytic co-pyrolysis of lignin and low-density polyethylene with HZSM-5 and MgO for improved bio-oil yield and quality. *Bioresource Technology* **225**: 199–205.
162. Veses, A., Aznar, M., Martinez, I. et al. (2014). Catalytic pyrolysis of wood biomass in an auger reactor using calcium-based catalysts. *Bioresource Technology* **162**: 250–258.
163. Lee, Y., Park, R.S., Hong, Y. et al. (2017). Catalytic pyrolysis of cork oak over supercritical hydrothermal synthesized nanosized ZrO_2. *Journal of Nanoscience and Nanotechnology* **17**: 2764–2767.
164. Lee, H., Ko, J.H., Kwon, W.H., and Park, Y.K. (2016). Conversion of acetic acid from the catalytic pyrolysis of xylan over CeO_2. *Journal of Nanoscience and Nanotechnology* **16**: 4480–4482.
165. Wang, K.G., Dayton, D.C., Peters, J.E., and Mante, O.D. (2017). Reactive catalytic fast pyrolysis of biomass to produce high-quality bio-crude. *Green Chemistry* **19**: 3243–3251.
166. Wang, W.L., Geng, J., Li, L.F., and Chang, J.M. (2016). Catalytic properties of fast pyrolysis char loaded with Cu-Zn on alkali lignin pyrolysis for monophenols. *Chemical Journal of Chinese Universities-Chinese* **37**: 736–744.
167. Murugappan, K., Mukarakate, C., Budhi, S. et al. (2016). Supported molybdenum oxides as effective catalysts for the catalytic fast pyrolysis of lignocellulosic biomass. *Green Chemistry* **18**: 5548–5557.

168. Aysu, T., Durak, H., Guner, S. et al. (2016). Bio-oil production via catalytic pyrolysis of Anchusa azurea: effects of operating conditions on product yields and chromatographic characterization. *Bioresource Technology* **205**: 7–14.
169. Wang, Z., Lu, Q.A., Zhu, X.F., and Zhang, Y. (2011). Catalytic fast pyrolysis of cellulose to prepare levoglucosenone using sulfated zirconia. *ChemSusChem* **4**: 79–84.
170. Lu, Q., Ye, X.-n., Zhang, Z.-b. et al. (2014). Catalytic fast pyrolysis of cellulose and biomass to produce levoglucosenone using magnetic SO2-4/TiO2-Fe3O4. *Bioresource Technology* **171**: 10–15.
171. Wang, S.Q., Xu, M.L., Wang, F., and Li, Z.H. (2016). Preparation of bio-oil by catalytic pyrolysis of corn stalks using red mud. *International Journal of Agricultural and Biological Engineering* **9**: 177–183.
172. Yathavan, B.K. and Agblevor, F.A. (2013). Catalytic pyrolysis of pinyon-juniper using red mud and HZSM-5. *Energy & Fuels* **27**: 6858–6865.
173. Gungor, A., Onenc, S., Ucar, S., and Yanik, J. (2012). Comparison between the "one-step" and "two-step" catalytic pyrolysis of pine bark. *Journal of Analytical and Applied Pyrolysis* **97**: 39–48.
174. Shadangi, K.P. and Mohanty, K. (2014). Production and characterization of pyrolytic oil by catalytic pyrolysis of Niger seed. *Fuel* **126**: 109–115.
175. Arteaga-Perez, L.E., Capiro, O.G., Romero, R. et al. (2017). In situ catalytic fast pyrolysis of crude and torrefied Eucalyptus globulus using carbon aerogel-supported catalysts. *Energy* **128**: 701–712.
176. Murata, K., Kreethawate, L., Larpkiattaworn, S., and Inaba, M. (2016). Evaluation of Ni-based catalysts for the catalytic fast pyrolysis of jatropha residues. *Journal of Analytical and Applied Pyrolysis* **118**: 308–316.
177. Griffin, M.B., Baddour, F.G., Habas, S.E. et al. (2016). Evaluation of silica-supported metal and metal phosphide nanoparticle catalysts for the hydrodeoxygenation of guaiacol under ex situ catalytic fast pyrolysis conditions. *Topics in Catalysis* **59**: 124–137.
178. Ye, X.N., Lu, Q., Li, W.T. et al. (2016). Selective production of nicotyrine from catalytic fast pyrolysis of tobacco biomass with Pd/C catalyst. *Journal of Analytical and Applied Pyrolysis* **117**: 88–93.
179. Zhang, Z.B., Lu, Q., Ye, X.N. et al. (2015). Selective production of 4-ethyl phenol from low-temperature catalytic fast pyrolysis of herbaceous biomass. *Journal of Analytical and Applied Pyrolysis* **115**: 307–315.
180. Lu, Q., Zhang, Z.B., Ye, X.N. et al. (2017). Selective production of 4-ethyl guaiacol from catalytic fast pyrolysis of softwood biomass using Pd/SBA-15 catalyst. *Journal of Analytical and Applied Pyrolysis* **123**: 237–243.
181. Lu, Q., Tang, Z., Zhang, Y., and Zhu, X.F. (2010). Catalytic upgrading of biomass fast pyrolysis vapors with Pd/SBA-15 catalysts. *Industrial & Engineering Chemistry Research* **49**: 2573–2580.
182. Lazdovica, K., Liepina, L., and Kampars, V. (2016). Catalytic pyrolysis of wheat bran for hydrocarbons production in the presence of zeolites and noble-metals by using TGA-FTIR method. *Bioresource Technology* **207**: 126–133.
183. Liu, W.J., Tian, K., Jiang, H. et al. (2012). Selectively improving the bio-oil quality by catalytic fast pyrolysis of heavy-metal-polluted biomass: take copper (Cu) as an example. *Environmental Science & Technology* **46**: 7849–7856.
184. J. Schaidle, K. Magrini, and H. Wang 2017. Catalytic fast pyrolysis, US DOE, BETO, Project Peer Reveiw, Denver, CO.
185. Wang, W.L., Ren, X.Y., Chang, J.M. et al. (2015). Characterization of bio-oils and bio-chars obtained from the catalytic pyrolysis of alkali lignin with metal chlorides. *Fuel Processing Technology* **138**: 605–611.
186. Paasikallio, V., Lindfors, C., Kuoppala, E. et al. (2014). Product quality and catalyst deactivation in a four day catalytic fast pyrolysis production run. *Green Chemistry* **16**: 3549–3559.
187. Carlson, T.R., Cheng, Y.T., Jae, J., and Huber, G.W. (2011). Production of green aromatics and olefins by catalytic fast pyrolysis of wood sawdust. *Energy & Environmental Science* **4**: 145–161.

188. Paasikallio, V., Agblevor, F., Oasmaa, A. et al. (2013). Catalytic pyrolysis of forest thinnings with ZSM-5 catalysts: effect of reaction temperature on bio-oil physical properties and chemical composition. *Energy & Fuels* **27**: 7587–7601.
189. Zhang, H.Y., Xiao, R., Jin, B.S. et al. (2013). Catalytic fast pyrolysis of straw biomass in an internally interconnected fluidized bed to produce aromatics and olefins: effect of different catalysts. *Bioresource Technology* **137**: 82–87.
190. Kim, S.W., Koo, B.S., and Lee, D.H. (2014). Catalytic pyrolysis of palm kernel shell waste in a fluidized bed. *Bioresource Technology* **167**: 425–432.
191. Murata, K., Liu, Y., Inaba, M., and Takahara, I. (2012). Catalytic fast pyrolysis of jatropha wastes. *Journal of Analytical and Applied Pyrolysis* **94**: 75–82.
192. Babich, I.V., van der Hulst, M., Lefferts, L. et al. (2011). Catalytic pyrolysis of microalgae to high-quality liquid bio-fuels. *Biomass and Bioenergy* **35**: 3199–3207.
193. Wang, Y., Dai, L., Fan, L. et al. (2017). Microwave-assisted catalytic fast co-pyrolysis of bamboo sawdust and waste tire for bio-oil production. *Journal of Analytical and Applied Pyrolysis* **123**: 224–228.
194. Zhang, B., Zhong, Z., Min, M. et al. (2015). Catalytic fast co-pyrolysis of biomass and food waste to produce aromatics: analytical Py–GC/MS study. *Bioresource Technology* **189**: 30–35.
195. Yildiz, G., Pronk, M., Djokic, M. et al. (2013). Validation of a new set-up for continuous catalytic fast pyrolysis of biomass coupled with vapour phase upgrading. *Journal of Analytical and Applied Pyrolysis* **103**: 343–351.
196. Xie, Q.L., Addy, M., Liu, S.Y. et al. (2015). Fast microwave-assisted catalytic co-pyrolysis of microalgae and scum for bio-oil production. *Fuel* **160**: 577–582.
197. Wang, J., Zhong, Z.P., Song, Z.W. et al. (2016). Modification and regeneration of HZSM-5 catalyst in microwave assisted catalytic fast pyrolysis of mushroom waste. *Energy Conversion and Management* **123**: 29–34.
198. Zhou, Y., Wang, Y.P., Fan, L.L. et al. (2017). Fast microwave-assisted catalytic co-pyrolysis of straw stalk and soapstock for bio-oil production. *Journal of Analytical and Applied Pyrolysis* **124**: 35–41.
199. Iisa, K., French, R.J., Orton, K.A. et al. (2017). Production of low-oxygen bio-oil via ex situ catalytic fast pyrolysis and hydrotreating. *Fuel* **207**: 413–422.
200. Mante, O.D. and Agblevor, F.A. (2011). Catalytic conversion of biomass to bio-syncrude oil. *Biomass Conversion and Biorefinery* **1**: 203.
201. Marker, T.L., Felix, L.G., Linck, M.B. et al. (2014). Integrated hydropyrolysis and hydroconversion (IH2®) for the direct production of gasoline and diesel fuels or blending components from biomass, Part 2: continuous testing. *Environmental Progress & Sustainable Energy* **33**: 762–768.
202. Resende, F.L.P. (2016). Recent advances on fast hydropyrolysis of biomass. *Catalysis Today* **269**: 148–155.
203. Stefanidis, S.D., Kalogiannis, K.G., Pilavachi, P.A. et al. (2016). Catalyst hydrothermal deactivation and metal contamination during the in situ catalytic pyrolysis of biomass. *Catalysis Science & Technology* **6**: 2807–2819.
204. Karanjkar, P.U., Coolman, R.J., Huber, G.W. et al. (2014). Production of aromatics by catalytic fast pyrolysis of cellulose in a bubbling fluidized bed reactor. *AICHE Journal* **60**: 1320–1335.
205. Jae, J., Coolman, R., Mountziaris, T.J., and Huber, G.W. (2014). Catalytic fast pyrolysis of lignocellulosic biomass in a process development unit with continual catalyst addition and removal. *Chemical Engineering Science* **108**: 33–46.
206. Zhang, H.Y., Carlson, T.R., Xiao, R., and Huber, G.W. (2012). Catalytic fast pyrolysis of wood and alcohol mixtures in a fluidized bed reactor. *Green Chemistry* **14**: 98–110.
207. Mullen, C.A. and Boateng, A.A. (2013). Accumulation of inorganic impurities on HZSM-5 zeolites during catalytic fast pyrolysis of switchgrass. *Industrial & Engineering Chemistry Research* **52**: 17156–17161.
208. Zhang, H.Y., Xiao, R., Jin, B.S. et al. (2013). Biomass catalytic pyrolysis to produce olefins and aromatics with a physically mixed catalyst. *Bioresource Technology* **140**: 256–262.

209. Choi, Y.S., Lee, K.H., Zhang, J. et al. (2015). Manipulation of chemical species in bio-oil using in situ catalytic fast pyrolysis in both a bench-scale fluidized bed pyrolyzer and micropyrolyzer. *Biomass and Bioenergy* **81**: 256–264.
210. Mendes, F.L., Ximenes, V.L., de Almeida, M.B.B. et al. (2016). Catalytic pyrolysis of sugarcane bagasse and pinewood in a pilot scale unit. *Journal of Analytical and Applied Pyrolysis* **122**: 395–404.
211. Paasikallio, V., Kalogiannis, K., Lappas, A. et al. (2017). Catalytic fast pyrolysis: influencing bio-oil quality with the catalyst-to-biomass ratio. *Energy Technology* **5**: 94–103.
212. Wang, K.G., Mante, O.D., Peters, J.E., and Dayton, D.C. (2017). Influence of the feedstock on catalytic fast pyrolysis with a solid acid catalyst. *Energy Technology* **5**: 183–188.
213. Dayton, D.C., Carpenter, J.R., Kataria, A. et al. (2015). Design and operation of a pilot-scale catalytic biomass pyrolysis unit. *Green Chemistry* **17**: 4680–4689.
214. Iisa, K., French, R.J., Orton, K.A. et al. (2016). Catalytic pyrolysis of pine over HZSM-5 with different binders. *Topics in Catalysis* **59**: 94–108.
215. Vitolo, S., Bresci, B., Seggiani, M., and Gallo, M.G. (2001). Catalytic upgrading of pyrolytic oils over HZSM-5 zeolite: behaviour of the catalyst when used in repeated upgrading–regenerating cycles. *Fuel* **80**: 17–26.
216. Mohammed, I.Y., Abakr, Y.A., Yusup, S. et al. (2017). Upgrading of Napier grass pyrolytic oil using microporous and hierarchical mesoporous zeolites: products distribution, composition and reaction pathways. *Journal of Cleaner Production* **162**: 817–829.
217. Wei, Y., Lei, H., Zhu, L. et al. (2016). Hydrocarbon produced from upgrading rich phenolic compound bio-oil with low catalyst coking. *Fuel* **178**: 77–84.
218. Pinho, A.d.R., de Almeida, M.B.B., Mendes, F.L. et al. (2015). Co-processing raw bio-oil and gasoil in an FCC Unit. *Fuel Processing Technology* **131**: 159–166.
219. Lindfors, C., Paasikallio, V., Kuoppala, E. et al. (2015). Co-processing of dry bio-oil, catalytic pyrolysis oil, and hydrotreated bio-oil in a micro activity test unit. *Energy & Fuels* **29**: 3707–3714.
220. de Miguel Mercader, F., Groeneveld, M.J., Kersten, S.R.A. et al. (2010). Production of advanced biofuels: co-processing of upgraded pyrolysis oil in standard refinery units. *Applied Catalysis B: Environmental* **96**: 57–66.
221. Fogassy, G., Thegarid, N., Toussaint, G. et al. (2010). Biomass derived feedstock co-processing with vacuum gas oil for second-generation fuel production in FCC units. *Applied Catalysis B: Environmental* **96**: 476–485.
222. Fogassy, G., Thegarid, N., Schuurman, Y., and Mirodatos, C. (2011). From biomass to bio-gasoline by FCC co-processing: effect of feed composition and catalyst structure on product quality. *Energy & Environmental Science* **4**: 5068–5076.
223. Fogassy, G., Thegarid, N., Schuurman, Y., and Mirodatos, C. (2012). The fate of bio-carbon in FCC co-processing products. *Green Chemistry* **14**: 1367–1371.
224. Ibarra, Á., Veloso, A., Bilbao, J. et al. (2016). Dual coke deactivation pathways during the catalytic cracking of raw bio-oil and vacuum gasoil in FCC conditions. *Applied Catalysis B: Environmental* **182**: 336–346.
225. Pinho, A.d.R., de Almeida, M.B.B., Mendes, F.L. et al. (2017). Fast pyrolysis oil from pinewood chips co-processing with vacuum gas oil in an FCC unit for second generation fuel production. *Fuel* **188**: 462–473.
226. Wang, C., Li, M., and Fang, Y. (2016). Coprocessing of catalytic-pyrolysis-derived bio-oil with VGO in a pilot-scale FCC riser. *Industrial & Engineering Chemistry Research* **55**: 3525–3534.
227. Chen, W., Luo, Z.Y., Yu, C.J. et al. (2014). Upgrading of bio-oil in supercritical ethanol: catalysts screening, solvent recovery and catalyst stability study. *Journal of Supercritical Fluids* **95**: 387–393.
228. Dang, Q., Luo, Z.Y., Zhang, J.X. et al. (2013). Experimental study on bio-oil upgrading over Pt/SO42-/ZrO2/SBA-15 catalyst in supercritical ethanol. *Fuel* **103**: 683–692.
229. Li, W., Pan, C.Y., Sheng, L. et al. (2011). Upgrading of high-boiling fraction of bio-oil in supercritical methanol. *Bioresource Technology* **102**: 9223–9228.

230. Li, W., Pan, C.Y., Zhang, Q.J. et al. (2011). Upgrading of low-boiling fraction of bio-oil in supercritical methanol and reaction network. *Bioresource Technology* **102**: 4884–4889.
231. Zhang, J.X., Luo, Z.Y., Dang, Q. et al. (2012). Upgrading of bio-oil over bifunctional catalysts in supercritical monoalcohols. *Energy & Fuels* **26**: 2990–2995.
232. Hilten, R.N., Bibens, B.P., Kastner, J.R., and Das, K.C. (2010). In-line esterification of pyrolysis vapor with ethanol improves bio-oil quality. *Energy & Fuels* **24**: 673–682.
233. Hu, X., Gunawan, R., Mourant, D. et al. (2012). Acid-catalysed reactions between methanol and the bio-oil from the fast pyrolysis of mallee bark. *Fuel* **97**: 512–522.
234. Hu, X., Gunawan, R., Mourant, D. et al. (2012). Esterification of bio-oil from mallee (Eucalyptus loxophleba ssp gratiae) leaves with a solid acid catalyst: conversion of the cyclic ether and terpenoids into hydrocarbons. *Bioresource Technology* **123**: 249–255.
235. Hu, X., Lievens, C., Mourant, D. et al. (2013). Investigation of deactivation mechanisms of a solid acid catalyst during esterification of the bio-oils from mallee biomass. *Applied Energy* **111**: 94–103.
236. Hu, X., Mourant, D., Gunawan, R. et al. (2012). Production of value-added chemicals from bio-oil via acid catalysis coupled with liquid-liquid extraction. *RSC Advances* **2**: 9366–9370.
237. Hu, X., Mourant, D., Wang, Y. et al. (2013). Acid-catalysed treatment of the mallee leaf bio-oil with methanol: effects of molecular structure of carboxylic acids and esters on their conversion. *Fuel Processing Technology* **106**: 569–576.
238. Hu, X., Westerhof, R.J.M., Wu, L.P. et al. (2015). Upgrading biomass-derived furans via acid-catalysis/hydrogenation: the remarkable difference between water and methanol as the solvent. *Green Chemistry* **17**: 219–224.
239. Liu, Y.C., Li, Z.L., Leahy, J.J., and Kwapinski, W. (2015). Catalytically upgrading bio-oil via esterification. *Energy & Fuels* **29**: 3691–3698.
240. Lu, J.X., Guo, S.J., Fu, Y., and Chang, J. (2017). Catalytic upgrading of bio-oil by simultaneous esterification and alkylation with azeotropic water removal. *Fuel Processing Technology* **161**: 193–198.
241. Wang, J.J., Chang, J., and Fan, J.A. (2010). Upgrading of bio-oil by catalytic esterification and determination of acid number for evaluating esterification degree. *Energy & Fuels* **24**: 3251–3255.
242. Ye, J., Liu, C.J., Fu, Y. et al. (2014). Upgrading bio-oil: simultaneous catalytic esterification of acetic acid and alkylation of acetaldehyde. *Energy & Fuels* **28**: 4267–4272.
243. Manayil, J.C., Osatiashtiani, A., Mendoza, A. et al. (2017). Impact of macroporosity on catalytic upgrading of fast pyrolysis bio-oil by esterification over silica sulfonic acids. *ChemSusChem* **10**: 3506–3511.
244. Milina, M., Mitchell, S., and Perez-Ramirez, J. (2014). Prospectives for bio-oil upgrading via esterification over zeolite catalysts. *Catalysis Today* **235**: 176–183.
245. Neves, P., Antunes, M.M., Russo, P.A. et al. (2013). Production of biomass-derived furanic ethers and levulinate esters using heterogeneous acid catalysts. *Green Chemistry* **15**: 3367–3376.
246. Qin, F., Cui, H.Y., Yi, W.M., and Wang, C.B. (2014). Upgrading the water-soluble fraction of bio-oil by simultaneous esterification and acetalation with online extraction. *Energy & Fuels* **28**: 2544–2553.
247. Reyhanitash, E., Tymchyshyn, M., Yuan, Z.S. et al. (2014). Upgrading fast pyrolysis oil via hydrodeoxygenation and thermal treatment: effects of catalytic glycerol pretreatment. *Energy & Fuels* **28**: 1132–1138.
248. Sundqvist, T., Oasmaa, A., and Koskinen, A. (2015). Upgrading fast pyrolysis bio-oil quality by esterification and azeotropic water removal. *Energy & Fuels* **29**: 2527–2534.
249. Tanneru, S.K., Parapati, D.R., and Steele, P.H. (2014). Pretreatment of bio-oil followed by upgrading via esterification to boiler fuel. *Energy* **73**: 214–220.
250. Zhang, Z.J., Wang, Q.W., Tripathi, P., and Pittman, C.U. (2011). Catalytic upgrading of bio-oil using 1-octene and 1-butanol over sulfonic acid resin catalysts. *Green Chemistry* **13**: 940–949.
251. Xu, X.M., Zhang, C.S., Zhai, Y.P. et al. (2014). Upgrading of bio-oil using supercritical 1-butanol over a Ru/C heterogeneous catalyst: role of the solvent. *Energy & Fuels* **28**: 4611–4621.

252. Zhang, X.H., Chen, L.G., Kong, W. et al. (2015). Upgrading of bio-oil to boiler fuel by catalytic hydrotreatment and esterification in an efficient process. *Energy* **84**: 83–90.
253. Ciddor, L., Bennett, J.A., Hunns, J.A. et al. (2015). Catalytic upgrading of bio-oils by esterification. *Journal of Chemical Technology and Biotechnology* **90**: 780–795.
254. Hu, X., Gunawan, R., Mourant, D. et al. (2017). Upgrading of bio-oil via acid-catalyzed reactions in alcohols — A mini review. *Fuel Processing Technology* **155**: 2–19.
255. McCormick, R.L., Ratcliff, M.A., Christensen, E. et al. (2015). Properties of oxygenates found in upgraded biomass pyrolysis oil as components of spark and compression ignition engine fuels. *Energy & Fuels* **29**: 2453–2461.
256. Christensen, E.D., Chupka, G.M., Luecke, J. et al. (2011). Analysis of oxygenated compounds in hydrotreated biomass fast pyrolysis oil distillate fractions. *Energy & Fuels* **25**: 5462–5471.
257. Elliott, D.C. and Schiefelbein, G.F. (1989). Liquid hydrocarbon fuels from biomass, preprints of papers. In: , 1160. Miami Beach, FL: American Chemical Society, Division of Fuel Chemistry. https://www.pnnl.gov/biobased/docs/liquid_fuels.pdf (accessed 28 November 2018).
258. Parisotto, G., Ferrão, M.F., Müller, A.L.H. et al. (2010). Total acid number determination in residues of crude oil distillation using ATR-FTIR and variable selection by chemometric methods. *Energy & Fuels* **24**: 5474–5478.
259. Alleman, T.L., McCormick, R.L., and Yanowitz, J. (2015). Properties of ethanol fuel blends made with natural gasoline. *Energy & Fuels* **29**: 5095–5102.
260. Christensen, E., Yanowitz, J., Ratcliff, M., and McCormick, R.L. (2011). Renewable oxygenate blending effects on gasoline properties. *Energy & Fuels* **25**: 4723–4733.
261. ASTM International 1958. Knocking Characterisitics of Pure Hydrocarbons, Philadelphia, PA,
262. Ratcliff, M.A., Burton, J., Sindler, P. et al. (2016). Knock resistance and fine particle emissions for several biomass-derived oxygenates in a direct-injection spark-ignition engine. *SAE International Journal of Fuels and Lubricants* **9**: 59–70.
263. M. Ajam, C. Woolard, and C.L. Viljoen 2013 Biomass Pyrolysis Oil as a Renewable Feedstocks for Bio-Jet Fuel, IASH 2013, the 13th International Symposium on Stability, Handling, and Use of LIquid Fuels, Rhodes, Greece.
264. Wang, J., Bi, P., Zhang, Y. et al. (2015). Preparation of jet fuel range hydrocarbons by catalytic transformation of bio-oil derived from fast pyrolysis of straw stalk. *Energy* **86**: 488–499.
265. Mueller, C.J. (2013). The feasibility of using raw liquids from fast pyrolysis of woody biomass as fuels for compression-ignition engines: a literature review. *SAE International Journal of Fuels and Lubricants* **6**: 251–262.
266. Martin, J.A., Mullen, C.A., and Boateng, A.A. (2014). Maximizing the stability of pyrolysis oil/diesel fuel emulsions. *Energy & Fuels* **28**: 5918–5929.
267. Baumgardner, M.E., Vaughn, T.L., Lakshminarayanan, A. et al. (2015). Combustion of lignocellulosic biomass based oxygenated components in a compression ignition engine. *Energy & Fuels* **29**: 7317–7326.
268. Covey, G., Allender, B., Laycock, B., and O'Shea, M. (2014). Biorefineries as sources of fuels and chemicals. *Appita Journal* **67**: 219–225.
269. Dang, Q., Hu, W., Rover, M. et al. (2016). Economics of biofuels and bioproducts from an integrated pyrolysis biorefinery. *Biofuels, Bioproducts & Biorefining* **10**: 790–803.
270. Brown, T.R., Zhang, Y., Hu, G., and Brown, R.C. (2012). Techno-economic analysis of biobased chemicals production via integrated catalytic processing. *Biofuels, Bioproducts & Biorefining* **6**: 73–87.
271. Rezaei, P.S., Shafaghat, H., and Daud, W.M.A.W. (2014). Production of green aromatics and olefins by catalytic cracking of oxygenate compounds derived from biomass pyrolysis: a review. *Applied Catalysis, A: General* **469**: 490–511.
272. Ma, Z., Custodis, V., Hemberger, P. et al. (2015). Chemicals from lignin by catalytic fast pyrolysis, from product control to reaction mechanism. *Chimia* **69**: 597–602.

273. Shen, D., Jin, W., Hu, J. et al. (2015). An overview on fast pyrolysis of the main constituents in lignocellulosic biomass to valued-added chemicals: structures, pathways and interactions. *Renewable and Sustainable Energy Reviews* **51**: 761–774.
274. X.-f. Zhu Q. Lu 2010. Production of Chemicals from Selective Fast Pyrolysis of Biomass, Sciyo, pp. 147–164.
275. Lu, D.-c., Liu, Y.-q., Wang, D., and Ye, Y.-y. (2013). Separation of chemicals from bio-oil and their application prospects. *Linchan Huaxue Yu Gongye* **33**: 137–143.
276. Ren, S., Ye, X.P., and Borole, A.P. (2017). Separation of chemical groups from bio-oil water-extract via sequential organic solvent extraction. *Journal of Analytical and Applied Pyrolysis* **123**: 30–39.
277. Wei, Y., Lei, H., Wang, L. et al. (2014). Liquid–liquid extraction of biomass pyrolysis bio-oil. *Energy & Fuels* **28**: 1207–1212.
278. Abou-Yousef, H. and Hassan, E.B. (2014). Efficient utilization of aqueous phase bio-oil to furan derivatives through extraction and sugars conversion in acid-catalyzed biphasic system. *Fuel* **137**: 115–121.
279. Bergem, H., Xu, R., Brown, R.C., and Huber, G.W. (2017). Low temperature aqueous phase hydrogenation of the light oxygenate fraction of bio-oil over supported ruthenium catalysts. *Green Chemistry* **19**: 3252–3262.
280. Park, L.K.-E., Ren, S., Yiacoumi, S. et al. (2016). Separation of switchgrass bio-oil by water/organic solvent addition and pH adjustment. *Energy & Fuels* **30**: 2164–2173.
281. Feng, Y. and Meier, D. (2015). Extraction of value-added chemicals from pyrolysis liquids with supercritical carbon dioxide. *Journal of Analytical and Applied Pyrolysis* **113**: 174–185.
282. Lu, D.-c., Liu, Y.-q., and Wang, D. (2013). Extracting acetic acid from bio-oil using TOA. *Huaxue Gongye Yu Gongcheng (Tianjin, China)* **30**: 32–36.
283. Bharti, A. and Banerjee, T. (2015). Enhancement of bio-oil derived chemicals in aqueous phase using ionic liquids: experimental and COSMO-SAC predictions using a modified hydrogen bonding expression. *Fluid Phase Equilibria* **400**: 27–37.
284. Chen, H.-W., Song, Q.-H., Liao, B., and Guo, Q.-X. (2011). Further separation, characterization, and upgrading for upper and bottom layers from phase separation of biomass pyrolysis oils. *Energy & Fuels* **25**: 4655–4661.
285. Zheng, A., Chen, T., Sun, J. et al. (2017). Toward fast pyrolysis-based biorefinery: selective production of platform chemicals from biomass by organosolv fractionation coupled with fast pyrolysis. *ACS Sustainable Chemistry & Engineering* **5**: 6507–6516.
286. Wilson, A.N., Price, M.J., Mukarakate, C. et al. (2017). Integrated biorefining: coproduction of renewable resol biopolymer for aqueous stream valorization. *ACS Sustainable Chemistry & Engineering* **5**: 6615–6625.
287. Jeong, J.-Y., Lee, U.-D., Chang, W.-S., and Jeong, S.-H. (2016). Production of bio-oil rich in acetic acid and phenol from fast pyrolysis of palm residues using a fluidized bed reactor: influence of activated carbons. *Bioresource Technology* **219**: 357–364.
288. Oh, S.-J., Choi, G.-G., and Kim, J.-S. (2017). Production of acetic acid-rich bio-oils from the fast pyrolysis of biomass and synthesis of calcium magnesium acetate deicer. *Journal of Analytical and Applied Pyrolysis* **124**: 122–129.
289. Oh, S.-J., Jung, S.-H., and Kim, J.-S. (2013). Co-production of furfural and acetic acid from corncob using $ZnCl_2$ through fast pyrolysis in a fluidized bed reactor. *Bioresource Technology* **144**: 172–178.
290. Zhang, X.-S., Yang, G.-X., Jiang, H. et al. (2013). Mass production of chemicals from biomass-derived oil by directly atmospheric distillation coupled with co-pyrolysis. *Scientific Reports* **3**: 1120.
291. Hoff, T.C., Holmes, M.J., Proano-Aviles, J. et al. (2017). Decoupling the role of external mass transfer and intracrystalline pore diffusion on the selectivity of HZSM-5 for the catalytic fast pyrolysis of biomass. *ACS Sustainable Chemistry & Engineering* **5** (10): 8766–8776. https://doi.org/10.1021/acssuschemeng.7b01578 (accessed 28 November 2018).

292. Ma, Z., Troussard, E., and van Bokhoven, J.A. (2012). Controlling the selectivity to chemicals from lignin via catalytic fast pyrolysis. *Applied Catalysis, A: General* **423-424**: 130–136.
293. Sorensen, C. (2017). *Chemicals and Fuel Blendstocks by a Catalytic Fast Pyrolysis Process*, 68. USA: Anellotech, Inc.
294. Zheng, A., Jiang, L., Zhao, Z. et al. (2017). Catalytic fast pyrolysis of lignocellulosic biomass for aromatic production: chemistry, catalyst and process. *Wiley Interdisciplinary Reviews: Energy and Environment* **6**.
295. Zhang, Z.-b., Lu, Q., Ye, X.-n. et al. (2015). Selective production of levoglucosenone from catalytic fast pyrolysis of biomass mechanically mixed with solid phosphoric acid catalysts. *Bioenergy Research* **8**: 1263–1274.
296. Budhi, S., Mukarakate, C., Iisa, K. et al. (2015). Molybdenum incorporated mesoporous silica catalyst for production of biofuels and value-added chemicals via catalytic fast pyrolysis. *Green Chemistry* **17**: 3035–3046.
297. Liu, J., Jiang, J., and Huang, H. (2011). Preparation of high-value chemicals from selective fast pyrolysis of bamboo directionally catalyzed by phosphoric acid. *Institute of Electrical and Electronics Engineers* 479–484.
298. Xia, H., Xu, S., and Yang, L. (2017). Efficient conversion of wheat straw into furan compounds, bio-oils, and phosphate fertilizers by a combination of hydrolysis and catalytic pyrolysis. *RSC Advances* **7**: 1200–1205.
299. Meng, X., Zhang, H., Liu, C., and Xiao, R. (2016). Comparison of acids and sulfates for producing levoglucosan and levoglucosenone by selective catalytic fast pyrolysis of cellulose using Py-GC/MS. *Energy & Fuels* **30**: 8369–8376.
300. Li, Q., Steele, P.H., Yu, F. et al. (2013). Pyrolytic spray increases levoglucosan production during fast pyrolysis. *Journal of Analytical and Applied Pyrolysis* **100**: 33–40.
301. Zheng, A., Zhao, K., Jiang, L. et al. (2016). Bridging the gap between pyrolysis and fermentation: improving anhydrosugar production from fast pyrolysis of agriculture and Forest residues by microwave-assisted organosolv pretreatment. *ACS Sustainable Chemistry & Engineering* **4**: 5033–5040.
302. Zheng, A., Zhao, Z., Huang, Z. et al. (2015). Overcoming biomass recalcitrance for enhancing sugar production from fast pyrolysis of biomass by microwave pretreatment in glycerol. *Green Chemistry* **17**: 1167–1175.
303. Lu, Q., Ye, X.-n., Zhang, Z.-b. et al. (2016). Catalytic fast pyrolysis of bagasse using activated carbon catalyst to selectively produce 4-ethyl phenol. *Energy & Fuels* **30**: 10618–10626.
304. Qu, Y.-C., Wang, Z., Lu, Q., and Zhang, Y. (2013). Selective production of 4-vinylphenol by fast pyrolysis of herbaceous biomass. *Industrial and Engineering Chemistry Research* **52**: 12771–12776.
305. Zhang, Z., Ye, X., Lu, Q. et al. (2014). Production of phenolic compounds from low temperature catalytic fast pyrolysis of biomass with activated carbon. *Applied Mechanics and Materials* **541-542**: 190–194, 6.
306. Zhang, Z.-b., Lu, Q., Ye, X.-n. et al. (2015). Production of phenolic-rich bio-oil from catalytic fast pyrolysis of biomass using magnetic solid base catalyst. *Energy Conversion and Management* **106**: 1309–1317.
307. Furusjö, E. and Pettersson, E. (2016). Mixing of fast pyrolysis oil and black liquor: preparing an improved gasification feedstock. *Energy & Fuels* **30**: 10575–10582.
308. Zheng, J.-L., Zhu, M.-Q., Wen, J.-L., and Sun, R.-c. (2016). Gasification of bio-oil: effects of equivalence ratio and gasifying agents on product distribution and gasification efficiency. *Bioresource Technology* **211**: 164–172.
309. Karlsruhe Institute of Technology 2016. The Karlsruhe bioliq Process. https://www.bioliq.de/downloads/Datenblatt_Bioliq_englisch_2016-a.pdf. Accessed 19 November 2018.
310. Choi, S.G., Chu, J., Brown, R.C. et al. (2017). Sustainable biocement production via microbially induced calcium carbonate precipitation: use of limestone and acetic acid derived from pyrolysis of lignocellulosic biomass. *ACS Sustainable Chemistry & Engineering* **5**: 5183–5190.

311. Caceres, L.A., McGarvey, B.D., Briens, C. et al. (2015). Insecticidal properties of pyrolysis bio-oil from greenhouse tomato residue biomass. *Journal of Analytical and Applied Pyrolysis* **112**: 333–340.
312. Rozhan, A.N., Cahyono, R.B., Yasuda, N. et al. (2012). Carbon deposition from biotar by fast pyrolysis using the chemical vapor infiltration process within porous low-grade iron ore for iron-making. *Energy & Fuels* **26**: 7340–7346.
313. Santhanaraj, D., Rover, M.R., Resasco, D.E. et al. (2014). Gluconic acid from biomass fast pyrolysis oils: specialty chemicals from the thermochemical conversion of biomass. *ChemSusChem* **7**: 3132–3137.
314. Uchimiya, M., Hiradate, S., and Antal, M.J. (2015). Dissolved phosphorus speciation of flash carbonization, slow pyrolysis, and fast pyrolysis biochars. *ACS Sustainable Chemistry & Engineering* **3**: 1642–1649.
315. Ye, J., Jiang, J., and Xu, J. (2017). Effect of alcohols on simultaneous bio-oil upgrading and separation of high value-added chemicals. *Waste and Biomass Valorization* Ahead of Print.
316. Zhang, R., Wang, H., You, Z. et al. (2017). Optimization of bio-asphalt using bio-oil and distilled water. *Journal of Cleaner Production* **165**: 281–289.
317. Elliott, D.C., Neuenschwander, G.G., and Hart, T.R. (2013). Hydroprocessing bio-oil and products separation for coke production. *ACS Sustainable Chemistry & Engineering* **1**: 389–392.
318. Ayalur Chattanathan, S., Adhikari, S., and Abdoulmoumine, N. (2012). A review on current status of hydrogen production from bio-oil. *Renewable and Sustainable Energy Reviews* **16**: 2366–2372.
319. Kong, H.-p., Zhou, H.-z., and Luo, T.-l. (2014). Progress of catalytic reforming of bio-oil steam for making hydrogen. *Henan Huagong* **31**: 25–29.
320. Kumar, A., Chakraborty, J.P., and Singh, R. (2017). Bio-oil: the future of hydrogen generation. *Biofuels* **8**: 663–674.
321. Chen, J., Sun, J., and Wang, Y. (2017). Catalysts for steam reforming of bio-oil: a review. *Industrial & Engineering Chemistry Research* **56**: 4627–4637.
322. Li, D., Li, X., and Gong, J. (2016). Catalytic reforming of oxygenates: state of the art and future prospects. *Chemical Reviews* **116**: 11529–11653.
323. Zeng, D., Xiao, R., Zhang, S. et al. (2017). Bio-oil heavy fraction as a feedstock for hydrogen generation via chemical looping process: reactor design and hydrodynamic analysis. *International Journal of Chemical Reactor Engineering* **15**.
324. Zeng, D.-W., Xiao, R., Zhang, S., and Zhang, H.-Y. (2015). Bio-oil heavy fraction for hydrogen production by iron-based oxygen carrier redox cycle. *Fuel Processing Technology* **139** (1–7).
325. Zeng, D.-W., Xiao, R., Huang, Z.-c. et al. (2016). Continuous hydrogen production from non-aqueous phase bio-oil via chemical looping redox cycles. *International Journal of Hydrogen Energy* **41**: 6676–6684.
326. Udomchoke, T., Wongsakulphasatch, S., Kiatkittipong, W. et al. (2016). Performance evaluation of sorption enhanced chemical-looping reforming for hydrogen production from biomass with modification of catalyst and sorbent regeneration. *Chemical Engineering Journal* **303**: 338.
327. Chen, L.W.A., Robles, J.A., Chow, J.C., and Hoekman, S.K. (2015). Renewable hydrogen production from bio-oil in an aerosol pyrolysis system. *Procedia Engineering* **102**: 1867–1876.
328. Bizkarra, K., Barrio, V.L., Arias, P.L., and Cambra, J.F. (2016). Sustainable hydrogen production from bio-oil model compounds (meta-xylene) and mixtures (1-butanol, meta-xylene and furfural). *Bioresource Technology* **216**: 281–293.
329. Valle, B., Aramburu, B., Remiro, A. et al. (2014). Effect of calcination/reduction conditions of Ni/La$_2$O$_3$–α-Al$_2$O$_3$ catalyst on its activity and stability for hydrogen production by steam reforming of raw bio-oil/ethanol. *Applied Catalysis, B: Environmental* **147**: 402.
330. Valle, B., Remiro, A., Aguayo, A.T. et al. (2013). Catalysts of Ni/α-Al$_2$O$_3$ and Ni/La$_2$O$_3$-αAl$_2$O$_3$ for hydrogen production by steam reforming of bio-oil aqueous fraction with pyrolytic lignin retention. *International Journal of Hydrogen Energy* **38**: 1307.

331. M. Chen, Y. Wang, J. Yang, et al. 2016 A regenerable catalyst for hydrogen production by bio-oil catalytic reforming and its preparation method, Anhui University of Science and Technology, Peoples Republic of China.
332. Dubey, V.R. and Vaidya, P.D. (2016). On the production of hydrogen from bio-oil: a representative study using propylene glycol. *Chemical Engineering Communications* **203**: 1234–1241.
333. Wang, S., Li, X., Zhang, F. et al. (2013). Bio-oil catalytic reforming without steam addition: application to hydrogen production and studies on its mechanism. *International Journal of Hydrogen Energy* **38**: 16038–16047.
334. Wang, S., Zhang, F., Cai, Q. et al. (2014). Catalytic steam reforming of bio-oil model compounds for hydrogen production over coal ash supported Ni catalyst. *International Journal of Hydrogen Energy* **39**: 2018–2025.
335. Xie, H., Yu, Q., Yao, X. et al. (2015). Hydrogen production via steam reforming of bio-oil model compounds over supported nickel catalysts. *Journal of Energy Chemistry* **24**: 299.
336. Xu, Y., Jiang, P.-w., and Li, Q.-x. (2013). Carbon nanofibers-supported Ni catalyst for hydrogen production from bio-oil through low-temperature reforming. *Wuli Huaxue Xuebao* **29**: 1041–1047.
337. Qiu, S., Gong, L., Liu, L. et al. (2011). Hydrogen production by low-temperature steam reforming of bio-oil over Ni/HZSM-5 catalyst. *Chinese Journal of Chemical Physics* **24**: 211.
338. Yuan, L.-x., Ding, F., Yao, J.-m. et al. (2013). Design of multiple metal doped Ni based catalyst for hydrogen generation from bio-oil reforming at mild-temperature. *Chinese Journal of Chemical Physics* **26**: 109–120.
339. Davidson, S.D., Spies, K.A., Mei, D. et al. (2017). Steam reforming of acetic acid over co-supported catalysts: coupling ketonization for greater stability. *ACS Sustainable Chemistry & Engineering* **5**: 9136–9149.
340. Xing, R., Dagle, V.L., Flake, M. et al. (2016). Steam reforming of fast pyrolysis-derived aqueous phase oxygenates over Co, Ni, and Rh metals supported on $MgAl_2O_4$. *Catalysis Today* **269**: 166.
341. Quan, C., Xu, S., and Zhou, C. (2016). Steam reforming of bio-oil from coconut shell pyrolysis over Fe/olivine catalyst. *Energy Conversion and Management* .
342. Chen, G., Yao, J., Liu, J. et al. (2016). Biomass to hydrogen-rich syngas via catalytic steam reforming of bio-oil. *Renewable Energy* **91**: 315.
343. Trane, R., Dahl, S., Skjøth-Rasmussen, M.S., and Jensen, A.D. (2012). Catalytic steam reforming of bio-oil. *International Journal of Hydrogen Energy* **37**: 6447.
344. Trane-Restrup, R. and Jensen, A.D. (2015). Steam reforming of cyclic model compounds of bio-oil over Ni-based catalysts: product distribution and carbon formation. *Applied Catalysis, B: Environmental* **165**: 117.
345. Montero, C., Oar-Arteta, L., Remiro, A. et al. (2015)). Thermodynamic comparison between bio-oil and ethanol steam reforming. *International Journal of Hydrogen Energy* **40**: 15963.
346. Zhang, F., Wang, N., Yang, L. et al. (2014). Ni-Co bimetallic MgO-based catalysts for hydrogen production via steam reforming of acetic acid from bio-oil. *International Journal of Hydrogen Energy* **39**: 18688.
347. Fu, P., Yi, W., Li, Z. et al. (2014). Investigation on hydrogen production by catalytic steam reforming of maize stalk fast pyrolysis bio-oil. *International Journal of Hydrogen Energy* **39**: 13962.
348. Zhang, Y.H., Li, W.Z., Zhang, S.P. et al. (2013). Hydrogen production by the catalytic reforming of volatile from biomass pyrolysis over a bimetallic catalyst. *Energy Sources, Part A: Recovery, Utilization, and Environmental Effects* **35**: 1975.
349. Medrano, J.A., Oliva, M., Ruiz, J. et al. (2011). Hydrogen from aqueous fraction of biomass pyrolysis liquids by catalytic steam reforming in fluidized bed. *Energy* **36**: 2215.
350. Remón, J., Medrano, J.A., Bimbela, F. et al. (2013). Ni/Al–Mg–O solids modified with Co or Cu for the catalytic steam reforming of bio-oil. *Applied Catalysis, B: Environmental* **132–133**: 433.

351. Salehi, E., Azad, F.S., Harding, T., and Abedi, J. (2011). Production of hydrogen by steam reforming of bio-oil over Ni/Al2O3 catalysts: effect of addition of promoter and preparation procedure. *Fuel Processing Technology* **92**: 2203.
352. Xie, H., Yu, Q., Zuo, Z. et al. (2016). Hydrogen production via sorption-enhanced catalytic steam reforming of bio-oil. *International Journal of Hydrogen Energy* **41**: 2345.
353. Gil, M.V., Esteban-Diez, G., Pevida, C. et al. (2014). H2 production by steam reforming with in situ CO_2 capture of biomass-derived bio-oil. *Energy Procedia* **63**: 6815–6823.
354. Sun, Z., Toan, S., Chen, S. et al. (2017). Biomass pyrolysis-gasification over Zr promoted CaO-HZSM-5 catalysts for hydrogen and bio-oil co-production with CO_2 capture. *International Journal of Hydrogen Energy* **42**: 16031–16044.
355. Hu, X. and Lu, G.-X. (2010). Bio-oil steam reforming, partial oxidation or oxidative steam reforming coupled with bio-oil dry reforming to eliminate CO_2 emission. *International Journal of Hydrogen Energy* **35**: 7169–7176.
356. Wiranarongkorn, K., Authayanun, S., Assabumrungrat, S., and Arpornwichanop, A. (2016). Analysis of thermally coupling steam and tri-reforming processes for the production of hydrogen from bio-oil. *International Journal of Hydrogen Energy* **41**: 18370–18379.
357. Jiang, P.-w., Wu, X.-p., Liu, J.-x., and Li, Q.-x. (2016). Preparation of bio-hydrogen and bio-fuels from lignocellulosic biomass pyrolysis-oil. *Chinese Journal of Chemical Physics* **29**: 635–643.
358. Pan, C., Li, W., Liu, Z. et al. (2012). Aqueous-phase reforming of low-boiling fraction of bio-oil for hydrogen production over Pt catalyst: effect of catalyst supports. *Huaxue Fanying Gongcheng Yu Gongyi* **28**: 149–153.
359. Kim, T.-W., Kim, H.-D., Jeong, K.-E. et al. (2011). Catalytic production of hydrogen through aqueous-phase reforming over platinum/ordered mesoporous carbon catalysts. *Green Chemistry* **13**: 1718–1728.
360. Lin, S.-b., Ye, T.-q., Yuan, L.-x. et al. (2010). Production of hydrogen from bio-oil using low-temperature electrochemical catalytic reforming approach over CoZnAl catalyst. *Chinese Journal of Chemical Physics* **23**: 451–458.

8

Solvent Liquefaction

Arpa Ghosh[1] and Martin R. Haverly[2]

[1]Chemical and Biological Engineering Department, Iowa State University, Ames, IA, USA
[2]Technical Services, Renewable Energy Group, Ames, IA, USA

8.1 Introduction

8.1.1 Definition of Solvent Liquefaction

Solvent liquefaction is a process by which carbonaceous feedstocks are converted to predominantly liquid or solubilized products in the presence of a liquid solvent at moderate temperatures and pressures: typically 105–400 °C and 2–20 MPa. Inert gases such as nitrogen are most often used to purge the system of oxygen and achieve reaction pressure. However, in some embodiments, reducing and oxidizing gases or recycled process gases are supplied to the reactor. Although this thermal deconstruction process superficially resembles pyrolysis in gaseous environments, most solvents significantly influence the course of reaction so resulting in important differences from pyrolysis. Also, operation at elevated pressures and the need for solvent recovery increases the capital costs of solvent liquefaction compared to fast pyrolysis. These costs can be offset by the improved liquid yields and product quality typically achieved by solvent liquefaction, as illustrated by Table 8.1.

8.1.2 History of Solvent Liquefaction

Continuous solvent liquefaction processes can be traced back to the early twentieth century and the development of the Bergius process. In 1914 Friedrich Bergius, who later won the Nobel Prize in Chemistry for his efforts in high pressure chemistry, developed a process for the liquefaction of coal mixed with heavy oil in the presence of hydrogen [1]. His initial

Table 8.1 Comparison of the yields and quality (as determined by oxygen content) of bio-oil produced from Southern Yellow Pine by fast pyrolysis and hydrocarbon solvent liquefaction.

Conversion technology	Moisture-free bio-oil yield (wt%)	Oxygen content (wt%)
Fast pyrolysis	40–50	40–45
Hydrocarbon solvent liquefaction	45–65	20–30

process was carried out in a 400 l batch reactor capable of treating 150 kg of coal at 400 °C and 20 MPa, but was converted to a continuous mode of operation by the mid-1920s [2].

Synthetic fuels from coal eventually fell out of favor as the global supply of crude oil increased. The energy crisis of the 1970s, when much of the world experienced escalating prices for crude oil as global supply could not meet demand, brought renewed interest to solvent liquefaction. The most notable effort was developed at the Pittsburgh Energy Research Center (PERC) of the US Bureau of Mines [3]. The PERC process, as it was known, reacted finely ground Douglas fir wood in recycled wood oil at 350–370 °C and 27 MPa. The process was eventually scaled up to a 3 ton/day continuous demonstration unit in Albany, Oregon [4].

The Albany project encountered numerous technical difficulties that prevented economic operation, and the Albany facility was shut down. Perhaps the foremost challenge faced by the PERC process was an inability to feed slurries with solids loading greater than 8% [3]. Researchers at the University of Arizona sought to remedy this challenge by adopting a modified single-screw extruder to feed biomass, which could convey slurries containing 60 wt% wood flour into a reactor operating at pressures up to 250 bar and temperatures of 400 °C [5]. Wood oil taken from the Albany facility was used as the solvent, while carbon monoxide and steam were used as the pressurizing gas [6]. This process was able to achieve bio-oil yields of approximately 48–58 wt%, which the researchers determined to be within 80–100% of the maximum theoretical yields. The bio-oil had an oxygen content of approximately 6–10 wt% [7].

At about the same time as the PERC process was being developed for biomass, several groups around the world were exploring coal liquefaction. The Exxon Donor Solvent (EDS) liquefaction system was unique in utilizing a hydrocarbon solvent not derived from the process itself (i.e. it was not a product of coal liquefaction). The process was modeled after the Bergius process, designed to liquefy coal in the presence of a hydrogen donor solvent (HDS) at approximately 400–450 °C and 0.3–2.1 MPa [8]. In order to provide hydrogen for upgrading the coal, the EDS system separated the hydrocarbon solvent from the product stream and hydrotreated it prior to recycling it back into the reactor, which made the EDS system unique among coal liquefaction efforts. The EDS process was largely successful, being scaled up to a pilot plant capable of processing 250 ton/day of dry feed in 1980 [9]. This process was also considered appropriate for the solvent liquefaction of biomass, although no significant progress was reported in the literature.

8.2 Feedstocks for Solvent Liquefaction

8.2.1 Feedstock Types

In solvent liquefaction, a variety of feedstocks can be converted into fuels and chemicals. Lignocellulosic biomass and algae are the two major types of feedstock that have been widely explored using liquefaction technologies. Other biomass feedstocks include microalgae, food waste, animal manure, paper-mill sludge, sewage sludge, and water hyacinth. Many biomass feedstocks are harvested with relatively high moisture content. High water content is thermodynamically a hindrance in the use of biomass for heat and power applications. In fact, the energy penalty for distillation and drying of the algae feedstock is almost 50% of the total energy required for the process [10]. Hence, processes that can directly use "wet" biomass are highly desirable. Solvent liquefaction has the potential to process wet feedstocks without extensive drying. Sludges and manures as feedstocks are likely to have bioactive contaminants, which can be sterilized during solvent liquefaction [11]. It is reported that solvent liquefaction can process feedstocks successfully with 65–80 wt% water content, thus it is possible to utilize wood and grasses (15–25 wt% moisture), food waste (69 wt% moisture), manure (20–76 wt% moisture), and paper sludge (50–62 wt%) as received in a biorefinery [12–17]. Although algal biomass as received can contain up to 90 wt% moisture, it can be dewatered through physical processes to solids content of 10–20 wt% solids [18, 19].

8.2.2 Benefits of Liquid Phase Processing

Solvent selection is a key parameter for optimal liquefaction performance. Solvents must be sufficiently stable at reaction conditions, miscible with the reactants stream, and facilitate the intended reaction chemistry. The process is usually configured for the solvent to remain in the liquid phase, although in some cases the solvent exists in a supercritical state. Often it is desirable to select a solvent that is easily recovered from the product stream. This can be achieved through use of a low-boiling point solvent than can be easily vaporized. Figure 8.1 shows the pressure–temperature curve for six common solvents used in solvent liquefaction. Ethanol, water, and tetrahydrofuran (THF) all reach their critical point below 400 °C, requiring substantial pressure to remain in the liquid phase. Conversely, phenol, tetralin, and γ-valerolactone (GVL) are relatively high boiling point solvents, requiring only moderate pressure to remain a liquid.

Processing in a solvent offers several advantages compared to liquid-free processing such as pyrolysis. The most palpable advantage is dilution of potentially reactive species. Biomass degradation products are notoriously reactive. Fast pyrolysis bio-oils for instance, are well documented to continue to undergo reactions for extended periods of time after collection [20, 21]. This reactivity reflects the non-equilibrium conditions under which liquids are produced in most thermochemical processes [22]. Another contributing factor is the hundreds of different compounds produced, with the result that many side-reactions occur during processing. Dilution of fast pyrolysis bio-oil with solvents has been shown to reduce the extent of undesired reactions in the condensed phase [23, 24]. Additionally, some

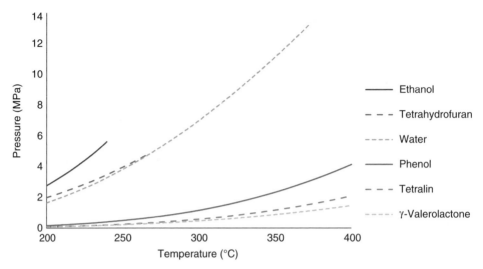

Figure 8.1 Pressure–temperature curves for several common solvents used in solvent liquefaction as determined by the Antoine equation for each pure species. The end point of the ethanol, tetrahydrofuran, and water curves denote their critical point. Phenol, tetralin, and γ-valerolactone do not reach their critical point within this window.

impurities found in biomass, such as alkali and alkaline earth metals, can catalyze undesirable reactions. These effects can be mitigated through dilution in a solvent phase [25]. For certain processing schemes wherein the solvent is not removed from the product stream, the benefits of dilution persist until the solvent is recovered or the product stream is utilized.

Liquid phase processing also provides an opportunity to recover solubilized non-volatile products. Recovery of these products is difficult through vapor-phase recovery as employed in pyrolysis because these molecules tend to have relatively low vapor pressures even at reaction temperatures. Therefore, these compounds are either entrained in the exiting gas stream or decompose to small enough molecules to be volatile. In the case of solvent liquefaction, however, these compounds can be recovered in the liquid product stream. To be sure, simply because these high-boiling point products are retained in the solvent stream does not mean that they are trivial to recover. Due to the inherent difficulty of using a thermal process to recover them suggests that more extensive separation processes are necessary; usually this results in implementation of potentially costly solvent extraction methods.

In addition to physical interactions, solvents can also participate in chemical reactions with the reactants and products, which is sometimes referred to as solvolysis. However, solvolysis is not necessarily the exclusive or even dominant kind of reaction in solvent liquefaction.

Solvolysis reactions are a type of nucleophilic substitution or elimination reaction wherein the nucleophile is a solvent molecule. Thus, the presence of a liquid phase in solvent liquefaction allows for these reactions to occur. Hydrolysis of polysaccharides in water to form monosaccharides and alcoholysis of triglycerides in methanol to form fatty acid methyl esters are two common examples of solvolysis reactions. Similar to these solvolysis reactions are hydrogenation or alkylation reactions wherein the solvent contributes a hydrogen atom or alkyl group to a solute molecule, respectively. These reactions are often desirable in many biomass applications due to the opportunity to decrease the oxygen content of biomass degradation products. The addition of a HDS is the most prevalent means of promoting this behavior through hydrodeoxygenation [26]. Although this often promotes formation of products with favorable C/O and C/H ratios, consumption of solvent in this manner can be detrimental to process economics. In most instances, the utilization of a solvent intended to participate in solvolysis reactions should be limited to yield only the optimal benefit for the cost. Complete hydrodeoxygenation or hydrogenation in this manner is not currently economically feasible for conversion of biomass to fuels.

Solvents do not need to be consumed during liquefaction for their chemical benefits to be realized. In certain schemes, a solvent can simply behave as a catalyst to promote desired reactions. Ionic liquids are perhaps the most impressive catalytic solvents. They have been demonstrated to be effective at decomposing biomass at very low temperatures [27]. Unfortunately, the cost of producing these solvents and the need for complete recovery challenges their utilization at any significant scale. Acid and base solvents exhibit similar catalytic behavior to more exotic ionic liquids, but at lower cost. The corrosivity and process handling concerns of these solvents can limit their adoption. Polar aprotic solvents are a potentially more cost effective and benign alternative to other catalytic solvents mentioned above. THF, GVL, and other similar solvents have been shown to significantly reduce the activation energy for cellulose decomposition and monosaccharide production from biomass molecules without being consumed during the reaction [28, 29].

Water is a particularly desirable solvent for many liquefaction applications. This is primarily due to its wide availability, low cost, and potentially favorable reaction chemistry. Extensive research has been conducted on a variety of processing conditions using water as a solvent, and has led to a secondary classification for aqueous versus non-aqueous solvent liquefaction systems [12, 30, 31]. Hydrothermal liquefaction (HTL) is defined as liquefaction processes in which water is the primary solvent. In this case solvent liquefaction can refer to liquefaction employing non-aqueous solvents.

HTL has advantages in processing wet feedstocks. Elevated pressures prevent vaporization of water in the system, reducing the thermal penalty associated with its high enthalpy of vaporization. HTL of high-moisture feedstocks such as microalgae, sewage sludge, and animal manure has been an area of recent interest [32, 33].

8.2.3 Reaction Types

A vast number of primary and secondary reactions are possible in solvent liquefaction. In some ways the reaction network can be more complex for solvent liquefaction than pyrolysis due to the presence of a solvent and the typically longer residence times, both of which increase opportunities for reaction. However, there are fewer side-reactions due to the solvation and separation of reactive products by the solvent. A few prevalent thermal decomposition reactions include:

- *thermolysis* – thermal decomposition of large molecules into smaller molecules
- *decarboxylation* – release of carbon dioxide from a base molecule, usually a carboxylic acid or ester functionality

- *decarbonylation* – release of carbon monoxide from a base molecule, usually a ketone functionality
- *dehydrogenation* – release of hydrogen from a base molecule
- *dehydration* – release of a water molecule from a base molecule, usually a carbohydrate species such as monosaccharide or oligosaccharide.

Thermal decomposition reactions can be assisted by homogeneous or heterogeneous catalysts, or by selection of a primary solvent that exhibits some degree of catalytic behavior. Nevertheless, it is crucial that the solvent(s) are not susceptible to thermal decomposition at reaction conditions, particularly if the solvent is to be recovered or recycled.

Some solvent liquefaction systems are intended to have the solvent(s) participate in the reaction chemistry through solvolytic reactions. These systems often require more targeted feedstock selection and pretreatment. This can result in greater product selectivities than systems where the solvent does not participate directly in the reactions. Solvolytic reactions consume the solvent and therefore require constant replenishment of the active solvent component to sustain reaction. This increases the operating cost of these systems relative to an inert solvent, but they also can result in higher value products that offset these costs. Several common solvolytic reactions include:

- *hydrolysis* – cleavage of ether linkages assisted by solvent water that results in two separate molecules, each with terminal hydroxyl groups
- *alcoholysis* – cleavage of ether linkages assisted by alcohol solvents that results in two separate molecules, one with a terminal hydroxide group and the other with a terminal alkyl group
- *hydrogenation* – donation of hydrogen atom(s) from the solvent that results in saturation of carbon chains
- *hydrodeoxygenation* – donation of hydrogen atom(s) from the solvent that results in removal of oxygen atoms from feedstock in the form of water.

Undesired reactions, such as polymerization, are common for thermochemical conversion of biomass. Due to the high concentration of reactive, oxygenated species the pyrolysis oils often undergo extensive polymerization reactions, with these reactions beginning immediately after production and extending long after storage [34]. This is also true for solvent liquefaction, but proper solvent selection and processing conditions can help to minimize this. One method of reducing the likelihood for secondary reactions is through the use of HDSs to cap reactive sites during thermal decomposition. Another method is to operate at solvent-to-biomass ratios sufficient to dilute reactive products to the point that secondary reactions are unlikely. This benefit can be maintained as long as the solvent is present in sufficient concentration with the products. Some researchers have demonstrated that adding diluent to pyrolysis oil can improve storage stability [24]. This scheme also extends to bio-oils produced from solvent liquefaction.

8.2.4 Processing Conditions

Broadly, solvent liquefaction conditions can be divided into two categories: subcritical and supercritical. Subcritical operation implies operation at sufficiently high pressure to prevent boiling of the solvent but operating below the critical point for the solvent.

Water is the most commonly used solvent in both subcritical and supercritical liquefaction processes for converting biomass in liquid products. Water chemistry under supercritical conditions differs significantly from subcritical conditions, which can be exploited to achieve unique product distributions.

Many organic solvents are employed as both subcritical and supercritical processing. If the solvent is strongly polar with a low boiling point it is often used in the supercritical state to exploit its exquisite solubilization properties. For example, methanol, ethanol, and acetone are commonly used for biomass conversion under supercritical conditions.

Conversely, it is common to use organic solvents with high boiling points for subcritical processing. The energy to heat solvents above their critical points is not insignificant, despite the absence of distinct phase change. Fortunately, high boiling point solvents usually exhibit excellent solubilization characteristics at temperatures well below their critical points. For example, high polarity solvents exhibit strong polar or hydrogen-donating interactions between solvent molecules and biomass-derived molecules even at moderate temperatures. This results in high yields of organic products without resorting to excessively high temperatures and pressures.

The solvent liquefaction of biomass is heavily influenced by several factors. The most common parameters are temperature and pressure. Generally speaking, high temperatures and pressures increase biomass liquid yields. Other important factors are residence time, type of primary solvent, biomass-to-solvent ratio, and the presence of co-solvent or catalyst.

8.3 Target Products

8.3.1 Bio-oil

Solvent liquefaction of whole biomass typically results in a wide range of products. These include solid residue, reaction water, non-condensable gases, and organic liquid containing a wide range of molecules. Choice of solvent has a strong impact on product selectivity and liquid yields for a given biomass feedstock [35]. Table 8.2 illustrates the effectiveness of various solvents in liquefying pine at 280–300 °C in a batch stirred reactor. Similar to bio-oils from fast pyrolysis, these are comprised of hundreds of chemical compounds, which makes targeted recovery of specific chemicals challenging.

Instead of attempting to recover specific compounds, many technologies fractionate bio-oils into only a few cuts based on crude boiling point ranges. In some embodiments,

Table 8.2 Liquid yields from the liquefaction of pine in various solvents.

Solvent	Liquid yield (wt%)	Reaction temperature (°C)	Source
Acetone	10%	300	[35]
Toluene	12%	280	
Ethanol	14%	300	[35]
Water	19%	300	[35]
GVL	60%	280	
Creosote	74%	280	
Tetralin	81%	280	

the entire liquid stream is collected together as whole oil. These bulk streams can be suitable as a replacement for heavy fuel oils due to a relatively high energy density and low oxygen content [36].

8.3.2 Production of Fuels and Chemicals

Careful choice of feedstock and operating conditions has potential to selectively produce desired products. A summary of the wide spectrum of useful products from solvent liquefaction of biomass is given in Figure 8.2.

Carbohydrate-rich biomass such as lignocellulose is converted into chemical intermediates that can be upgraded into value-added fuels and chemicals using strong polar solvents with or without catalyst. Cellulose as well as herbaceous and woody biomass could be used to produce solubilized carbohydrates including glucose, xylose, and anhydrosugars at high yields and selectivities in aqueous acid, supercritical water, GVL, THF, and ionic liquids, just to name a few [37–41]. Solubilized carbohydrates can also be upgraded to liquid fuels, platform biochemicals, and commodity chemicals such as ethanol, liquid alkanes, 2-dimethylfuran, 5-hydroxymethylfurfural (5-HMF), furfural, levulinic acid, acetic acid, levoglucosenone, and sorbitol by different biological and chemical reaction pathways

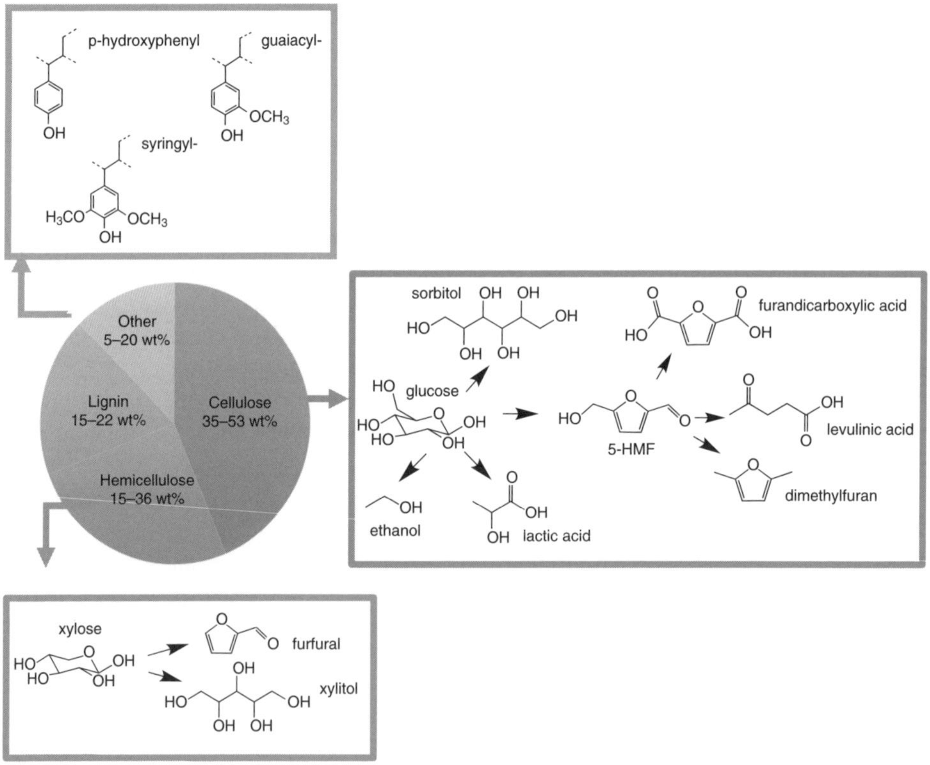

Figure 8.2 Schematic of target biochemicals, biofuels, and their precursors produced from biomass by solvent liquefaction.

including fermentation, isomerization, dehydration, hydrogenation, and hydrodeoxygenation [39, 42–44]. Many of these chemicals are expensive to manufacture by other means, and can be used as building blocks for a variety of molecules used in plastics and pharmaceuticals.

Another category of valuable chemicals from biomass via solvent liquefaction is aromatic fuel, chemicals, or their precursors. Solvent liquefaction of lignin in primary alcohols, aromatic hydrocarbons, aromatic alcohols, or water as solvent are common pathways for production of mixtures of phenolic monomers and oligomers. A variety of aromatic monomers can be produced from soft and hardwood lignin such as guaiacol, coniferyl alcohol, 2,6-dimethoxyphenol, sinapyl alcohol, and the γ-methyl ethers as guaiacyl or syringyl lignin-derived products [45]. In the presence of hydrogen, the product distribution could be shifted toward alkylated phenols from lignin [46]. Application of formic acid and tetralin as the hydrogen-donating solvent results in production of significantly high yields of low molecular weight aromatics, vanillin, guaiacyl, and syringyl monomers [47–49]. Phenolic products have to undergo several types of catalytic upgrading mechanisms such as oxidation and hydrodeoxygenation to produce valuable aromatic fuels and chemicals, including benzene, toluene, xylene, alkylbenzenes, and vanillin [50–52].

8.3.3 Co-products

In general, non-condensable gas and solid residue co-products from solvent liquefaction and fast pyrolysis are similar in composition and yields. Perhaps the most significant difference in co-products is the nature of the solid residue. Unless extensive solvent extraction or washing procedures are employed, the solid residue from solvent liquefaction is often laden with bio-oil and high-boiling point solvents. Effective solvent liquefaction reduces biomass feedstock to such small residual particles that this co-product has the characteristics of sludge or filter cake.

8.4 Processing Solvents

As illustrated in Figure 8.3, solvents can be broadly classified as either organic or inorganic. Inorganic solvents can be further divided into aqueous or non-aqueous solvents. Non-aqueous solvents may include a wide range of acids, bases, or oxidizing agents. When water is used for thermochemical decomposition of biomass, the process is referred to as hydrothermal liquefaction, which is described in Section 8.4.1.2. Non-aqueous inorganic solvents used widely in pretreatment and fractionation of biomass include concentrated acids, concentrated alkali, and oxidizing agents.

Organic solvents are used both in thermochemical decomposition and pretreatment of biomass. These solvents can be further divided into three major categories: polar solvents, non-polar solvents, and ionic liquids. Polar solvents are defined as liquids with electric dipoles or multi-pole moments [53]. Depending on the hydrogen atom donating capability, polar solvents bifurcate into two types. Solvents having a hydrogen atom attached to a highly electronegative atom such as oxygen (hydroxyl group) or a nitrogen (amine group) are called polar protic solvents [54]. These solvents can form strong hydrogen bonds with a suitable acceptor. On the other hand, solvents that lack hydrogen atoms attached to electronegative atoms (e.g. oxygen and nitrogen) are known as polar aprotic solvents. Usually,

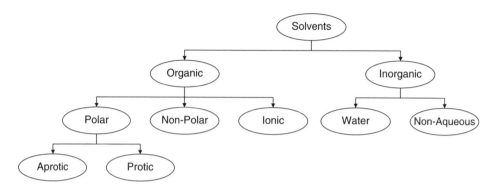

Figure 8.3 Types of solvents used in solvent liquefaction.

polar protic solvents are capable of accepting a hydrogen bond due to the presence of a lone pair of electrons and can donate hydrogen or participate in hydrogen bonding due to presence of an acidic proton. Alcohols and carboxylic acids are strongly polar protic solvents. Commonly used protic solvents in biomass conversion include methanol, ethanol, isopropanol, and formic acid.

Polar aprotic solvents can accept a hydrogen bond through a lone pair electron, but are unable to donate acidic protons or form hydrogen bonds [55]. Biomass conversions in various polar aprotic solvents such as acetone, THF, acetonitrile, 1,4-dioxane, ethyl acetate, dimethylacetamide (DMA), dimethylformamide (DMF), dimethylsulfoxide (DMSO), sulfolane, and GVL are prevalent in the literature.

Non-polar solvents have negligible electric dipole or multi-pole moments [53]. This type of solvent does not possess or accept acidic protons or create hydrogen bonds with other molecules. For example, most hydrocarbon solvents are non-polar in nature, such as hexane, toluene, xylene, and tetralin. Generally, non-polar solvents are immiscible in strongly polar compounds but can dissolve other non-polar compounds.

Ionic liquids (IL) are liquid organic salts that are the subject of increasing research in biomass pretreatment and conversion technologies [56]. The ions in IL are poorly coordinated, which contributes to their liquid state at temperatures below 100 °C, sometimes as low as room temperature [57]. ILs have at least one ion with a delocalized charge and one organic component, which prevents the formation of a stable crystal lattice [58]. Ionic liquids have attractive properties for biomass conversion including high solvation strength and solubility characteristics, strong coordination ability, high boiling point, low flammability, high thermal stability, and low environmental impact [59]. IL can solvate a wide range of polar and non-polar compounds and exhibit high solubility toward insoluble components of biomass such as cellulose and lignin. The most commonly investigated ILs for biomass utilization are 1-ethyl-3-methylimidazolium chloride ([EMIM]Cl), 1-ethyl-3-methylimidazolium acetate ([EMIM]OAc), 1-butyl-3-methylimidazolium chloride ([BMIM]Cl), 1-butyl-4-methylpyridinium chloride, cholinium acetate ([Ch][OAc]), and cholinium lysinate ([Ch][Lys]) [55].

Table 8.3 summarizes major applications of polar protic, polar aprotic, non-polar, and ionic liquid solvents in biomass conversion for the production of crude bio-oils, fuel and chemical precursors, platform chemicals, and commodity chemicals.

Table 8.3 Major applications of various types of solvents in biomass conversion processes.

Process	Biomass	Solvent class	Solvent used	Product yields	Ref.
Pretreatment	Lignocellulose[a]	Inorganic	Dilute acid[b]	Glucose: 51–90%	[40, 60–70]
		Ionic liquid	[Ch][Lys][c], [EMIM]Cl[d]	Glucose: 80–96%	
		Polar protic	Methanol, ethanol	Glucose: 70–97%	
		Polar aprotic	GVL, THF	Glucose: 95–99%	
Saccharification	Lignocellulose	Inorganic	Dilute acid, concentrated acid	Glucose: 50–99%, xylose: 78–88%	[37, 39, 55, 71–83]
		Ionic liquid	[BMIM]Cl[d], [EMIM]Cl	Glucose: 53–89%, xylose: 75–88%	
		Polar aprotic	GVL	lucose: 65–69%, xylose: 70–73%	
Carbohydrate alkylation	Cellulose	Polar protic	Methanol, ethanol, ethylene glycol	Alkylated C_6 sugars: 25–90%	[84–88]
Carbohydrate depolymerization	Cellulose	Polar aprotic	1,4-Dioxane, THF, acetone, GVL, acetonitrile, sulfolane	Levoglucosan: 15–41%	[29, 89, 90]
Dehydration[e]	Glucose	Inorganic	Water	5-HMF: 10–27%	[55, 91–93]
		Ionic liquid	[BMIM]Cl, [EMIM]Cl	5-HMF: 68–91%	
		Polar aprotic	THF, DMSO, MIBK, DMA	5-HMF: 34–76%	
	Xylose	Inorganic	Dilute acid	Furfural: 30–48%	
		Polar aprotic	THF, MIBK	Furfural: 50–75%	
Lignin depolymerization	Organosolv, kraft lignin	Inorganic	Alkali, acid	Phenolic monomer: 5–48%	[55, 94–97]
		Non-polar	Phenol	Phenolic products: 7–19 wt%	
			p-cresol	Phenolic products: 80% C	
		Polar protic	Methanol, ethanol, butanol, formic acid	Phenolic monomer: 5–35%	
Reductive fractionation ("lignin-first")	Lignocellulose	Polar protic	Methanol, ethanol, 2-propanol, formic acid	Phenolic monomer: 23–55%	[98–103]

Table 8.3 Continued

Process	Biomass	Solvent class	Solvent used	Product yields	Ref.
Wet oxidation	Technical lignin	Inorganic	Water	Aromatic aldehyde: 4–21%	[104]
Liquefaction	Lignocellulose	Inorganic	Water	Bio-oil: 19–35 wt%	[7, 18, 32, 35, 105, 106]
		Non-polar	Tetralin, creosote	Bio-oil: 47–81 wt%	
		Polar protic	Methanol, ethanol, propanol, butanol	Bio-oil: 21–44 wt%	
		Polar aprotic	Acetone, GVL	Bio-oil: 10–60 wt%	
	Algal biomass, manure	Inorganic	Water	Bio-oil: 27–64 wt%	

[a] Glucose yields for pretreatment are from enzymatic hydrolysis of solvent-pretreated biomass.
[b] Catalyst information is not included.
[c] [Ch][Lys] is the abbreviation for cholinium lysinate.
[d] [EMIM]Cl and [BMIM]Cl are abbreviations for 1-ethyl-3-methylimidazolium chloride and 1-butyl-3-methylimidazolium chloride, respectively.
[e] The solvent is either the primary reaction solvent or the extraction solvent in a biphasic reaction.

8.4.1 Inorganic Solvents

8.4.1.1 Non-aqueous

Non-aqueous inorganic solvents are primarily used to pretreat biomass ahead of fractionation or saccharification. Many concentrated acids and alkaline solutions are effective in opening the structure of biomass by solubilizing hemicellulose and dislodging lignin, making it more susceptible to deconstruction by subsequent enzymatic or thermochemical processing. Use of 55–75% sulfuric acid, 79–86% phosphoric acid, 60–88% nitric acid, and 59–61% perchloric acid for pretreating biomass are described in the literature [107]. Common alkaline solutions used for pretreatment include sodium hydroxide, potassium hydroxide, calcium hydroxide and amines. One of the most well-known alkaline pretreatments is the Kraft process for woody biomass, which uses a mixture of water, sodium hydroxide, and sodium sulfide to solubilize lignin and liberate cellulose fibers [108]. Leading pretreatments are based on dilute sulfuric acid, liquid ammonia, ammonia fiber expansion, and lime [109]. Oxidizing agents such as dissolved molecular oxygen, hydrogen peroxide, sodium hypochlorite, peracetic acid are also employed for thermochemical pretreatment of biomass [107, 110, 111].

8.4.1.2 Aqueous Hydrothermal Liquefaction

HTL is the name given to solvent liquefaction with water as the solvent. Hydrothermal processing includes operation at both subcritical and supercritical conditions. Subcritical HTL occurs below the critical point of water at 374 °C and 22 MPa. In order to maintain water in the liquid phase, system pressure must equal or exceed the vapor pressure of water. As the temperature of supercritical water increases, most literature acknowledges a shift to supercritical hydrothermal gasification due to the tendency for the reaction products to decompose to small, non-condensable gases [30].

Hydrothermal carbonization exists as another subset of hydrothermal processing, although it is not a liquefaction process. In hydrothermal carbonization, water and carbonaceous feedstocks are reacted at temperatures up to 250 °C with the goal of producing a carbon-rich solid product [112]. These processes are currently being evaluated for the production of renewable carbon products such as nanotubes and composites [113].

Properties of liquid water vary greatly with temperature and even more significantly across the critical point. Figure 8.4 illustrates the effect of temperature on the density, dielectric constant, and ion dissociation constant of water. Significant reductions in each of these properties occur at the critical point. Unlike most other properties, the ion dissociation constant of water demonstrates a maximum at approximately 280 °C. At these conditions water takes on an ionic nature that changes the reaction chemistry with many feedstocks to favor acid- or base-catalyzed reactions [114]. This also contributes to high corrosion rates with many stainless steels and requires higher alloys to prevent excessive deterioration [115].

Reduction in the dielectric constant indicates another shift in the solvent behavior of water. As the temperature of water increases, the structure of water flattens and its molecular polarity decreases. Once water reaches the supercritical regime it is effectively non-polar and begins to have solvating ability similar to non-polar solvents.

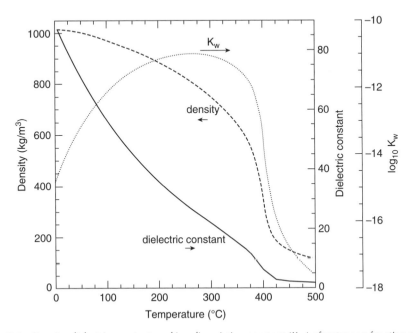

Figure 8.4 Density, dielectric constant, and ion dissociation constant (K_W) of water as a function of temperature. Source: taken from Peterson et al. [30] with permission.

8.4.1.2.1 Hydrothermal Liquefaction of Whole Biomass

HTL of whole biomass has been widely studied. It has primarily been evaluated as a means of processing high-moisture feedstocks with the goal of eliminating the drying of feedstock. Common examples of high-moisture feedstocks suitable for HTL include sewage sludge, animal manure, and microalgae [32]. Despite this energy saving from the absence of drying, the moisture in the feedstock nevertheless incurs a significant energy penalty to heat it to reaction temperature. Economization of heat recovery is a crucial component to the development of HTL systems to mitigate this.

Product yields from HTL are highly dependent upon the feedstock type and composition [31]. HTL is well suited to convert saccharides, lipids, and proteins to liquid products due to the tendency for these feedstocks to undergo hydrolysis. HTL of lipids is a relatively well developed technology often referred to as "fat splitting" or "oil splitting." The Colgate-Emery process is the most common example. Originally developed in the 1940s, this non-catalytic process involves reacting lipids with compressed water or steam at temperatures of approximately 250 °C and pressures up to 5 MPa to produce glycerol and free fatty acids [116]. This technique is still used today in the oleo-chemical and feed industries.

Microalgae is a particularly attractive feedstock for hydrothermal liquefaction because of its absence of lignin. Liquid yields as high as 60 wt% have been reported [117, 118]. Relative to fast pyrolysis bio-oil, the liquid products obtained from liquefaction of microalgae are characterized by low oxygen content, moderate energy content, and high nitrogen content due to the high concentration of proteins [119].

Because lignins are largely recalcitrant to hydrothermal processing and often contribute to high solid yields, they pose a challenge for hydrothermal liquefaction of lignocellulosic biomass. For example, Demirbas [120] found a positive correlation between lignin content and solid yields after testing eight lignocellulosic feedstocks with lignin contents ranging from 15 to 54 wt%. Solid yields ranged from around 20 to 50 wt%, respectively. The addition of a KOH catalyst reduced the solid yields on average, but a correlation between lignin and solid yields was still observed. Pretreatment and removal of lignin prior to hydrothermal liquefaction of lignocellulosic feedstocks is a promising alternative to adding catalysts to assist in its decomposition.

8.4.1.2.2 Hydrothermal Liquefaction of Saccharides

Cellulose decomposition has been extensively studied under acid-catalyzed condition in subcritical and supercritical water [41, 121–123]. Subcritical water encouraged the formation of dehydration products of glucose, namely, 5-HMF, due to catalytic behavior of water in the range of 200–300 °C [123–125]. On the other hand, supercritical water dissolves cellulose easily due to its exquisite solubilization properties and results in an enhanced rate of hydrolysis, forming higher yields of glucose [121].

The addition of dilute acid catalyst to subcritical water is a promising approach to hydrolysis of polysaccharides. In particular, acid-catalyzed solubilization and hydrolysis of hemicellulose is an environmentally friendly and cost-effective alternative to severe pretreatments with organic solvents. The primary product from hemicellulose hydrolysis is xylose. Small amounts of arabinose, galactose, and glucose are likely to also be present in the products. Ideal reaction conditions are in the range of 120–140 °C for 0.25–1 hour residence time and 1.5–3.5% sulfuric acid as catalyst [126]. It is important to conduct hemicellulose hydrolysis at optimal conditions to avoid the formation of undesirable degradation products such as acetic acid and furfural.

Current research has raised concerns over some of the adverse effects of water being employed as a solvent for biomass conversion. It has been shown that subcritical water in the presence of acid catalyst facilitates dehydration of monosaccharides into 5-HMF and furfural, which are precursors to humins, an undesirable by-product [127–129].

8.4.1.2.3 Hydrothermal Liquefaction of Lignin

Despite its relative recalcitrance, lignin can be decomposed into soluble products via hydrothermal processing in supercritical water. Sasaki and Goto [130] deconstructed alkaline lignin in supercritical water. Identified products were mainly catechol, phenol, and cresol. Subcritical alkaline solutions (4–5 wt% NaOH, KOH, $Ca(OH)_2$) can also be useful for lignin conversion as they inhibit char formation and thus increase yields of phenolic oil [94]. Alkali-assisted hydrothermal liquefaction is typically conducted at 300–330 °C and 3.5–90 MPa, resulting in phenol and phenol derivatives such as guaiacol, catechol, vanillin, pyrocatechol, and syringol as major products from Kraft lignin, organosolv lignin, and steam explosion extracted lignin [131–134]. The mechanism involves cleavage of aryl-alkyl β-O-4 bonds with the transition state species of this reaction forming a cation adduct with the help of the base cation of the solvent [132]. This process, however, leads to low selectivity of monophenols due to high temperature operation.

HTL can also be used for wet oxidation of lignin. Since lignin has abundant hydroxyphenyl groups, it can be oxidized with molecular oxygen or oxidants such as hydrogen peroxide, metal oxides, and nitrobenzene and catalytically cracked to produce valuable aromatic aldehydes and carboxylic acid products such as vanillin, syringaldehyde, p-hydroxybenzaldehyde, syringic acid, and oxalic acid [104, 135].

8.4.2 Polar Protic Solvents

The hydrogen-donating capability of polar protic solvents is particularly useful in hydrogenating products of lignin depolymerization. Generally, hydrogenation of the products of polysaccharide deconstruction is not desirable, with the goal to simply depolymerize carbohydrate to monosaccharides or anhydrosugars. Thus, application of polar protic solvents in biomass liquefaction mostly revolves around lignin depolymerization.

8.4.2.1 Processing Whole Biomass in Polar Protic Solvents

Although less widely investigated than hydrothermal processing, many protic solvents can effectively convert biomass into a liquid bio-oil at high temperature. For example, aromatic alcohols can enhance bio-oil yields relatively to primary alcohols. As illustrated in Table 8.2, while creosote achieved bio-oil yield up to 74 wt% from lignocellulosic biomass, methanol, ethanol, 1-propanol, and 1-butanol could only reach 21–44 wt% of bio-oil yield.

Protic solvents have been traditionally explored for pretreating biomass to separate lignin and so to improve the digestibility of polysaccharide via enzymatic hydrolysis. In 1893, Klason [136] demonstrated the effectiveness of organosolv pretreatment to isolate lignin from wood. Since then methanol, ethanol, glycerol, and ethylene glycol have been extensively used with and without catalysts for extracting lignin and hemicellulose from biomass ahead of enzymatic hydrolysis of the cellulose pulp. Alcohol-mediated pretreatment can achieve delignification of 65–93% at 170–235 °C within 30–180 minutes. The lignin obtained is a high-quality precursor for aromatic chemical production [137–143].

Carboxylic acids can be good choices for solvent-based pretreatment of biomass owing to their operability at relatively low pressures. Formic acid and acetic acid can be utilized at atmospheric pressure for biomass pretreatment. Nevertheless, their use might be limited by their corrosivity and tendency to acetylate cellulose [144].

Alcohols have been used to simultaneously deconstruct biomass and separately extract carbohydrates and aromatic monomers. Compared to HTL, alcoholysis has several potential benefits including augmentation of reaction rate and product yields, suppression of humins and chars, and reducing waste water treatment.

Historically, biomass fractionation technologies have used harsh reaction conditions of high temperature and pressure to efficiently extract lignin. However, this process leads to significant modification of the native structure of lignin and decreases the maximum achievable monomer yield from lignin. Ether linkages that are cleaved during lignin extraction tend to recondense to recalcitrant C—C bonds. These bonds are difficult to further depolymerize to phenolic monomers. A recent approach has emerged in the biorefinery literature, namely "lignin-first strategy," which aims at mitigating this problem by dissembling recovered lignin from biomass prior to carbohydrate valorization [145].

The lignin-first approach has two manifestations. The first couples depolymerization and stabilization while the second preserves β-O-4 bonds, resulting in higher monomer yields from lignin. The depolymerization/stabilization pathway employs polar protic solvents to produce stable phenolic monomers and oligomers from woody biomass at impressively high yields and selectivity directly [98, 100–103, 146]. Yields as high as 23–55% of monomeric phenol (phenol, 4-propylguaiacol, 4-propylsyringol) have been attained from birch and poplar wood processed in methanol at 180–250 °C with redox catalysts. The remaining solid material is a carbohydrate pulp that can be hydrolyzed or otherwise upgraded into fuels and chemicals. Although this approach results in insignificant condensation of lignin-derived species, the key challenge is finding a suitable window of operation that can solubilize and depolymerize lignin without damaging hemicellulose. Additionally, solid catalyst separation from the delignified pulp can be challenging.

The second pathway usually employs polar aprotic solvents or ionic liquids to preserve ether linkages. However, Luterbacher et al. [49] has recently demonstrated beech wood pretreatment in a polar protic solvent, formaldehyde, for production of stabilized lignin monomers from biomass. A high phenolic monomer yield of 78% could be achieved at mild conditions (80 °C, 5 hours) from the above pretreatment method. Formaldehyde could prevent lignin condensation by forming 1,3-dioxane structures with the side-chain hydroxyl groups in lignin. This pathway extracts soluble and stabilized lignin prior to hydrogenolysis. Hemicellulose is preserved by reaction with formaldehyde to form diformyl-xylose (75%), which can be readily hydrolyzed to xylose in an aqueous acid solution, making this pathway particularly attractive for carbohydrate recovery. The cellulose pulp is extracted as a highly susceptible feedstock for enzymatic hydrolysis, achieving 73% glucose yield.

8.4.2.2 Processing of Saccharides in Polar Protic Solvents

Only a few studies have undertaken the use of polar protic solvents to convert cellulose into solubilized carbohydrates. Ishikawa [87] employed supercritical methanol to deconstruct cellulose and obtained methylated glucosides as the major products. Other alcohols that could be used for the same purpose include ethanol and ethylene glycol [84, 88]. It is intriguing that alcoholysis of cellulose results in 57–63% yields of the methyl and ethyl glucosides in presence of strong acid catalyst compared to only 5–10% yields in pure water [85]. Furthermore, a remarkably high yield of 99.5% of n-decyl pentosides was obtained from conversion of wheat bran hemicellulose in n-decanol at 100 °C in presence of sulfuric acid catalyst [147]. However, alkylated sugars are not a particularly desirable product if the goal is to produce fermentable sugars. They have been proposed as precursors for production of cosmetics, detergents, emulsifiers, and nonionic surfactants [147, 148].

8.4.2.3 Processing of Lignin in Polar Protic Solvents

Thermal depolymerization of lignin has been investigated extensively in polar protic solvents. Supercritical alcohols are well known for their capability of effectively depolymerizing lignin. Alcohols have excellent solubilization ability under supercritical conditions. Low molecular weight primary and secondary alcohols are frequently used for solvolysis of lignin [149]. They offer a range of practical benefits including low cost and low boiling points. These alcohols can be readily recovered and recycled from the reaction mixture by simple distillation.

Researchers have found that the chemistry of alcohols is particularly favorable for depolymerization of lignin, serving as hydrogen donors. Studies have shown that primary alcohols promote cleavage of ether linkages in lignin and hydrogen donation to the products [146, 150, 151]. This phenomenon leads to alkylation of the aromatic monomers and oligomers as produced from lignin depolymerization. Most likely, the alcohols inhibit repolymerization of reactive free radicals produced at high temperatures [152]. Thus, phenolic oil yield can be increased in the presence of supercritical alcohols [146].

Methanol, ethanol, 2-propanol, and 1-butanol are commonly employed in lignin decomposition via solvolysis [149, 153, 154]. Depolymerization of lignin usually occurs in these solvents at supercritical conditions to improve bio-oil yield. Additionally, a high reaction temperature and high lignin-to-solvent ratio are helpful for increasing liquid oil yield from lignin in the alcohols. Often an acid or base catalyst is added to enhance lignin conversion and selectivity of phenolic monomer yields [94].

Using alcohol to depolymerize lignin typically results in consumption of the solvent. It is sometimes perceived that the alkyl group originates from the solvent and might attach to lignin by a nucleophilic substitution, thus solvolysis is often used to describe lignin depolymerization in alcohols [155]. Evidence for this is that the process yields more mass in the products than in the starting lignin, indicating that solvent was consumed during reaction [156].

Many aromatic alcohols could be used as solvent for production of phenolic oil and phenolic monomers from lignin. The improvement in conversion and selectivity occurs through the reaction of the alcoholic solvent with the reactive phenolic monomers and oligomers derived from cracking the lignin polymer. Usually, the aromatic alcohol serves as capping agent for preventing re-polymerization of lignin-derived fragments via free radical mechanisms. This type of solvent effect leads to production of biphenyl dimer structures, which is formed by addition of one phenolic moiety from lignin and the other from the solvent itself. Aromatic alcohols such as *p*-cresol and phenol are common choices for effective depolymerization of organosolv lignin [96, 97, 157–159]. When blended with supercritical water at high temperature, these solvents can produce high yields of monomeric phenolic compounds in addition to achieving a high phenolic oil yield [97, 160].

Formic acid is another widely used protic solvent for selectively producing hydrodeoxygenated aromatic chemicals from lignin depolymerization. Similar to primary alcohols, formic acid participates in the reaction by donating hydrogen during lignin fragmentation to form stabilized phenolic monomers. With added acid or metallic catalyst, conversion of lignin into alkoxyphenol monomers in formic acid can be further improved. Acid-catalyzed depolymerization in the presence of formic acid produces methoxyphenols and catechols as major products while metal-catalyzed reactions produce guaiacol, pyrocatechol, and resorcinol as primary monomer products [94]. Acid-catalyzed lignin solvent liquefaction in formic acid requires high temperatures (>300 °C) to provide the activation energy for cleavage of C—O and C—C bonds in lignin. Using metal catalysts such as Ni, Pt, and Pd on carbon or nafion supports dramatically reduced the activation energies of the depolymerization reactions [146, 161, 162]. Products observed were mainly guaiacol, 4-methylguaiacol, 4-propylguaiacol, 4-propenylguaiacol, and 4-ethylphenol from various technical and biomass lignin sources.

8.4.3 Polar Aprotic Solvents

Polar aprotic solvents do not directly participate in decomposing biomass but may influence the energetics and thermodynamics of reaction equilibrium or facilitate heat and mass transfer during the process. Recently, considerable attention has been paid to polar aprotic solvents due to their unique properties, making them attractive alternatives to water for hydrolysis and depolymerization of polysaccharides, as explored in the following sections.

8.4.3.1 Polar Aprotic Solvents for Saccharides

Several polar aprotic solvents, such as GVL, 1,4-dioxane, and THF, have been shown to both reduce the apparent activation energy of acid-catalyzed hydrolysis of cellulose to form monosaccharides and increase the activation energy of secondary reactions [163]. This influence on cellulose depolymerization makes polar aprotic solvents highly selective toward sugar production from carbohydrate-rich biomass. In the absence of water, polar aprotic solvents often promote high rates of cellulose solubilization and production of anhydrosugars. It has been shown that 1,4-dioxane, sulfolane, and acetone readily produce levoglucosan at high yields from cellulose [89, 164, 165]. Furthermore, it was found that the type of polar aprotic solvent can govern the yield and distribution of levoglucosan and anhydro-oligosaccharides from cellulose in the absence of catalysts [29]. As shown in Figure 8.5, it appears that high polarity solvents such as acetonitrile and GVL can

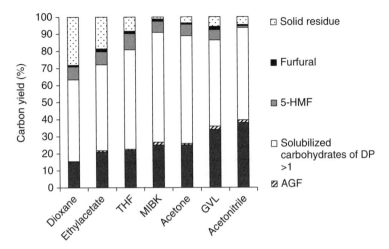

Figure 8.5 Product distribution of cellulose thermal depolymerization in various polar aprotic solvents. Source: taken from [29] with permission.

result in higher yields of solubilized carbohydrates, i.e. the sum of levoglucosan and anhydro-oligosaccharides from cellulose, with very little solid residue.

Solvent-mediated dehydration is another promising pathway to produce valuable products from polysaccharides. Cellulose can be transformed directly into 5-HMF and levoglucosenone by dehydration of its intermediate monosaccharide derivatives including glucose, fructose, and levoglucosan in a polar aprotic solvent. THF, MIBK, DMSO, DMA, and acetone are effective polar aprotic media for most acid-catalyzed dehydration reactions because they accelerate dehydration reactions while improving extraction of furanic products, thus inhibiting formation of undesired humins [166, 167]. Polar aprotic solvents such as THF can easily convert hemicellulose to furfural via dehydration of xylose in the presence of an acid catalyst [128, 168]. Cellulose can also yield levulinic acid, a high-value product of 5-HMF hydrolysis, by controlling water content and catalyst concentration in a GVL system [169].

8.4.3.2 Polar Aprotic Solvents for Lignin

Polar aprotic solvents are generally not suited for lignin depolymerization due to their inability to participate in free radical shuttling and hydrogen donation. However, a few cyclic ethers, such as 1,4-dioxane, 2-methyltetrahydrofuran, and THF, have been shown to promote solubilization and depolymerization of lignin in the presence of metal or base catalysts at high temperatures. Phenolic monomer yields observed in these solvents can range from 13 to 61% when reaction occurs at 150–250 °C [170]. These yields are surprising considering none of them are HDSs [171]. It was proposed that the oxygen center of a cyclic ether can coordinate to Lewis acids, and therefore enhance the basicity of the catalyst [170]. Major products from lignin depolymerization in cyclic ethers were phenol, guaiacol, syringol, 4-propylcyclohexanol, 4-propylcyclohexane-1,2-diol, 4-propylsyrignol, and dihydroxysinapyl alcohol.

Supercritical polar aprotic solvents can be effective in deconstructing lignin. Gosselink et al. [172] demonstrated the use of acetone and CO_2/water at supercritical conditions to convert organosolv hardwood and wheat straw lignin at 300–370 °C and 10 MPa into primarily syringol and guaiacol, respectively. In this type of solvent process, the yield of depolymerization products likely depends on the structure of lignin, which is different in woody and herbaceous biomass.

8.4.3.3 Polar Aprotic Solvents for Whole Biomass

Luterbacher et al. [173] demonstrated conversion of whole biomass (corn stover and maple wood) using a mixture of GVL, water, and dilute sulfuric acid to solubilized C_6 and C_5 carbohydrates with 69–73% yields. A semi-continuous GVL process for saccharification of biomass into fermentable sugars was proposed by Luterbacher et al. [173]. In their work, the solvent/acid mixture flowed through a fixed bed reactor loaded with biomass and the solubilized sugars were extracted using liquid CO_2 or NaCl solution. This type of sugar extraction unit for polar aprotic solvent does not require high pressure due to the very low vapor pressure of GVL. Bai et al. [25] have produced soluble monosaccharides from acid-infused switchgrass in a co-solvent of 1,4-dioxane and water at 300 °C to attain total yield of monomeric sugars up to 19.8 wt% based on the biomass fed although no information on oligosaccharides was included.

Polar aprotic solvents are good candidates for biomass pretreatment and fractionation. Organosolv extraction techniques using 1,4-dioxane, acetone, and dimethyl formamide have widely been used to delignify biomass. However, these processes are typically conducted at high temperatures (190–210 °C) for prolonged times (25–180 minutes), which has the potential to significantly alter the structure of lignin [174, 175]. Instead, lignin isolation can be achieved that preserves β-O-4 bonds through mild fractionation, qualifying as a "lignin-first" strategy. Some polar aprotic solvents, such as GVL and THF, demonstrate excellent solubilization properties and can selectively extract lignin from woody biomass. Cleavage of ether linkages and condensation of reactive aromatic species are discouraged by the mild reaction conditions (117–150 °C and 0.5–2 wt% H_2SO_4). The mild conditions for this process ensures lignin of near-native structure, offering high monomer yields during hydrogenolysis [176]. Furthermore, the cellulose-rich pulp shows high digestibility [39, 40, 65]. Unfortunately, complete delignification is still difficult.

8.4.4 Ionic Liquids

Ionic liquids have been explored for solubilizing cellulose, lignin, and whole biomass at low temperatures to produce phenolic monomers or fermentable carbohydrates [37, 153]. One advantage of ionic liquids is a significant difference in volatility between the solvent and phenolic products, which makes distillation feasible. Ionic liquids are also non-flammable and thermally stable, which makes them a good choice for biomass liquefaction. Unfortunately, ionic liquids are very costly, which demands very high recovery efficiency for these solvents [177].

8.4.4.1 Ionic Liquids for Conversion of Saccharides

Chlorine- and acetate-based imidazolium ionic liquids are reported to be highly effective solvents for decrystallization and dissolution of cellulose [57, 59, 178]. These basic anions can easily disrupt the hydrogen bonding network of cellulose and therefore require only a small amount of thermal energy to deconstruct the polymer matrix. Moreover, ionic liquids such as 1-ethyl-3-methylimidazolium facilitate hydrolysis of cellulose to glucose [179]. Rinaldi et al. [180] reported that ionic liquids can also effectively depolymerize cellulose in presence of solid catalyst. The ionic liquids [EMIM]Cl and [BMIM]Cl can produce furans from saccharides. Although use of ionic liquids are mostly centered around conversion of glucose and fructose to 5-HMF at very high yields, a few studies have shown that cellulose can be directly converted at 100–120 °C to 5-HMF at yields of 15–21% [91].

8.4.4.2 Ionic Liquids for Conversion of Lignin

Ionic liquids can depolymerize lignin at remarkably low temperatures. Depolymerization of organosolv lignin has been investigated in the presence of 1-ethyl-3-methylimidazolium trifluoromethansulfonate ([EMIM][CF$_3$SO$_3$]) at 100 °C and 8.4 MPa [181]. 2,6-Dimethoxy-1,4-benzoquinone (DMBQ) was produced at 11.5% overall yield. Other ionic liquids, including 1H-3-methylimidazolium chloride methylsulfate, butyl-1,8-diazabicyclo[5.4.0]undec-7-enium chloride, 1-ethyl-3-methylimidazolium triflate, have been used to convert lignin into small phenolic monomers [182–184].

8.4.4.3 Ionic Liquids for Converting Whole Biomass

In 2009, Binder et al. [37] produced high yields of glucose from corn stover using a chloride-based ionic liquid by gradually introducing water into the system during the hydrolysis reaction. More recently, the Joint BioEnergy group at Sandia National Laboratory has developed a number of biocompatible ionic liquids or "bionic liquids" for liquefaction of whole biomass [185, 186]. Cholinium lysinate ([Ch][Lys]) is one example, which has potential for one-pot conversion of biomass. The ionic liquid enables simultaneous removal of lignin, enzymatic hydrolysis of the polysaccharides and fermentation of the resulting sugars in one vessel without requiring separation of IL. The authors indicate that this novel process could significantly improve the economics of processing whole biomass in ionic liquids for cellulosic bioethanol production.

8.4.5 Non-Polar Solvents

The transition from non-polar to polar solvents is commonly taken to occur when the molecular dipole moment exceeds 1.6. Molecules classified as non-polar includes hydrocarbons and phenolic compounds and excludes most oxygenated solvents and water (the latter having a dipole moment of 1.87). Table 8.4 summarizes several non-polar solvents that meet this definition and the associated dipole moment of these molecules.

Non-polar solvents have historically been used in coal liquefaction due to the low oxygen content of coal [9]. Non-polar solvents are less effective at penetrating and solubilizing the polymer matrix of oxygen-rich biomass components like cellulose and hemicellulose. On the other hand, low polarity solvents are more promising for solubilizing the aromatic structure of lignin.

Regardless of the type of biomass or feedstock used, non-polar solvents offer some distinct advantages. Non-polar solvents typically have very low solubility in water, and therefore tend to phase separate in the presence of water to form separate organic and aqueous phases. This reduces the energy requirements of separating water from the product

Table 8.4 Non-polar solvents and associated dipole moment commonly found in the literature.

Molecule	Dipole moment
Naphthalene	0.00
Toluene	0.31
Tetralin	0.61
o-Cresol	1.44
Phenol	1.55

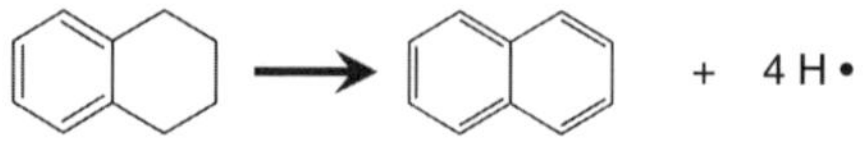

Figure 8.6 At elevated temperatures tetralin (left) releases four hydrogen atoms to form naphthalene (middle).

streams. Non-polar solvents also tend to have very low reactivity and high thermal stability, reducing decomposition even for long reaction times.

Non-polar solvents can participate as hydrogen donors to stabilize biomass degradation products. Tetralin is perhaps the foremost example of a HDS. As shown in Figure 8.6, tetralin has four hydrogen atoms that can be cleaved to form naphthalene and hydrogen atoms [187]. The hydrogen atoms attach to free radicals released during thermal deconstruction of feedstock molecules, promoting liquid yields by reducing product polymerization [188].

Non-polar HDSs are generally stable enough to donate one or more hydrogen atoms without significant alterations to the molecular structure. In some cases the expended molecule can then be separated from the product stream and hydrogenated for further use as a HDS. This is in contrast to most polar HDSs, such as ethanol and formic acid, which are altered to the extent that re-use is difficult. Some non-polar solvents have been demonstrated to catalyze the cleavage of heteroatoms, particularly oxygen and nitrogen [189]. Lastly, non-polar solvents, particularly hydrocarbons, are generally soluble in crude petroleum and most refinery cuts. Therefore, product streams from hydrocarbon solvent liquefaction can often be easily mixed with petroleum for upgrading or end use.

8.4.6 Influence of Process Conditions

8.4.6.1 Temperature

Reaction temperature is one of the most important variables for controlling solvent liquefaction. Biomass liquefaction primarily occurs over the range of 150–350 °C [12]. As reaction temperature progresses through this temperature range, bio-oil yields tend to increase, but eventually decrease as solid and gas yields increase. In hydrothermal liquefaction of lignocellulose, for example, the following three behaviors can be observed as a function of temperature: biomass hydrolysis, bio-oil formation, and bio-oil decomposition. Biomass is first hydrolyzed to small molecular weight carbohydrates (glucose and fructose) by 250 °C. Below 280 °C, incomplete decomposition of biomass components keeps bio-oil yield low,

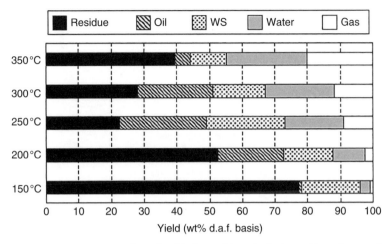

Figure 8.7 Effect of temperature on bio-oil yield obtained from hydrothermal liquefaction of Eucalyptus HTL. Source: taken from [190] with permission.

as shown in Figure 8.7. In fact, although hemicellulose can be completely solubilized at 250 °C, cellulose and lignin are only 22% and 60% solubilized, respectively, at this temperature [191]. Beyond 250 °C, degradation of cellulose and hemicellulose starts to dominate and bio-oil yield increases significantly [192]. At 300 °C, lignin begins to depolymerize very rapidly. Nevertheless, depolymerization is accompanied by condensation of lignin-derived products [193]. Maximum bio-oil yield occurs near 300 °C. Temperatures higher than 300 °C generally promote degradation of carbohydrate products and condensation of lignin products, which in turn decreases the bio-oil yield and increases solid and gaseous products [194].

8.4.6.2 Pressure

In the subcritical regime, the main role of pressure is to maintain solvents as liquids. Given that most liquids are essentially incompressible, pressure has little effect on liquid density or other liquid properties that might affect reactions. Gibbins et al. [195] found that pressure had a negligible effect on the conversion of coal in tetralin over the wide range of 5 to 10 MPa. However, through manipulation of pressure, the selective vaporization of water, light oxygenates, and other undesirable products can be achieved, increasing the concentration of desirable products in the solvent. In some instances, however, increasing the reaction pressure to maintain a higher solubility of light gases has been beneficial. Appell et al. [196] found that increasing the pressure from 6.9 to 12.4 MPa improved oil yields and conversion of cellulose in water at 250 °C by about 15 and 10 wt%, respectively. These improvements were attributed to increases in the initial charge pressure of carbon monoxide.

8.4.6.3 Residence Time

Reaction time is an important parameter in controlling product composition and yield. The example of converting cellulose into 5-HMF is illustrated in Figure 8.8. Solvent liquefaction

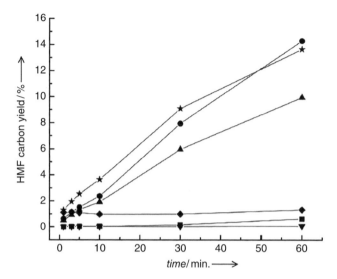

Figure 8.8 Time-dependent yields of 5-HMF from cellulose in various protic and aprotic solvent at 170 °C with 5 mM sulfuric acid as catalyst. Source: taken from [197] with permission.

reactions can occur over seconds to hours depending upon desired product distributions. The optimal residence time for desired products depends upon solvent, catalyst, feedstock, and, especially, reaction temperature. For example, cellulose hydrolysis in near-critical water (320 °C, 25 MPa) occurs in less than 10 seconds while cellulose dehydration at 170 °C can take up to 60 minutes.

Yields typically show non-monotonic behavior with time. As shown in Figure 8.9, cellulose hydrolyzes to monosaccharides and oligosaccharides following distinctly different time evolutions. While oligosaccharide production dominates for the first four seconds, it is eventually overtaken by monosaccharide production. Total carbohydrate yields also

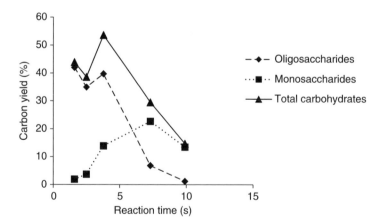

Figure 8.9 Product yields versus reaction time for cellulose hydrolysis in near critical water (320 °C, 25 MPa). Source: reproduced from [41].

diminish after the first four seconds of reaction, possibly due to secondary reactions such as dehydration of monosaccharides to furanic derivatives [197]. Thus, control of reaction time is critical for maximizing yield of desired products.

8.4.6.4 Feedstock-to-Solvent Ratio

Higher solids loadings are always desired for improving process throughput and economics, as long as it does not adversely affect the desired product yields. An example of a positive outcome of increased solids loading is illustrated in Figure 8.10a for hydrothermal liquefaction of microalgae [198]. A counter example is shown in Figure 8.10b for solvent liquefaction of lignin in a mixture of o-cresol and HDS tetralin. Yields of guaiacol decreased as solids loading increased, especially for high HDS blend ratios [47].

8.4.6.5 Water Content

High moisture content in feedstock can have either a positive or negative impact on liquid yields. In many biomass liquefaction processes, feedstock moisture acts as co-solvent to enhance solubilization of products, such as saccharides. Water can also promote decomposition of certain kinds of feedstocks. For example, the rate of cellobiose conversion in

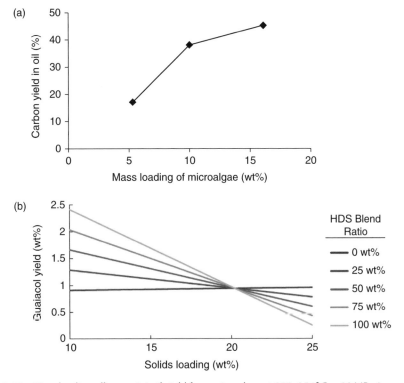

Figure 8.10 Mass loading effect on (a) oil yield from microalgae at 350–354 °C, ~20 MPa (reproduced) and (b) guaiacol yield from lignin liquefaction in tetralin. Source: taken from [47, 198] with permission.

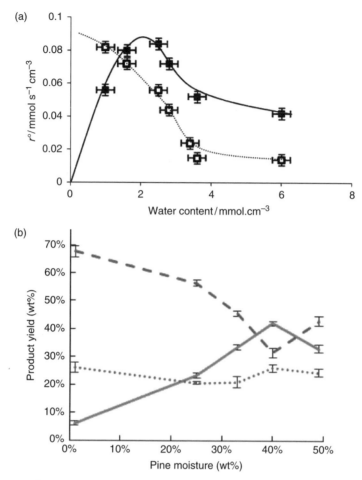

Figure 8.11 (a) Effect of water on rate of cellobiose hydrolysis (solid squares) and glucose dehydration (hollow squares) in ionic liquid; and (b) effect of water on liquefaction of pine in tetralin at 280 °C, where dashed line is liquid yield, solid line is solids yield, dotted line is gas yield. Source: taken from [199, 200] with permission.

ionic liquid increases with increasing water content, leading to enhanced glucose production via hydrolysis, as shown in Figure 8.11a. An example of the deleterious effect of water is shown in Figure 8.11b, where water enhances glucose dehydration to 5-HMF during solvent liquefaction of pine in tetralin [199]. Conversely, the negative effect of water has also been evaluated for tetralin-assisted liquefaction [200].

8.4.6.6 Catalysts

Table 8.5 summarizes the use of catalysts in liquefaction.

Base catalysts have been used in solvent liquefaction of whole biomass and lignin. Addition of alkaline or metal catalysts can significantly improve the liquid oil yields from

Table 8.5 Use of catalysts in solvent-mediated biomass conversion processes.

Feedstock	Solvent/ condition	Catalyst	Product	Catalyst concentration	Product yield	Ref.
Woody biomass	Water, 280 °C, 15 min	K_2CO_3	Bio-oil	0.235 M 0.47 M 0.94 M	17.8 wt% 35.9 wt% 33.7 wt%	[201]
Potato peel	Water, 150 °C	H_3PO_4	Glucose	5.0% 10.0%	25.1 wt% 46.0 wt%	[75]
Cellulose	[BMIM]Cl[a], 160 °C, 10 min	None $CrCl_3$/LiCl $CrCl_3$	5-HMF	Zero 50/50 mol% 100 mol%	0.9% 61.9% 52.6%	[202]
Alcell® hardwood lignin	Iso-propanol/ formic acid, 400 °C, 240 min	None Ru/C	Phenolic oil	Zero 1.4 wt% on feed	18.2 wt% 71.2 wt%	[203]
Waste poultry fat	Methanol, 62 °C, 60 min	NaOH	Bio-diesel	0.6 wt% 1.0 wt% 1.4 wt%	60 wt% 88 wt% 85 wt%	[204]

[a] [BMIM]Cl is the abbreviation for 1-butyl-3-methylimidazolium chloride.

biomass and technical lignin. Additionally, use of noble metal catalysts can selectively produce targeted phenolic monomers from lignin depolymerization.

Acid-based homogeneous and solid catalysts have been widely used for saccharification, thermal depolymerization, and dehydration of carbohydrates in biomass. Usually, increasing concentration of mineral acid leads to higher yields of monomeric sugars due to an enhanced hydrolysis rate of cellulose and hemicellulose. However, high acid concentrations can also increase dehydration of monosaccharides. Selection of optimum acid catalyst concentration is important in maximizing sugar yields. It is important to note that the type of solvent system could shift the optimum of monosaccharide production significantly due to variable kinetics of both saccharification and dehydration reactions in different solvents [28, 128].

8.5 Solvent Effects

In order to facilitate biomass decomposition to liquid products, it is important to identify physical and chemical solvent properties that influence liquefaction reactions. This section outlines the major physical and chemical effects in solvent liquefaction.

8.5.1 Physical Effects

Although solvent selection is based primarily on its influence on reaction chemistry, thermophysical properties should also be taken into consideration. Because solvent is the primary constituent of the process, its physical properties will be the major determinant of the physical behavior of the system. Volatility of a solvent relative to products affects their separation. Thermal conductivity and viscosity influence heat and mass transfer in the reactor, determining the extent that a reaction is isothermal and homogeneous. Specific heat determines how much thermal energy is required to reach reaction temperature.

8.5.2 Solubility Effects

8.5.2.1 Solubility Parameters

High solubility or degree of interaction of biomass polymer with a solvent can contribute to faster reaction and higher degree of depolymerization [205]. Comparing the solubility parameter of solvent with that of reactants (and products) is a good indicator of the degree of interaction between the two [206–208]. Liquids with similar solubility parameters will be miscible, and reactants (and products) will dissolve in solvents with similar solubility parameters. Solubility parameters are sometimes called cohesion energy parameters, as they are derived from the energy required to convert a liquid to a gas. The energy of vaporization is a direct measure of the total (cohesive) energy holding molecules together in a solvent.

8.5.2.1.1 Hildebrand Solubility Parameter

Hildebrand and Scott [209] introduced *solubility parameter* δ in 1950. It is defined as the square root of the cohesive energy density:

$$\delta = \sqrt{\frac{E}{V}} \tag{8.1}$$

where E is the enthalpy of vaporization and V is the molar volume of the solvent and is measured in units of $MPa^{1/2}$.

For a solute to dissolve spontaneously in a solvent at a given temperature T, the Gibbs free energy, ΔG, of mixing must be negative. Gibbs energy can be expressed in terms of the enthalpy of mixing (ΔH) and entropy of mixing (ΔS):

$$\Delta G = \Delta H - T\Delta S \tag{8.2}$$

Since entropy of mixing is always positive and T (in K) is positive, ΔH must be smaller than $T\Delta S$ for solubilization to occur. Hildebrand proposed the following equation for ΔH,

$$\Delta H = \varphi_1 \varphi_2 V_M (\delta_1 - \delta_2)^2 \tag{8.3}$$

where φ_1 and φ_2 are volume fractions of solvent and solute in the solution, respectively, V_M is the volume of the mixture, and δ_1 and δ_2 are the solubility parameters of solvent and solute, respectively. Clearly, ΔH would be small or zero when the solubility parameters are of similar values, in which case the Gibbs free energy is negative and the solute would be soluble in the solvent.

The solubility parameters of both reactant and products can affect the extent of reaction during solvent liquefaction. For example, the Hildebrand solubility parameter can be used to determine the relative solubilization of lignin-derived phenolic monomers in various solvents. Solvation of phenolic products is critical in discouraging interaction between reactive products from lignin. The difference in the Hildebrand solubility parameter for solvent and solute is typically less than $3\ MPa^{1/2}$ for a good solvent [210]. Table 8.6 lists the estimated Hildebrand solubility parameters at 300 °C for several phenolic monomers commonly derived from solvent liquefaction of lignin. *o*-Cresol and phenol are excellent solvents for solubilizing phenolic monomers. Tetralin is another solvent often used to solubilize lignin. However, the difference in the Hildebrand solubility parameter for phenol

Table 8.6 Hildebrand solubility parameters of solvents and common phenolic monomers at reaction temperature of 300 °C.

Compound	Hildebrand solubility parameter (MPa$^{1/2}$)
Phenol	20.99
o-Cresol	19.54
Guaiacol	19.17
Syringol	18.70
p-Xylenol	18.64
2-Ethylphenol	18.46
Cresol	18.26

(20.99 MPa$^{1/2}$) and tetralin (16.70 MPa$^{1/2}$) exceeds the recommended maximum difference of 3 MPa$^{1/2}$. Accordingly, it was found that the solubilization of a wide range of phenolic monomer products of lignin in tetralin was low [47].

The use of the Hildebrand solubility parameter is limited to regular solutions, which were defined by Hildebrand and Scott [211] as non-ideal solutions whose entropy of mixing is the same as that of an ideal solution. Practically speaking, the use of the Hildebrand solubility parameter does not account for associative molecular interactions such as polar and hydrogen bonding interactions. Thus, this model is well suited for non-polar solvents such as hydrocarbon and phenolic solvents (e.g. pentane, heptane, cyclohexane, toluene, tetralin, and o-cresol).

8.5.2.1.2 Hansen Solubility Parameter

The Hansen solubility parameter (δ_{Tot}) is based on the total energy of vaporization of a liquid that may arise from three types of atomic and molecular interactions. These contributions are (atomic) dispersion forces, (molecular) permanent dipole–permanent dipole forces, and (molecular) hydrogen bonding (electron exchange). These interaction forces give rise to three individual parameters for calculating the total solubility parameter: dispersive solubility parameter (δ_D), polar solubility parameter (δ_P) and the hydrogen bonding solubility parameter (δ_H). The Hansen total solubility parameter is calculated as:

$$\delta_{Tot} = \sqrt{(\delta_D)^2 + (\delta_P)^2 + (\delta_H)^2} \tag{8.4}$$

For highly polar solvents, such as acetonitrile, sulfolane, GVL, and DMSO, the contribution of δ_P can be significant, which in turn indicates good solubilization of highly polar biomass components. Similarly, polar solvents with hydrogen bonding capabilities, such as methanol, ethanol, and water, generally show significantly higher polar solubility parameter and hydrogen bonding solubility parameter. These solvents could potentially participate in strong hydrogen bonding interactions in addition to polar–polar interactions with biomass. Both polysaccharides and lignin have several hydroxyl groups in their structure, which could easily interact with the solvent via interactive forces of hydrogen bonding and dipole moments.

In some cases the solubility parameters calculated at ambient conditions will be significantly lower than for the biomass components. For instance, solubility parameters of commonly employed polar aprotic solvents used in cellulose depolymerization are in the range of 17.0–26.3 MPa$^{1/2}$ at room temperature and atmospheric pressure, which is significantly lower than the solubility parameter for cellulose (39.3 MPa$^{1/2}$). Nevertheless, these solvents can produce very high yields of solubilized products from cellulose at reaction conditions [29], the result of solubility parameters being strong functions of temperature and pressure. Individual and total solubility parameters for a range of polar aprotic solvents were estimated using thermophysical properties of the solvents at hot and pressurized solvent conditions [29]. Total solubility parameters of the tested solvents were within 25.7–33.8 MPa$^{1/2}$, as shown in Table 8.7, which approaches the solubility parameter of cellulose.

Generally, increasing temperature reduced solubility parameters while increasing pressure increased solubility parameters. The individual solubility parameters are functions of temperature and pressure according to the following equations:

$$\delta_D = \frac{\delta_{Dref}}{\left(\frac{v_{ref}}{v}\right)^{-1.25}} \tag{8.5}$$

$$\delta_P = \frac{\delta_{Pref}}{\left(\frac{v_{ref}}{v}\right)^{-0.5}} \tag{8.6}$$

$$\delta_H = \frac{\delta_{Href}}{\exp\left(-1.32 \times 10^{-3}(T_{ref} - T) - \ln\left(\frac{v_{ref}}{v}\right)^{0.5}\right)} \tag{8.7}$$

where δ_{Dref}, δ_{Pref}, and δ_{Href} represent the values of dispersion parameter, polar interaction parameter, and hydrogen bonding parameter at T_{ref} and P_{ref}, respectively. v_{ref} and v are equal to the molar volumes of the solvent at reference and reaction conditions, respectively. The reference temperature is 25 °C and to the reference pressure is 0.1 MPa.

8.5.2.2 Kamlet-Taft Parameters

The Kamlet-Taft parameter is often used to characterize solvents that have specific acidic and basic interactions with biomolecules. This parameter expresses combined solubility

Table 8.7 Solubility parameters of the polar aprotic solvents at near or supercritical state [29].

Solvent	δ_D (MPa$^{1/2}$)	δ_P (MPa$^{1/2}$)	δ_H (MPa$^{1/2}$)	δ_{Tot} (MPa$^{1/2}$)
1,4-Dioxane	26.6	2.1	0.5	26.7
Ethyl acetate	27.0	6.6	0.4	27.7
THF	27.7	7.0	0.4	28.6
MIBK	24.6	7.4	0.2	25.7
Acetone	27.5	13.1	0.4	30.4
Acetonitrile	25.6	22.1	0.3	33.8
GVL[a]	25.6	18.7	0.4	31.7

[a]Values approximated from gamma-butyrolactone (GBL) due to unavailability of actual data for GVL [212].

effects arising from polarity and acidity and basicity associated with hydrogen bonding [56, 213], which can be determined by UV-Vis spectra of specific dyes in the solvent. Ionic liquids are example of a class of solvents where Kamlet-Taft could be highly useful as a predictive model for solvent performance [214].

8.5.3 Structural Effects

Pretreatment via solvent processing, by changing the structure of biomass, can increase the susceptibility of cellulose to subsequent deconstruction by enzymes or thermochemical processes. Water at subcritical conditions (300 °C, 27.6 MPa) was found to increase enzymatic digestibility of microcrystalline cellulose [215]. It was proposed that formation of cellulose I-II polymorphs and generation of surface crack and trenches contributed to increased reactivity of enzymes. Luterbacher et al. [39] demonstrated that both water and GVL initially increase the crystallinity index (CrI) of cellulose by removing a fraction of amorphous cellulose by solvent-assisted hydrolysis. For longer times (40 minutes at 448 K), the CrI of cellulose decreases in the presence of GVL, indicating the disruption of crystalline parts of cellulose.

Significant structural changes are also observed during solvent liquefaction of lignin. Nguyen et al. [40] have developed a biomass pretreatment using THF and water as co-solvents that solubilize lignin and enhance subsequent production of fermentable sugars. Using molecular dynamics simulation, it was shown that lignin was preferentially solvated by THF [216]. This in turn shifted the equilibrium configurational distribution of lignin from a crumpled globule to coil over a wide range of temperature. It was further revealed that water was a poor solvent for lignin while THF–water mixtures equalized the strength of solvent–lignin and lignin–lignin interactions. Lignin polymer tends to aggregate much less in THF–water mixtures, thus promoting high degree of solubilization of lignin. This mechanism also may help in reducing the recalcitrance of lignocellulosic biomass.

THF is more effective than water in approaching the surface of lignin at high concentration [216]. Steric hindrance of water molecules in approaching lignin prevents formation of lignin–lignin contacts and intra-polymeric hydrogen bonds that lead to collapsed coil structure of the biopolymer. At high ratio of THF to water, the end-to-end distances of lignin polymer monotonically increases as a function of number of separated monomers, indicating an extended chain conformation of the polymer. With pure water as the pretreatment medium, this end–end distance remains plateaued with respect to the number of monomers, leading to a crumpled and densely packed structure of lignin in water.

As a whole, solvent can contribute significantly to the alteration of biomass structure at the molecular level. In fact, certain ionic liquids could completely disrupt the structure of plant cell wall after pretreatment [217]. This could have resulted from several physical effects between the solvent and biomass components, leading to disruption of the polymeric matrix of lignocellulose.

8.5.4 Chemical Effects

Solvents can have strong influences on the thermodynamic equilibrium of a chemical reaction. The extent of this effect depends on the type and degree of interactions between the solvent and the chemical species present in the reaction media. These interactions may lead to significant alterations to the free energy of solvation of reactant, product, or even the catalyst and transition state. This can subsequently govern the direction of chemical equilibrium and be used to influence reaction rates, yields, and selectivities of target products.

8.5.4.1 Solvent Effects on Reaction Energetics

The apparent activation energy of representative model compounds have been determined for several solvents. Mellmer et al. [28] reported that polar aprotic solvents could remarkably increase the rate of monosaccharide formation from cellobiose compared to a reaction medium of pure water. This was attributed to reduction in the activation energy for acid-catalyzed hydrolysis of cellobiose from 131 to 81 kJ/mol by using GVL as the primary solvent in comparison to pure water. Figure 8.12 illustrates that increasing the proportion of GVL in water resulted in a significant drop in the activation energy of hydrolysis of cellobiose to glucose. Glucose was produced at higher rate with increasing GVL content. This kind of solvent effect is also evident in thermal depolymerization of cellulose in polar aprotic solvents in absence of both water and acid catalysts. Kawamoto et al. [89] and Ghosh et al. [29] both demonstrated that sulfolane, THF, acetonitrile, and GVL can reduce activation energy for the depolymerization of crystalline cellulose to 82–111 kJ/mol. These values are significantly lower than that reported for pyrolytic depolymerization of cellulose in the absence of either a solvent or direct hydrolysis of crystalline cellulose [218–220]. Furthermore, Ghosh et al. [29] found an inverse correlation between solvent polarity and the activation energy of cellulose depolymerization. The authors found that increasing solvent polarity decreased activation energy. A similar correlation between yields of levoglucosan and polar solubility parameter of the solvents was also observed as depicted in Figure 8.13.

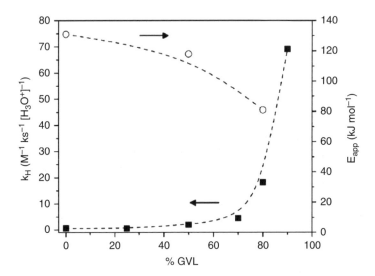

Figure 8.12 Rate constant (square; left axis) and apparent activation energy (circle; right axis) of cellobiose hydrolysis versus increasing GVL content in GVL/water mixture performed in presence of sulfuric acid as catalyst. Source: taken from [28] with permission.

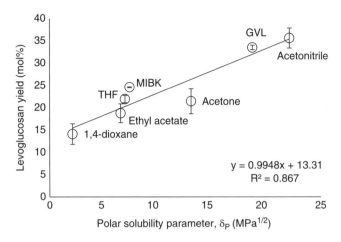

Figure 8.13 Relationship of levoglucosan yields with polar solubility parameter of aprotic solvents from non-catalytic processing cellulose at 350 °C for 8–16 minutes. Source: taken from [29] with permission.

The type of solvent seemed to play a major role in driving chemical equilibrium toward monomer production by changing the activation energy of the primary reaction involved.

Solvents can influence the activation energy of secondary reactions that convert biomass-derived products into undesirable products. Mellmer et al. [28] determined that a 4 : 1 GVL and water mixture had an activation energy of 138 kJ/mol compared to 135 kJ/mol in pure water for acid-catalyzed glucose dehydration. This may explain the relative stability of glucose in GVL and its higher degradation rates in water. A summary of the activation energies of various biomass feedstocks in aqueous and non-aqueous organic solvents is given in Table 8.8 to demonstrate the effect of solvents on energetics of biomass conversion.

8.5.4.2 Proposed/Simulated Solvent Effects on Reaction Thermodynamics

8.5.4.2.1 Solvent–Reactant Interactions

Solvents can interact with the molecules involved in biomass conversion by creating solvation shells around molecules. This may facilitate or prevent interactions between two reacting species of biomass and thus affect the reaction rate. Vasudevan and Mushrif [222] investigated the hydrogen bonding interactions of glucose in water in the presence of several organic solvents. The authors suggested that when organic solvent was used with water, the mobility of glucose molecules was hindered due to a higher degree of interaction with the organic solvent. For example, increasing the proportion of THF in water resulted in increased interaction between THF and glucose, which pushed water molecules away from glucose. This is shown by a volumetric spatial density map of glucose in solvent mixtures (Figure 8.14). Thus, stronger hydrogen bonds could be formed between the organic solvent and C_3 and C_4 positions of glucose, causing less chance of protonation of these carbon atoms with water. The authors argued that this phenomenon leads to reduced humin formation, which is believed to start by protonation of the above carbon atoms. The

Table 8.8 Apparent activation energy of biomass reactions in various solvents.

Reactant	Reaction	Product	Solvent	Apparent activation energy (kJ/mol)	Ref.
Cellobiose	Hydrolytic depolymerization	Glucose	Water	131	[28, 199]
			GVL : water 4 : 1	81	
			IL	84	
Cellulose (crystalline)	Hydrolytic depolymerization	Glucose, C_6 oligosaccharides	Water	170–190	[73, 199, 221]
			IL	92–96	
Cellulose (crystalline)	Depolymerization	Levoglucosan, C_6 anhydro-oligosaccharides	No solvent	126–251	[29, 89, 218–220]
			THF	111	
			Acetonitrile	82	
			GVL	85	
			Sulfolane	88	
Glucose	Dehydration	5-HMF	Water	135	[28]
			GVL:Water 4 : 1	138	
Xylose	Dehydration	Furfural	Water	145	[128]
			GVL:Water 4 : 1	114	

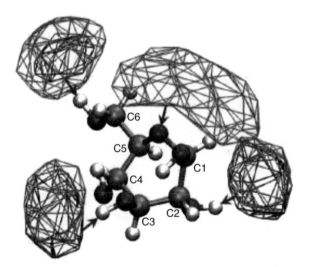

Figure 8.14 Volumetric spatial density maps (isosurfaces) of the time averaged distribution of THF (blue surfaces) and water (red surfaces) around glucose molecules in mixtures of 90/10 wt% THF and water. Source: taken from [222] with permission. See the plate section for a color representation of this figure.

net effect was a lower rate of sugar condensation reactions and increased sugar yields in organic solvents compared to pure water.

8.5.4.2.2 Solvent–Catalyst Interactions

Researchers have also explored the role of catalysts and transition states in solvent–solute interactions during biomass deconstruction. In a study by Phan et al. [223], the reaction rate of glycosidic bond hydrolysis was enhanced by the presence of Br^- and Cl^- ions. This effect was more pronounced in 1,4-dioxane/water systems. They proposed that hydrolysis occurred by nucleophilic attack of the anions. They further attributed the increased yields of glucose to the hindered solvation of anions in 1,4-dioxane and water. As the proportion of 1,4-dioxane in water increased, the extent of solvation of anions decreased, and therefore the probability for nucleophilic attack of the ions was increased.

The Gibbs free energy of solvation of ions for organic solvents has been compared to water. It has been shown that polar aprotic solvents, such as acetonitrile, can solvate protons more easily than water due to a lower Gibbs energy of proton solvation [224]. This can lead to higher proton reactivity in polar aprotic solvents compared to water, and thus show enhanced rates of acid-catalyzed reactions. In fact, this hypothesis is in good agreement with the enhanced hydrolysis rate of cellobiose in GVL. Mellmer et al. [128] proposed that destabilization of the acidic proton in these solvents relative to the protonated transition state leads to accelerated rates of acid-catalyzed reactions. This solvation behavior of acidic protons predicts that the activation energy barrier in water (ΔG_{act}) is higher than ΔG_{act} in a polar aprotic solvents in the presence of a stronger acid while the above difference in ΔG_{act} should be minimal in a weak acid environment. This whole effect is shown in the free energy diagram in Figure 8.15.

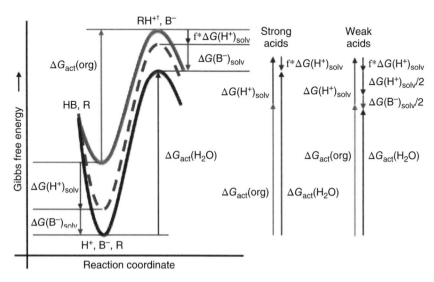

Figure 8.15 Energy diagram for destabilization of proton in water and organic solvent. Source: taken from [128] with permission. See the plate section for a color representation of this figure.

Table 8.9 Partition coefficient and selectivity of 5-HMF in various solvents.

Feedstock	Biphasic solvent	Partition coefficient	Selectivity (%)	Conversion (%)	Ref.
Fructose	Water	0.0	57	89	[227]
	Water/MIBK	0.9–1.1	65–91	47–73	[225]
	Water/THF	7.1	83	53	[166]
	Water/acetone	3.6	79	62	[166]
	Water/1-propanol	3.9	80	73	[166]
	Water/2-butanol	3.7	85	67	[166]
Glucose	Water	0.0	51	21	[228]
	Water/MIBK	–	85	90	[229]
	Water/THF	–	72	79	[230]
	Water/2-butanol	–	83	70	[231]

8.5.4.2.3 Solvent–Product Interactions

Many researchers have incorporated an extraction solvent directly in the reaction medium to increase the selectivity of a target product during biomass processing. For example, selectivity for 5-HMF in the conversion of biomass-derived saccharides is reported to be significantly higher in a biphasic reaction system consisting of a polar aprotic solvent such as DMSO, MIBK, or THF as the extraction solvent and water as the reaction solvent compared to water as a monophasic solvent [166, 225]. This effect is likely due to enhanced stabilization of 5-HMF in non-aqueous solvents. Density functional theory calculations revealed that DMSO preferentially solvates the 5-HMF carbonyl group. This can reduce susceptibility to nucleophilic attack and minimize undesirable reactions that would lead to the formation of polymeric humin species from 5-HMF [226]. As shown in Table 8.9, the enhanced selectivity of 5-HMF in polar organic solvents during fructose and glucose dehydration can be attributed to high partition coefficients of the product in these solvents.

8.6 Engineering Challenges

Solvent liquefaction is a promising technology for the conversion of biomass to higher-value products. However, the technology has rarely been developed beyond laboratory scale, and is generally at a low technology readiness level because of several engineering challenges, which are described below.

8.6.1 High Pressure Feed Systems

Feeding solids into high pressure reactors is a significant challenge to the development of practical solvent liquefaction systems, especially at high solids loadings. The higher the solids loading, the smaller the reactor, the higher the concentration of products, and the smaller the solvent inventory for the plant, which translates into reduced capital and operating expenses. Researchers working on the PERC process in Albany, OR, found that wood chips had to be diluted to 10–12 wt% to avoid clogging of the pumping system [232]. Wet milling procedures and other pretreatment methods can reduce biomass particle size and help densify the feedstock, but these operations can be expensive [233, 234].

For decades, the pulp and paper industry has used special "chip pumps" capable of pumping woodchips up to several centimeters in size. However, this is at considerably reduced pressures compared to what is required for most solvent liquefaction systems. Development of chip pumps to reach greater pressures is necessary to make this option viable.

Extruder systems have been explored at the pilot scale for feeding biomass slurries into high pressure reactors [6, 235, 236]. Although these machines can perform well at high pressures and temperatures, they require stable and predictable feedstock size, consistency, and solids concentration. In the polyolefin and food manufacturing industries, where extruders are often used, these tight feedstock controls are achievable, but it will be challenging for biomass-processing facilities to profitably meet these stringent conditions.

Progressive cavity, piston, and plunger pumps have also been explored in the literature to feed biomass slurries into high pressure reactors [237–239]. These pumps are often used in series, with the progressive cavity pump providing the suction pressure needed for the high-pressure piston pump to deliver the slurry to pressures up to 8000 psig and higher. In general these pumps have limited scalability, and are considerably more expensive than most other types of pumps. Furthermore, both piston and plunger pumps deliver pulsed or discontinuous flow since the cylinder has to refill periodically. Pulseless flow requires operating at least two pumps in tandem, which further adds to the cost.

8.6.2 Separation of Solid Residue

Separation of solid residue from solvent and liquefied products is considerably more difficult than for "dry" thermochemical processes like pyrolysis and gasification where solids can be removed with inertial separators such as cyclones or baffles. In some cases, density differences are enough to allow solid residues to settle out of solution for easier recovery. More commonly, solids must be removed via filters, screens, or possibly hydrocyclones. Continuous operation of these removal methods can be challenging due to high likelihood for plugging.

Biomass residues are often highly porous, retaining significant amounts of bio-oil and solvent. In this case, removal of liquid from recovered solid residues using mechanical pressing or solvent extraction might be important for economical operation of a solvent liquefaction plant.

8.6.3 Solvent Recovery and Recycle

The high cost of solvents is one of the most significant barriers to economical operation of a solvent liquefaction biorefinery. Table 8.10 illustrates that it would cost approximately $3.9 M annually to provide fresh solvent to a 1000 ton/day hydrothermal liquefaction biorefinery operating at 75% solvent (water) recovery and recycle. This does not factor in wastewater disposal and treatment costs that could potentially add another $0.10/gal or $19.8 M annually. Even after recovering and recycling 75% of the solvent, the fresh solvent cost is still an onerous component to biorefinery economics. Of course, processes employing non-aqueous solvents would simply exacerbate the cost of solvent recovery and recycling compared to water.

Despite the significant economic incentives, solvent recovery and recycle has received little attention in the literature [173, 240–244]. Recycle streams can accumulate products

Table 8.10 Annual fresh solvent cost to operate a 1000 ton/day hydrothermal liquefaction biorefinery assuming 75% effective solvent recovery and recycle and 330 day per year operation.

Nameplate capacity	1000	ton/day
Biomass loading	10%	wt%
Solvent recovery	75%	wt%
Fresh solvent demand	600 962	gal/day
Solvent unit cost	$0.02	USD/gal
Daily solvent cost	$12 019	USD
Annual solvent cost	$3 966 346	USD

or impurities in the solvent which may adversely affect reaction chemistry. Purification of recycle streams through thermal separation technologies, such as distillation or flash separation, may have an impact on bio-oil quality due to rapid heating. van Rossum et al. [244] found that recycle oils continued to increase in viscosity and molecular weight over time, approaching that of vacuum residual oils. However, van Rossum suggests that in some instances, recycle oils were more effective than the starting solvent at solubilizing feedstocks.

It is also conceivable to have solvent liquefaction systems in which the solvent is produced either directly or indirectly from the feedstock. This would reduce the demand for solvent recovery assuming a net production of solvent molecules in the process. Similarly, if a petroleum cut were selected as solvent, co-location of the solvent liquefaction biorefinery and petroleum refinery would allow for on-site production of the solvent and therefore reduce solvent costs.

8.7 Conclusions

Solvent liquefaction is a promising thermochemical platform for production of renewable fuels and chemicals from biomass. The unique physicochemical effects of solvent liquefaction make it a highly versatile platform compared to other thermochemical processes for biomass conversion. Solvent interactions with biomass can improve reaction energetics and enable selective production of target molecules. The solvent can further act as a diluent or capping agent for suppressing undesirable secondary reactions. Solvent liquefaction has been observed to significantly deoxygenate the liquid product, an advantage when the goal is to produce hydrocarbon fuels. The special case of hydrothermal processing, in which water is the solvent, is particularly attractive for processing wet biomass. Despite processing flexibility and excellent quality of the resulting products, solvent liquefaction has several technical barriers to overcome before commercialization. These include feeding biomass under high pressure, separation of solid residue, and solvent recovery and recycle, problems that are best addressed through pilot-scale evaluations in continuous reactor systems.

References

1. Stranges, A.N. (1984). Friedrich Bergius and the rise of the German synthetic fuel industry. *Isis* **75** (4): 643–667.
2. Stranges, A.N. (1983). Synthetic petroleum from coal hydrogenation: its history and present state of development in the United States. *Journal of Chemical Education* **60** (8): 617.

3. Chornet, E. and Overend, R.P. (1985). Biomass liquefaction: an overview. In: *Fundamentals of Thermochemical Biomass Conversion*, 967–1002. Springer.
4. Ergun, S. (1980). *Biomass Liquefaction Efforts in the United States*. Lawrence Berkeley Laboratory.
5. White, D., Schott, N., and Wolf, D. (1989). Experimental study of an extruder-feeder for biomass direct liquefaction. *The Canadian Journal of Chemical Engineering* **67** (6): 969–977.
6. White, D.H., Wolf, D., and Zhao, Y. (1987). Biomass liquefaction utilizing extruder-feeder reactor system. *American Chemical Society, Division of Fuel Chemistry Preprints* **32** (2): 106.
7. Elliott, D.C., Beckman, D., Bridgwater, A.V. et al. (1991). Developments in direct thermochemical liquefaction of biomass: 1983–1990. *Energy & Fuels* **5** (3): 399–410.
8. Neavel, R.C. (1976). Liquefaction of coal in hydrogen-donor and non-donor vehicles. *Fuel* **55** (3): 237–242.
9. Neavel, R.C., Knights, C.F., and Schulz, H. (1981). *Exxon Donor Solvent Liquefaction Process [and Discussion]*. *Philosophical Transactions of the Royal Society of London A: Mathematical, Physical and Engineering Sciences* **300** (1453): 141–156.
10. Guo, Y., Yeh, T., Song, W. et al. (2015). A review of bio-oil production from hydrothermal liquefaction of algae. *Renewable & Sustainable Energy Reviews* **48**: 776–790.
11. Pham, M., Schideman, L., Scott, J. et al. (2013). Chemical and biological characterization of wastewater generated from hydrothermal liquefaction of Spirulina. *Environmental Science & Technology* **47** (4): 2131–2138.
12. Akhtar, J. and Amin, N.A.S. (2011). A review on process conditions for optimum bio-oil yield in hydrothermal liquefaction of biomass. *Renewable and Sustainable Energy Reviews* **15** (3): 1615–1624.
13. Boushy, A.E., Klaassen, G., and Ketelaars, E. (1985). Biological conversion of poultry and animal waste to a feedstuff for poultry. *World's Poultry Science Journal* **41** (2): 133–145.
14. Mahmood, T. and Elliott, A. (2006). A review of secondary sludge reduction technologies for the pulp and paper industry. *Water Research* **40** (11): 2093–2112.
15. Xiu, S., Shahbazi, A., and Wang, L. (2016). Co-liquefaction of swine manure with waste vegetable oil for enhanced bio-oil production. *Energy Sources, Part A: Recovery, Utilization, and Environmental Effects* **38** (4): 459–465.
16. Yin, S., Dolan, R., Harris, M. et al. (2010). Subcritical hydrothermal liquefaction of cattle manure to bio-oil: effects of conversion parameters on bio-oil yield and characterization of bio-oil. *Bioresource Technology* **101** (10): 3657–3664.
17. Zastrow, D.J. and Jennings, P.A. (2013). *Hydrothermal Liquefaction of Food Waste and Model Food Waste Compounds*. Florida Institute of Technology.
18. Elliott, D.C., Biller, P., Ross, A.B. et al. (2015). *Hydrothermal liquefaction of biomass: developments from batch to continuous process*. *Bioresource Technology* **178**: 147–156.
19. Sambusiti, C., Bellucci, M., Zabaniotou, A. et al. (2015). *Algae as promising feedstocks for fermentative biohydrogen production according to a biorefinery approach: a comprehensive review*. *Renewable and Sustainable Energy Reviews* **44**: 20–36.
20. Czernik, S., Johnson, D.K., and Black, S. (1994). *Stability of wood fast pyrolysis oil*. *Biomass and Bioenergy* **7** (1–6): 187–192.
21. Adjaye, J.D., Sharma, R.K., and Bakhshi, N.N. (1992). *Characterization and stability analysis of wood-derived bio-oil*. *Fuel Processing Technology* **31** (3): 241–256.
22. Chen, D., Zhou, J., Zhang, Q. et al. (2014). *Evaluation methods and research progresses in bio-oil storage stability*. *Renewable and Sustainable Energy Reviews* **40**: 69–79.
23. Diebold, J. and Czernik, S. (1997). *Additives to lower and stabilize the viscosity of pyrolysis oils during storage*. *Energy & Fuels* **11** (5): 1081–1091.
24. Boucher, M.E., Chaala, A., and Roy, C. (2000). *Bio-oils obtained by vacuum pyrolysis of softwood bark as a liquid fuel for gas turbines. Part I: Properties of bio-oil and its blends with methanol and a pyrolytic aqueous phase*. *Biomass and Bioenergy* **19** (5): 337–350.

25. Bai, X., Brown, R.C., Fu, J. et al. (2014). *The influence of alkali and alkaline earth metals and the role of acid pretreatments in production of sugars from switchgrass based on solvent liquefaction*. Energy & Fuels **28** (2): 1111–1120.
26. Vasilakos, N.P. and Austgen, D.M. (1985). *Hydrogen-donor solvents in biomass liquefaction*. Industrial & Engineering Chemistry Process Design and Development **24** (2): 304–311.
27. Honglu, X. and Tiejun, S. (2006). *Wood liquefaction by ionic liquids*. Holzforschung **60** (5): 509–512.
28. Mellmer, M.A., Alonso, D.M., Luterbacher, J.S. et al. (2014). *Effects of γ-valerolactone in hydrolysis of lignocellulosic biomass to monosaccharides*. Green Chemistry **16** (11): 4659–4662.
29. Ghosh, A., Brown, R.C., and Bai, X. (2016). *Production of solubilized carbohydrate from cellulose using non-catalytic, supercritical depolymerization in polar aprotic solvents*. Green Chemistry **18** (4): 1023–1031.
30. Peterson, A.A., Vogel, F., Lachance, R.P. et al. (2008). Thermochemical biofuel production in hydrothermal media: A review of sub-and supercritical water technologies. *Energy & Environmental Science* **1** (1): 32–65.
31. Toor, S.S., Rosendahl, L., and Rudolf, A. (2011). Hydrothermal liquefaction of biomass: a review of subcritical water technologies. *Energy* **36** (5): 2328–2342.
32. Huang, H.-J. and Yuan, X.-Z. (2015). *Recent progress in the direct liquefaction of typical biomass*. Progress in Energy and Combustion Science **49** (Suppl. C): 59–80.
33. Dimitriadis, A. and Bezergianni, S. (2017). *Hydrothermal liquefaction of various biomass and waste feedstocks for biocrude production: A state of the art review*. Renewable and Sustainable Energy Reviews **68** (Part 1): 113–125.
34. Diebold, J.P. (1999). *A review of the chemical and physical mechanisms of the storage stability of fast pyrolysis bio-oils*. Golden, CO: National Renewable Energy Lab.
35. Liu, Z. and Zhang, F.-S. (2008). *Effects of various solvents on the liquefaction of biomass to produce fuels and chemical feedstocks*. Energy Conversion and Management **49** (12): 3498–3504.
36. Xu, C. and Lad, N. (2007). *Production of heavy oils with high caloric values by direct liquefaction of woody biomass in sub/near-critical water*. Energy & Fuels **22** (1): 635–642.
37. Binder, J.B. and Raines, R.T. (2010). *Fermentable sugars by chemical hydrolysis of biomass*. Proceedings of the National Academy of Sciences **107** (10): 4516–4521.
38. Kim, J.S., Lee, Y., and Torget, R.W. (2001). Cellulose hydrolysis under extremely low sulfuric acid and high-temperature conditions. In: *Twenty-Second Symposium on Biotechnology for Fuels and Chemicals* (ed. B.H. Davison, J. McMillan and M. Finkelstein), 331–340. Springer.
39. Luterbacher, J.S., Rand, J.M., Alonso, D.M. et al. (2014). *Nonenzymatic sugar production from biomass using biomass-derived γ-valerolactone*. Science **343** (6168): 277–280.
40. Nguyen, T.Y., Cai, C.M., Kumar, R. et al. (2015). *Co-solvent pretreatment reduces costly enzyme requirements for high sugar and ethanol yields from lignocellulosic biomass*. ChemSusChem **8** (10): 1716–1725.
41. Sasaki, M., Fang, Z., Fukushima, Y. et al. (2000). *Dissolution and hydrolysis of cellulose in subcritical and supercritical water*. Industrial & Engineering Chemistry Research **39** (8): 2883–2890.
42. Bond, J.Q., Upadhye, A.A., Olcay, H. et al. (2014). *Production of renewable jet fuel range alkanes and commodity chemicals from integrated catalytic processing of biomass*. Energy & Environmental Science **7** (4): 1500–1523.
43. Cao, F., Schwartz, T.J., McClelland, D.J. et al. (2015). *Dehydration of cellulose to levoglucosenone using polar aprotic solvents*. Energy & Environmental Science **8** (6): 1808–1815.
44. Chheda, J.N., Huber, G.W., and Dumesic, J.A. (2007). *Liquid-phase catalytic processing of biomass-derived oxygenated hydrocarbons to fuels and chemicals*. Angewandte Chemie International Edition **46** (38): 7164–7183.
45. Minami, E. and Saka, S. (2003). *Comparison of the decomposition behaviors of hardwood and softwood in supercritical methanol*. Journal of Wood Science **49** (1): 0073–0078.

46. Kim, J.-Y., Oh, S., Hwang, H. et al. (2013). *Effects of various reaction parameters on solvolytical depolymerization of lignin in sub-and supercritical ethanol. Chemosphere* **93** (9): 1755–1764.
47. Haverly, M.R. (2016). Optimization of phenolic monomer production from solvent liquefaction of ligin. In: *An Experimental Study on Solvent Liquefaction*, 93.
48. Rahimi, A., Ulbrich, A., Coon, J.J. et al. (2014). *Formic-acid-induced depolymerization of oxidized lignin to aromatics. Nature* **515** (7526): 249.
49. Shuai, L., Amiri, M.T., Questell-Santiago, Y.M. et al. (2016). *Formaldehyde stabilization facilitates lignin monomer production during biomass depolymerization. Science* **354** (6310): 329–333.
50. Jiang, X., Zhou, J., Zhao, J. et al. (2017). *Catalytic conversion of guaiacol as a model compound for aromatic hydrocarbon production. Biomass and Bioenergy*.
51. Mathias, A.L., Lopretti, M.I., and Rodrigues, A.E. (1995). *Chemical and biological oxidation of Pinus pinaster lignin of the production of vanillin. Journal of Chemical Technology and Biotechnology* **64** (3): 225–234.
52. Olcese, R., Bettahar, M., Petitjean, D. et al. (2012). *Gas-phase hydrodeoxygenation of guaiacol over Fe/SiO_2 catalyst. Applied Catalysis B: Environmental* **115**: 63–73.
53. Reichardt, C. and Welton, T. (2011). *Solvents and Solvent Effects in Organic Chemistry*. Wiley.
54. Parker, A.J. (1969). *Protic-dipolar aprotic solvent effects on rates of bimolecular reactions. Chemical Reviews* **69** (1): 1–32.
55. Shuai, L. and Luterbacher, J. (2016). *Organic solvent effects in biomass conversion reactions. ChemSusChem* **9** (2): 133–155.
56. Brandt, A., Gräsvik, J., Hallett, J.P. et al. (2013). *Deconstruction of Lignocellulosic Biomass with Ionic Liquids. Green Chemistry* **15** (3): 550–583.
57. Pinkert, A., Marsh, K.N., Pang, S. et al. (2009). *Ionic liquids and their interaction with cellulose. Chemical Reviews* **109** (12): 6712–6728.
58. Wasserscheid, P. and Keim, W. (2000). *Ionic liquids – new "solutions" for transition metal catalysis. Angewandte Chemie International Edition* **39** (21): 3772–3789.
59. Zhu, S., Wu, Y., Chen, Q. et al. (2006). *Dissolution of cellulose with ionic liquids and its application: a mini-review. Green Chemistry* **8** (4): 325–327.
60. Akpinar, O., Levent, O., Sabanci, S. et al. (2011). *Optimization and comparision of dilute acid pretreatment of selected agricultural residues for recovery of xylose. BioResources* **6** (4): 4103–4116.
61. Fu, D. and Mazza, G. (2011). *Optimization of processing conditions for the pretreatment of wheat straw using aqueous ionic liquid. Bioresource Technology* **102** (17): 8003–8010.
62. Hsu, T.-C., Guo, G.L., Chen, W.H. et al. (2010). *Effect of dilute acid pretreatment of rice straw on structural properties and enzymatic hydrolysis. Bioresource Technology* **101** (13): 4907–4913.
63. Kootstra, A.M.J., Beeftink, H.H., Scott, E.L. et al. (2009). *Comparison of dilute mineral and organic acid pretreatment for enzymatic hydrolysis of wheat straw. Biochemical Engineering Journal* **46** (2): 126–131.
64. Li, C., Knierim, B., Manisseri, C. et al. (2010). *Comparison of dilute acid and ionic liquid pretreatment of switchgrass: biomass recalcitrance, delignification and enzymatic saccharification. Bioresource Technology* **101** (13): 4900–4906.
65. Shuai, L., Questell-Santiago, Y.M., and Luterbacher, J.S. (2016). *A mild biomass pretreatment using γ-valerolactone for concentrated sugar production. Green Chemistry* **18** (4): 937–943.
66. Shuai, L., Yang, Q., Zhu, J.Y. et al. (2010). *Comparative study of SPORL and dilute-acid pretreatments of spruce for cellulosic ethanol production. Bioresource Technology* **101** (9): 3106–3114.
67. Sun, N., Parthasarathi, R., Socha, A.M. et al. (2014). *Understanding pretreatment efficacy of four cholinium and imidazolium ionic liquids by chemistry and computation. Green Chemistry* **16** (5): 2546–2557.
68. Xin, D., Yang, Z., Liu, F. et al. (2015). *Comparison of aqueous ammonia and dilute acid pretreatment of bamboo fractions: structure properties and enzymatic hydrolysis. Bioresource Technology* **175**: 529–536.

69. Yan, L., Zhang, L., and Yang, B. (2014). *Enhancement of total sugar and lignin yields through dissolution of poplar wood by hot water and dilute acid flowthrough pretreatment*. Biotechnology for Biofuels **7** (1): 76.
70. Zhao, X., Cheng, K., and Liu, D. (2009). *Organosolv pretreatment of lignocellulosic biomass for enzymatic hydrolysis*. Applied Microbiology and Biotechnology **82** (5): 815.
71. Bergius, F. (1937). *Conversion of wood to carbohydrates*. Industrial & Engineering Chemistry **29** (3): 247–253.
72. Fagan, R.D., Grethlein, H.E., Converse, A.O. et al. (1971). *Kinetics of the acid hydrolysis of cellulose found in paper refuse*. Environmental Science & Technology **5** (6): 545–547.
73. Franzidis, J.-P., Porteous, A., and Anderson, J. (1982). *The acid hydrolysis of cellulose in refuse in a continuous reactor*. Conservation & Recycling **5** (4): 215–225.
74. Harris, J.F., Baker, A.J., Conner, A.H., et al. 1985. Two-stage dilute sulfuric acid hydrolysis of wood: an investigation of fundamentals. General technical report, Forest Products Laboratory, US Department of Agriculture, p. 45. https://doi.org/10.2737/FPL-GTR-45 (accessed 20November 2018).
75. Lenihan, P., Orozco, A., O'Neill, E. et al. (2010). *Dilute acid hydrolysis of lignocellulosic biomass*. Chemical Engineering Journal **156** (2): 395–403.
76. Liu, Z.-S., Wu, X.L., Kida, K. et al. (2012). *Corn stover saccharification with concentrated sulfuric acid: effects of saccharification conditions on sugar recovery and by-product generation*. Bioresource Technology **119**: 224–233.
77. Malester, I.A., Green, M., and Shelef, G. (1992). *Kinetics of dilute acid hydrolysis of cellulose originating from municipal solid wastes*. Industrial & Engineering Chemistry Research **31** (8): 1998–2003.
78. Moe, S.T., Janga, K.K., Hertzberg, T. et al. (2012). *Saccharification of lignocellulosic biomass for biofuel and biorefinery applications – a renaissance for the concentrated acid hydrolysis?* Energy Procedia **20**: 50–58.
79. Saeman, J.F. (1945). *Kinetics of wood saccharification-hydrolysis of cellulose and decomposition of sugars in dilute acid at high temperature*. Industrial & Engineering Chemistry **37** (1): 43–52.
80. Selke, S.M., Hawley, M.C., Hardt, H. et al. (1982). *Chemicals from wood via HF*. Industrial & Engineering Chemistry Product Research and Development **21** (1): 11–16.
81. Sun, N., Liu, H., Sathitsuksanoh, N. et al. (2013). *Production and extraction of sugars from switchgrass hydrolyzed in ionic liquids*. Biotechnology for Biofuels **6** (1): 39.
82. Wijaya, Y.P., Putra, R.D.D., Widyaya, V.T. et al. (2014). *Comparative study on two-step concentrated acid hydrolysis for the extraction of sugars from lignocellulosic biomass*. Bioresource Technology **164**: 221–231.
83. Ioelovich, M. (2012). *Study of cellulose interaction with concentrated solutions of sulfuric acid*. ISRN Chemical Engineering **2012** 428974: 1–7.
84. Deng, W., Liu, M., Zhang, Q. et al. (2010). *Acid-catalysed direct transformation of cellulose into methyl glucosides in methanol at moderate temperatures*. Chemical Communications **46** (15): 2668–2670.
85. Deng, W., Liu, M., Zhang, Q. et al. (2011). *Direct transformation of cellulose into methyl and ethyl glucosides in methanol and ethanol media catalyzed by heteropolyacids*. Catalysis Today **164** (1): 461–466.
86. Dora, S., Bhaskar, T., Singh, R. et al. (2012). *Effective catalytic conversion of cellulose into high yields of methyl glucosides over sulfonated carbon based catalyst*. Bioresource Technology **120**: 318–321.
87. Ishikawa, Y. and Saka, S. (2001). *Chemical conversion of cellulose as treated in supercritical methanol*. Cellulose **8** (3): 189–195.
88. Yamada, T. and Ono, H. (2001). *Characterization of the products resulting from ethylene glycol liquefaction of cellulose*. Journal of Wood Science **47** (6): 458–464.
89. Kawamoto, H., Hatanaka, W., and Saka, S. (2003). *Thermochemical conversion of cellulose in polar solvent (sulfolane) into levoglucosan and other low molecular-weight substances*. Journal of Analytical and Applied Pyrolysis **70** (2): 303–313.

90. Köll, P. and Metzger, J. (1978). *Thermal degradation of cellulose and chitin in supercritical acetone*. Angewandte Chemie International Edition in English **17** (10): 754–755.
91. van Putten, R.-J., Van Der Waal, J.C., De Jong, E.D. et al. (2013). *Hydroxymethylfurfural, a versatile platform chemical made from renewable resources*. Chemical Reviews **113** (3): 1499–1597.
92. Weingarten, R., Cho, J., Conner, W.C. Jr. et al. (2010). *Kinetics of furfural production by dehydration of xylose in a biphasic reactor with microwave heating*. Green Chemistry **12** (8): 1423–1429.
93. Yang, Y., Hu, C.W., and Abu-Omar, M.M. (2012). *Synthesis of furfural from xylose, xylan, and biomass using AlCl3·6 H2O in biphasic media via xylose isomerization to xylulose*. ChemSusChem **5** (2): 405–410.
94. Wang, H., Tucker, M., and Ji, Y. (2013). *Recent development in chemical depolymerization of lignin: a review*. Journal of Applied Chemistry **2013**: 1–9.
95. Narani, A., Chowdari, R.K., Cannilla, C. et al. (2015). *Efficient catalytic hydrotreatment of Kraft lignin to alkylphenolics using supported NiW and NiMo catalysts in supercritical methanol*. Green Chemistry **17** (11): 5046–5057.
96. Okuda, K., Man, X., Umetsu, M. et al. (2004). *Efficient conversion of lignin into single chemical species by solvothermal reaction in water–p-cresol solvent*. Journal of Physics: Condensed Matter **16** (14): S1325.
97. Saisu, M., Sato, T., Watanabe, M. et al. (2003). *Conversion of lignin with supercritical water – phenol mixtures*. Energy & Fuels **17** (4): 922–928.
98. Ferrini, P. and Rinaldi, R. (2014). *Catalytic biorefining of plant biomass to non-pyrolytic lignin bio-oil and carbohydrates through hydrogen transfer reactions*. Angewandte Chemie International Edition **53** (33): 8634–8639.
99. Galkin, M.V. and Samec, J.S. (2014). *Selective route to 2-propenyl aryls directly from wood by a tandem organosolv and palladium-catalysed transfer hydrogenolysis*. ChemSusChem **7** (8): 2154–2158.
100. Huang, X., Gonzalez, O.M.M., Zhu, J. et al. (2017). *Reductive fractionation of woody biomass into lignin monomers and cellulose by tandem metal triflate and Pd/C catalysis*. Green Chemistry **19** (1): 175–187.
101. Huang, X., Zhu, J., Korányi, T.I. et al. (2016). *Effective release of lignin fragments from lignocellulose by Lewis acid metal triflates in the lignin-first approach*. ChemSusChem **9** (23): 3262–3267.
102. Renders, T., Schutyser, W., Van den Bosch, S. et al. (2016). *Influence of acidic (H3PO4) and alkaline (NaOH) additives on the catalytic reductive fractionation of lignocellulose*. ACS Catalysis **6** (3): 2055–2066.
103. Van den Bosch, S., Schutyser, W., Vanholme, R. et al. (2015). *Reductive lignocellulose fractionation into soluble lignin-derived phenolic monomers and dimers and processable carbohydrate pulps*. Energy & Environmental Science **8** (6): 1748–1763.
104. Kang, S., Li, X., Fan, J. et al. (2013). *Hydrothermal conversion of lignin: A review*. Renewable and Sustainable Energy Reviews **27**: 546–558.
105. Cemek, M. and Küçük, M.M. (2001). *Liquid products from Verbascum stalk by supercritical fluid extraction*. Energy Conversion and Management **42** (2): 125–130.
106. Li, Q., Liu, D., Song, L. et al. (2014). *Direct liquefaction of sawdust in supercritical alcohol over ionic liquid nickel catalyst: effect of solvents*. Energy & Fuels **28** (11): 6928–6935.
107. Wyman, C.E. (2013). *Aqueous Pretreatment of Plant Biomass for Biological and Chemical Conversion to Fuels and Chemicals*. Wiley.
108. Gellerstedt, G. (2015). *Softwood kraft lignin: raw material for the future*. Industrial Crops and Products **77**: 845–854.
109. Mosier, N., Wyman, C., Dale, B. et al. (2005). *Features of promising technologies for pretreatment of lignocellulosic biomass*. Bioresource Technology **96** (6): 673–686.
110. Chang, V.S., Nagwani, M., Kim, C.H. et al. (2001). *Oxidative lime pretreatment of high-lignin biomass*. Applied Biochemistry and Biotechnology **94** (1): 1–28.

111. Bjerre, A.B., Olesen, A.B., Fernqvist, T. et al. (1996). *Pretreatment of wheat straw using combined wet oxidation and alkaline hydrolysis resulting in convertible cellulose and hemicellulose. Biotechnology and Bioengineering* **49** (5): 568–577.
112. Libra, J.A., Ro, K.S., Kammann, C. et al. (2011). *Hydrothermal carbonization of biomass residuals: a comparative review of the chemistry, processes and applications of wet and dry pyrolysis. Biofuels* **2** (1): 71–106.
113. Hu, B., Wang, K., Wu, L. et al. (2010). *Engineering carbon materials from the hydrothermal carbonization process of biomass. Advanced Materials* **22** (7): 813–828.
114. Akiya, N. and Savage, P.E. (2002). *Roles of water for chemical reactions in high-temperature water. Chemical Reviews* **102** (8): 2725–2750.
115. Kritzer, P. (2004). *Corrosion in high-temperature and supercritical water and aqueous solutions: a review. The Journal of Supercritical Fluids* **29** (1): 1–29.
116. Barnebey, H. and Brown, A. (1948). *Continuous fat splitting plants using the Colgate-Emery process. Journal of the American Oil Chemists' Society* **25** (3): 95–99.
117. Dote, Y., Sawayama, S., Inoue, S. et al. (1994). *Recovery of liquid fuel from hydrocarbon-rich microalgae by thermochemical liquefaction. Fuel* **73** (12): 1855–1857.
118. Duan, P. and Savage, P.E. (2010). *Hydrothermal liquefaction of a microalga with heterogeneous catalysts. Industrial & Engineering Chemistry Research* **50** (1): 52–61.
119. Barreiro, D.L., Prins, W., Ronsse, F. et al. (2013). *Hydrothermal liquefaction (HTL) of microalgae for biofuel production: state of the art review and future prospects. Biomass and Bioenergy* **53**: 113–127.
120. Demirbaş, A. (2000). *Effect of lignin content on aqueous liquefaction products of biomass. Energy Conversion and Management* **41** (15): 1601–1607.
121. Adschiri, T., Hirose, S., Malaluan, R. et al. (1993). *Noncatalytic conversion of cellulose in supercritical and subcritical water. Journal of Chemical Engineering of Japan* **26** (6): 676–680.
122. Saka, S. and Ueno, T. (1999). *Chemical conversion of various celluloses to glucose and its derivatives in supercritical water. Cellulose* **6** (3): 177–191.
123. Ehara, K. and Saka, S. (2005). *Decomposition behavior of cellulose in supercritical water, subcritical water, and their combined treatments. Journal of Wood Science* **51** (2): 148–153.
124. Sasaki, M., Kabyemela, B., Malaluan, R. et al. (1998). *Cellulose hydrolysis in subcritical and supercritical water. Journal of Supercritical Fluids* **13** (1–3): 261–268.
125. Meyer, C., McClintock, R.B., Silvestri, G.J. et al. (1993). *ASME Steam Tables: Thermodynamic and Transport Properties of Steam*. New York: American Society of Mechanical Engineers.
126. Chandel, A.K., Antunes, F.A., De Arruda, P.V. et al. (2012). Dilute acid hydrolysis of agro-residues for the depolymerization of hemicellulose: state-of-the-art. In: *D-Xylitol* (ed. S.S. da Silva and A.K. Chandel), 39–61. Springer.
127. Gurbuz, E.I., Gallo, J.M.R., Alonso, D.M. et al. (2013). *Conversion of hemicellulose into furfural using solid acid catalysts in gamma-Valerolactone. Angewandte Chemie-International Edition* **52** (4): 1270–1274.
128. Mellmer, M.A., Sener, C., Gallo, J.M.R. et al. (2014). *Solvent effects in acid-catalyzed biomass conversion reactions. Angewandte Chemie International Edition* **53** (44): 11872–11875.
129. Alonso, D.M., Gallo, J.M.R., Mellmer, M.A. et al. (2013). *Direct conversion of cellulose to levulinic acid and gamma-valerolactone using solid acid catalysts. Catalysis Science & Technology* **3** (4): 927–931.
130. Sasaki, M. and Goto, M. (2008). *Recovery of phenolic compounds through the decomposition of lignin in near and supercritical water. Chemical Engineering and Processing: Process Intensification* **47** (9–10): 1609–1619.
131. Lavoie, J.-M., Baré, W., and Bilodeau, M. (2011). *Depolymerization of steam-treated lignin for the production of green chemicals. Bioresource Technology* **102** (7): 4917–4920.
132. Roberts, V., Stein, V., Reiner, T. et al. (2011). *Towards quantitative catalytic lignin depolymerization. Chemistry-A European Journal* **17** (21): 5939–5948.

133. Beauchet, R., Monteil-Rivera, F., and Lavoie, J. (2012). *Conversion of lignin to aromatic-based chemicals (L-chems) and biofuels (L-fuels)*. Bioresource Technology **121**: 328–334.
134. Toledano, A., Serrano, L., and Labidi, J. (2012). *Organosolv lignin depolymerization with different base catalysts*. Journal of Chemical Technology and Biotechnology **87** (11): 1593–1599.
135. Pandey, M.P. and Kim, C.S. (2011). *Lignin depolymerization and conversion: a review of thermochemical methods*. Chemical Engineering & Technology **34** (1): 29–41.
136. Klason, P. (1893). *Bidrag till kannedomen om sammansattningen af granens ved samt de kemiska processerna vid framstallning af cellulosa darur*. Teknisk Tidskrift, Afdelningen for Kemi och Metallurgi **23** (2): 17–22.
137. Brosse, N., El Hage, R., Sannigrahi, P. et al. (2010). *Dilute sulphuric acid and ethanol organosol pretreatment of Miscanthus x Giganteus*. Cellulose Chemistry & Technology **44** (1): 71.
138. Goh, C.S., Tan, H.T., Lee, K.T. et al. (2011). *Evaluation and optimization of organosolv pretreatment using combined severity factors and response surface methodology*. Biomass and Bioenergy **35** (9): 4025–4033.
139. Hörmeyer, H., Tailliez, P., Millet, J. et al. (1988). *Ethanol production by Clostridium thermocellum grown on hydrothermally and organosolv-pretreated lignocellulosic materials*. Applied Microbiology and Biotechnology **29** (6): 528–535.
140. Jimenez, L., Rodríguez, A., Díaz, M.J. et al. (2004). *Organosolv pulping of olive tree trimmings by use of ethylene glycol/soda/water mixtures*. Holzforschung **58** (2): 122–128.
141. Obama, P., Ricochon, G., Muniglia, L. et al. (2012). *Combination of enzymatic hydrolysis and ethanol organosolv pretreatments: effect on lignin structures, delignification yields and cellulose-to-glucose conversion*. Bioresource Technology **112**: 156–163.
142. Shimizu, K. and Usami, K. (1978). *Enzymatic hydrolysis of wood. III. Pretreatment of woods with acidic methanol-water mixture*. Mokuzai Gakkaishi **24**: 632–637.
143. Sun, F. and Chen, H. (2008). *Enhanced enzymatic hydrolysis of wheat straw by aqueous glycerol pretreatment*. Bioresource Technology **99** (14): 6156–6161.
144. Espinoza-Acosta, J.L., Torres-Chávez, P.I., Carvajal-Millán, E. et al. (2014). *Ionic liquids and organic solvents for recovering lignin from lignocellulosic biomass*. BioResources **9** (2): 3660–3687.
145. Renders, T., Van den Bosch, S., Koelewijn, S.F. et al. (2017). *Lignin-first biomass fractionation: the advent of active stabilisation strategies*. Energy & Environmental Science **10** (7): 1551–1557.
146. Song, Q., Wang, F., Cai, J. et al. (2013). *Lignin depolymerization (LDP) in alcohol over nickel-based catalysts via a fragmentation–hydrogenolysis process*. Energy & Environmental Science **6** (3): 994–1007.
147. Marinkovic, S. and Estrine, B. (2010). *Direct conversion of wheat bran hemicelluloses into n-decyl-pentosides*. Green Chemistry **12** (11): 1929–1932.
148. Villandier, N. and Corma, A. (2010). *One pot catalytic conversion of cellulose into biodegradable surfactants*. Chemical Communications **46** (24): 4408–4410.
149. Kim, J.-Y., Park, J., Kim, U.J. et al. (2015). *Conversion of lignin to phenol-rich oil fraction under supercritical alcohols in the presence of metal catalysts*. Energy & Fuels **29** (8): 5154–5163.
150. Kuznetsov, B., Sharypov, V.I., Kuznetsova, S.A. et al. (2009). *The study of different methods of bio-liquids production from wood biomass and from biomass/polyolefine mixtures*. International Journal of Hydrogen Energy **34** (16): 7051–7056.
151. Zhao, W., Xu, W.J., Lu, X.J. et al. (2009). *Preparation and property measurement of liquid fuel from supercritical ethanolysis of wheat stalk*. Energy & Fuels **24** (1): 136–144.
152. Huang, X., Korányi, T.I., Boot, M.D. et al. (2014). *Catalytic depolymerization of lignin in supercritical ethanol*. ChemSusChem **7** (8): 2276–2288.
153. Kozliak, E.I., Kubátová, A., Artemyeva, A.A. et al. (2016). *Thermal liquefaction of lignin to aromatics: efficiency, selectivity, and product analysis*. ACS Sustainable Chemistry & Engineering **4** (10): 5106–5122.

154. Nielsen, J.B., Jensen, A., Madsen, L.R. et al. (2017). *Noncatalytic direct liquefaction of biorefinery lignin by ethanol*. Energy & Fuels **31** (7): 7223–7233.
155. Løhre, C., Barth, T., and Kleinert, M. (2016). *The effect of solvent and input material pretreatment on product yield and composition of bio-oils from lignin solvolysis*. Journal of Analytical and Applied Pyrolysis **119**: 208–216.
156. Nielsen, J., Jensen, A., Schandel, C.B. et al. (2017). *Solvent consumption in non-catalytic alcohol solvolysis of biorefinery lignin*. Sustainable Energy & Fuels **1** (9): 2006–2015.
157. Matsumura, Y., Sasaki, M., Okuda, K. et al. (2006). *Supercritical water treatment of biomass for energy and material recovery*. Combustion Science and Technology **178** (1–3): 509–536.
158. Okuda, K., Umetsu, M., Takami, S. et al. (2004). *Disassembly of lignin and chemical recovery – rapid depolymerization of lignin without char formation in water – phenol mixtures*. Fuel Processing Technology **85** (8–10): 803–813.
159. Yoshikawa, T., Yagi, T., Shinohara, S. et al. (2013). *Production of phenols from lignin via depolymerization and catalytic cracking*. Fuel Processing Technology **108**: 69–75.
160. Takami, S., Okuda, K., Man, X. et al. (2012). *Kinetic study on the selective production of 2-(Hydroxybenzyl)-4-methylphenol from organosolv lignin in a mixture of supercritical water and p-cresol*. Industrial & Engineering Chemistry Research **51** (13): 4804–4808.
161. Liguori, L. and Barth, T. (2011). *Palladium-Nafion SAC-13 catalysed depolymerisation of lignin to phenols in formic acid and water*. Journal of Analytical and Applied Pyrolysis **92** (2): 477–484.
162. Xu, W., Miller, S.J., Agrawal, P.K. et al. (2012). *Depolymerization and hydrodeoxygenation of switchgrass lignin with formic acid*. ChemSusChem **5** (4): 667–675.
163. Mellmer, M.A., Alonso, D.M., Luterbacher, J.S. et al. (2014). *Effects of gamma-valerolactone in hydrolysis of lignocellulosic biomass to monosaccharides*. Green Chemistry **16** (11): 4659–4662.
164. Bao, G., Shiro, S., and Wang, H. (2008). *Cellulose decomposition behavior in hot-compressed aprotic solvents*. Science in China Series B: Chemistry **51** (5): 479–486.
165. Koll, P. and Metzger, J. (1978). *Thermal-degradation of cellulose and chitin in super-critical acetone*. Angewandte Chemie-International Edition in English **17** (10): 754–755.
166. Román-Leshkov, Y. and Dumesic, J.A. (2009). *Solvent effects on fructose dehydration to 5-hydroxymethylfurfural in biphasic systems saturated with inorganic salts*. Topics in Catalysis **52** (3): 297–303.
167. Saha, B. and Abu-Omar, M.M. (2014). *Advances in 5-hydroxymethylfurfural production from biomass in biphasic solvents*. Green Chemistry **16** (1): 24–38.
168. Cai, C.M., Zhang, T., Kumar, R. et al. (2013). *THF co-solvent enhances hydrocarbon fuel precursor yields from lignocellulosic biomass*. Green Chemistry **15** (11): 3140–3145.
169. Wettstein, S.G., Alonso, D.M., Chong, Y. et al. (2012). *Production of levulinic acid and gamma-valerolactone (GVL) from cellulose using GVL as a solvent in biphasic systems*. Energy & Environmental Science **5** (8): 8199–8203.
170. Wang, X. and Rinaldi, R. (2012). *Solvent effects on the hydrogenolysis of diphenyl ether with Raney nickel and their implications for the conversion of lignin*. ChemSusChem **5** (8): 1455–1466.
171. Long, J., Zhang, Q., Wang, T. et al. (2014). *An efficient and economical process for lignin depolymerization in biomass-derived solvent tetrahydrofuran*. Bioresource Technology **154**: 10–17.
172. Gosselink, R.J., Teunissen, W., Van Dam, J.E. et al. (2012). *Lignin depolymerisation in supercritical carbon dioxide/acetone/water fluid for the production of aromatic chemicals*. Bioresource Technology **106**: 173–177.
173. Luterbacher, J.S., Rand, J.M., Alonso, D.M. et al. (2014). *Nonenzymatic sugar production from biomass using biomass-derived gamma-Valerolactone*. Science **343** (6168): 277–280.
174. Araque, E., Parra, C., Freer, J. et al. (2008). *Evaluation of organosolv pretreatment for the conversion of Pinus radiata D. Don to ethanol*. Enzyme and Microbial Technology **43** (2): 214–219.
175. Ghozatloo, A., Mohammadi-Rovshandeh, J., and Hashemi, S. (2006). *Optimization of pulp properties by dimethyl formamide pulping of rice straw*. Cellulose Chemistry and Technology **40** (8): 659–667.

176. Luterbacher, J.S., Azarpira, A., Motagamwala, A.H. et al. (2015). *Lignin monomer production integrated into the γ-valerolactone sugar platform.* Energy & Environmental Science **8** (9): 2657–2663.
177. Mai, N.L., Ahn, K., and Koo, Y.-M. (2014). *Methods for recovery of ionic liquids – a review.* Process Biochemistry **49** (5): 872–881.
178. Swatloski, R.P., Spear, S.K., Holbrey, J.D. et al. (2002). *Dissolution of cellulose with ionic liquids.* Journal of the American Chemical Society **124** (18): 4974–4975.
179. Li, C. and Zhao, Z.K. (2007). *Efficient acid-catalyzed hydrolysis of cellulose in ionic liquid.* Advanced Synthesis & Catalysis **349** (11–12): 1847–1850.
180. Rinaldi, R., Palkovits, R., and Schüth, F. (2008). *Depolymerization of cellulose using solid catalysts in ionic liquids.* Angewandte Chemie International Edition **47** (42): 8047–8050.
181. Stärk, K., Taccardi, N., Bösmann, A. et al. (2010). *Oxidative depolymerization of lignin in ionic liquids.* ChemSusChem **3** (6): 719–723.
182. Binder, J.B., Gray, M.J., White, J.F. et al. (2009). *Reactions of lignin model compounds in ionic liquids.* Biomass and Bioenergy **33** (9): 1122–1130.
183. Diop, A., Jradi, K., Daneault, C. et al. (2015). *Kraft lignin depolymerization in an ionic liquid without a catalyst.* BioResources **10** (3): 4933–4946.
184. Jia, S., Cox, B.J., Guo, X., et al. (2010). Catalysis of lignin depolymerization in ionic liquids. ACS National Meeting Book of Abstracts.
185. Liszka, M.J., Kang, A., Konda, N.M. et al. (2016). *Switchable ionic liquids based on di-carboxylic acids for one-pot conversion of biomass to an advanced biofuel.* Green Chemistry **18** (14): 4012–4021.
186. Xu, F., Sun, J., Konda, N.M. et al. (2016). *Transforming biomass conversion with ionic liquids: process intensification and the development of a high-gravity, one-pot process for the production of cellulosic ethanol.* Energy & Environmental Science **9** (3): 1042–1049.
187. Curran, G.P., Struck, R.T., and Gorin, E. (1967). *Mechanism of hydrogen-transfer process to coal and coal extract.* Industrial & Engineering Chemistry Process Design and Development **6** (2): 166–173.
188. Kim, K.H., Brown, R.C., Kieffer, M. et al. (2014). *Hydrogen-donor-assisted solvent liquefaction of lignin to short-chain alkylphenols using a micro reactor/gas chromatography system.* Energy & Fuels **28** (10): 6429–6437.
189. King, H.-H. and Stock, L.M. (1982). *Aspects of the chemistry of donor solvent coal dissolution. The role of phenol in the reaction.* Fuel **61** (11): 1172–1174.
190. Sugano, M., Takagi, H., Hirano, K. et al. (2008). *Hydrothermal liquefaction of plantation biomass with two kinds of wastewater from paper industry.* Journal of Materials Science **43** (7): 2476–2486.
191. Mok, W.S.L. and Antal, M.J. Jr. (1992). *Uncatalyzed solvolysis of whole biomass hemicellulose by hot compressed liquid water.* Industrial & Engineering Chemistry Research **31** (4): 1157–1161.
192. Osada, M., Sato, T., Watanabe, M. et al. (2006). *Catalytic gasification of wood biomass in subcritical and supercritical water.* Combustion Science and Technology **178** (1–3): 537–552.
193. Boocock, D.G. and Porretta, F. (1986). *Physical aspects of the liquefaction of poplar chips by rapid aqueous thermolysis.* Journal of Wood Chemistry and Technology **6** (1): 127–144.
194. Bobleter, O. and Concin, R. (1979). Degradation of poplar lignin by hydrothermal treatment. *Cellulose Chemistry and Technology* **13**: 583–593.
195. Gibbins, J.R. and Kandiyoti, R. (1990). *Development of a flowing-solvent liquefaction apparatus.* Fuel Processing Technology **24** (0): 237–243.
196. Appell, H.R., Fu, Y.C., Illig, E.G., et al. 1975. Conversion of cellulosic wastes to oil. NASA STI/Recon Technical Report N,. 75: p. 27572.
197. Weingarten, R., Rodriguez-Beuerman, A., Cao, F. et al. (2014). *Selective conversion of cellulose to hydroxymethylfurfural in polar aprotic solvents.* ChemCatChem **6** (8): 2229–2234.
198. Elliott, D.C., Hart, T.R., Neuenschwander, G.G. et al. (2013). *Hydrothermal processing of macroalgal feedstocks in continuous-flow reactors.* ACS Sustainable Chemistry & Engineering **2** (2): 207–215.

199. Vanoye, L., Fanselow, M., Holbrey, J.D. et al. (2009). *Kinetic model for the hydrolysis of lignocellulosic biomass in the ionic liquid, 1-ethyl-3-methyl-imidazolium chloride*. Green chemistry **11** (3): 390–396.
200. Haverly, M.R. (2016). The effect of moisture on hydrocarbon-based solvent liquefaction of biomass. *An Experimental Study on Solvent Liquefaction* **1001**: 58.
201. Karagöz, S., Bhaskar, T., Muto, A. et al. (2006). *Hydrothermal upgrading of biomass: effect of K_2CO_3 concentration and biomass/water ratio on products distribution*. Bioresource Technology **97** (1): 90–98.
202. Wang, P., Yu, H., Zhan, S. et al. (2011). *Catalytic hydrolysis of lignocellulosic biomass into 5-hydroxymethylfurfural in ionic liquid*. Bioresource Technology **102** (5): 4179–4183.
203. Kloekhorst, A., Shen, Y., Yie, Y. et al. (2015). *Catalytic hydrodeoxygenation and hydrocracking of Alcell® lignin in alcohol/formic acid mixtures using a Ru/C catalyst*. Biomass and Bioenergy **80**: 147–161.
204. Gürü, M., Koca, A., Can, Ö. et al. (2010). *Biodiesel production from waste chicken fat based sources and evaluation with Mg based additive in a diesel engine*. Renewable Energy **35** (3): 637–643.
205. Su, Y., Brown, H.M., Huang, X. et al. (2009). *Single-step conversion of cellulose to 5-hydroxymethylfurfural (HMF), a versatile platform chemical*. Applied Catalysis A: General **361** (1–2): 117–122.
206. Archer, W.L. (1991). *Determination of hansen solubility parameters for selected cellulose ether derivatives*. Industrial & Engineering Chemistry Research **30** (10): 2292–2298.
207. Barton, A.F. (1991). *CRC Handbook of Solubility Parameters and Other Cohesion Parameters*. CRC press.
208. Hansen, C.M. (1967). *The Three Dimensional Solubility Parameter and Solvent Diffusion Coefficient: Their Importance in Surface Coating Formalation*. Danish Technical Press.
209. Hildebrand, J.H. and Scott, R.L. (1950). *The Solubility of Nonelectrolytes*, 3e. New York, NY: Reinhold Publishing Corporation.
210. Vanasse, C., Chornet, E., and Overend, R. (1988). *Liquefaction of lignocellulosics in model solvents: creosote oil and ethylene glycol*. The Canadian Journal of Chemical Engineering **66** (1): 112–120.
211. Hildebrand, J.H. and Scott, R.L. (1962). *Regular Solutions*. Prentice-Hall.
212. Hansen, C.M. (2012). *Hansen Solubility Parameters: A User's Handbook*. CRC press.
213. Lu, J., Brown, J.S., Liotta, C.L. et al. (2001). *Polarity and hydrogen-bonding of ambient to near-critical water: Kamlet–Taft solvent parameters*. Chemical Communications (7): 665–666.
214. Crowhurst, L., Mawdsley, P.R., Perez-Arlandis, J.M. et al. (2003). *Solvent–solute interactions in ionic liquids*. Physical Chemistry Chemical Physics **5** (13): 2790–2794.
215. Kumar, S., Gupta, R., Lee, Y.Y. et al. (2010). *Cellulose pretreatment in subcritical water: effect of temperature on molecular structure and enzymatic reactivity*. Bioresource Technology **101** (4): 1337–1347.
216. Smith, M.D., Mostofian, B., Cheng, X. et al. (2016). *Cosolvent pretreatment in cellulosic biofuel production: effect of tetrahydrofuran-water on lignin structure and dynamics*. Green Chemistry **18** (5): 1268–1277.
217. Singh, S., Simmons, B.A., and Vogel, K.P. (2009). *Visualization of biomass solubilization and cellulose regeneration during ionic liquid pretreatment of switchgrass*. Biotechnology and Bioengineering **104** (1): 68–75.
218. Agrawal, R.K. (1988). *Kinetics of reactions involved in pyrolysis of cellulose II. The modified kilzer-bioid model*. The Canadian Journal of Chemical Engineering **66** (3): 413–418.
219. Bradbury, A.G., Sakai, Y., and Shafizadeh, F. (1979). *A kinetic model for pyrolysis of cellulose*. Journal of Applied Polymer Science **23** (11): 3271–3280.
220. Varhegyi, G., Jakab, E., and Antal, M.J. Jr. (1994). *Is the Broido-Shafizadeh model for cellulose pyrolysis true?* Energy & Fuels **8** (6): 1345–1352.

221. Dee, S.J. and Bell, A.T. (2011). *A study of the acid-catalyzed hydrolysis of cellulose dissolved in ionic liquids and the factors influencing the dehydration of glucose and the formation of humins.* ChemSusChem **4** (8): 1166–1173.
222. Vasudevan, V. and Mushrif, S.H. (2015). *Insights into the solvation of glucose in water, dimethyl sulfoxide (DMSO), tetrahydrofuran (THF) and N, N-dimethylformamide (DMF) and its possible implications on the conversion of glucose to platform chemicals.* RSC Advances **5** (27): 20756–20763.
223. Phan, H.D., Yokoyama, T., and Matsumoto, Y. (2012). *Direct participation of counter anion in acid hydrolysis of glycoside.* Organic & Biomolecular Chemistry **10** (36): 7382–7391.
224. Kelly, C.P., Cramer, C.J., and Truhlar, D.G. (2007). *Single-ion solvation free energies and the normal hydrogen electrode potential in methanol, acetonitrile, and dimethyl sulfoxide.* The Journal of Physical Chemistry B **111** (2): 408–422.
225. Román-Leshkov, Y., Chheda, J.N., and Dumesic, J.A. (2006). *Phase modifiers promote efficient production of hydroxymethylfurfural from fructose.* Science **312** (5782): 1933–1937.
226. Tsilomelekis, G., Josephson, T.R., Nikolakis, V. et al. (2014). *Origin of 5-hydroxymethylfurfural stability in water/dimethyl sulfoxide mixtures.* ChemSusChem **7** (1): 117–126.
227. Li, Y., Lu, X., Yuan, L. et al. (2009). *Fructose decomposition kinetics in organic acids-enriched high temperature liquid water.* Biomass and Bioenergy **33** (9): 1182–1187.
228. Chareonlimkun, A., Champreda, V., Shotipruk, A. et al. (2010). *Reactions of C5 and C6-sugars, cellulose, and lignocellulose under hot compressed water (HCW) in the presence of heterogeneous acid catalysts.* Fuel **89** (10): 2873–2880.
229. Fan, C., Guan, H., Zhang, H. et al. (2011). *Conversion of fructose and glucose into 5-hydroxymethylfurfural catalyzed by a solid heteropolyacid salt.* Biomass and Bioenergy **35** (7): 2659–2665.
230. Nikolla, E., Román-Leshkov, Y., Moliner, M. et al. (2011). *"One-pot" synthesis of 5-(hydroxymethyl)furfural from carbohydrates using tin-beta zeolite.* ACS Catalysis **1** (4): 408–410.
231. Yang, F., Liu, Q., Bai, X. et al. (2011). *Conversion of biomass into 5-hydroxymethylfurfural using solid acid catalyst.* Bioresource Technology **102** (3): 3424–3429.
232. Thigpen, P.L. and Berry, W.L. Jr. (1982). Liquid fuels from wood by continuous operation of the Albany, Oregon biomass liquefaction facility. In: *Energy from Biomass and Wastes VI* (ed. D.L. Klass), 1057–1095. Institute of Gas Technology.
233. Baker, E.G. and Elliott, D.C. (1988). Catalytic upgrading of biomass pyrolysis oils. In: *Research in Thermochemical Biomass Conversion* (ed. A.V. Bridgwater and J.L. Kuester), 883–895. Springer.
234. Elliott, D.C., Sealock Jr., L., Butner, R.S., et al. (1989). *Low-temperature conversion of high-moisture biomass: continuous reactor system results.* No. PNL-7126. Washington, USA: Pacific Northwest National Laboratory, Richland.
235. White, D., Coates, W., and Wolf, D. (1996). *Conversion of cotton plant and cotton gin residues to fuels by the extruder feeder liquefaction process.* Bioresource Technology **56** (1): 117–123.
236. Haverly, M.R., Schulz, T.C., Whitmer, L.E. et al. (2018). *Continuous solvent liquefaction of biomass in a hydrocarbon solvent.* Fuel **211** (Suppl. C): 291–300.
237. Figueroa, C., Schaleger, L.L., and Davis, H.G. (1982). *LBL continuous bench-scale liquefaction unit, operation and results.* Energy from Biomass and Wastes **VI**: 1097–1112.
238. Jazrawi, C., Biller, P., Ross, A.B. et al. (2013). *Pilot plant testing of continuous hydrothermal liquefaction of microalgae.* Algal Research **2** (3): 268–277.
239. Elliott, D.C., Hart, T.R., Schmidt, A.J. et al. (2013). *Process development for hydrothermal liquefaction of algae feedstocks in a continuous-flow reactor.* Algal Research **2** (4): 445–454.
240. Derbyshire, F.J., Varghese, P., and Whitehurst, D.D. (1982). *Synergistic effects between light and heavy solvent components during coal liquefaction.* Fuel **61** (9): 859–864.
241. Hörnell, C. and Björnbom, P. (1989). *Dissolution of peat in simulated recycle solvents.* Fuel **68** (4): 491–497.

242. Aiura, M., Masunaga, T., Kageyama, Y. et al. (1986). *Chemistry of solvents in brown coal liquefaction: estimation of steady state recycle solvent quality*. Fuel Processing Technology **14**: 13–22.
243. Ogi, T. and Yokoyama, S.-y. (1993). *Liquid fuel production from woody biomass by direct liquefaction*. Journal of the Japan Petroleum Institute **36** (2): 73–84.
244. van Rossum, G., Zhao, W., Castellvi Barnes, M. et al. (2014). *Liquefaction of lignocellulosic biomass: solvent, process parameter, and recycle oil screening*. ChemSusChem **7** (1): 253–259.

9

Hybrid Processing

Zhiyou Wen[1] and Laura R. Jarboe[2]

[1]Department of Food Science and Human Nutrition, Iowa State University, Ames, IA, USA
[2]Department of Chemical and Biological Engineering, Iowa State University, Ames, IA, USA

9.1 Introduction

Human societies have converted biomass into energy and products for millennia using both biochemical and thermochemical processes. Familiar examples of biochemical processing includes fermentation of sugar- or starch-rich crops and milk into sauerkraut, beer, wine, yogurt, and cheese. Familiar examples of thermochemical processing include baking and cooking of food and burning wood for heat and power.

Biochemical and thermochemical processes are also employed for the production of advanced biofuels and bio-based products. The biochemical platform uses pretreatments and enzymatic hydrolysis to release reducing sugars from biomass followed by microbial fermentation to fuels or chemicals. The thermochemical platform uses heat and catalysts via the processes of pyrolysis, solvent liquefaction, and gasification to produce liquid products (bio-oil) or gaseous products (syngas), which are further upgraded to fuels and chemicals. Combining biochemical and thermochemical processes in a single system is known as hybrid processing, and provides unique opportunities for improving selectivity and efficiency in the production of fuels and chemicals [1, 2]. In general, hybrid processing encompasses a wide combination of biological, thermal, and/or catalytic processes. In this chapter, we focus on the sequence of thermochemical deconstruction of biomass followed by biochemical upgrading to final products. Two prominent examples of hybrid thermochemical–biochemical processing are (i) fast pyrolysis of biomass into pyrolytic substrates followed by microbial fermentation and (ii) gasification of biomass into synthesis gas (syngas) followed by syngas fermentation.

Hybrid processing captures the benefits of thermochemical and biochemical processes while mitigating their deficiencies. For example, thermochemical processing eliminates complex and costly pretreatment and enzymatic hydrolysis steps. It rapidly and economically converts the carbohydrate in whole biomass (cellulose and hemicellulose) into the fermentable intermediates irrespective of the biomass type and composition. For the pyrolysis-based hybrid processing, the thermochemical deconstruction of biomass can be performed in close proximity to biomass production sites to produce crude bio-oil suitable for transportation to a central upgrading facility. In this manner, low-density biomass is converted to high-density feedstock, reducing transportation costs. With proper deconstruction and fractionation, lignin in the biomass is converted to a phenolic oil suitable for upgrading to drop-in hydrocarbon fuel [3], while the carbohydrate fraction of the biomass can be used as fermentation substrates for producing ethanol fuel and oxygenated chemicals [4]. For the gasification-based hybrid processing, biomass is converted into syngas, a uniform substrate for fermentation that includes carbon from both the carbohydrate and lignin content of biomass. Unlike Fischer–Tropsch (FT) synthesis usually envisioned for processing syngas into fuels and chemicals, syngas fermentation does not require a strictly controlled hydrogen-to-carbon monoxide ratio and the biocatalysts are more tolerant towards sulfur and chloride contaminants in syngas than are metal catalysts. Syngas fermentation is also more selective of desired products than FT synthesis [5, 6].

Although the research on syngas fermentation [7] and microbial utilization of bio-oil components [8] dates back to the early 1990s, the categorization as hybrid processing was first proposed in 1999 by Brown [9]. To date, the hybrid process based on fast pyrolysis of biomass followed by fermentation of pyrolytic substrates has been studied by relatively few researchers [1, 4, 10]. The gasification–syngas fermentation route has been more widely studied, including research on gas-to-liquid mass transfer [6, 11], strain development [12, 13], and commercialization [14–16]. This chapter explores the unique features, technical challenges, and future perspectives of these two approaches to hybrid processing.

9.2 Thermochemical Conversion of Lignocellulosic Biomass for Fermentative Substrates

9.2.1 Fast Pyrolysis for Production of Pyrolytic Substrates

Fast pyrolysis of lignocellulosic biomass is the thermal decomposition of biomass in the absence of oxygen. This process produces an energy-rich liquid (bio-oil), a flammable gas mixture (syngas), and a carbon-rich/nutrient-rich solid (biochar) [17, 18]. The composition of bio-oil varies depending on biomass properties and pyrolysis type and operating conditions. An example of bio-oil composition from woody biomass (based on dry weight of biomass) is as follows: 15 wt% carboxylic acids, 25 wt% sugars, 4 wt% alcohols, 10 wt% aldehydes, 2 wt% esters, 7 wt% furans, 5 wt% ketones, and 20 wt% aromatics [19]. The use of these compounds as substrates for microbial fermentation [1, 4], is discussed in the Sections 9.2.1.1–9.2.1.3.

9.2.1.1 Pyrolytic Sugars

Pyrolytic sugars are produced from depolymerization of cellulose and hemicellulose during fast pyrolysis [1, 20]. The major sugar product of the fast pyrolysis of cellulose is the anhydrosugar levoglucosan (1,6-anhydro-β-D-glucopyranose) with small amounts of cellobiosan [21]. The yield of levoglucosan from cellulose can be as high as 60 wt%. The alkali and alkaline earth metals (AAEM, e.g. K, Ca, and Mg) in biomass decrease levoglucosan yields due to the fragmentation of anhydrosugar rings, generating oxygenated compounds such as formic acid, glycolaldehyde, and acetol [22]. Mayes et al. [23] elucidated from the molecular level that the metal ions can decrease levoglucosan formation by up to 41.6 times, depending on the adjacency of the cation to the reaction center. The major sugar products from fast pyrolysis of hemicellulose are xylose and dianhydrose xylose [24]. Acid washing of biomass to remove AAEM has been shown to increase anhydrosugar yield [25–29]. Alternatively, acid infusion to passivate AAEM through formation of thermally stable salts allows glycosidic bond cleavage to dominate pyranose and furanose ring fragmentation [30, 31]. Kim et al. [32] reported a sugar yield of 20.62 g/100 g biomass when acid-infused red oak was partially oxidized with nitrogen sweep gas containing 2.1 vol% oxygen.

9.2.1.2 Acetic Acid

Acetic acid is the predominant carboxylic acid produced by deacetylation of hemicellulose during the fast pyrolysis of lignocellulosic biomass. In general, acetic acid is recovered in the aqueous phase of bio-oil [33, 34]. Due to its low heating value and corrosiveness, acetic acid is commonly regarded as an undesirable by-product.

9.2.1.3 Lignin Derivatives

Lignin is a major component in lignocellulosic biomass (15–30% by dry weight and 20–40% by energy density) and is responsible for the biomass structural integrity [35]. Because of the difficulty of its depolymerization and the heterogeneous nature of the depolymerization products, lignin valorization has long been a key challenge of traditional biochemical conversion processes [36]. During fast pyrolysis, lignin depolymerizes to various phenolic monomers and oligomers [37], which can be used as precursors for production of fuels and chemicals, although the subsequent repolymerization reduces their yield among the pyrolysis products [37, 38].

9.2.2 Gasification of Biomass for Syngas Production

Compared to the fast pyrolysis, gasification of lignocellulosic biomass occurs at higher temperatures (usually 800–1000 °C) with a limited amount of oxygen, producing syngas as the main product. Syngas is a gaseous mixture primarily comprised of 6–59 vol% carbon monoxide (CO), 29–76 vol% hydrogen (H_2), and 1–16 vol% carbon dioxide (CO_2), with a small amount of methane (CH_4). Depending on the biomass type and composition, the raw syngas also contains trace amounts of sulfur compounds (H_2S, COS), nitrogen compounds

(NH$_3$, HCN), tars (i.e. condensable hydrocarbons), alkali metals, and chlorine [39]. Traditionally, syngas has been upgraded to alcohols and liquid hydrocarbon fuels through FT synthesis, in which iron, cobalt, or ruthenium are used as catalysts for converting the gaseous compounds into liquid products at temperatures of 200–350 °C [5]. FT synthesis usually requires a strict H$_2$ to CO ratio and is vulnerable to CO$_2$; the metal catalysts have lower selectivity and are very sensitive to some of the impurities (e.g. sulfur) in the syngas [6].

9.3 Biological Conversion of Fermentative Substrates into Fuels and Chemicals

9.3.1 Fermentation of Pyrolytic Substrates

9.3.1.1 Pyrolytic Sugars

Some microorganisms can utilize levoglucosan as source of both carbon and energy (Table 9.1) [48]. Many of these microorganisms use levoglucosan kinase (*lgk*) to convert levoglucosan into glucose-6-phosphate (G6P), which then enters the glycolysis pathway [46]. The yield of the products produced from levoglucosan is similar to that based on glucose [49, 50], indicating that levoglucosan to G6P is not a rate-limiting step. Some workhorse strains such as *Escherichia coli* can be engineered for direct utilization of levoglucosan. For example, Zhuang and Zhang [44] expressed a cDNA library of *lgk* from

Table 9.1 Microbial utilization of levoglucosan as carbon source.

Microorganism	Key enzyme	Intermediate	Product	Reference
Aspergillus terreus K-26	Levoglucosan kinase	Glucose-6-phosphate	Itaconic acid	[40]
Aspergillus awamori[a]	Levoglucosan kinase	Glucose-6-phosphate	N/A	[41]
Arthrobacter sp. I-552	Levoglucosan dehydrogenase	Glucose	N/A	[42]
Aspergillus niger CBX-209	Levoglucosan kinase	Glucose-6-phosphate	Citric acid	[43]
Escherichia coli, expressing cDNA of *lgk*[b] from *A. niger* CBX-209	Levoglucosan kinase	Glucose-6-phosphate	N/A	[44]
E. coli BL21, expressing *lgk* from *Lipomyces starkeyi* YZ-215	Levoglucosan kinase	Glucose-6-phosphate	Ethanol	[45]
E. coli KO11, expressing *lgk* from *L. starkeyi*	Levoglucosan kinase	Glucose-6-phosphate	Ethanol	[46]
Rhodosporidium toruloides *Rhodotorula glutinis*	Levoglucosan kinase	Glucose-6-phosphate	Lipids	[47]

[a]Multiple strains were reported including *Aspergillus fonsecaeus, Aspergillus luchuensis, Aspergillus niger, Aspergillus oryzae, Aspergillus sojae, Cryptococcus albidus, Fusarium solami, Neurospora crassa, Penicillium citrinum, Penicillium cyclopium, Penicillium expansum, Penicillium granulatum, Penicillium griseolum, Penicillium italicum, Rhizopus niveus, Rhizopus oryzae, Sporobolomyces salmonicolor.*
[b]*lgk*, levoglucosan kinase.
Source: adapted from Ref. [2].

fungus *Aspergillus niger* CBX-209 in *E. coli*, although the resulting enzyme activity was only one-third of that in the wild strain. Dai et al. [45] expressed the *lgk* gene from yeast *Lipomyces starkeyi* YZ-215 in *E. coli* BL21 and the resulting recombinant *E. coli* strain could grow on levoglucosan. More recently, ethanologenic *E. coli* KO11 was engineered to express *lgk* from *L. starkeyi* YZ-215; the engineered strain was able to use levoglucosan as a sole carbon source for ethanol production without additional antibiotics or inducers [46].

In addition to being directly utilized, levoglucosan can also be acid hydrolyzed into reducing sugars for yeasts and fungi to produce ethanol or lipids (Table 9.2). This method, however, can result in some sugar loss during the neutralization of the acid hydrolysate [53].

Cellobiosan (1,6-anhydro-β-cellobiose) is another anhydrosugar in bio-oil [51, 54–56]. Lian et al. [57] characterized soil samples and isolated six microbial species that could utilize cellobiosan as sole carbon source: *Sphingobacterium multivorum*, *Acinetobacter oleivorans* JC3-1, *Enterobacter* sp. SJZ-6, *Microbacterium* spx FXJ8.207 and 203, and *Cryptococcus* sp. Each of these organisms was able to use levoglucosan as sole carbon source. However, to date, direct microbial fermentation of cellobiosan into a fuel or chemical product has not been reported. Cellobiosan can be hydrolyzed into reducing sugars including glucose, levoglucosan, and/or cellobiose [58] that can be potentially utilized as

Table 9.2 Microbial utilization of levoglucosan that was acid hydrolyzed into glucose.

Pyrolysis feedstock	Post-pyrolysis treatment	Strain(s)	Product/Yield	Reference
Wood	Aqueous extraction: water-to-oil 1:2 Detoxification: activated charcoal treatment Hydrolysis: 2% (v/v) H_2SO_4 at 100°C for 120 min Neutralization: lime	Yeasts; Fungi	Ethanol, 0.43 g/g substrate	[8]
Cotton cellulose	Hydrolysis: 0.2 M H_2SO_4 at 121°C for 20 min Neutralization: $Ca(OH)_2$ to pH 6.0 Detoxification: 10% (w/v) absorbent diatomite Filtration: 0.45 μm membrane	*Saccharomyces cerevisiae*	Ethanol, 0.45 g/g substrate	[51]
Softwood	Aqueous extraction: 62% (w/w) water Hydrolysis: 0.5 M H_2SO_4 at 125°C for 44 min	*S. cerevisiae*	Ethanol, 0.46 g/g substrate	[52]
Hardwood	Solvent extraction: ethyl acetate/biodiesel blends Hydrolysis: 0.5 M H_2SO_4 at 120°C for 42 min Detoxification: 100% (v/v) activated carbon Neutralization: $Ba(OH)_2$ to pH 7.0 Filtration	*S. cerevisiae*, *Cryptococcus curvatus*	Ethanol, 0.47 g/g substrate Lipids, 0.1 g/g substrate	[10]

Source: adapted from Ref. [2].

fermentative substrates. It has also been shown that exogenously provided beta-glucosidase enzyme can cleave cellobiosan into levoglucosan and glucose [59].

9.3.1.2 Acetic Acid

Acetic acid is the predominant carboxylic acid in bio-oil and is produced from deacetylation of hemicellulose [20]. Some microorganisms can metabolize acetic acid to produce acetyl-CoA, a central intermediate for biosynthesis of a wide variety of compounds, including fatty acids [60]. A number of oleaginous yeasts (*Cryptococcus albidus*, *Cryptococcus curvatus*, *Yarrowia lipolytica*) [61–63] and microalgae (*Chlorella protothecoides*) [64] have been reported to produce lipids through the utilization of acetic acid as the sole carbon source, with resulting lipid content as high as 55% (g/g biomass). The main lipids produced by these oleaginous strains are C_{16}–C_{18} fatty acids, ready for biodiesel production. However, there are only a few studies on lipid production using pyrolysis-derived acetic acid [65–68]. Unlike pyrolytic sugar fermentation based on the anaerobic glycolysis pathway, lipid biosynthesis from acetate is an aerobic process. Therefore, maintaining an appropriate dissolved oxygen level is important for the carbon metabolism [62, 63, 69]. Important engineering issues to consider include aeration efficiency and power consumption.

9.3.1.3 Lignin Derivatives

Two approaches are commonly used in the microbial utilization of lignin and its derivatives. One approach targets transformation of lignin-derived monomers into a specific target product. For example, vanillin can be produced by a number of specialized microorganisms from aromatic molecules such as eugenol [70–74], isoeugenol [74–85], ferulic acid [86–98], and vanillic acid [99] or by solid-state fermentation of green coconut husk [100]. The other approach is to funnel various lignin derivatives through the central metabolism of a microorganism to a single metabolic node, such as acetyl-CoA, and tune the target product based on industrial relevance [101].

A number of phenolic monomers [37] can be utilized by aromatic-catabolizing microorganisms to produce the central intermediates (e.g. catechol and protocatechuate) via the aerobic peripheral pathway [102]. These central intermediates are then subjected to O_2-dependent aromatic-ring cleavages by dioxygenase enzymes found in many bacteria and fungi [102]. This aromatic catabolism serves as a "biological funnel" to reduce the heterogeneity of the lignin derivatives [101]. Linger et al. [101] reported utilization of a natural aromatic-catabolizing bacterium, *Pseudomonas putida* KT2440, to medium chain-length polyhydroxyalkanoates (*mcl*-PHAs) from lignin-derived phenolic liquor. Therefore, *P. putida* KT2440 could be used as a biological platform to produce muconate (13.5 g/l) from diverse lignin-derived aromatic monomers (e.g. phenol, vanillin, ferulic acid) via protocatechuate and catechol branches of the β-ketoadipate pathway. The muconic acid was recovered with high efficiency (74%) and high purity (>97%) and then subjected to catalytic hydrogenation to produce adipic acid, the most commercially important dicarboxylic acid. The aromatic-catabolizing pathways in *P. putida* KT2440 can be further modulated to optimize the yield of the desired product via the central intermediates (pyruvate, succinate, acetyl-CoA). For example, Johnson and Beckham [103] attained a fivefold increase in pyruvate production by replacing the protocatechuate *ortho* pathway in *P. putida* KT2440 with a xenogeneic *meta*-cleavage pathway from *Sphingobium* sp. SYK-6. These advances in biological utilization of lignin derivatives enhance economic viability of the hybrid processing approach [35]. More recently, Salvachúa et al. [104] reported that a bacterial consortium of *P. putida* strains, *Amycolatopsis* sp., *Acinetobacter* ADP1, and *Rhodococcus jostii* were able to depolymerize lignin with high molecular weight and catabolize the resulting aromatic monomers simultaneously. This may enable biological utilization of lignin-derived phenolic oligomers in bio-oil, thus alleviating aforementioned monomer-repolymerization issues. However, since biological upgrading of lignin derivatives depends on aromatic catabolism with oxygen consumption, developing cost-efficient aeration will be crucial for economic viability and scalability of lignin valorization.

9.3.2 Fermentation of Syngas

9.3.2.1 Metabolic Pathway in Syngas Fermentation

A variety of microorganisms are capable of performing syngas fermentation. Based on the end products, those microorganisms can be hydrogenogens (Table 9.3) and acetogens (Table 9.4). The hydrogenogens produce H_2 from proton reduction coupled with CO oxidation to CO_2, which is also referred to as the biological water-gas shift reaction [170]:

$$CO + H_2O \rightarrow CO_2 + 2H^+ + 2e^- \rightarrow CO_2 + H_2$$

The above reactions are catalyzed by two key enzymes: nickel-CO dehydrogenase (Ni-CODH) and CO-induced hydrogenase [109, 171].

Table 9.3 Hydrogenogenic microorganisms in syngas fermentation and the optimal culture conditions[a].

Microorganism	T (°C)	pH	Genome sequence	Reference
Citrobacter sp. Y19	30–40	5.0–8.0	N/A	[105, 106]
Rhodopseudomonas palustris	30	7.0	N/A	[107, 108]
Rhodospirillum rubrum	30	6.8	Available	[109, 110]
Rubrivivax gelatinosus	35	7.5	Available	[111–113]
Caldanaerobacter subterraneus	70	6.8–7.1	Draft	[114]
Carboxydothermus hydrogenoformans	70–72	6.8–7.0	Available	[115]
Carboxydocella sporoproducens	60	6.8	N/A	[116]
Carboxydocella thermautotrophica	58	7.0	N/A	[117]
Thermincola carboxydiphila	55	8.0	N/A	[118]
Thermincola ferriacetica	57–60	7.0–7.2	N/A	[119]
Thermolithobacter carboxydivorans	70	7.0	N/A	[120]
Thermosinus carboxydivorans	60	6.8–7.0	Draft	[121]
Desulfotomaculum carboxydivorans[b]	55	6.8–7.2	Available	[122]
Thermococcus onnurineus NA1	80	6.5	Available	[123–125]
Thermococcus sp. strain AM4	82	6.8	Available	[126, 127]

[a] Unless noted otherwise, CO was used as substrate and H_2 was the product.
[b] CO and sulfate were reported as substrates. H_2 and H_2S were reported as products.
Source: adapted from Ref. [2].

Table 9.4 Acetogenic microorganisms used for syngas fermentation, their optimal culture condition, and end products.

Microorganism	Substrate	T (°C)	pH	Product(s)	Genome sequence	Reference
Acetobacterium woodii	CO_2/H_2, CO	30.0	6.8	Acetate	Available	[128]
Acetogenium kivui	CO_2/H_2, CO	N/A[a]	6.6	Acetate	N/A	[129]
Acetonema longum	CO_2/H_2	30–33	7.8	Acetate, butyrate	N/A	[130, 131]
Alkalibaculum bacchi	CO_2/H_2, CO	37	8.0–8.5	Acetate, ethanol	N/A	[132, 133]
Blautia producta/Peptostreptococcus productus	CO_2/H_2, CO	37	7.0	Acetate	N/A	[134, 135]
Butyribacterium methylotrophicum	CO_2/H_2, CO	37	5.5–7.4	Acetate, ethanol, butyrate, butanol	N/A	[136, 137]
Clostridium aceticum	CO_2/H_2, CO	30	8.3	Acetate	Available	[138, 139]
Clostridium autoethanogenum	CO_2/H_2, CO	37	5.8–6.0	Acetate, ethanol, lactate, 2,3-butanediol	Available	[140–143]
Clostridium carboxidivorans P7	CO_2/H_2, CO	37	5.8–6.2	Acetate, ethanol, butyrate, butanol, lactate	Draft	[144–146]
Clostridium drakei	CO_2/H_2, CO	25–30	5.8–6.9	Acetate, ethanol, butyrate	N/A	[144, 147]
Clostridium formicoaceticum	CO	37	NA	Acetate, formate	N/A	[148]
Clostridium glycolicum	CO_2/H_2	37–40	7.0–7.5	Acetate	Draft	[149, 150]
Clostridium ljungdahlii	CO_2/H_2, CO	37	6.0	Acetate, ethanol, lactate, 2,3-butanediol	Available	[13, 151]
Clostridium magnum	CO_2/H_2	30–32	7.0	Acetate	N/A	[152]
Clostridium mayombei	CO_2/H_2	33	7.3	Acetate	N/A	[153]

Organism	Substrate	Temp	pH	Products	Genome	Ref
Clostridium methoxybenzovorans	CO_2/H_2	37	7.4	Acetate	Draft	[154]
Clostridium ragsdalei P11	CO_2/H_2, CO	37	6.3	Acetate, ethanol, lactate, 2,3-butanediol	Under construction	[141, 155]
Eubacterium limosum	CO_2/H_2, CO	38–39	7.0–7.2	Acetate, ethanol, butyrate	Available	[156, 157]
Oxobacter pfennigii	CO_2/H_2, CO	36–38	7.3	Acetate, butyrate	N/A	[158]
Moorella thermoacetica	CO_2/H_2, CO	55	6.5–6.8	Acetate	Available	[159]
Moorella sp. HUC22-1	CO_2/H_2	55	5.8–6.2	Acetate, ethanol	N/A	[160]
Moorella thermoautotrophica	CO_2/H_2, CO	58	6.1	Acetate	N/A	[161, 162]
Thermoanaerobacter kivui	CO_2/H_2	66	6.4	Acetate	N/A	[163, 164]
Desulfotomaculum kuznetsovii	CO, sulfate	60	7.0	Acetate, H_2S	Available	[165, 166]
Desulfotomaculum thermobenzoicum subsp. thermosyntrophicum	CO, sulfate	55	7.0	Acetate, H_2S	N/A	[165, 167]
Archaeoglobus fulgidus	CO, sulfate	83	6.4	Acetate, formate, H_2S	Available	[168, 169]

[a]NA, not available.
Source: adapted from Ref. [2].

Acetogens are facultative autotrophs capable of $CO/H_2/CO_2$ metabolism via the Wood-Ljungdahl (WL) pathway [172, 173]. The carbonyl branch and the methyl branch are two branches of the WL pathway [15]. The overall stoichiometry of the pathway is presented as:

$$2CO_2 + 4H_2 + ATP \rightarrow Acetyl - CoA + 2H_2O + nATP$$

where n is the ATP conservation coefficient; acetyl-CoA serves as a central intermediate and an ATP source, which is utilized to produce metabolites such as acetate and ethanol, with the supply of electron donors (H_2 or CO).

9.3.2.2 Metabolic Engineering of Acetogens in Syngas Fermentation

Most acetogens produce acetate as the sole product (Table 9.4). Some organisms are also capable of producing products such as ethanol, butanol, butyrate, and 2,3-butanediol, but the yields are usually low. Native acetogens have been engineered to divert carbon flow to the desired products with improved yields. Table 9.5 summarizes recent advances in genetic manipulation of acetogens. Among those strains, *Clostridium ljungdahlii* has been widely used as a platform for heterologous gene expression. Six butanol pathway genes from *Clostridium acetobutylicum* were expressed in *C. ljungdahlii* via a pIMP1 plasmid-based shuttle vector, with up to 2 mM butanol produced by the recombinant *C. ljungdahlii* [13]. Leang et al. [174] reported a more efficient electroporation protocol to perform chromosomal gene deletion for *C. ljungdahlii* via double-crossover homologous recombination with suicide vector. With this toolkit, Banerjee et al. [175] adapted the $bgaR$-P_{bgaL} plasmid-based lactose-inducible system originally developed for *Clostridium perfringens* to *C. ljungdahlii* for acetone production. The inducible system redirected the carbon and electron flow for biosynthesis of the desired products other than acetate, producing 13 mM acetone with 25% carbon yield based on syngas input. The same lab later transformed eight genes (*thl*, *crt*, *bcd*, *etfA*, *etfB*, *hbd*, *ptb*, and *buk*) encoding the key enzymes for butyrate pathway from *C. acetobutylicum* to *C. ljungdahlii* [176]. The authors inactivated *pta*-dependent acetate, *adhE1*-dependent ethanol, and *ctf*-dependent fatty acid synthesis pathways to increase the butyrate titer to 17 mM, with 68% carbon yield and 73% electron yield [176]. Genetic manipulation of other acetogens has also been reported (Table 9.5). For example, LanzaTech reported up to 25.6 mM butanol produced from steel mill waste gas (mainly CO) by overexpression of butanol synthesis pathway genes from *C. acetobutylicum* in *Clostridium autoethanogenum* [177]. LanzaTech researchers identified two key enzymes, 2,3-butanediol dehydrogenase (2,3-BDH) and NADPH-dependent primary-secondary alcohol dehydrogenase (CaADH), contributing to 2,3-butanediol production during *C. autoethanogenum* syngas fermentation [179]. CaADH was demonstrated *in vitro* to convert acetoin, acetone, and butanone to 2,3-butanediol, isopropanol, and 2-butanol, respectively. This indicates the potential of using *C. autoethanogenum* as a platform for producing higher alcohols (C_3, C_4). Straub et al. [178] increased the acetate titer of *Acetobacterium woodii* to 51 g/l by overexpression of the native genes encoding the four tetrahydrofuran (THF)-dependent enzymes, phosphotransacetylase and acetate kinase.

Overall, development of metabolic engineering strategies for acetogens is still in its infancy. Kopke et al. [13] reported low butanol titer of 2 mM by engineered *C. ljungdahlii*;

Table 9.5 Engineered acetogens in syngas fermentation, target products, and metabolic engineering strategies.

Host strain	Products	Metabolic engineering strategy	Results	Reference
Clostridium ljungdahlii	Butanol	Plasmid overexpression of butanol synthesis pathway genes thlA, hbd, crt, bcd, adhE, and bdhA from Clostridium acetobutylicum	2 mM butanol production from syngas	[13]
Clostridium ljungdahlii	Acetate	Deletion of the bifunctional aldehyde/alcohol dehydrogenases adhE1 and adhE2	Increased acetate production (64.4 mM) at expense of ethanol (4.7 mM)	[174]
Clostridium ljungdahlii	Ethanol, acetone	Adapting the bgaR-P$_{bgaL}$ plasmid-based lactose-inducible system developed for Clostridium perfringens to C. ljungdahlii	25% carbon flux redirected to acetone (~13 mM) as a result of reduced acetate yield	[175]
Clostridium ljungdahlii	Butyrate	Integration of Clostridium acetobutylicum butyrate pathway genes thl, crt, bcd, etfA, etfB, hbd, ptb, and buk at pta locus of chromosome, and inactivation of adhE1 and ctf genes encoding CoA transferase	Up to 1.5 g/l butyrate production from H_2/CO_2 or CO/CO_2	[176]
Clostridium autoethanogenum	Butanol	Plasmid overexpression of butanol synthesis pathway genes thlA, hbd, crt, bcd, etfA, and etfB from C. acetobutylicum	25.66 mM butanol production from steel mill waste gas	[177]
Acetobacterium woodii	Acetate	Plasmid overexpression of genes encoding the formyl-THF synthetase, methenyl-THF-cyclohydrolase, methylene-THF dehydrogenase, methylene-THF reductase, phosphotransacetylase and acetate kinase	Increased acetate titer from H_2/CO_2 to 51 g/l	[178]

Source: adapted from Ref. [2].

by the end of the batch fermentation, butanol was no longer detected, possibly having been consumed by the cells. In another project, Ueki et al. [176] developed a genetic knock out system for *C. ljungdahlii* to interrupt acetate kinase and CoA transferase. However, this attempt did not completely eliminate acetate and ethanol production; the majority of carbon flux in the WL pathway still flowed to acetate, indicating the existence of other unidentified genes controlling their metabolism. Further efforts in metabolic engineering of acetogens are needed.

9.4 Challenges of Hybrid Processing and Mitigation Strategies

9.4.1 Pyrolysis–Fermentation Process

9.4.1.1 Fractionation of Bio-oil

Bio-oil fractionation is commonly used to enrich specific pyrolytic substrate(s) in a distinct fraction. Bio-oil can be phase separated by addition of water to produce two fractions: a light fraction containing oxygenated compounds such as carboxylic acids, acetol, and aldehydes, and a heavy fraction containing lignin-derivatives [180]. Pollard et al. [33] reported a unique multi-stage fractionation system containing a series of condensers and electrostatic precipitators to recover bio-oil as five distinct stage fractions (SFs), with each fraction having different physiochemical properties. With this innovative system, over 85 wt% of the sugar (mainly levoglucosan) in the crude bio-oil was recovered in SF1 and SF2, whereas acetic acid-rich aqueous fraction was condensed in SF4 and SF5. Both the sugar-rich phase and the acetic acid-rich phase have been tested as fermentation substrates [4, 66, 67]. Conventional distillation usually causes polymerization of bio-oil due to thermal and chemical instability of several components in the bio-oil. Vacuum distillation at lower temperatures is more successful [181–183] but still generates distillation residue of little value. Zhang et al. [184] reported a more efficient fractionation process with no waste generated. The authors separated bio-oil by atmospheric distillation, followed by co-pyrolysis of the distillation residue into acetic acid, propionic acid, and furfural.

9.4.1.2 Toxicity of Crude Pyrolytic Substrates

The toxicity of contaminants in crude pyrolytic substrate is a major challenge for pyrolysis-based hybrid processing. Crude pyrolytic substrate is highly heterogeneous, containing hundreds of chemical compounds [185]. Some of these can severely inhibit microbial biocatalysts. For example, organic acids, furfural, and 5-hydroxymethylfurfural (5-HMF) are well known for inhibiting growth and fermentation of ethanologenic *E. coli* [186, 187]. Furanic compounds can deactivate cell replication, inducing DNA damage, and inhibit key enzymes in central carbon metabolism [188], while phenolics cause inhibition by altering permeability of cell membranes and/or generating reactive oxygen species [189]. Acetol, furfural, 5-HFM, and phenolics are toxic to oleaginous microalgae [66–68]. In addition to the well-known mechanisms described above, crude pyrolytic bio-oil also contains numerous unidentified compounds with unknown inhibitory effects [55, 67].

9.4.1.3 Detoxification of Pyrolytic Substrates

Detoxification is a promising method to increase the fermentability of pyrolytic substrates. Various physical and chemical methods have been developed to remove or reduce inhibitory compounds present in crude bio-oil or its fractions, including water extraction [190], air stripping [191], activated carbon adsorption and/or absorption [8, 10, 47, 51, 65, 66, 191], biochar adsorption [192], liquid–liquid extraction [191, 193], and alkali/overliming treatment [51, 193, 194]. Treatments can be combined into highly efficient detoxification processes. For example, Rover et al. [195] used hot water washes (80–90 °C) to recover 93 wt% of the pyrolytic sugar in the heavy fraction of fractionated bio-oil, followed by alkali (NaOH) treatment to remove most of the acetol, furans, and phenolic compounds without sugar loss. Ethanologenic *E. coli* were able to grow on up to 2 wt% treated pyrolytic sugars with cell densities comparable to that obtained for glucose at the same concentration; in contrast, untreated sugars strongly inhibited *E. coli* growth beyond 1 wt% [195]. It should be noted that challenges still exist for implementing industrial-scale detoxification processes. For example, overliming procedures consume large quantities of base and acid, which increases processing cost. Gypsum ($CaSO_4$) formed during the overliming will precipitate with sugars at alkaline pH, reducing sugar yield [196]. Humbird et al. [197] reported that sugar loss during precipitation of gypsum can be as high as 13 wt%. Stanford et al. [198] recently reported the efficacy of resins to remove contaminants from pyrolytic sugars (such as levoglucosan), which can be adapted to simulated moving bed filters.

Microbial and enzymatic treatments are alternative approaches for bio-oil detoxification. In this approach, microorganisms and/or enzymes selectively degrade specific inhibitory compounds while leaving fermentable substrates intact [199]. For example, laccases and peroxidases are capable of degrading phenolics and other aromatics [200–203]. Microorganisms including fungi (*Trichoderma reesei, Coniochaeta ligniaria*) [203–206], bacteria (*Ureibacillus thermophaericus, Cupriavidus basilensis, Rhodopseudomonas palustris*) [207–209], and yeasts (*Issatchenkia occidentalis, Iris orienalis*) [210, 211] can also detoxify various toxic compounds. The microbial/enzymatic-based detoxification has been used to treat acid hydrolysate in the lignocellulosic biomass biochemical conversion platform to remove phenolic compounds [200, 202, 203], furfural and 5-HMF [204, 205, 208, 212], and carboxylic acids [210]. As most of the inhibitors derived from lignocellulose pretreatment and/or hydrolysis also exist in pyrolytic bio-oil [33], this method can be adapted to the detoxification of the pyrolysis product. For example, Khiyami et al. [213] reported detoxification of corn stover pyrolysis liquors by white-rot fungi *Phanerochaete chrysosporium* capable of secreting ligninolytic enzymes. The biological detoxification approaches have advantages of mild reacting conditions, elimination of extra complicated separation procedures, and less waste generation. However, a long incubation period, high enzyme costs, and sugar loss are the challenges to be overcome [203].

9.4.1.4 Enhancing Tolerance of Microorganisms to Toxic Compounds in Crude Pyrolytic Substrates

Another strategy for solving the toxicity problems in pyrolytic substrate fermentation is to enhance microbial tolerance to toxic compounds [214]. When the inhibition mechanism of a toxic compound is known, rational strain engineering can sometimes be used to increase tolerance. For example, ethanologenic *E. coli* possesses an NADPH-dependent furfural reductase capable of converting furfural to a less toxic product, furfuryl alcohol [215], but this reaction depletes the NADPH needed for cysteine biosynthesis, which is especially important when cells are grown in low-cost defined mineral salts media. Miller et al. [216] enhanced furfural tolerance of *E. coli* LY180 by silencing two NADPH-dependent oxidoreductase genes (*yqhD* and *dkgA*). These gene deletions decreased the ability of the cells to convert furfural to the less toxic alcohol, but the associated increase in NADPH availability resulted in increased growth and ethanol production in the presence of furfural. The tolerance of *E. coli* LY180 to furfural was also improved to 10 mM by plasmid-based expression of *thyA* gene from *Bacillus subtilis* [217]. Cell growth and ethanol production (30 g/l) were restored after an initial 48-hours lag phase, during which time furfural was completely converted to less toxic furfuryl alcohol. As NADH represents a favorable alternative for reducing furfural to less toxic alcohol instead of NADPH, Wang et al. [218] overexpressed *fucO*, an NADH-dependent propanediol reductase, in *E. coli* LY180 to enhance furfural tolerance. The engineered strain could tolerate 15 mM furfural and produced ethanol at a titer of 45 g/l, comparable to the titer observed when cells were grown in furfural-free medium. Wang et al. [219] further investigated the epistatic interactions among the four beneficial genetic traits ($\Delta yqhD$, *pntAB*, *fucO*, and *ucpA*) for furfural tolerance in *E. coli* LY180, leading to substantial biomass growth and ethanol productivity during xylose fermentation in the presence of 15 mM furfural.

Random mutagenesis, such as directed evolution, is another strategy to improve microbial tolerance to toxic compounds [66, 193]. This approach is more commonly used in pyrolytic substrate fermentation as the toxic compounds in crude substrate solution are very complex and the inhibition mechanisms are usually unknown. Directed evolution mimics natural evolution in the laboratory environment. Microorganisms are exposed to an environment containing small amounts of the toxic compounds so that they acquire mutations that improve growth in the environment. The concentration of the toxic compounds is gradually increased during the evolution process, and strains with beneficial mutations will eventually dominate the population. The mutant strains can then be characterized by, for example, DNA sequencing. A typical example of using directed evolution for enhancing a microorganism's tolerance of crude pyrolytic substrates is reported by Liang et al. [66], in which the microalga *Chlamydomonas reinhardtii* was cultured with acetic acid-rich bio-oil fraction for producing lipid-containing biomass. After multiple generations, the evolved algal strain was able to grow in media containing substantial amounts of pyrolytic acetate.

9.4.2 Gasification–Syngas Fermentation Process

9.4.2.1 *Inhibitory Compounds in Syngas and Gas Cleanup*

Syngas produced by gasification of lignocellulosic biomass typically contains small amounts of sulfur gases (H_2S, COS), nitrogen gases (NH_3, NO, HCN), tars, and particulate matter (ash and char) [6]. Although syngas fermentation does not require as strict a clean-gas composition as metal-catalyzed Fischer-Tropsch synthesis, contaminants nevertheless can suppress product yield, or even cause process failure if contaminants exist in too high a concentration [220–223]. Indeed, it has been reported that INEOS Bio, the first commercial cellulosic ethanol producer based on syngas fermentation, experienced severe disruption in its process resulting from high levels (~15 ppm) of HCN contaminant [224].

Syngas contaminants can be minimized by feedstock pretreatment and/or gasification process optimization. For example, Broer et al. [225] used steam/oxygen-blown gasification with optimized equivalence ratio (ER) to reduce contaminant production. Contaminants that formed were effectively removed though multiple gas-cleaning steps post-gasification [225]. Cyclones, electrostatic separation, and barrier filtration are commonly used in hot syngas cleanup, while water/liquid absorption is used for up cold-gas cleaning [39].

It should be noted that the composition of major syngas species (H_2, CO, CO_2) varies with biomass feedstocks and gasification conditions, which can impact performance of syngas fermentation. For example, excess H_2 was reported to promote ethanol production by *C. ljungdahlii* [226]. The electrons and protons required by the strain can be obtained from H_2 oxidation and/or CO oxidation/CO_2 reduction. Electrons from H_2 oxidation are preferred, so that CO can be utilized as a carbon source for metabolites synthesis rather than be sacrificed as an energy source for electron donation. In addition to H_2 content, the ratio of CO to CO_2 was found to affect syngas fermentation in certain strains such as *Clostridium carboxidivorans* P7 [227].

9.4.2.2 Mass Transfer Limitation and Bioreactor Design

Gas-to-liquid mass transfer is a main bottleneck for syngas fermentation due to the low solubility of gas species such as CO and H_2 [11, 228]. Reactor configuration is crucial for enhancing the volumetric mass transfer coefficient ($k_L a$). A continuous stirred tank reactor (CSTR) is the most commonly used configuration. Increasing the gas flow rate and agitation speed enhances $k_L a$, but at high energy cost [229]. Other strategies to enhance $k_L a$ are the use of microbubble dispersion [230] and/or porous particles with large Brunauer, Emmett, and Teller (BET) surface areas and specific surface properties [231]. Zhu et al. [232] reported an increase of 190% in CO $k_L a$ using mesoporous silica nanoparticles. In *Rhodospirillum rubrum* fermentation, H_2 yield was improved by 200% when the medium was dosed with MCM41 functionalized nanoparticles [232]. Applying this strategy to commercial-scale syngas fermentation, however, is still challenging due to the high cost of purchasing and recovering these particles.

Bubble column reactors (BCRs) can also be used in syngas fermentation. BCR can achieve high $k_L a$ values with low power consumption [233]. The maintenance and operational costs of BCR are low. However, back-mixing and gas bubble coalescence may reduce mass transfer efficiency [234]. Using BCR, Chang et al. [235] reported CO $k_L a$ of $72\,h^{-1}$ with acetate productivity of 5.8 g/l/day in *Eubacterium limosum* fermentation. Rajagopalan et al. [236] achieved ethanol concentrations of 0.16 wt% and productivity of 1 g/l/day for syngas fermentation with *C. carboxidivorans* P7 [234]. A BCR containing a monolithic column improved ethanol production during syngas fermentation [237]. The monolith is composed of thousands of parallel micro channels separated by thin walls that support biofilm growth. Slug flow of gas and liquid inside each microchannel occurs under certain operating conditions [238]. Formation of a very thin liquid boundary layer sandwiched between the gas and biofilm greatly increases mass transfer efficiency [237].

Trickle bed reactors (TBRs) have also been used for syngas fermentation. Klasson et al. [239] reported CO $k_L a$ value of $55.5\,h^{-1}$ in a TBR, considerably higher than reported for a packed-bed reactor ($k_L a$ $2.1\,h^{-1}$). Orgill et al. [240] reported the effects of bead size, gas, and liquid flow rates on O_2 $k_L a$ in a TBR, achieving $k_L a$ of $421\,h^{-1}$ at liquid flow rate of 50 ml/min for 6 mm beads. The authors applied the same condition during semi-continuous fermentation of *Clostridium ragsdalei* P11 and achieved 90% and 70% conversion of CO and H_2, respectively [241].

Hollow fiber membrane (HFM) bioreactors are promising for syngas fermentation, characterized by efficient mass transfer [240, 242–246]. Syngas is fed into the lumen of the HFM bundle and diffuses through the membrane. Cells growing on the outer wall of the membrane are able to consume dissolved gas and excrete metabolites into the medium. Mass transfer efficiency depends on numerous factors, such as hydrophobicity, porosity, pore size, fiber diameter and thickness of membrane material [240], internal or external placement of the HFM module [244, 245], membrane surface area [242], gas pressure [245], and gas flow rate through the lumen and liquid flow rate through the shell-side [242, 246]. Shen et al. [246] used HFM for continuous syngas fermentation of *C. carboxidivorans*, reporting a 10-fold improvement in CO $k_L a$ compared to a CSTR and achieved maximum ethanol productivity of 3.44 g/l/day. Major challenges in using HFM are fouling and loss of microbial viability during long-term operation. Cost-efficient strategies have been developed to maintain biomass viability and to minimize membrane fouling such as in situ physical cleaning [247, 248].

Biologically active polymeric coatings represent an alternative approach to overcome mass transfer limitations in syngas fermentation. In this approach, microbial cells are concentrated and embedded in a thin, adhesive, and nanoporous latex coating as an integral component of the bio-reactive structure. Cells are preserved under desiccated conditions, rehydrated to regain activity [249]. Applications include microbial photosynthesis, air pollution control [250–252], and more recently, syngas fermentation by immobilizing *C. ljungdahlii* on latex paper coatings [253].

9.5 Efforts in Commercialization of Hybrid Processing

Pyrolysis-based hybrid processing remains in its infancy. However, recent progress in the production of pyrolytic sugar [254] and continuing advances in microbial engineering [255] give promise for the economical production of fuels and chemicals via fermentation of pyrolysis-derived sugars.

Gasification-based hybrid processing has a longer history of development, with several commercialization efforts emerging in the last few years [256]. For example, Coskata Inc., an Illinois, USA based company, envisioned the steam reforming of natural gas to syngas followed by fermentation to ethanol. It reportedly operated a demonstration plant in Madison, PA, producing 40 000 gal per year of ethanol before the company was sold to Synata Bio in 2016 [257]. LanzaTech (www.lanzatech.com) entered a joint venture with the two largest steel companies in China to produce ethanol via microbial fermentation of CO-rich steel flue gas with a capacity of 300 million tons annually [14, 16]. Later, the company collaborated with Concord Blue (www.concordblueenergy.com) to produce ethanol and 2,3-butanediol (a jet fuel precursor) via gasification of woody biomass or solid waste (municipal solid waste, sewage sludge, agricultural residues) into high-quality syngas, followed by syngas fermentation. The pilot-scale facility began operation in 2015 to produce ethanol and value-added chemicals with a capacity of 250 l/day [258]. Another start-up company in syngas fermentation was INEOS Bio. They entered a joint venture with New Planet Energy Florida known as INEOS New Planet Bioenergy, LLC, to construct a commercial plant near Vero Beach, FL [259]. The plant was designed to produce 8 million gallons per year of ethanol and 6 MW of renewable electricity via gasification of vegetative waste followed by syngas fermentation [260]. Unexpectedly high concentrations of HCN (15 ppm) in syngas caused production delays and eventually wet scrubber towers were installed to remove HCN to less than 1 ppm [261]. The company was sold to a

Chinese investor in 2017 and renamed Jupeng Bio [262] while the biofuels plant was sold to Texas-based Frankens Energy LLC [263].

9.6 Conclusion and Perspectives

Hybrid processing employs a thermochemical process, pyrolysis, or gasification, to deconstruct biomass into fermentation substrates. Fast pyrolysis produces fermentable pyrolytic sugars, carboxylic acids, and lignin derivatives, which can be utilized by microorganisms to produce advanced hydrocarbon biofuels. Currently, the inhibition from the toxic compounds is the biggest hurdle for the viability of pyrolytic substrate fermentation. Further research should focus on the development of more efficient detoxification methods for less toxic substrates and robust strains highly tolerant to toxicity. It is important to use both of the two strategies in order to enhance the economic feasibility of the pyrolytic substrate fermentation process. Moreover, it is necessary to elucidate the mechanism of toxicity so that a reverse engineering of evolved strains can be developed for an expanded genetic toolbox to be applied to other microorganisms.

Gasification-based hybrid processing is more advanced than the pyrolysis-based platform, having been developed over several decades. It is attractive for its ability to utilize both the carbohydrate and lignin in biomass feedstocks. Nevertheless, it faces several challenges to successful commercial development. Although many microorganisms are more robust to contaminants in syngas, gas clean-up is still required and can be both capital intensive and entail high operating costs. Gas-to-liquid mass transfer is rate limiting to the fermentation process and requires further improvement. Finally, the number of products from syngas fermentation are currently more limited than traditional sugar-based fermentations, which reduces market opportunities. For example, as the most prominent product of syngas fermentation, ethanol is not an optimal fuel due to its low energy density and the "blend wall" limitation [264].

Despite these challenges, hybrid processing remains an attractive approach for advanced biofuels and bio-based products because of the robustness of thermochemical deconstruction of biomass and opportunities to improve fermentation of the resulting substrates through advances in biotechnology. Research directed toward these challenges in concert with growing demand for low-carbon fuels and bio-based products will improve the commercial prospects of hybrid processing in the coming years.

References

1. Brown, R.C. (2007). Hybrid thermochemical/biological processing. *Applied Biochemistry and Biotechnology* **137**: 947–956.
2. Shen, Y.W., Jarboe, L., Brown, R. et al. (2015). A thermochemical-biochemical hybrid processing of lignocellulosic biomass for producing fuels and chemicals. *Biotechnology Advances* **33**: 1799–1813.
3. Bridgwater, A.V. (2012). Review of fast pyrolysis of biomass and product upgrading. *Biomass and Bioenergy* **38**: 68–94.
4. Jarboe, L.R., Wen, Z., Choi, D. et al. (2011). Hybrid thermochemical processing: fermentation of pyrolysis-derived bio-oil. *Applied Microbiology and Biotechnology* **91**: 1519–1523.
5. Dry, M.E. (2002). The Fischer-Tropsch process: 1950–2000. *Catalysis Today* **71**: 227–241.
6. Munasinghe, P.C. and Khanal, S.K. (2010). Biomass-derived syngas fermentation into biofuels: opportunities and challenges. *Bioresource Technology* **101**: 5013–5022.

7. Phillips, J.R., Klasson, K.T., Clausen, E.C. et al. (1993). Biological production of ethanol from coal synthesis gas – medium development studies. *Applied Biochemistry and Biotechnology* **39–40**: 559–571.
8. Prosen, E.M., Radlein, D., Piskorz, J. et al. (1993). Microbial utilization of levoglucosan in wood pyrolysate as a carbon and energy source. *Biotechnology and Bioengineering* **42**: 538–541.
9. So, K. and Brown, R.C. (1999). Economic analysis of selected lignocellulose-to-ethanol conversion technologies. *Applied Biochemistry and Biotechnology* **77–79**: 633–640.
10. Lian, J., Chen, S., Zhou, S. et al. (2010). Separation, hydrolysis and fermentation of pyrolytic sugars to produce ethanol and lipids. *Bioresource Technology* **101**: 9688–9699.
11. Bredwell, M.D., Srivastava, P., and Worden, R.M. (1999). Reactor design issues for synthesis-gas fermentations. *Biotechnology Progress* **15**: 834–844.
12. Henstra, A.M., Sipma, J., Rinzema, A. et al. (2007). Microbiology of synthesis gas fermentation for biofuel production. *Current Opinion in Biotechnology* **18**: 200–206.
13. Kopke, M., Held, C., Hujer, S. et al. (2010). *Clostridium ljungdahlii* represents a microbial production platform based on syngas. *Proceedings of the National Academy of Sciences* **107**: 13087–13092.
14. Daniell, J., Köpke, M., and Simpson, S. (2012). Commercial biomass syngas fermentation. *Energies* **5**: 5372–5417.
15. Kopke, M., Mihalcea, C., Bromley, J.C. et al. (2011). Fermentative production of ethanol from carbon monoxide. *Current Opinion in Biotechnology* **22**: 320–325.
16. Liew, M.F., Köpke, M., and Simpson, S.D. (2013). Gas fermentation for commercial biofuels production. In: *Liquid, Gaseous and Solid Biofuels – Conversion Techniques* (ed. Z. Fang), 125–173. Rijeka, Croatia: InTech Inc.
17. Bridgwater, A.V. and Peacocke, G.V.C. (2000). Fast pyrolysis processes for biomass. *Renewable & Sustainable Energy Reviews* **4**: 1–73.
18. Oasmaa, A. and Czernik, S. (1999). Fuel oil quality of biomass pyrolysis oils – state of the art for the end users. *Energy & Fuels* **13**: 914–921.
19. Huber, G.W., Iborra, S., and Corma, A. (2006). Synthesis of transportation fuels from biomass: chemistry, catalysts, and engineering. *Chemical Reviews* **106**: 4044–4098.
20. Mohan, D., Pittman, C.U.J., and Steele, P.H. (2006). Pyrolysis of wood/biomass for bio-oil: a critical review. *Energy & Fuels* **20**: 848–889.
21. Patwardhan, P.R., Satrio, J.A., Brown, R.C. et al. (2009). Product distribution from fast pyrolysis of glucose-based carbohydrates. *Journal of Analytical and Applied Pyrolysis* **86**: 323–330.
22. Patwardhan, P.R., Satrio, J.A., Brown, R.C. et al. (2010). Influence of inorganic salts on the primary pyrolysis products of cellulose. *Bioresource Technology* **101**: 4646–4655.
23. Mayes, H.B., Nolte, M.W., Beckham, G.T. et al. (2015). The alpha-bet(a) of salty glucose pyrolysis: computational investigations reveal carbohydrate pyrolysis catalytic action by sodium ions. *American Chemical Society Catalysis* **5**: 192–202.
24. Patwardhan, P., Brown, R., and Shanks, B. (2011). Product distribution from the fast pyrolysis of hemicellulose. *ChemSusChem* **5**: 636–643.
25. Dobele, G., Dizhbite, T., Rossinskaja, G. et al. (2003). Pre-treatment of biomass with phosphoric acid prior to fast pyrolysis – a promising method for obtaining 1,6-anhydrosaccharides in high yields. *Journal of Analytical and Applied Pyrolysis* **68–69**: 197–211.
26. Dobele, G., Rossinskaja, G., Dizhbite, T. et al. (2005). Application of catalysts for obtaining 1,6-anhydrosaccharides from cellulose and wood by fast pyrolysis. *Journal of Analytical and Applied Pyrolysis* **74**: 401–405.
27. Dobele, G., Rossinskaja, G., Telysheva, G. et al. (1999). Cellulose dehydration and depolymerization reactions during pyrolysis in the presence of phosphoric acid. *Journal of Analytical and Applied Pyrolysis* **49**: 307–317.
28. Li, Q., Steele, P.H., Yu, F. et al. (2013). Pyrolytic spray increases levoglucosan production during fast pyrolysis. *Journal of Analytical and Applied Pyrolysis* **100**: 33–40.

29. Piskorz, J., Radlein, D.A.G., Scott, D.S. et al. (1989). Pretreatment of wood and cellulose for production of sugars by fast pyrolysis. *Journal of Analytical and Applied Pyrolysis* **16**: 127–142.
30. Dalluge, D.L., Daugaard, T., Johnston, P. et al. (2014). Continuous production of sugars from pyrolysis of acid-infused lignocellulosic biomass. *Green Chemistry* **16**: 4144.
31. Kuzhiyil, N., Dalluge, D., Bai, X. et al. (2012). Pyrolytic sugars from cellulosic biomass. *ChemSusChem* **5**: 2228–2236.
32. Kim, K.H., Bai, X., Rover, M. et al. (2014). The effect of low-concentration oxygen in sweep gas during pyrolysis of red oak using a fluidized bed reactor. *Fuel* **124**: 49–56.
33. Pollard, A.S., Rover, M.R., and Brown, R.C. (2012). Characterization of bio-oil recovered as stage fractions with unique chemical and physical properties. *Journal of Analytical and Applied Pyrolysis* **93**: 129–138.
34. Westerhof, R.J.M., Brilman, D.W.F., Garcia-Perez, M. et al. (2011). Fractional condensation of biomass pyrolysis vapors. *Energy & Fuels* **25**: 1817–1829.
35. Ragauskas, A.J., Beckham, G.T., Biddy, M.J. et al. (2014). Lignin valorization: improving lignin processing in the biorefinery. *Science* **344**: 1246843.
36. Tuck, C.O., Perez, E., Horvath, I.T. et al. (2012). Valorization of biomass: deriving more value from waste. *Science* **337**: 695–699.
37. Bai, X., Kim, K.H., Brown, R.C. et al. (2014). Formation of phenolic oligomers during fast pyrolysis of lignin. *Fuel* **128**: 170–179.
38. Patwardhan, P.R., Brown, R.C., and Shanks, B.H. (2011). Understanding the fast pyrolysis of lignin. *ChemSusChem* **4**: 1629–1636.
39. Woolcock, P.J. and Brown, R.C. (2013). A review of cleaning technologies for biomass-derived syngas. *Biomass and Bioenergy* **52**: 54–84.
40. Mitsutoshi, N., Yoshiom, S., and Tsuneo, Y. (1984). Itaconic acid fermentation of levaglucosan. *Journal of Fermentation Technology* **62**: 201–203.
41. Kitamura, Y., Abe, Y., and Yasui, T. (1991). Metabolism of levoglucosan (1,6-anhydro-β-D-glucopyranose) in microorganisms. *Agricultural and Biological Chemistry* **55**: 515–521.
42. Nakahara, K., Kitamura, Y., Yamagishi, Y. et al. (1994). Levoglucosan dehydrogenase involved in the assimilation of levoglucosan in *Arthrobacter* sp. I-552. *Bioscience, Biotechnology and Biochemistry* **58**: 2193–2196.
43. Zhuang, X.L., Zhang, H.X., and Yang, J.Z. (2001). Preparation of levoglucosan by pyrolysis of cellulose and its citric acid fermentation. *Bioresource Technology* **79**: 63–66.
44. Zhuang, X. and Zhang, H. (2002). Identification, characterization of levoglucosan kinase, and cloning and expression of levoglucosan kinase cDNA from *Aspergillus niger CBX-209* in *Escherichia coli*. *Protein Expression and Purification* **26**: 71–81.
45. Dai, J., Yu, Z., He, Y. et al. (2009). Cloning of a novel levoglucosan kinase gene from *Lipomyces starkeyi* and its expression in *Escherichia coli*. *World Journal of Microbiology and Biotechnology* **25**: 1589–1595.
46. Layton, D.S., Ajjarapu, A., Choi, D.W. et al. (2011). Engineering ethanologenic *Escherichia coli* for levoglucosan utilization. *Bioresource Technology* **102**: 8318–8322.
47. Lian, J., Garcia-Perez, M., and Chen, S. (2013). Fermentation of levoglucosan with oleaginous yeasts for lipid production. *Bioresource Technology* **133**: 183–189.
48. Bacik, J.P. and Jarboe, L.R. (2016). Bioconversion of anhydrosugars: emerging concepts and strategies. *IUBMB Life* **68**: 700–708.
49. Nakagawa, M., Sakai, Y., and Yasui, T. (1984). Itaconic acid fermentation of levoglucosan. *Journal of Fermentation Technology* **62**: 201–203.
50. Zhuang, X.L., Zhang, H.X., and Tang, J.J. (2001). Levoglucosan kinase involved in citric acid fermentation by *Aspergillus niger CBX-209* using levoglucosan as sole carbon and energy source. *Biomass and Bioenergy* **21**: 53–60.

51. Yu, Z. and Zhang, H. (2003). Pretreatment of cellulose pyrolysate for ethanol production by *Saccharomyces cerevisiae*, *Pichia* sp. YZ-1 and *Zymomonas mobilis*. *Biomass and Bioenergy* **24**: 257–262.
52. Bennett, N.M., Helle, S.S., and Duff, S.J. (2009). Extraction and hydrolysis of levoglucosan from pyrolysis oil. *Bioresource Technology* **100**: 6059–6063.
53. Sluiter, A., Hames, B., Ruiz, R. et al. (2012). *Determination of Structural Carbohydrates and Lignin in Biomass*. National Renewable Energy Laboratory.
54. Johnston, P.A. and Brown, R.C. (2014). Quantitation of sugar content in pyrolysis liquids after acid hydrolysis using high-performance liquid chromatography without neutralization. *Journal of Agricultural and Food Chemistry* **62**: 8129–8133.
55. Rover, M.R., Johnston, P.A., Lamsal, B.P. et al. (2013). Total water-soluble sugars quantification in bio-oil using the phenol-sulfuric acid assay. *Journal of Analytical and Applied Pyrolysis* **104**: 194–201.
56. Tessini, C., Vega, M., Muller, N. et al. (2011). High performance thin layer chromatography determination of cellobiosan and levoglucosan in bio-oil obtained by fast pyrolysis of sawdust. *Journal of Chromatography A* **1218**: 3811–3815.
57. Lian, J., Choi, J., Tan, Y.S. et al. (2016). Identification of soil microbes capable of utilizing cellobiosan. *PLoS One* **11**: 2.
58. Helle, S., Bennett, N.M., Lau, K. et al. (2007). A kinetic model for production of glucose by hydrolysis of levoglucosan and cellobiosan from pyrolysis oil. *Carbohydrate Research* **342**: 2365–2370.
59. Linger, J.G., Hobdey, S.E., Franden, M.A. et al. (2016). Conversion of levoglucosan and cellobiosan by *Pseudomonoas putida* KT2440. *Metabolic Engineering Communications* **3**: 24–29.
60. Ratledge, C. (2004). Fatty acid biosynthesis in microorganisms being used for single cell oil production. *Biochimie* **86**: 807–815.
61. Christophe, G., Deo, J.L., Kumar, V. et al. (2012). Production of oils from acetic acid by the oleaginous yeast *Cryptococcus curvatus*. *Applied Biochemistry and Biotechnology* **167**: 1270–1279.
62. Fei, Q., Chang, H.N., Shang, L. et al. (2012). The effect of volatile fatty acids as a sole carbon source on lipid accumulation by *Cryptococcus albidus* for biodiesel production. *Bioresource Technology* **102**: 2695–2701.
63. Fontanille, P., Kumar, V., Christophe, G. et al. (2012). Bioconversion of volatile fatty acids into lipids by the oleaginous yeast *Yarrowia lipolytica*. *Bioresource Technology* **114**: 443–449.
64. Fei, Q., Fu, R., Shang, L. et al. (2015). Lipid production by microalgae *Chlorella protothecoides* with volatile fatty acids (VFAs) as carbon sources in heterotrophic cultivation and its economic assessment. *Bioprocess and Biosystems Engineering* **38**: 691–700.
65. Lian, J., Garcia-Perez, M., Coates, R. et al. (2012). Yeast fermentation of carboxylic acids obtained from pyrolytic aqueous phases for lipid production. *Bioresource Technology* **118**: 177–186.
66. Liang, Y., Zhao, X., Chi, Z. et al. (2013). Utilization of acetic acid-rich pyrolytic bio-oil by microalga *Chlamydomonas reinhardtii*: reducing bio-oil toxicity and enhancing algal toxicity tolerance. *Bioresource Technology* **133**: 500–506.
67. Zhao, X., Chi, Z., Rover, M. et al. (2013). Microalgae fermentation of acetic acid-rich pyrolytic bio-oil: reducing bio-oil toxicity by alkali treatment. *Environmental Progress & Sustainable Energy* **32**: 955–961.
68. Zhao, X., Davis, K., Brown, R. et al. (2015). Alkaline treatment for detoxification of acetic acid-rich pyrolytic bio-oil for microalgae fermentation: effects of alkaline species and the detoxification mechanisms. *Biomass and Bioenergy* **80**: 203–212.
69. Rakicka, M., Lazar, Z., Dulermo, T. et al. (2015). Lipid production by the oleaginous yeast *Yarrowia lipolytica* using industrial by-products under different culture conditions. *Biotechnology for Biofuels* **8**: 104.
70. Overhage, J., Steinbuchel, A., and Priefert, H. (2003). Highly efficient biotransformation of eugenol to ferulic acid and further conversion to vanillin in recombinant strains of Escherichia coli. *Applied and Environmental Microbiology* **69**: 6569–6576.

71. Plaggenborg, R., Overhage, J., Loos, A. et al. (2006). Potential of *Rhodococcus* strains for biotechnological vanillin production from ferulic acid and eugenol. *Applied Microbiology and Biotechnology* **72**: 745–755.
72. Overhage, J., Steinbuchel, A., and Priefert, H. (2006). Harnessing eugenol as a substrate for production of aromatic compounds with recombinant strains of *Amycolatopsis* sp. hr167. *Journal of Biotechnology* **125**: 369–376.
73. Srivastava, S., Luqman, S., Khan, F. et al. (2010). Metabolic pathway reconstruction of eugenol to vanillin bioconversion in *Aspergillus niger*. *Bioinformation* **4**: 320–325.
74. Unno, T., Kim, S.J., Kanaly, R.A. et al. (2007). Metabolic characterization of newly isolated *Pseudomonas nitroreducens* Jin1 growing on eugenol and isoeugenol. *Journal of Agricultural and Food Chemistry* **55**: 8556–8561.
75. Ashengroph, M., Nahvi, I., Zarkesh-Esfahani, H. et al. (2012). Conversion of isoeugenol to vanillin by *Psychrobacter* sp. strain CSW4. *Applied Biochemistry and Biotechnology* **166**: 1–12.
76. Shimoni, E., Ravid, U., and Shoham, Y. (2000). Isolation of a *Bacillus* sp. capable of transforming isoeugenol to vanillin. *Journal of Biotechnology* **78**: 1–9.
77. Furukawa, H., Morita, H., Yoshida, T. et al. (2003). Conversion of isoeugenol into vanillic acid by *Pseudomonas putida* 158 cells exhibiting high isoeugenol-degrading activity. *Journal of Bioscience and Bioengineering* **96**: 401–403.
78. Shimoni, E., Baasov, T., Ravid, U. et al. (2003). Biotransformations of propenylbenzenes by an *Arthrobacter* sp. and its t-anethole blocked mutants. *Journal of Biotechnology* **105**: 61–70.
79. Zhao, L.Q., Sun, Z.H., Zheng, P. et al. (2005). Biotransformation of isoeugenol to vanillin by a novel strain of Bacillus fusiformis. *Biotechnology Letters* **27**: 1505–1509.
80. Zhang, Y.M., Xu, P., Han, S. et al. (2006). Metabolism of isoeugenol via isoeugenol-diol by a newly isolated strain of *Bacillus subtilis* hs8. *Applied Microbiology and Biotechnology* **73**: 771–779.
81. Zhao, L.Q., Sun, Z.H., Zheng, P. et al. (2006). Biotransformation of isoeugenol to vanillin by *Bacillus fusiformis* CGMCC1347 with the addition of resin HD-8. *Process Biochemistry* **41**: 1673–1676.
82. Kasana, R.C., Sharma, U.K., Sharma, N. et al. (2007). Isolation and identification of a novel strain of *Pseudomonas chlororaphis* capable of transforming isoeugenol to vanillin. *Current Microbiology* **54**: 457–461.
83. Hua, D.L., Ma, C.Q., Lin, S. et al. (2007). Biotransformation of isoeugenol to vanillin by a newly isolated Bacillus pumilus strain: identification of major metabolites. *Journal of Biotechnology* **130**: 463–470.
84. Yamada, M., Okada, Y., Yoshida, T. et al. (2007). Biotransformation of isoeugenol to vanillin by *Pseudomonas putida* IE27 cells. *Applied Microbiology and Biotechnology* **73**: 1025–1030.
85. Ashengroph, M., Nahvi, I., Zarkesh-Esfahani, H. et al. (2011). Candida galli strain PGO6: a novel isolated yeast strain capable of transformation of isoeugenol into vanillin and vanillic acid. *Current Microbiology* **62**: 990–998.
86. Achterholt, S., Priefert, H., and Steinbuchel, A. (2000). Identification of Amycolatopsis sp strain hr167 genes, involved in the bioconversion of ferulic acid to vanillin. *Applied Microbiology and Biotechnology* **54**: 799–807.
87. Plaggenborg, R., Overhage, J., Steinbuchel, A. et al. (2003). Functional analyses of genes involved in the metabolism of ferulic acid in Pseudomonas putida KT2440. *Applied Microbiology and Biotechnology* **61**: 528–535.
88. Yoon, S.H., Li, C., Lee, Y.M. et al. (2005). Production of vanillin from ferulic acid using recombinant strains of *Escherichia coli*. *Biotechnology and Bioprocess Engineering* **10**: 378–384.
89. Alvarado, I.E., Lomascolo, A., Navarro, D. et al. (2001). Evidence of a new biotransformation pathway of p-coumaric acid into p-hydroxybenzaldehyde in *Pycnoporus cinnabarinus*. *Applied Microbiology and Biotechnology* **57**: 725–730.
90. Agrawal, R., Seetharam, Y.N., Kelamani, R.C. et al. (2003). Biotransformation of ferulic acid to vanillin by locally isolated bacterial cultures. *Indian Journal of Biotechnology* **2**: 610–612.

91. Brunati, M., Marinelli, F., Bertolini, C. et al. (2004). Biotransformations of cinnamic and ferulic acid with actinomycetes. *Enzyme and Microbial Technology* **34**: 3–9.
92. Torre, P., de Faveri, D., Perego, P. et al. (2004). Bioconversion of ferulate into vanillin by Escherichia coli strain JM109/pBB1 in an immobilized-cell reactor. *Annual of Microbiology* **54**: 517–527.
93. Martinez-Cuesta, M.D., Payne, J., Hanniffy, S.B. et al. (2005). Functional analysis of the vanillin pathway in a vdh-negative mutant strain of *Pseudomonas fluorescens* AN103. *Enzyme and Microbial Technology* **37**: 131–138.
94. Bloem, A., Bertrand, A., Lonvaud-Funel, A. et al. (2007). Vanillin production from simple phenols by wine-associated lactic acid bacteria. *Letters in Applied Microbiology* **44**: 62–67.
95. Hua, D.L., Ma, C.Q., Song, L.F. et al. (2007). Enhanced vanillin production from ferulic acid using adsorbent resin. *Applied Microbiology and Biotechnology* **74**: 783–790.
96. Lee, E.G., Yoon, S.H., Das, A. et al. (2009). Directing vanillin production from ferulic acid by increased acetyl-CoA consumption in recombinant *Escherichia coli*. *Biotechnology and Bioengineering* **102**: 200–208.
97. Sarangi, P.K. and Sahoo, H.P. (2010). Enhancing the rate of ferulic acid bioconversion utilizing glucose as carbon source. *Science* **6**: 115–117.
98. Tilay, A., Bule, M., and Annapure, U. (2010). Production of biovanillin by one-step biotransformation using fungus *Pycnoporus cinnabarinus*. *Journal of Agricultural and Food Chemistry* **58**: 4401–4405.
99. Lesage-Meessen, L., Lomascolo, A., Bonnin, E. et al. (2002). A biotechnological process involving filamentous fungi to produce natural crystalline vanillin from maize bran. *Applied Biochemistry and Biotechnology* **102**: 141–153.
100. Barbosa, E.D., Perrone, D., Vendramini, A.L.D., and Leite, S.G.F. (2008). Vanillin production by Phanerochaete hrysosorium grown on green coconut agro-industrial husk in solid state fermentation. *BioResources* **3**: 1042–1050.
101. Linger, J.G., Vardon, D.R., Guarnieri, M.T. et al. (2014). Lignin valorization through integrated biological funneling and chemical catalysis. *Proceedings of the National Academy of Sciences* **111**: 12013–12018.
102. Fuchs, G., Boll, M., and Heider, J. (2011). Microbial degradation of aromatic compounds – from one strategy to four. *Nature Reviews Microbiology* **9**: 803–816.
103. Johnson, C.W. and Beckham, G.T. (2015). Aromatic catabolic pathway selection for optimal production of pyruvate and lactate from lignin. *Metabolic Engineering* **28**: 240–247.
104. Salvachúa, D., Karp, E.M., Nimlos, C.T. et al. (2015). Towards lignin consolidated bioprocessing: simultaneous lignin depolymerization and product generation by bacteria. *Green Chemistry* **17**: 4951–4967.
105. Jung, G.Y., Kim, J.R., Jung, H.O. et al. (1999). A new chemoheterotrophic bacterium catalyzing water-gas shift reaction. *Biotechnology Letters* **21**: 869–873.
106. Oh, Y.-K., Seol, E.-H., Kim, J.R. et al. (2003). Fermentative biohydrogen production by a new chemoheterotrophic bacterium *Citrobacter* sp. Y19. *International Journal of Hydrogen Energy* **28**: 1353–1359.
107. Jung, G.Y., Jung, H.O., Kim, J.R. et al. (1999). Isolation and characterization of *Rhodopseudomonas palustris* P4 which utilizes CO with the production of H_2. *Biotechnology Letters* **21**: 525–529.
108. Oh, Y.-K., Seol, E.-H., Kim, M.-S. et al. (2004). Photoproduction of hydrogen from acetate by a chemoheterotrophic bacterium *Rhodopseudomonas palustris* P4. *International Journal of Hydrogen Energy* **29**: 1115–1121.
109. Kerby, R.L., Ludden, P.W., and Roberts, G.P. (1995). Carbon monoxide-dependent growth of *Rhodospirillum rubrum*. *Journal of Bacteriology* **177**: 2241–2244.
110. Munk, A.C., Copeland, A., Lucas, S. et al. (2011). Complete genome sequence of *Rhodospirillum rubrum* type strain (S1^T). *Standards in Genomic Sciences* **4**: 293–302.
111. Maness, P.C. and Weaver, P.F. (2002). Hydrogen production from a carbon-monoxide oxidation pathway in *Rubrivivax gelatinosus*. *International Journal of Hydrogen Energy* **27**: 1407–1411.

112. Hu, P., Lang, J., Wawrousek, K. et al. (2012). Draft genome sequence of *Rubrivivax gelatinosus* CBS. *Journal of Bacteriology* **194**: 3262.
113. Nagashima, S., Kamimura, A., Shimizu, T. et al. (2012). Complete genome sequence of phototrophic betaproteobacterium *Rubrivivax gelatinosus* IL144. *Journal of Bacteriology* **194**: 3541–3542.
114. Sokolova, T.G., Gonzalez, J.M., Kostrikina, N.A. et al. (2001). *Carboxydobrachium pacificum* gen. nov., sp nov., a new anaerobic, thermophilic, CO-utilizing marine bacterium from Okinawa Trough. *International Journal of Systematic Evolution Microbiology* **51**: 141–149.
115. Wu, M., Ren, Q., Durkin, A.S. et al. (2005). Life in hot carbon monoxide: the complete genome sequence of *Carboxydothermus hydrogenoformans* Z-2901. *PLoS Genetics* **1**: 563–574.
116. Slepova, T.V., Sokolova, T.G., Lysenko, A.M. et al. (2006). *Carboxydocella sporoproducens* sp. nov., a novel anaerobic CO-utilizing/H_2-producing thermophilic bacterium from a Kamchatka hot spring. *International Journal of Systematic Evolution Microbiology* **56**: 797–800.
117. Sokolova, T.G., Kostrikina, N.A., Chernyh, N.A. et al. (2002). *Carboxydocella thermautotrophica* gen. nov., sp. nov., a novel anaerobic, CO-utilizing thermophile from a Kamchatkan hot spring. *International Journal of Systematic Evolution Microbiology* **52**: 1961–1967.
118. Sokolova, T.G., Kostrikina, N.A., Chernyh, N.A. et al. (2005). *Thermincola carboxydiphila* gen. nov., sp. nov., a novel anaerobic, carboxydotrophic, hydrogenogenic bacterium from a hot spring of the Lake Baikal area. *International Journal of Systematic Evolution Microbiology* **55**: 2069–2073.
119. Zavarzina, D.G., Sokolova, T.G., Tourova, T.P. et al. (2007). *Thermincola ferriacetica* sp. nov., a new anaerobic, thermophilic, facultatively chemolithoautotrophic bacterium capable of dissimilatory Fe(III) reduction. *Extremophiles* **11**: 1–7.
120. Sokolova, T., Hanel, J., Onyenwoke, R.U. et al. (2007). Novel chemolithotrophic, thermophilic, anaerobic bacteria *Thermolithobacter ferrireducens* gen. nov., sp. nov. and *Thermolithobacter carboxydivorans* sp. nov. *Extremophiles* **11**: 145–157.
121. Sokolova, T.G., Gonzalez, J.M., Kostrikina, N.A. et al. (2004). *Thermosinus carboxydivorans* gen. nov., sp. nov., a new anaerobic, thermophilic, carbon-monoxide-oxidizing, hydrogenogenic bacterium from a hot pool of Yellowstone National Park. *International Journal of Systematic Evolution Microbiology* **54**: 2353–2359.
122. Parshina, S.N., Sipma, J., Nakashimada, Y. et al. (2005). *Desulfotomaculum carboxydivorans* sp. nov., a novel sulfate-reducing bacterium capable of growth at 100% CO. *International Journal of Systematic Evolution Microbiology* **55**: 2159–2165.
123. Lee, H.S., Kang, S.G., Bae, S.S. et al. (2008). The complete genome sequence of *Thermococcus onnurineus* NA1 reveals a mixed heterotrophic and carboxydotrophic metabolism. *Journal of Bacteriology* **190**: 7491–7499.
124. Kim, Y.J., Lee, H.S., Kim, E.S. et al. (2010). Formate-driven growth coupled with H_2 production. *Nature* **467**: 352–355.
125. Kim, M., Bae, S., Kim, Y. et al. (2013). CO-dependent H_2 production by genetically engineered *Thermococcus onnurineus* NA1. *Applied and Environmental Microbiology* **79**: 2048–2053.
126. Sokolova, T.G., Jeanthon, C., Kostrikina, N.A. et al. (2004). The first evidence of anaerobic CO oxidation coupled with H_2 production by a hyperthermophilic archaeon isolated from a deep-sea hydrothermal vent. *Extremophiles* **8**: 317–323.
127. Oger, P., Sokolova, T.G., Kozhevnikova, D.A. et al. (2011). Complete genome sequence of the hyperthermophilic archaeon *Thermococcus* sp. strain AM4, capable of organotrophic growth and growth at the expense of hydrogenogenic or sulfidogenic oxidation of carbon monoxide. *Journal of Bacteriology* **193**: 7019–7020.
128. Poehlein, A.S., Kaster, A.-K., Goenrich, M. et al. (2012). An ancient pathway combining carbon dioxide fixation with the generation and utilization of a sodium ion gradient for ATP synthesis. *PLoS One* **7**: e33439.
129. Gaddy, J.L. (2000). Biological production of ethanol from waste gases with Clostridium ljungdahlii. US Patent No. 6136577 A.

130. Kane, M.D. and Breznak, J.A. (1991). *Acetonema longum* gen. nov. sp. nov., an H_2/CO_2 acetogenic bacterium from the termite, *Pterotermes occidentis*. *Archives of Microbiology* **156**: 91–98.
131. Tocheva, E.I., Dekas, A.E., McGlynn, S.E. et al. (2013). Polyphosphate storage during sporulation in the gram-negative bacterium *Acetonema longum*. *Journal of Bacteriology* **195**: 3940–3946.
132. Allen, T.D., Caldwell, M.E., Lawson, P.A. et al. (2010). *Alkalibaculum bacchi* gen. nov., sp. nov., a CO-oxidizing, ethanol-producing acetogen isolated from livestock-impacted soil. *International Journal of Systematic Evolution Microbiology* **60**: 2483–2489.
133. Liu, K., Atiyeh, H.K., Tanner, R.S. et al. (2012). Fermentative production of ethanol from syngas using novel moderately alkaliphilic strains of *Alkalibaculum bacchi*. *Bioresource Technology* **104**: 336–341.
134. Misoph, M. and Drake, H.L. (1996). Effect of CO_2 on the fermentation capacities of the acetogen *Peptostreptococcus productus* U-1. *Journal of Bacteriology* **178**: 3140–3145.
135. Liu, C., Finegold, S.M., Song, Y. et al. (2008). Reclassification of *Clostridium coccoides*, *Ruminococcus hansenii*, *Ruminococcus hydrogenotrophicus*, *Ruminococcus luti*, *Ruminococcus productus* and *Ruminococcus schinkii* as *Blautia coccoides* gen. nov., comb. nov., *Blautia hansenii* comb. nov., *Blautia hydrogenotrophica* comb. nov., *Blautia luti* comb. nov., *Blautia producta* comb. nov., *Blautia schinkii* comb. nov. and description of *Blautia wexlerae* sp. nov., isolated from human faeces. *International Journal of Systematic Evolution Microbiology* **8**: 1896–1902.
136. Worden, R.M., Grethlein, A.J., Zeikus, J.G. et al. (1989). Butyrate production from carbon monoxide by *Butyribacterium methylotrophicum*. *Applied Biochemistry and Biotechnology* **20–21**: 687–698.
137. Grethlein, A.J., Worden, R.M., Jain, M.K. et al. (1991). Evidence of production of n-butanol from carbon monoxide by *Butyribacterium methylotrophicum*. *Journal of Fermentation and Bioengineering* **72**: 58–60.
138. Braun, M. and Gottschalk, G. (1981). *Clostridium aceticum* (Wieringa), a microorganism producing acetic acid from molecular hydrogen and carbon dioxide. *Archives of Microbiology* **128**: 288–293.
139. Poehlein, A., Cebulla, M., Ilg, M.M. et al. (2015). The complete genome sequence of *Clostridium aceticum*: a missing link between Rnf- and cytochrome-containing autotrophic acetogens. *MBio* **6**: e01168–e01115.
140. Abrini, J.N.H. and Nyns, E.-J. (1994). *Clostridium autoethanogenum*, sp. nov., an anaerobic bacterium that produces ethanol from carbon monoxide. *Archives of Microbiology* **161**: 345–351.
141. Kopke, M., Mihalcea, C., Liew, F. et al. (2011). 2,3-butanediol production by acetogenic bacteria, an alternative route to chemical synthesis, using industrial waste gas. *Applied and Environmental Microbiology* **77**: 5467–5475.
142. Bruno-Barcena, J.M., Chinn, M.S., and Grunden, A.M. (2013). Genome sequence of the autotrophic acetogen *Clostridium autoethanogenum* JA strain DSM 10061, a producer of ethanol from carbon monoxide. *Genome Announcements* **1**: e00628–e00613.
143. Utturkar, S.M., Klingeman, D.M., Bruno-Barcena, J.M. et al. (2015). Sequence data for *Clostridium autoethanogenum* using three generations of sequencing technologies. *Scientific Data* **2**: 150014.
144. Liou, J.S., Balkwill, D.L., Drake, G.R. et al. (2005). *Clostridium carboxidivorans* sp. nov., a solvent-producing clostridium isolated from an agricultural settling lagoon, and reclassification of the acetogen *Clostridium scatologenes* strain SL1 as *Clostridium drakei* sp. nov. *International Journal of Systematic Evolution Microbiology* **55**: 2085–2091.
145. Bruant, G., Levesque, M.J., Peter, C. et al. (2010). Genomic analysis of carbon monoxide utilization and butanol production by *Clostridium carboxidivorans* strain P7[T]. *PLoS One* **5**: e13033.
146. Paul, D., Austin, F.W., Arick, T. et al. (2010). Genome sequence of the solvent-producing bacterium *Clostridium carboxidivorans* strain P7[T]. *Journal of Bacteriology* **192**: 5554–5555.
147. Gossner, A.S., Picardal, F., Tanner, R.S. et al. (2008). Carbon metabolism of the moderately acid-tolerant acetogen *Clostridium drakei* isolated from peat. *FEMS Microbiology Letters* **287**: 236–242.
148. Diekert, G.B. and Thauer, R.K. (1978). Carbon monoxide oxidation by *Clostridium thermoaceticum* and *Clostridium formicoaceticum*. *Journal of Bacteriology* **136**: 597–606.

149. Ohwaki, K. and Hungate, R.E. (1977). Hydrogen utilization by clostridia in sewage sludge. *Applied and Environmental Microbiology* **33**: 1270–1274.
150. Kusel, K., Karnholz, A., Trinkwalter, T. et al. (2001). Physiological ecology of *Clostridium glycolicum* RD-1, an aerotolerant acetogen isolated from sea grass roots. *Applied and Environmental Microbiology* **67**: 4734–4741.
151. Tanner, R.S., Miller, L.M., and Yang, D. (1993). *Clostridium ljungdahlii* sp. nov., an acetogenic species in *Clostridial* rRNA homology group I. *International Journal of Systematic Evolution Microbiology* **43**: 232–236.
152. Schink, B. (1984). *Clostridium magnum* sp. nov., a non-autotrophic homoacetogenic bacterium. *Archives of Microbiology* **137**: 250–255.
153. Kane, M.D., Brauman, A., and Breznak, J.A. (1991). *Clostridium mayombei* sp. nov., an H_2/CO_2 acetogenic bacterium from the gut of the African soil-feeding termite, *Cubitermes speciosus*. *Archives of Microbiology* **156**: 99–104.
154. Mechichi, T., Labat, M., Patel, B.K.C. et al. (1999). *Clostridium methoxybenzovorans* sp. nov., a new aromatic o-demethylating homoacetogen from an olive wastewater treatment digester. *International Journal of Systematic Evolution Microbiology* **49**: 1201–1209.
155. Kundiyana, D.K., Huhnke, R.L., Maddipati, P. et al. (2010). Feasibility of incorporating cotton seed extract in *Clostridium* strain P11 fermentation medium during synthesis gas fermentation. *Bioresource Technology* **101**: 9673–9680.
156. Genthner, B.R.S. and Bryant, M.P. (1987). Additional characteristics of one-carbon-compound utilization by *Eubacterium limosum* and *Acetobacterium woodii*. *Applied and Environmental Microbiology* **53**: 471–476.
157. Roh, H., Ko, H.J., Kim, D. et al. (2011). Complete genome sequence of a carbon monoxide-utilizing acetogen, *Eubacterium limosum* KIST612. *Journal of Bacteriology* **193**: 307–308.
158. Krumholz, L.R. and Bryant, M.P. (1985). *Clostridium pfennigii* sp. nov. uses methoxyl groups of monobezenoids and produces butyrate. *International Journal of Systematic Evolution Microbiology* **35**: 454–456.
159. Drake, H.L. and Daniel, S.L. (2004). Physiology of the thermophilic acetogen *Moorella thermoacetica*. *Research in Microbiology* **155**: 869–883.
160. Sakai, S., Nakashimada, Y., Inokuma, K. et al. (2005). Acetate and ethanol production from H_2 and CO_2 by *Moorella* sp. using a repeated batch culture. *Journal of Bioscience and Bioengineering* **99**: 252–258.
161. Wiegel, J., Braun, M., and Gottschalk, G. (1981). *Clostridium thermoautotrophicum* species novum, a thermophile producing acetate from molecular hydrogen and carbon dioxide. *Current Microbiology* **5**: 255–260.
162. Savage, M.D., Wu, Z.G., Daniel, S.L. et al. (1987). Carbon monoxide-dependent chemolithotrophic growth of *Clostridium thermoautotrophicum*. *Applied and Environmental Microbiology* **53**: 1902–1906.
163. Leigh, J.A. and Wolfe, R.S. (1981). *Acetogenium kivui*, a new thermophilic hydrogen-oxidizing, acetogenic bacterium. *Archives of Microbiology* **129**: 275–280.
164. Daniel, S.L., Hsu, T., Dean, S.I. et al. (1990). Characterization of the H_2- and CO-dependent chemolithotrophic potentials of the acetogens *Clostridium thermoaceticum* and *Acetogenium kivui*. *Journal of Bacteriology* **172**: 4464–4471.
165. Parshina, S.N., Kijlstra, S., Henstra, A.M. et al. (2005). Carbon monoxide conversion by thermophilic sulfate-reducing bacteria in pure culture and in co-culture with *Carboxydothermus hydrogenoformans*. *Applied Microbiology and Biotechnology* **68**: 390–396.
166. Visser, M., Worm, P., Muyzer, G. et al. (2013). Genome analysis of *Desulfotomaculum kuznetsovii* strain 17^T reveals a physiological similarity with *Pelotomaculum thermopropionicum* strain SI^T. *Standards in Genomic Sciences* **8**: 69–87.

167. Plugge, C.M., Balk, M., and Stams, A.J.M. (2002). *Desulfotomaculum thermobenzoicum* subsp *thermosyntrophicum* subsp nov., a thermophilic, syntrophic, propionate-oxidizing, spore-forming bacterium. *International Journal of Systematic Evolution Microbiology* **52**: 391–399.
168. Klenk, H.P., Clayton, R.A., Tomb, J.F. et al. (1997). The complete genome sequence of the hyperthermophilic, sulphate-reducing archaeon *Archaeoglobus fulgidus*. *Nature* **390**: 364–375.
169. Henstra, A.M., Dijkema, C., and Stams, A.J. (2007). *Archaeoglobus fulgidus* couples CO oxidation to sulfate reduction and acetogenesis with transient formate accumulation. *Environmental Microbiology* **9**: 1836–1841.
170. Do, Y.S., Smeenk, J., Broer, K.M. et al. (2007). Growth of *Rhodospirillum rubrum* on synthesis gas: conversion of CO to H_2 and poly-beta-hydroxyalkanoate. *Biotechnology and Bioengineering* **97**: 279–286.
171. Drennan, C.L., Heo, J., Sintchak, M.D. et al. (2001). Life on carbon monoxide: X-ray structure of *Rhodospirillum rubrum* Ni-Fe-S carbon monoxide dehydrogenase. *Proceedings of the National Academy of Sciences* **98**: 11973–11978.
172. Ljungdahl, L.G. (1986). The autotrophic pathway of acetate synthesis in acetogenic bacteria. *Annual Review of Microbiology* **40**: 415–450.
173. Wood, H.G., Ragsdale, S.W., and Pezacka, E. (1986). The acetyl-CoA pathway of autotrophic growth. *FEMS Microbiology Reviews* **39**: 345–362.
174. Leang, C., Ueki, T., Nevin, K.P. et al. (2013). A genetic system for *Clostridium ljungdahlii*: a chassis for autotrophic production of biocommodities and a model homoacetogen. *Applied and Environmental Microbiology* **79**: 1102–1109.
175. Banerjee, A., Leang, C., Ueki, T. et al. (2014). Lactose-inducible system for metabolic engineering of *Clostridium ljungdahlii*. *Applied and Environmental Microbiology* **80**: 2410–2416.
176. Ueki, T., Nevin, K.P., Woodard, T.L. et al. (2014). Converting carbon dioxide to butyrate with an engineered strain of *Clostridium ljungdahlii*. *MBio* **5**: e01636–e01614.
177. Kopke, M. and Liew, F. (2011). Recombinant microorganism and methods of production thereof. US Patent 13/049,263.
178. Straub, M., Demler, M., Weuster-Botz, D. et al. (2014). Selective enhancement of autotrophic acetate production with genetically modified *Acetobacterium woodii*. *Journal of Biotechnology* **178**: 67–72.
179. Kopke, M., Gerth, M.L., Maddock, D.J. et al. (2014). Reconstruction of an acetogenic 2,3-butanediol pathway involving a novel NADPH-dependent primary-secondary alcohol dehydrogenase. *Applied and Environmental Microbiology* **80**: 3394–3403.
180. Czernik, S. and Bridgwater, A.V. (2004). Overview of applications of biomass fast pyrolysis oil. *Energy & Fuels* **18**: 590–598.
181. Capunitan, J.A. and Capareda, S.C. (2013). Characterization and separation of corn stover bio-oil by fractional distillation. *Fuel* **112**: 60–73.
182. Wang, S., Gu, Y., Liu, Q. et al. (2009). Separation of bio-oil by molecular distillation. *Fuel Processing Technology* **90**: 738–745.
183. Zheng, J.-L. and Wei, Q. (2011). Improving the quality of fast pyrolysis bio-oil by reduced pressure distillation. *Biomass and Bioenergy* **35**: 1804–1810.
184. Zhang, X.S., Yang, G.X., Jiang, H. et al. (2013). Mass production of chemicals from biomass-derived oil by directly atmospheric distillation coupled with co-pyrolysis. *Scientific Reports* **3**: 1120.
185. Ruddy, D.A., Schaidle, J.A., Ferrell, J.R. et al. (2014). Recent advances in heterogeneous catalysts for bio-oil upgrading via "ex situ catalytic fast pyrolysis": catalyst development through the study of model compounds. *Green Chemistry* **16**: 454–490.
186. Zaldivar, J. and Ingram, L.O. (1999). Effect of organic acids on the growth and fermentation of ethanologenic *Escherichia coli* LY01. *Biotechnology and Bioengineering* **66**: 203–210.
187. Zaldivar, J., Martinez, A., and Ingram, L.O. (1999). Effect of selected aldehydes on the growth and fermentation of ethanologenic *Escherichia coli*. *Biotechnology and Bioengineering* **65**: 24–33.

188. Palmqvist, E. and Hahn-Hagerdal, B. (2000). Fermentation of lignocellulosic hydrolysates. II: inhibitors and mechanisms of inhibition. *Bioresource Technology* **74**: 25–33.
189. Monlau, F., Sambusiti, C., Barakat, A. et al. (2014). Do furanic and phenolic compounds of lignocellulosic and algae biomass hydrolyzate inhibit anaerobic mixed cultures? A comprehensive review. *Biotechnology Advances* **32**: 934–951.
190. Oasmaa, A., Kuoppala, E., Gust, S. et al. (2003). Fast pyrolysis of forestry residue. 1. Effect of extractives on phase separation of pyrolysis liquids. *Energy & Fuels* **17**: 1–12.
191. Wang, H., Livingston, D., Srinivasan, R. et al. (2012). Detoxification and fermentation of pyrolytic sugar for ethanol production. *Applied Biochemistry and Biotechnology* **168**: 1568–1583.
192. Li, Y., Shao, J., Wang, X. et al. (2013). Upgrading of bio-oil: removal of the fermentation inhibitor (furfural) from the model compounds of bio-oil using pyrolytic char. *Energy & Fuels* **27**: 5975–5981.
193. Chan, J.K. and Duff, S.J. (2010). Methods for mitigation of bio-oil extract toxicity. *Bioresource Technology* **101**: 3755–3759.
194. Chi, Z., Rover, M., Jun, E. et al. (2013). Overliming detoxification of pyrolytic sugar syrup for direct fermentation of levoglucosan to ethanol. *Bioresource Technology* **150**: 220–227.
195. Rover, M.R., Johnston, P.A., Jin, T. et al. (2014). Production of clean pyrolytic sugars for fermentation. *ChemSusChem* **7**: 1662–1668.
196. Kazi, F.K., Fortman, J.A., Anex, R.P. et al. (2010). Techno-economic comparison of process technologies for biochemical ethanol production from corn stover. *Fuel* **89**: S20–S28.
197. Humbird, D., Davis, R., Tao, L. et al. (2011). *Process Design and Economics for Biochemical Conversion of Lignocellulosic Biomass to Ethanol: Dilute-Acid Pretreatment and Enzymatic Hydrolysis of Corn Stover*. Golden, Colorado: National Renewable Energy Laboratory.
198. Stanford, J.P., Hall, P.H., Rover, M.R. et al. (2018). Separation of sugars and phenolics from the heavy fraction of bio-oil using polymeric resin adsorbents. *Separation and Purification Technology* **194**: 170–180.
199. Parawira, W. and Tekere, M. (2011). Biotechnological strategies to overcome inhibitors in lignocellulose hydrolysates for ethanol production: review. *Critical Review in Biotechnology* **31**: 20–31.
200. Jonsson, L.J., Palmqvist, E., Nilvebrant, N.-O. et al. (1998). Detoxification of wood hydrolysates with laccase and peroxidase from the white-rot fungus *Trametes versicolor*. *Applied Microbiology and Biotechnology* **49**: 691–697.
201. Kapoor, R.K., Rajan, K., and Carrier, D.J. (2015). Applications of *Trametes versicolor* crude culture filtrates in detoxification of biomass pretreatment hydrolyzates. *Bioresource Technology* **189**: 99–106.
202. Ludwig, D., Amann, M., Hirth, T. et al. (2013). Development and optimization of single and combined detoxification processes to improve the fermentability of lignocellulose hydrolyzates. *Bioresource Technology* **133**: 455–461.
203. Larsson, S., Reimann, A., Nilvebrant, N.-O. et al. (1999). Comparison of different methods for the detoxification of lignocellulosic hydrolyzates of spruce. *Applied Biochemistry and Biotechnology* **77–79**: 91–103.
204. Lopez, M.J., Nichols, N.N., Dien, B.S. et al. (2004). Isolation of microorganisms for biological detoxification of lignocellulosic hydrolysates. *Applied Microbiology and Biotechnology* **64**: 125–131.
205. Nichols, N.N., Sharma, L.N., Mowery, R.A. et al. (2008). Fungal metabolism of fermentation inhibitors present in corn stover dilute acid hydrolysate. *Enzyme and Microbial Technology* **42**: 624–630.
206. Palmqvist, E., Hahn-Hagerdal, B., Szengyel, Z. et al. (1997). Simultaneous detoxification and enzyme production of hemicellulose hydrolysates obtained after steam pretreatment. *Enzyme and Microbial Technology* **20**: 286–293.
207. Austin, S., Kontur, W.S., Ulbrich, A. et al. (2015). Metabolism of multiple aromatic compounds in corn stover hydrolysate by *Rhodopseudomonas palustris*. *Environmental Science & Technology* **49**: 8914–8922.

208. Okuda, N., Soneura, M., Ninomiya, K. et al. (2008). Biological detoxification of waste house wood hydrolysate using *Ureibacillus thermosphaericus* for bioethanol production. *Journal of Bioscience and Bioengineering* **106**: 128–133.
209. Wierckx, N., Koopman, F., Bandounas, L. et al. (2010). Isolation and characterization of *Cupriavidus basilensis* HMF14 for biological removal of inhibitors from lignocellulosic hydrolysate. *Microbial Biotechnology* **3**: 336–343.
210. Fonseca, B.G., Moutta, R.O., Ferraz, F.O. et al. (2011). Biological detoxification of different hemicellulosic hydrolysates using *Issatchenkia occidentalis* CCTCC M 206097 yeast. *Journal of Industrial Microbiology and Biotechnology* **38**: 199–207.
211. Zhang, H.-R., Qin, X.-X., Silva, S.S. et al. (2009). Novel isolates for biological detoxification of lignocellulosic hydrolysate. *Applied Biochemistry and Biotechnology* **152**: 199–212.
212. Zhang, J., Zhu, Z., Wang, X. et al. (2010). Biodetoxification of toxins generated from lignocellulose pretreatment using a newly isolated fungus, *Amorphotheca resinae* ZN1, and the consequent ethanol fermentation. *Biotechnology for Biofuels* **3**: 26.
213. Khiyami, M., Pometto, A.L., and Brown, R.C. (2005). Detoxification of corn stover and corn starch pyrolysis liquors by ligninolytic enzymes of *Phanerochaete chrysosporium*. *Journal of Agricultural and Food Chemistry* **53**: 2969–2977.
214. Jarboe, L.R., Liu, P., and Royce, L.A. (2011). Engineering inhibitor tolerance for the production of biorenewable fuels and chemicals. *Current Opinion in Chemical Engineering* **1**: 38–42.
215. Gutierrez, T., Ingram, L.O., and Preston, J.F. (2006). Purification and characterization of a furfural reductase (FFR) from *Escherichia coli* strain LYO1–an enzyme important in the detoxification of furfural during ethanol production. *Journal of Biotechnology* **121**: 154–164.
216. Miller, E.N., Jarboe, L.R., Yomano, L.P. et al. (2009). Silencing of NADPH-dependent oxidoreductase genes (*yqhD* and *dkgA*) in furfural-resistant ethanologenic *Escherichia coli*. *Applied and Environmental Microbiology* **75**: 4315–4323.
217. Zheng, H., Wang, X., Yomano, L.P. et al. (2012). Increase in furfural tolerance in ethanologenic *Escherichia coli* LY180 by plasmid-based expression of *thyA*. *Applied and Environmental Microbiology* **78**: 4346–4352.
218. Wang, X., Miller, E.N., Yomano, L.P. et al. (2011). Increased furfural tolerance due to overexpression of NADH-dependent oxidoreductase FucO in *Escherichia coli* strains engineered for the production of ethanol and lactate. *Applied and Environmental Microbiology* **77**: 5132–5140.
219. Wang, X., Yomano, L.P., Lee, J.Y. et al. (2013). Engineering furfural tolerance in *Escherichia coli* improves the fermentation of lignocellulosic sugars into renewable chemicals. *Proceedings of the National Academy of Sciences* **110**: 4021–4026.
220. Ahmed, A., Cateni, B.G., Huhnke, R.L. et al. (2006). Effects of biomass-generated producer gas constituents on cell growth, product distribution and hydrogenase activity of *Clostridium carboxidivorans* P7^T. *Biomass and Bioenergy* **30**: 665–672.
221. Ahmed, A. and Lewis, R.S. (2007). Fermentation of biomass-generated synthesis gas: effects of nitric oxide. *Biotechnology and Bioengineering* **97**: 1080–1086.
222. Datar, R.P., Shenkman, R.M., Cateni, B.G. et al. (2004). Fermentation of biomass-generated producer gas to ethanol. *Biotechnology and Bioengineering* **86**: 587–594.
223. Xu, D., Tree, D.R., and Lewis, R.S. (2011). The effects of syngas impurities on syngas fermentation to liquid fuels. *Biomass and Bioenergy* **35**: 2690–2696.
224. Greenwood, A. (2014). INEOS Bio making little cellulosic ethanol at US plant. http://www.icis.com/resources/news/2014/09/05/9818462/ineos-bio-making-little-cellulosic-ethanol-at-us-plant. Houston, TX: ICIS. Accessed: Jul 3, 2018.
225. Broer, K.M., Woolcock, P.J., Johnston, P.A. et al. (2015). Steam/oxygen gasification system for the production of clean syngas from switchgrass. *Fuel* **140**: 282–292.
226. Gaddy, J.L., Arora, D.K., and Ko, C.-W., et al. (2007). Methods for increasing the production of ethanol from microbial fermentation. US Patent No. 7, 285,402 B2.

227. Hurst, K.M. and Lewis, R.S. (2010). Carbon monoxide partial pressure effects on the metabolic process of syngas fermentation. *Biochemical Engineering Journal* **48**: 159–165.
228. Worden, R.M., Bredwell, M.D., and Grethlein, A.J. (1997). Engineering issues in synthesis-gas fermentations. In: *Fuels and Chemicals from Biomass*, ACS Symposium Series, vol. **666** (ed. B.C. Saha and J. Woodward), 320–335. Washington DC: American Chemical Society.
229. Riggs, S.S. and Heindel, T.J. (2006). Measuring carbon monoxide gas-liquid mass transfer in a stirred tank reactor for syngas fermentation. *Biotechnology Progress* **22**: 903–906.
230. Bredwell, M.D. and Worden, R.M. (1998). Mass-transfer properties of microbubbles. 1. Experimental studies. *Biotechnology Progress* **14**: 31–38.
231. Olle, B., Bucak, S., Holmes, T.C. et al. (2006). Enhancement of oxygen mass transfer using functionalized magnetic nanoparticles. *Industrial & Engineering Chemistry Research* **45**: 4355–4363.
232. Zhu, H., Shanks, B.H., Choi, D.W. et al. (2010). Effect of functionalized MCM41 nanoparticles on syngas fermentation. *Biomass and Bioenergy* **34**: 1624–1627.
233. Kantarci, N., Borak, F., and Ulgen, K.O. (2005). Bubble column reactors. *Process Biochemistry* **40**: 2263–2283.
234. Shetty, S.A. and Kantak, M.V. (1992). Gas-phase backmixing in bubble-colmn reactors. *AIChE Journal* **38**: 1013–1026.
235. Chang, I.S., Kim, B.H., Lovitt, R.W. et al. (2001). Effect of CO partial pressure on cell-recycled continuous CO fermentation by *Eubacterium limosum* KIST612. *Process Biochemistry* **37**: 411–421.
236. Rajagopalan, S., Datar, R.P., and Lewis, R.S. (2002). Formation of ethanol from carbon monoxide via a new microbial catalyst. *Biomass and Bioenergy* **23**: 487–493.
237. Shen, Y., Brown, R., and Wen, Z. (2014). Enhancing mass transfer and ethanol production in syngas fermentation of *Clostridium carboxidivorans* P7 through a monolithic biofilm reactor. *Applied Energy* **136**: 68–76.
238. Ebrahimi, S., Kleerebezem, R., Kreutzer, M.T. et al. (2006). Potential application of monolith packed columns as bioreactors, control of biofilm formation. *Biotechnology and Bioengineering* **93**: 238–245.
239. Klasson, K.T., Ackerson, M.D., Clausen, E.C. et al. (1992). Bioconversion of synthesis gas into liquid or gaseous fuels. *Enzyme and Microbial Technology* **14**: 602–608.
240. Orgill, J.J., Atiyeh, H.K., Devarapalli, M. et al. (2013). A comparison of mass transfer coefficients between trickle-bed, hollow fiber membrane and stirred tank reactors. *Bioresource Technology* **133**: 340–346.
241. Atiyeh, H.K., Devarapalli, M., Lewis, R.S., et al. (2013). Semi-continuous syngas fermentation in a trickle bed reactor. AIChE Annual Meeting. San Francisco, CA.
242. Lee, P.-H., Ni, S.-Q., Chang, S.-Y. et al. (2012). Enhancement of carbon monoxide mass transfer using an innovative external hollow fiber membrane (HFM) diffuser for syngas fermentation: experimental studies and model development. *Chemical Engineering Journal* **184**: 268–277.
243. Liu, K., Atiyeh, H.K., Stevenson, B.S. et al. (2014). Continuous syngas fermentation for the production of ethanol, *n*-propanol and *n*-butanol. *Bioresource Technology* **151**: 69–77.
244. Munasinghe, P.C. and Khanal, S.K. (2010). Syngas fermentation to biofuel: evaluation of carbon monoxide mass transfer coefficient ($k_L a$) in different reactor configurations. *Biotechnology Progress* **26**: 1616–1621.
245. Munasinghe, P.C. and Khanal, S.K. (2012). Syngas fermentation to biofuel: evaluation of carbon monoxide mass transfer and analytical modeling using a composite hollow fiber (CHF) membrane bioreactor. *Bioresource Technology* **122**: 130–136.
246. Shen, Y., Brown, R.C., and Wen, Z. (2014). Syngas fermentation of *Clostridium carboxidivorans* P7 in a hollow fiber membrane biofilm reactor: evaluating the mass transfer coefficient and ethanol production performance. *Biochemical Engineering Journal* **85**: 21–29.
247. Cote, P., Alam, Z., and Penny, J. (2012). Hollow fiber membrane life in membrane bioreactors (MBR). *Desalination* **288**: 145–151.

248. Robles, A., Ruano, M.V., Ribes, J. et al. (2013). Factors that affect the permeability of commercial hollow-fibre membranes in a submerged anaerobic MBR (HF-SAnMBR) system. *Water Research* **47**: 1277–1288.
249. Flickinger, M.C., Schottel, J.L., Bond, D.R. et al. (2007). Painting and printing living bacteria: engineering nanoporous biocatalytic coatings to preserve microbial viability and intensify reactivity. *Biotechnology Progress* **23**: 2–17.
250. Bernal, O.I., Mooney, C.B., and Flickinger, M.C. (2014). Specific photosynthetic rate enhancement by cyanobacteria coated onto paper enables engineering of highly reactive cellular biocomposite "leaves". *Biotechnology and Bioengineering* **111**: 1993–2008.
251. Estrada, J.M., Bernal, O.I., Flickinger, M.C. et al. (2015). Biocatalytic coatings for air pollution control: a proof of concept study on VOC biodegradation. *Biotechnology and Bioengineering* **112**: 263–271.
252. Gosse, J.L., Engel, B.J., Hui, J.C. et al. (2010). Progress toward a biomimetic leaf: 4,000 h of hydrogen production by coating-stabilized nongrowing photosynthetic *Rhodopseudomonas palustris*. *Biotechnology Progress* **26**: 907–918.
253. Gosse, J.L., Chinn, M.S., Grunden, A.M. et al. (2012). A versatile method for preparation of hydrated microbial-latex biocatalytic coatings for gas absorption and gas evolution. *Journal of Industrial Microbiology and Biotechnology* **39**: 1269–1278.
254. Zhang, Y., Brown, T.R., Hu, G. et al. (2013). Techno-economic analysis of monosaccharide production via fast pyrolysis of lignocellulose. *Bioresource Technology* **127**: 358–365.
255. Peralta-Yahya, P.P., Zhang, F., del Cardayre, S.B. et al. (2012). Microbial engineering for the production of advanced biofuels. *Nature* **488**: 320–328.
256. Brown, T.R. and Brown, R.C. (2013). A review of cellulosic biofuel commercial-scale projects in the United States. *Biofuels, Bioproducts and Biorefining* **7**: 235–245.
257. Lane, J. (2016). Coskata's technology re-emerges as Synata Bio. Biofuels Digest, 24 January 2016. http://www.biofuelsdigest.com/bdigest/2016/01/24/coskatas-technology-re-emerges-as-synata-bio (accessed 3 July 2018).
258. Anon. Facilities. http://www.lanzatech.com/facilities (accessed 3 July 2018).
259. US Department of Energy (2010). INEOS Bio commercializes bioenergy technology in Florida, 2010. http://www1.eere.energy.gov/bioenergy/pdfs/ibr_arraprojects_ineos.pdf (accessed 3 July 2018).
260. INEOS Bio (2013). INEOS Bio announces operational progress at Florida plant. Ethanol Producer Magazine, 9 December 2013. http://ethanolproducer.com/articles/10528/ineos-bio-announces-operational-progress-at-florida-plant (accessed 3 July 2018).
261. Lane, J. (2014). One the mend: Why INEOS Bio isn't producing ethanol in Florida. BiofuelsDigest, 5 September 2014. http://www.biofuelsdigest.com/bdigest/2014/09/05/on-the-mend-why-ineos-bio-isnt-reporting-much-ethanol-production (accessed 3 July 2018).
262. Anon (2017). Jupeng Bio has acquired INEOS Bio for the manufacture of biofuels from renewable carbon sources, 19 August 2017. http://www.jupengbio.com/blog/jupeng-bio-has-acquired-ineos-bio-for-the-manufacture-of-biofuels-from (accessed 3 July 2018).
263. Wixon, C. (2018). Former INEOS bioenergy plant sold; county in talks with new owner over possible partnership. TCPalm, 24 January 2018. https://www.tcpalm.com/story/news/local/indian-river-county/2018/01/24/former-ineos-bioenergy-plant-sold-county-talks-new-owner-over-possible-partnership-promises-great-pl/1060002001 (accessed 3 July 2018).
264. Peplow, M. (2014). Cellulosic ethanol fights for life. *Nature* **507**: 152–153.

10

Costs of Thermochemical Conversion of Biomass to Power and Liquid Fuels

Mark M. Wright[1] and Tristan Brown[2]

[1] Department of Mechanical Engineering, Iowa State University, Ames, IA, USA
[2] Department of Forest and Natural Resources Management, SUNY ESF, Syracuse, NY, USA

10.1 Introduction

This chapter describes capital and operating costs of various thermochemical conversion technology for biomass to power and liquid fuel applications. These technologies range from direct combustion to fast pyrolysis and gasification to fuels. All of these technologies can contribute energy to a low-carbon transportation fuel sector including combustion engines and battery-powered electric vehicles.

Capital and operating costs are gathered from techno-economic analysis studies, which employ a wide range of assumptions to estimate costs of mature commercial plants. Thus, the accuracy of these estimates depends on the availability of reliable data. The typical uncertainty for these costs is +100%/−30%, underscoring the potential risks of novel technologies. Uncertainty analysis can mitigate some of these risks. Alternatively, comparative analysis using a common basis can be useful for selecting among competing technologies. Estimates summarized below have not been corrected for different basis years, plant scales, or assumed rates of return among the different sources of data, which can strongly affect cost estimates. Other assumptions that vary widely in the literature include the cost of feedstock, equipment, and utilities. An improved comparison would adjust results found in the literature for the appropriate cost basis year, plant capacity, and depreciation method among other considerations [1].

10.2 Electric Power Generation

10.2.1 Direct Combustion to Power

Biomass combustion power plants have applications dating to the nineteenth century. While the pathway has evolved over that time, a typical system involves the combustion of biomass to produce hot gases that exchange heat with a water cycle to yield steam. The steam is then used to either provide process heat or to drive a turbine to generate electricity, or both in a combined heat and power (CHP) system. Figure 10.1 shows a general schematic for a biomass combustion power plant. A relative lack of process steps, high capacity factors, and the development of widespread practical experience with biomass combustion systems over time results in high boiler efficiencies of 70% or more [2].

A 2007 report by the US Environmental Protection Agency (EPA) identified the presence of large economies of scale for both stoker boiler integrated steam plants and fluidized bed integrated steam plants [2]. Unit costs of $121/lb steam and $151/lb steam were reported for the two systems, respectively, at a 900 tons/day capacity. These unit costs increased to $232/lb steam and $480/lb steam, respectively, when the capacity was reduced to 100 tons/day, and the fluidized bed boiler equipment cost was determined to incur comparatively high capital costs at the lower scale. Non-fuel O&M costs for the stoker boiler and fluidized bed systems were also found to be almost identical at $0.73 and $0.74/1000 lb steam, respectively for the 900 tons/day capacity systems, but much higher at $3.55 and $4.19/1000 lb steam, respectively for the 100 tons/day systems. Small-scale systems can be economically viable despite these higher costs in rural locations with limited natural gas supply and forested areas with low-cost biomass feedstocks (such as forest residues from forestry operations).

Surveys of techno-economic analyses of biomass combustion power plants in different countries have found capital costs to be roughly comparable across both locations and systems. A comparison of steam boiler, Stirling engine, and organic Rankine cycle (ORC) systems in Austria and Denmark calculated specific investment costs in the range of $284–$381/kW$_{th}$ (€1 – $1.18) [3]. The survey identified the lowest production costs for the Stirling engine system of $0.02/kWh$_{th}$, followed by the ORC system at $0.03/kWh$_{th}$ and the steam boiler system at $0.04/kWh$_{th}$. Within CHP systems the steam boiler and Stirling engine systems were found to incur the lowest production costs at approximately

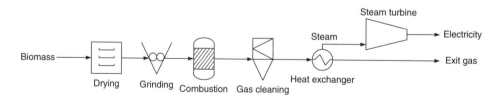

Figure 10.1 Process diagram for direct combustion to power.

$0.12/kWh_e$ each, compared to $0.15/kWh_e$ for the ORC. The latter's comparatively low electric efficiency was identified as the cause of its higher electricity production cost.

10.2.2 Gasification to Power

Biomass gasification systems are capable of achieving overall thermodynamic efficiencies superior to those of biomass combustion systems when combined with integrated gasification combined cycle (IGCC) power systems. While both types of systems convert the biomass to electricity via steam, IGCC systems include an additional bottoming cycle that uses the high temperatures of the gases as they leave the gasifier to drive a gas turbine as well (see Figure 10.2). IGCC power systems are characterized by power-to-heat ratios of 47.3% versus 18.7% for ORC systems with comparable thermal capacities [4].

The higher complexity and equipment needs of IGCC power systems results in higher capital costs; a comparison of European ORC and IGCC power systems calculated capital costs of $3.1 million and $4.1 million, respectively, for 2.1 MW_{th} capacities [4]. The IGCC systems' higher electricity output results in higher revenues, however, and internal rates of return of 14.8% and 18.1% for the ORC and IGCC power systems, respectively.

Biomass gasification power systems are very sensitive to differences in capacity. A comparison of two different gasification power systems, one employing an externally fired gas turbine and the other employing a gas engine, found that both achieved IRRs in excess of 17.5% at a capacity of 300 kW_e but below 6% at a capacity of 100 kW_e [5]. The gas engine system was especially sensitive to capacity, with IRRs of 19.7% and 1.2% reported for the two capacities, respectively.

10.2.3 Fast Pyrolysis to Power

There are several possible configurations for fast pyrolysis power generation. Pyrolysis products can be combusted either separately or in combinations [6, 7]. Gas turbines can combust pyrolysis vapors. Bio-oil can fuel conventional oil boilers [8]. Bio-oil can be treated to form solid Lignocol which pulverizes into fine particles [9]. Biochar can substitute for coal in conventional boilers. The versatility of fast pyrolysis to power configurations means that they can be adapted to meet economic, environmental, or logistical needs.

Distributed power generation is the most common scenario for fast pyrolysis systems. Small-scale and mobile fast pyrolysis units can be deployed in remote locations to convert low-value biomass into pyrolysis products, which can be combusted directly or stored and transported to a centralized plant [10]. Combined fast pyrolysis and diesel engine generation plants have capital costs of $2400–$6200/kW for units with capacities of

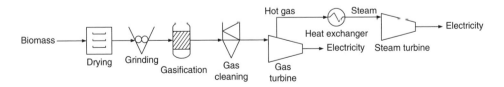

Figure 10.2 Process diagram for gasification to power.

1–20 MW, respectively [6]. Operating costs for these systems range between $0.08/kWh and $0.25/kWh with costs decreasing at larger scales.

Coal-fired boilers can combust bio-oil and biochar. However, conventional boilers are not designed to use liquid feeds or tolerate the high ash content of biochar. Feeding the bio-oil and biochar as a slurry partially addresses these challenges because the mixture still contains a high ash content. A novel approach is to thermally treat the pyrolysis liquids to form a friable material known as Lignocol that shares similar mechanical and thermal properties as coal [9]. This approach allows co-feeding of bio-oil in conventional coal power plants. Co-feeding Lignocol with coal in an existing power plant yields electricity at the cost of $0.088–$0.149/kWh [11]. In this scenario, the biochar could be combusted to reduce costs or sequestered as a form of biomass energy and carbon capture and sequestration (BECCS) agent providing significant carbon reductions.

10.3 Liquid Fuels via Gasification

Gasification has a long history dating back to at least the 1940s as a pathway for the conversion of carbonaceous feedstocks to liquid fuels. Subsequent decades have seen the development of several additional methods for producing liquid fuels from biomass via the reaction of syngas with various metallic or biocatalysts. These include Fischer-Tropsch synthesis (FTS), acetic acid synthesis (AAS), mixed alcohols synthesis (MAS), methanol-to-gasoline synthesis (MTG), syngas-to-distillates (S2D), and syngas fermentation (SF). All of these pathways involve the thermal decomposition of biomass at temperatures of up to 1500 °C into syngas, which is a gaseous mixture of carbon monoxide, carbon dioxide, hydrogen, methane, and trace amounts of light hydrocarbons. The syngas is then cleansed of contaminants to minimize catalyst poisoning, although the degree of cleaning that is necessary varies by process according to the resistance of the catalyst to poisoning, with biocatalysts being more resistant than metal catalysts. At this point, the clean syngas stream is upgraded to one or more liquid fuels.

10.3.1 Gasification to Fischer-Tropsch Liquids

Figure 10.3 shows the primary steps involved in gasification and FTS. Following the production of biomass syngas, the first step of the FTS process involves the reacting of the clean syngas over a cobalt, iron, or ruthenium catalyst to yield high molecular weight alkanes and

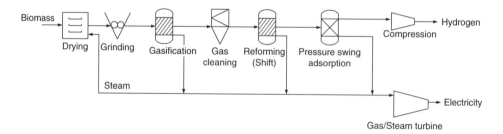

Figure 10.3 Process diagram for gasification and FTS.

waxes. The waxes are hydrocracked to fuel-range hydrocarbon molecules, allowing the process to yield refinery blendstocks capable of further conversion to bio-based versions of refined fuels such as diesel fuel and gasoline.

The FTS process has been the subject of a relatively large number of techno-economic assessments due to its history of use at commercial-scale volumes (albeit with fossil feedstocks). An early analysis [12] that was the basis for multiple subsequent studies calculated a minimum selling price (MSP) for liquid fuels produced by the pathway of $2.70 per gallon of gasoline equivalent (gge) ($0.71/l of gasoline equivalent, lge). Subsequent analyses of the pathway at larger capacities have calculated higher MSPs ranging from $4.63/gge ($1.22/lge) to $7.13/gge ($1.88/lge) [13]. An important contributor to these high production costs are the large capital expenditures of up to $16/gge capacity necessary for a pathway that includes a depolymerization step followed by repolymerization and yet another depolymerization step, reducing the process efficiency while increasing the equipment requirements [14]. The pathway has been found to have unattractive financial returns compared to other cellulosic biofuel pathways under harmonized operating assumptions as a result [15].

10.3.2 Gasification to Mixed Alcohols

Reacting clean syngas with a ZnO/CuO catalyst yields methanol. While methanol is capable of being blended with gasoline for use in the existing transportation infrastructure, the alcohol's high toxicity and potential for groundwater contamination has limited its use as such. However, distilling the raw methanol and reacting it over iodide- and iridium-based catalysts as part of the AAS process yields acetic acid that is hydrogenated to produce fuel ethanol after an additional distillation step [16].

The techno-economics of the AAS process have not been extensively studied. A 2009 analysis of two different system configurations identified capital costs of up to $752 million at a feedstock capacity of 2000 metric tons per day (MTPD) [14], resulting in MSPs of up to $4.41/gge ($1.16/lge). A subsequent analysis that compared the AAS process with other cellulosic biofuel pathways under uncertainty calculated a very high probability that the process MSP would fall no lower than $4/gge ($1.06/lge) even under optimistic operating conditions [17].

The MTG and S2D processes also convert syngas to methanol. They diverge by dehydrating the methanol to produce dimethyl ether, however, which is reacted in the MTG process over a zeolite catalyst to yield a mix of alkanes and aromatics with characteristics that are very similar to gasoline blendstock. The S2D process reduces the number of process steps by reacting the methanol produced with catalysts capable of accomplishing both dehydration and hydrocarbon synthesis in a single reactor.

When the MAS process is employed, the syngas is first compressed and then combined with methanol, at which point the mixture is reacted with a metal sulfide catalyst to produce a stream of mixed alcohols. The stream is separated and divided into its methanol, ethanol, and high molecular weight alcohol components. The methanol is recycled back into the process while the ethanol is distilled into fuel ethanol.

Techno-economic studies of the MAS process resemble those of the MTG process in reporting a wide range of MSPs for 2000 MTPD facilities, from a low of $1.74/gge ($0.46/lge) [18] to a high of $3.90/gge ($1.03/lge) [19]. A more recent analysis calculates

Table 10.1 Capital costs for 72.6 MMGPY gasification-to-mixed-alcohols plant.

Section (2000 dry tonnes/d)	Capital cost (US$ million)
Feed handling and drying	23.2
Gasification	12.9
Tar reforming and quench	38.4
Acid gas and sulfur removal	14.5
Alcohol synthesis – compression	16.0
Alcohol synthesis – other	4.60
Alcohol separation	7.20
Steam system and power generation	16.8
Cooling water and other utilities	3.60
Total	137

Table 10.2 Operating costs for 72.6 MMGPY gasification-to-mixed-alcohols plant.

Operating costs (2000 dry tons/day)	Annual cost (US$ million)
Feedstock	27.0
Catalysts	0.20
Olivine	0.40
Other raw materials	0.30
Waste disposal	0.30
Electricity	0.00
Fixed costs	12.1
Co-product credits	(12.8)
Capital depreciation	9.50
Average income tax	7.30
Average return on investment	17.6
Total	34.9

a pathway MSP of $3.33/gge ($0.88/lge) for a facility with the same capacity [20]. The MSP result exhibits a strong sensitivity to the pathway capital cost, estimates of which range from $220 million to $560 million for 2000 MTPD facilities [13]. A harmonized analysis calculated a very low probability that the pathway would be capable of achieving a positive 20-year net present value (NPV) despite the comparatively low bottom estimate for pathway MSP, illustrating the sensitivity of techno-economic analyses of the pathway to analytical assumptions [13].

Researchers at the National Renewable Energy Laboratory (NREL) estimated MAS capital costs of $137 million and operating costs of $1.01/gal for a 72.6 MMGPY mixed-alcohol biomass plant [21]. Tables 10.1 and 10.2 show capital and operating cost breakdowns for these estimates.

10.3.3 Gasification to Gasoline

The MTG process has been the subject of several techno-economic studies that have arrived at very different results for MSP: from a low of $2.12/gge ($0.56/lge) [22] for a 2000 MTPD facility to a high of $6.26/gge ($1.65/lge) for a 2472 MTPD facility that only includes syngas upgrading equipment and does not account for gasification equipment costs [23]. A primary distinction between these results is the installation factors employed in the calculation of the process capital costs, with lower factors being associated with low capital costs and vice versa. A harmonized analysis has found that the MSP of the process is very sensitive to this factor, although its process economics remain attractive compared to other cellulosic biofuel pathways even when a higher install factor is employed [24].

10.3.4 Gasification and Syngas Fermentation to Ethanol

The SF process differs from the other upgrading processes in that it employs a biological step in the form of fermentation. The fermentation of the cleaned syngas is accomplished by *Clostridium* bacteria that metabolize the hydrogen and carbon monoxide in the syngas to yield ethanol. This process has undergone less development than the other processes but is attracting the interest of both researchers and industry due to the reduced number of process steps.

An analysis of ethanol production via the SF process calculated a capital cost for a 2030 MTPD facility of $562 million [25]. While this figure is approximately in the middle of capital cost estimates for comparable cellulosic biofuel production facilities [13], the analysis calculates a MSP of $7.29/gge ($1.93/l) that is among the highest on the same basis. Low conversion yields to ethanol, high capital costs, and an energy-intensive ethanol recovery step are major drivers of the high MSP.

A comparative analysis of the syngas fermentation process considering multiple lignocellulosic feedstocks and countries calculated a MSP of up to $1.03/l, although this was found to be sensitive to both feedstock and production location [26]. The main distinction between the two SF process analyses was an assumed ethanol yield in the latter that was more than twice that used in the former [17, 18]. Both analyses agreed that the ethanol recovery step was an important contributor to MSP, however, illustrating both the challenges faced by the SF process and a potential area for improvement.

10.3.5 Gasification and Syngas Fermentation to PHA and Co-product Hydrogen

Syngas fermentation is a hybrid thermochemical/biochemical process that can produce a variety of bio-based products [27]. Although it is being commercially developed for biofuels production by several companies, the published literature contains very little on the costs of the process. One prominent exception is a techno-economic study by Choi et al. [28] that considers the use of *Rhodospirillum rubrum*, a purple non-sulfur bacterium, to simultaneously convert the carbon monoxide in syngas to the biopolymer polyhydroxyalkonate (PHA) and enrich the hydrogen content of the gas through a biologically mediated water-gas shift reaction. The process of gasifying biomass followed by steam reforming to remove organic contaminants and syngas fermentation to PHA and co-product hydrogen is illustrated in Figure 10.4.

Choi et al. [28] assumed the syngas fermentation biorefinery would have a daily production output of 12 tonne of PHA and 50 tonne of hydrogen gas. Grassroots capital for the plant was estimated to be $55.5 million, with annual net operating cost of $6.7 million based on a credit of $2.00/kg for hydrogen co-product. Assuming a plant capacity factor of

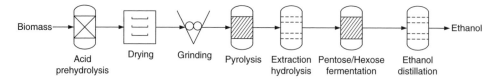

Figure 10.4 Process diagram for bio-oil fermentation to ethanol.

Table 10.3 Operating costs for 40 MTPY gasification and syngas fermentation-to-PHA and co-product hydrogen plant.

Operating costs	Annual costs (US$ million)
Raw materials	16.1
Credit for H_2	(32.3)
Labor, utilities, maintenance	9.1
Indirect costs	6.4
Annual capital charges	7.4
Total	6.7
PHA production costs	$1.65/kg

93%, the unit cost of PHA was estimated to be $1.65/kg (see Table 10.3). Co-production of hydrogen was essential to this attractive production cost.

10.4 Liquid Fuels via Fast Pyrolysis

There are several pathways for converting fast pyrolysis liquids into transportation fuels: these include (i) non-catalytic fast pyrolysis and hydroprocessing, (ii) catalytic fast pyrolysis and hydroprocessing, and (iii) fast pyrolysis and FTS. Each of these pathways can be implemented as centralized facilities or in hub and spoke configurations with multiple distributed fast pyrolysis units feeding to a centralized upgrading site. Furthermore, these pathways can have different configurations depending on the use of pyrolysis by-products. Fast pyrolysis is an interesting option for BECCS. By sequestering biochar, fast pyrolysis could yield carbon-negative biofuel while improving soil carbon and nutrient content [29].

10.4.1 Fast Pyrolysis and Hydroprocessing

Conventional, non-catalytic fast pyrolysis has been studied extensively. This approach is depicted in Figure 10.5. Biomass drying, grinding, pyrolysis, gas cleaning, and oil collection can be done either at small scales or large scales. Following bio-oil collection, the pyrolysis liquids can either be hydroprocessed on site or transported to a centralized refinery.

Bio-oil could be co-fed with petroleum in existing crude oil refineries, reducing equipment and infrastructure costs. Operating costs depend primarily on the use of natural gas for heating and hydrogen input needs. A standalone fast pyrolysis facility could rely solely on biomass, but natural gas provides a lower cost heat and hydrogen resource. Tables 10.4

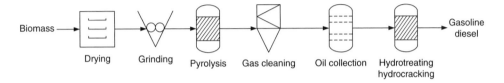

Figure 10.5 Diesel and gasoline production from bio-oil upgrading.

Table 10.4 Capital costs for 35 MMGPY fast pyrolysis and bio-oil upgrading plant.

Section	Capital costs (US$ million)
Hydroprocessing	48.7
Combustion	47.3
Pyrolysis and oil recovery	28.0
Pretreatment	20.2
Utilities	9.1
Storage	5.8
Installed equipment cost	159.1
Total	276.6

Table 10.5 Operating costs for 35 MMGPY fast pyrolysis and bio-oil upgrading plant.

Operating costs	Annual costs (US$ million)
Feedstock	54.4
Electricity	5.8
Solids disposal	1.8
Catalyst	1.8
Fixed costs	11.2
Co-product credits	(11.3)
Capital depreciation	11.9
Average income tax	9.3
Average return on investment	22.5
Total	107.4

and 10.5 show capital and operating costs respectively for fast pyrolysis and bio-oil upgrading. Costs for conventional fast pyrolysis biofuels range between $1.49 and $3.69/gal [30].

10.4.2 Catalytic Fast Pyrolysis and Hydroprocessing

Catalytic fast pyrolysis increases the compatibility of pyrolysis liquids with the petroleum infrastructure. Catalysts can be employed in situ or ex situ depending on whether they are placed within the pyrolysis reactor or downstream [31].

The advantage of in situ catalytic pyrolysis is that it enables direct contact between the biomass and catalyst. Direct contact with biomass is attractive for promoting desirable primary reactions. However, in situ catalysts are exposed to alkali and alkaline earth metals in the biomass that can lead to their deactivation. In situ catalyst regeneration requires periodical removal and regeneration of the catalyst that can add to the costs.

Ex situ catalytic pyrolysis feeds the pyrolysis products into a downstream reactor either in the vapor phase or as a condensed liquid. The advantage of ex situ catalytic pyrolysis is that it processes a relatively clean pyrolysis stream, which could increase the catalyst lifetime. The biofuel costs are similar for both approaches, with prices of $4.20 and $4.27/gal for in situ and ex situ respectively.

10.4.3 Fast Pyrolysis and Gasification to Fuels

Gasification provides an alternative approach to upgrading bio-oil. Slurries made of bio-oil and biochar can be fed into entrained flow gasifiers to generate syngas. The syngas can then be upgraded to transportation fuels via the various pathways described in the gasification section.

Wright and co-worker [10] conducted a production-scale analysis of the costs to produce FTL using a distributed biomass processing system. This study estimates higher capital costs than a conventional single-facility system, but lower fuel costs offset the capital costs due to reduced feedstock transportation costs. Figure 10.6 illustrates the concept of distributed biomass processing and centralized gasification to fuels. Capital and operating costs for a 550 MMGPY FTL scenario are estimated at $4.03 billion and $852 million, respectively. The costs of upgrading syngas to gasoline via dimethyl ether synthesis are estimated to be $4.32/gal [32].

Fast pyrolysis benefits from the ability to operate as a biorefinery, producing chemicals and other products in addition to fuels. A comparison of various pyrolysis biorefinery configurations shows that switching between hydrocarbons and biochemicals can improve profitability over different market conditions [33, 34].

10.4.4 Fast Pyrolysis and Bio-oil Fermentation to Ethanol

In the absence of inorganic contaminants, especially alkali and alkaline earth metals, thermal depolymerization of cellulose and hemicelluloses yields anhydrosugars [35], which can be hydrolyzed to fermentable sugars [36]. Pretreating biomass before pyrolysis to either remove or otherwise deactivate alkali and alkaline earth metals can increase yields of fermentable sugars in bio-oil to over 25% [37]. A scheme for accomplishing this so-called bio-oil fermentation is illustrated in Figure 10.7. In many respects, it resembles conventional biochemical production of cellulosic ethanol production, including an acid pretreatment step, except that acid or enzymatic hydrolysis is replaced with pyrolysis for depolymerizing plant carbohydrates.

So and Brown [38] have performed one of the only techno-economic analyses of bio-oil fermentation. Tables 10.6 and 10.7 summarize the capital and operating costs for bio-oil fermentation for 50 MMGPY production. Total capital investment was estimated to be $142 million in 1997 dollars. Annual operating costs were estimated to be $118 million. Assuming a plant capacity factor of 90%, this translates to a unit cost of production of $2.35/gal. Costs can be expected to be significantly higher in 2010 dollars.

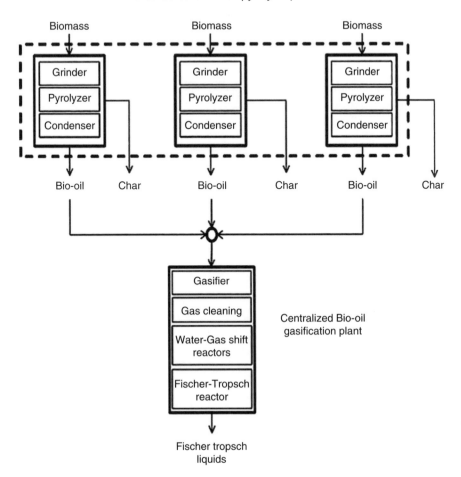

Figure 10.6 Distributed bio-oil production and centralized gasification upgrading to FTL [10].

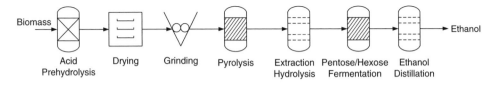

Figure 10.7 Process diagram for bio-oil fermentation to ethanol.

Table 10.6 Capital costs for 35 MMGPY fast pyrolysis and bio-oil upgrading plant.

Section	Capital costs (US$ million)
Hydroprocessing	48.7
Combustion	47.3
Pyrolysis and oil recovery	28.0
Pretreatment	20.2
Utilities	9.1
Storage	5.8
Installed equipment cost	159.1
Total	276.6

Table 10.7 Operating costs for 35 MMGPY fast pyrolysis and bio-oil upgrading plant.

Operating costs	Annual costs (US$ million)
Feedstock	54.4
Electricity	5.8
Solids disposal	1.8
Catalyst	1.8
Fixed costs	11.2
Co-product credits	(11.3)
Capital depreciation	11.9
Average income tax	9.3
Average return on investment	22.5
Total	107.4

10.5 Liquid Fuels via Direct Liquefaction

Direct liquefaction involves the thermal conversion of biomass within a solvent. Hydrothermal processing employs water as a solvent whereas solvent liquefaction can employ a variety of solvents including gamma-valerolactone (GVL), tetrahydrofuran (THF), and acetone [39]. Direct liquefaction costs are similar to fast pyrolysis but differ primarily in higher capital costs for biomass conversion. The higher pressures employed in liquefaction result in higher material costs, which are compensated by a higher yield and lower oxygen content in the biocrude product.

Biomass hydrothermal processing costs range between $2.52/gal and $4.44/gal, depending on the maturity of the technology [40]. Hydrothermal processing is attractive for converting high moisture-containing feedstocks such as algae. Assuming that the lipid portion can be sold to the animal feed market, the algae remnants could be upgraded into fuels at a price of $2.57/gal [41].

Solvent liquefaction can improve the conversion of biomass to sugars and other chemical products. The sugars can be fermented into ethanol or hydrocarbons. This approach could yield ethanol at the cost of $4.33–$4.99/gal [42].

10.6 Liquid Fuels via Esterification

The production of liquid fuels via esterification differs from the pathways covered above in that it has been widely commercialized in a large number of countries. The process also utilizes lipid feedstocks, primarily soybean oil and palm oil, rather than lignocellulosic biomass. The growing use of biodiesel to meet carbon intensity reduction mandates has prompted interest in the use of waste lipid feedstocks, such as used cooking oil and animal processing waste, with the esterification process. These spent lipids are characterized by their high free fatty acid contents, however, that necessitate the use of different catalysts than are used with conventional lipid feedstocks.

A comparative techno-economic analysis of three different esterification process facilities with capacities of 36 000 metric tons of biodiesel per year employing different catalysts (homogenous catalyst with preesterification, homogenous catalyst without preesterification, and heterogeneous catalyst) in conjunction with spent lipid feedstock found substantial differences in process economics [43]. A biodiesel production cost of approximately $0.57/l was calculated for all three catalyst scenarios. The heterogeneous catalyst scenario achieved an IRR of 32% compared to 11% and 7% for the two homogenous catalyst scenarios, however, due to its ability to produce a higher volume of biodiesel despite incurring increased catalyst costs.

Microalgae as lipid feedstock for esterification has been the subject of several high-profile research efforts due to the limited availability of waste lipid feedstocks. A recent review of techno-economic assessments of microalgal-based esterification processes identified a vast range of MSPs in the refereed literature, from a low of $1.64/gge ($0.43/l) to a high of more than $15/gge ($3.96/l) [44]. This lack of agreement between assessments was attributed by the authors to differences in technology statuses, financial assumptions, lipid yields, and microalgae production systems. More fundamentally, the range reflects the lack of commercialization of microalgae production systems to date.

10.7 Summary and Conclusions

Table 10.8 summarizes capital and operating costs for the technologies described in this chapter. Direct comparisons among the technologies are not possible because of differences in cost basis and underlying assumptions. Capital costs are a strong function of plant capacity, and feedstock costs can have a major impact on the final cost of fuel from biomass. Also, comparisons of different kinds of fuels on a volumetric basis, which is common in the techno-economic literature on biofuels, can be misleading because of the different energy content of fuels. For example, methanol and ethanol have only 50% and 66% of the energy content of gasoline on a volumetric basis, which affects the range of vehicles fueled on these alcohols. Similarly, liquefied hydrogen has only 25% of the energy content of gasoline. Finally, many of the gasification scenarios appear to assume very high capacity factors (90% or higher) without including sufficient equipment redundancy to assure this level of operational reliability. In practice, the capacity factors for gasifiers can be as low as 70%; thus, plant capacity factors of 90% would require two gasifiers with one serving as back-up.

Table 10.8 Summary of costs for thermochemical biomass conversion technologies.

Technology	Capital cost (US$/kW)	Electricity cost (US$/kWh)
Direct combustion to power	600	0.075
Pyrolysis to power	2400	0.080
Pyrolysis (Lignocol) to power	1200[a]	0.088
Gasification to power	1600	0.050

	Capital cost (US$ million)	Operating cost (US$ million)	Product cost (US$)
Fast pyrolysis and hydroprocessing	287	109	$3.09/gal
Bio-oil gasification to liquid fuels	4029	852	$1.55/gal
Bio-oil fermentation to ethanol	69	39.20	$1.57/gal
Direct liquefaction (human toxicity potential [HTP])	424	158	$2.57/gal
Gasification to hydrogen	282	52.10	$0.29/gal
Gasification to methanol	224	60.60	$0.70/gal
Gasification to mixed alcohols	137	34.90	$1.01/gal
Gasification to FTL	341	50.80	$1.45/gal
Gasification and fermentation to PHA with co-product H_2	103	18.20	$2.80/kg PHA

[a] Brownfield capital costs.

Similar redundancy would be required for gas compressors, which at the scale of these plants are custom manufactured.

There are numerous pathways to the production of biorenewable fuels. Projected economic costs are an important factor in the selection of biomass conversion technologies for investment in further research or commercial enterprises. Given the relative scarcity of established commercial data for biomass thermochemical processes, techno-economic analysis will continue to play a major role in the research of these technologies.

Investigations into new pathways for the conversion of biomass into fuel continue to discover intriguing possibilities for the use of biomass as a renewable source of energy. Unfortunately, some of these efforts have yet to publish detailed economic analysis.

There are various key opportunities and bottlenecks inherent in the different biomass conversion pathways. All biomass conversion technologies present attractive environmental benefits, and technological breakthroughs can provide the economic incentive for widespread consumer adoption. Therefore, it is important to continue exploring all potential pathways to the utilization of biomass for power and fuel applications.

References

1. Wright, M.M. and Brown, R.C. (2007). Comparative economics of biorefineries based on the biochemical and thermochemical platforms. *Biofuels, Bioproducts & Biorefining* **1**: 49–56.
2. EPA (2007). *Biomass Combined Heat and Power Catalog of Technologies*. Chapter 5:. Washington, DC: Biomass Conversion Technologies/U.S. Environmental Protection Agency Combined Heat and Power Partnership.
3. Obernberger, I. and Thek, G. (2004). Techno-economic evaluation of selected decentralized CHP applications based on biomass combustion in IEA partner countries. Final report from BIOS Bioenergiesysteme GmbH, Graz, Austria.

4. Rentizelas, A., Karellas, S., Kakaras, E., and Tatsiopoulos, I. (2009). Comparative techno-economic analysis of ORC and gasification for bioenergy applications. *Energy Conversion and Management* **50** (3): 674–681.
5. Arena, U., Di Gregorio, F., and Santonastasi, M. (2010). A techno-economic comparison between two design configurations for a small scale, biomass-to-energy gasification based system. *Chemical Engineering Journal* **162**: 580–590.
6. Bridgwater, A.V., Toft, A.J., and Brammer, J.G. (2002). A techno-economic comparison of power production by biomass fast pyrolysis with gasification and combustion. *Renewable Sustainable Energy Reviews* **6**: 181–248.
7. Chiaramonti, D., Oasmaa, A., and Solantausta, Y. (2007). Power generation using fast pyrolysis liquids from biomass. *Renewable Sustainable Energy Reviews* **11**: 1056–1086.
8. Martin, J.A. and Boateng, A.A. (2014). Combustion performance of pyrolysis oil/ethanol blends in a residential-scale oil-fired boiler. *Fuel* **133**: 34–44.
9. Rover, M., Smith, R., and Brown, R.C. (2018). Enabling biomass combustion and co-firing through the use of Lignocol. *Fuel* **211**: 312–317.
10. Wright, M.M., Brown, R.C., and Boateng, A.A. (2008). Distributed processing of biomass to bio-oil for subsequent production of Fischer-Tropsch liquids. *Biofuels, Bioproducts & Biorefining* 229–238. https://doi.org/10.1002/bbb.73.
11. Dang, Q., Wright, M., and Brown, R.C. (2015). Ultra-low carbon emissions from coal-fired power plants through bio-oil co-firing and biochar sequestration. *Environmental Science & Technology* **49**: 14688–14695.
12. Faaij, P.C., Hamelinck, C.N., Hardeveld, M.R.M. et al. (2002). Exploration of the possibilities for production of Fischer Tropsch liquids and power via biomass gasification. *Biomass and Bioenergy* **23**: 129–152.
13. Brown, T.R. (2015). A techno-economic review of thermochemical cellulosic biofuel pathways. *Bioresource Technology* **178**: 166–176.
14. Swanson, R.M., Platon, A., Satrio, J.A., and Brown, R.C. (2010). Techno-economic analysis of biomass-to-liquids production based on gasification. *Fuel* **89**: S11–S19.
15. Brown, T.R. and Wright, M.M. (2014). Techno-economic impacts of shale gas on cellulosic biofuel pathways. *Fuel* **117**: 989–995.
16. Zhu, Y. and Jones, S.B. (2009). *Techno-Economic Analysis for the Thermochemical Conversion of Lignocellulosic Biomass to Ethanol via Acetic Acid Synthesis*. Richland: Pacific Northwest National Laboratory.
17. Zhao, X., Brown, T.R., and Tyner, W.E. (2016). Stochastic techno-economic evaluation of cellulosic biofuel pathways. *Bioresource Technology* **198**: 755–763.
18. Phillips, S.D. (2007). Technoeconomic analysis of a lignocellulosic biomass indirect gasification process to make ethanol via mixed alcohols synthesis. *Industrial and Engineering Chemistry Research* **46**: 8887–8897.
19. Dutta, A., Bain, R.L., and Biddy, M.J. (2010). Techno-economics of the production of mixed alcohols from lignocellulosic biomass via high-temperature gasification. *Environmental Progress and Sustainable Energy* **29** (2): 163–174.
20. Dutta, A., Talmadge, M., Hensley, J. et al. (2012). Techno-economics for conversion of lignocellulosic biomass to ethanol by indirect gasification and mixed alcohol synthesis. *Environmental Progress & Sustainable Energy* **31**: 182–190.
21. Phillips, S., Aden, A., Jechura, J., Dayton, D., and Eggeman, T. (2007). Thermochemical Ethanol via Indirect Gasification and Mixed Alcohol Synthesis of Lignocellulosic Biomass. National Renewable Energy Laboratory. https://www.nrel.gov/docs/fy07osti/41168.pdf (accessed 27 November 2018).
22. Phillips, S.D., Tarud, J.K., Biddy, M.J., and Dutta, A. (2011). Gasoline from woody biomass via thermochemical gasification, methanol synthesis, and methanol-to-gasoline technologies: a technoeconomic analysis. *Industrial and Engineering Chemistry Research* **50**: 11734–11745.

23. Trippe, F., Frohling, M., Schultmann, F. et al. (2013). Comprehensive techno-economic assessment of dimethyl ether (DME) synthesis and Fischer-Tropsch synthesis as alternative process steps within biomass-to-liquid production. *Fuel Processing Technology* **106**: 577–586.
24. Brown, T.R. (2015). A critical analysis of thermochemical cellulosic biorefinery capital cost estimates. *Bioresource Technology* **9** (4): 412–421.
25. Piccolo, C. and Bezzo, F. (2009). A techno-economic comparison between two technologies for bioethanol production from lignocellulose. *Biomass and Bioenergy* **33**: 478–491.
26. Benalcazar, E.A. and Deynoot, B.G. (2017). Production of bulk chemicals from lignocellulosic biomass via thermochemical conversion and syngas fermentation: a comparative techno-economic and environmental assessment of different site-specific supply chain configurations. *Biofuels, Bioproducts and Biorefining* **11**: 861–886.
27. Brown, R.C. (2005). Hybrid thermochemical/biological processing: an overview. In: *Biorefineries, Biobased Industrial Processes and Products*. Weinheim, Germany: Wiley.
28. Choi, D.W., Chipman, D.C., Bents, S.C., and Brown, R.C. (2010). A techno-economic analysis of polyhydroxyalkanoate and hydrogen production from syngas fermentation of gasified biomass. *Applied Biochemistry and Biotechnology* **160**: 1032–1046.
29. Li, W., Dang, Q., Brown, R.C. et al. (2017). The impacts of biomass properties on pyrolysis yields, economic and environmental performance of the pyrolysis-bioenergy-biochar platform to carbon negative energy. *Bioresource Technology* **241**: 959–968.
30. Li, W., Dang, Q., Smith, R. et al. (2017). Techno-economic analysis of the stabilization of bio-oil fractions for insertion into petroleum refineries. *ACS Sustainable Chemistry & Engineering* **5**: 1528–1537.
31. Li, B., Ou, L., Dang, Q. et al. (2015). Techno-economic and uncertainty analysis of in situ and ex situ fast pyrolysis for biofuel production. *Bioresource Technology* **196**: 49–56.
32. Haro, P., Trippe, F., Stahl, R., and Henrich, E. (2013). Bio-syngas to gasoline and olefins via DME – -a comprehensive techno-economic assessment. *Applied Energy* **108**: 54–65.
33. Hu, W., Dang, Q., Rover, M. et al. (2016). Comparative techno-economic analysis of advanced biofuels, biochemicals, and hydrocarbon chemicals via the fast pyrolysis platform. *Biofuels* **7**: 57–67.
34. Dang, Q., Hu, W., Rover, M. et al. (2016). Economics of biofuels and bioproducts from an integrated pyrolysis biorefinery. *Biofuels, Bioproducts and Biorefining* **10**: 790–803.
35. Patwardhan, P.R., Satrio, J.A., Brown, R.C., and Shanks, B.H. (2009). Product distribution from fast pyrolysis of glucose-based carbohydrates. *Journal of Analytical and Applied Pyrolysis* **86**: 323–330.
36. Scott, D.S., Paterson, L., Piskorz, J., and Radlein, D. (2001). Pretreatment of poplar wood for fast pyrolysis: rate of cation removal. *Journal of Analytical and Applied Pyrolysis* **57**: 169–176.
37. Brown, R.C., Radlein, D., and Piskorz, J. (2001). Pretreatment processes to increase pyrolytic yield of levoglucosan from herbaceous feedstocks. *ACS Symposium Series* **784**: 123–132.
38. So, K.S. and Brown, R.C. (1999). Economic analysis of selected lignocellulose-to-ethanol conversion technologies. *Applied Biochemistry and Biotechnology* **77**: 633–640.
39. Ghosh, A., Brown, R.C., and Bai, X. (2016). Production of solubilized carbohydrate from cellulose using non-catalytic, supercritical depolymerization in polar aprotic solvents. *Green Chemistry* **18**: 1023–1031.
40. Zhu, Y., Biddy, M.J., Jones, S.B. et al. (2014). Techno-economic analysis of liquid fuel production from woody biomass via hydrothermal liquefaction (HTL) and upgrading. *Applied Energy* **129**: 384–394.
41. Ou, L., Thilakaratne, R., Brown, R.C., and Wright, M.M. (2015). Techno-economic analysis of transportation fuels from defatted microalgae via hydrothermal liquefaction and hydroprocessing. *Biomass and Bioenergy* **72**: 45–54.

42. Han, J., Luterbacher, J.S., Alonso, D.M. et al. (2015). A lignocellulosic ethanol strategy via nonenzymatic sugar production: Process synthesis and analysis. *Bioresource Technology* **182**: 258–266.
43. Marchetti, J.M., Miguel, V.U., and Errazu, A.F. (2008). Techno-economic study of different alternatives for biodiesel production. *Fuel Processing Technologies* **89** (8): 740–748.
44. Quinn, J.C. and Davis, R. (2015). The potentials and challenges of algae based biofuels: A review of the techno-economic, life cycle, and resource assessment modeling. *Bioresource Technology* **184**: 444–452.

11

Life Cycle Assessment of the Environmental Performance of Thermochemical Processing of Biomass

Eskinder Demisse Gemechu, Adetoyese Olajire Oyedun, Edson Norgueira, Jr., and Amit Kumar

Department of Mechanical Engineering, Donadeo Innovation Centre for Engineering, University of Alberta, Edmonton, Alberta, Canada

11.1 Introduction

Climate change is a global issue [1]. Global emissions of greenhouse gas (GHG) caused by human activities have increased by 10 Gt (CO_2eq) between 2000 and 2010. GHG emissions from fossil fuel combustion are responsible for 88% of this increase [1]. Governments around the globe have been urged to adopt low-carbon energy technologies to reduce our dependence on fossil fuels.

Adoption of green energy technologies is acknowledged as important to achieve the Paris Agreement, keeping the rise of global average temperature well below 2 °C [2]. The European Union has established a renewable energy directive with the aim of fulfilling 20% of the EU's energy demand with renewables by 2020 [3]. The United States federal government has set a target of 30% electricity from renewable energy within the federal government by 2025 [4]. Renewable energy is acknowledged by 145 of the 194 parties to the Paris Convention as a way to mitigate climate change, and more than 56% of these

---

Thermochemical Processing of Biomass: Conversion into Fuels, Chemicals and Power, Second Edition.
Edited by Robert C. Brown.
© 2019 John Wiley & Sons Ltd. Published 2019 by John Wiley & Sons Ltd.

countries have a specific renewable energy target [5]. Recent trends show that investments in renewable energy technologies increased by 28% between 2010 and 2015 [6]. In 2014, renewable energy reached an estimated share of 19.2% of global energy consumption. Biomass, geothermal, and solar heat together made up about 4.2% of that share [7]. According to a recent study, renewable energy made up 27.7% of global power generation capacity and 22.8% of global electricity demand in 2015 [8]. In that same year, the total primary energy demand from biomass was approximately 16 250 TWh [8].

Thermochemical processing is one pathway to convert biomass into valuable products such as heat, power, transportation fuels, syngas, and chemicals [9, 10]. Direct combustion, torrefaction, carbonization, pyrolysis, gasification, hydrothermal liquefaction (HTL), and hydrothermal gasification are among thermochemical technologies described in the literature for processing biomass [9, 11–16]. While all of these technologies aim to reduce fossil fuel dependence and associated climate change impacts, several issues need to be addressed before they are implemented at a commercial scale. These issues include resource availability in a particular region, technical challenges in processing biomass, and the environmental, economic, and social impacts associated with processing biomass. This chapter reviews the environmental impacts of thermochemical processing of biomass.

Life cycle assessment (LCA) can be used to evaluate the environmental sustainability and socioeconomic implications of thermochemical processes. LCA is an internationally standardized tool used to evaluate the environmental performance of a product across its entire life cycle, including resource extraction, manufacture, use, and final disposal. This chapter describes the procedure for performing an LCA and reviews the literature on environmental performance of bioenergy systems. Specific objectives are:

- provide an overview of the concept of LCA as an environmental assessment tool and its methodological foundation,
- perform an extensive literature review on the use of LCA on thermochemical biomass processing pathways,
- identify the research gaps and the environmental bottlenecks relevant to the large-scale implementation of different thermochemical processes, and
- suggest ways to address the gaps.

11.2 Life Cycle Assessment

11.2.1 Introduction to LCA and Life Cycle Thinking

Environmental and socioeconomic impacts of human activities are assessed using system-based approaches. LCA is among the most widely used environmental assessment tools. According to ISO 14040, "life cycle" is defined as the "consecutive and interlinked stages of a product system, from raw material acquisition or generation from natural resources to final disposal" [17]. LCA is defined as the "compilation and evaluation of the inputs, outputs and the potential environmental impacts of a product system throughout its life cycle" [17].

LCA is a stand-alone tool for measuring and evaluating the environmental performance of a product system. LCA is based on the concept of life cycle thinking, which goes beyond the traditional approach of assessing a product system from only the manufacturing process to include the entire life cycle. The life cycle of a product system starts from the extraction of natural resources to manufacturing, packaging, distribution to end consumers, use, reuse, recycling, and final disposal in nature. At each stage of the product's life, energy and material inputs are taken into account and translated to associated emission values. By considering the entire life cycle of a product system and applying several environmental criteria, LCA avoids burden shifting between life cycle stages and different environmental problems while helping to identify environmental hotspots that allow consumers, producers, and policy authorities to make informed decisions.

LCA is an internationally harmonized and standardized tool. The International Organization for Standardization (ISO) provides the principles and frameworks (ISO 14040:2006) and requirements and guidelines (14044:2006) for the goal and scope definitions, inventory analysis, life cycle impact assessment (LCIA), and interpretation to have a common ground for individuals, industries, and governments performing an LCA [17, 18]. Each stage is discussed below.

11.2.2 Goal and Scope Definition

Defining the goal and scope is the first step of an LCA. The goal sets the context of the study by answering questions such as: What is the main purpose of conducting an LCA and the reasons behind it? Who is the target audience? How can the main findings of the study be used? The goal is the basis of scope definition. The scope of the LCA defines a product system and its boundary, identifies the main function of the product, defines the functional unit, and sets an allocation, cut-off, and other procedures. In addition, the scope also clearly specifies time and place, modeling approach, and assumptions and limitations, all of which are taken into account when the results are interpreted.

The functional unit is a crucial element and must be defined properly, especially when the LCA compares the environmental performance of different products. All the energy and material inputs to the product system and emissions outputs are matched with the functional unit.

The system boundary defines the inclusion and exclusion of sub-processes or sub-systems in an LCA. It should match the goal and scope. Each unit process within the system boundary is usually described in a process flow diagram as it helps to easily understand the product system and the interlinks among processes or sub-processes. The desired output from the system boundary is known as a reference flow, which is a measured output that is required in order to provide a certain function as defined in the functional unit [17, 18]. All the inputs and outputs in the product system should be scaled up to the reference flow.

11.2.3 Life Cycle Inventory

In a life cycle inventory (LCI) analysis, data on elementary flows in physical units are collected from all the processes in a product system. These include resource inputs, energy, materials, and the associated emissions and waste outputs. The data compilation and modeling should be in accordance with the goal and scope. The LCI analysis calculates the inventory results using the method set in the goal and scope, and the results are used to model potential impacts.

An important aspect of the LCI is the handling of the multifunctional process, a process that delivers more than one product. If a product system is multifunctional, an allocation of energy and material flows and emissions and waste outputs between the primary product and its co-product is required. The ISO defines allocation as "partitioning the input or output flows of a process or a product system between the product system under study and one or more other product systems" [17]. The allocation procedure depends on the product or system analyzed. There are different ways of performing allocation, for example, partitioning, mass or economic allocation, system expansion, or substitution.

LCI modeling can be based on attribution or consequential LCA approaches, depending on the questions addressed in the LCA [19]. Sonnemann et al. [20] define an attributional LCA as a system modeling approach in which inputs and outputs are attributed to the defined functional unit of the product system by linking or partitioning each unit process based on a normative rule. The aim of an attributional LCA is, using market average data and an allocation procedure, to trace the environmental impacts of a product system and determine the allocated shares of the activities that contributed throughout the product's life cycle (production, consumption, and end of life). A consequential LCA is a modeling approach that determines changes in a product system in response to a change in demand for the functional unit [21]). Using marginal data and the system expansion allocation procedure, a consequential LCA aims to determine the environmental consequences of a change in demand. It is mainly used as a decision support tool.

11.2.4 Life Cycle Impact Assessment

LCIA translates the inventory results on elementary flows into impacts on human health, ecosystem services, and natural resources. Environmental emissions and resource uses quantified in the inventory analysis are characterized with the aim of understanding and measuring the magnitude of the environmental impacts of each input and output. Important stages in an impact assessment are the selection and identification of relevant impact categories, relating inventory results to the identified impact categories (classification), and quantifying and aggregating the inventory results in given impact categories by multiplying by the characterization factor to have a common unit for each impact category (characterization). The results from different impact categories reflect the environmental performance of the system under study. Normalization, grouping, and weighting are optional elements according to the ISO, and they could be applied depending on the goal and scope of the study [17, 18].

Figure 11.1 shows how LCI results are translated to environmental impact categories. The environmental mechanism pathway, based on a cause–effect chain, is developed for each emission and natural resource use. The cause–effect chain determines the fate, exposure, effect, and potential damage. There are two ways of defining the environmental impacts associated with each elementary flow, through the midpoint or the endpoint. The midpoint impact assessment is a problem-oriented approach that translates impacts into environmental themes such as acidification, climate change, eutrophication, human toxicity, and so on. Environmental impact categories at the midpoint level can be further linked to damage-oriented endpoint impacts to reflect the potential damage of each elementary flow on human health, ecosystem services, and natural resources, the three areas of protection defined by Hauschild and Huijbregts [22] as a "cluster of category endpoints of recognizable

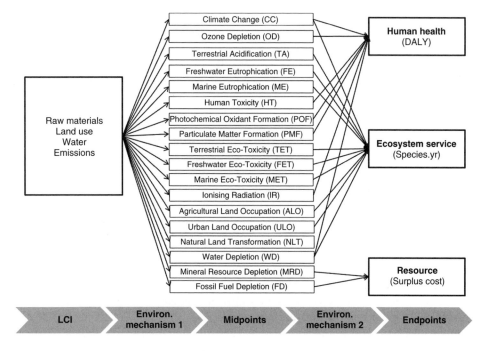

Figure 11.1 Life cycle impact assessment process.

value to society." Endpoint impact assessment results are easier to understand by decision makers than midpoint level results but have a higher level of uncertainty. There are several midpoint and endpoint LCIA methods, for example, CML [23], Eco-Indicator 99 [24], Impact 2002+ [25], ReCiPe [26], TRACI 2.0 [27], and USEtox [28].

11.2.5 Life Cycle Interpretation

The final phase of an LCA, in which results from the LCIA are analyzed, is the life cycle interpretation. In this phase the most relevant processes or subsystems are identified from the results of the inventory analysis and impact assessment. The completeness, consistency, and sensitivity of the information considered in various stages of the LCA processes are evaluated. Conclusions and recommendations are the main outcomes, and they help the intended audience easily understand the results and make informed decisions. The study's main limitations are also stated in this stage.

11.2.6 Sensitivity and Uncertainty Analyses

Since its first use, LCA has had a prominent role as a decision support tool for authorities at different organizational levels. However, there are concerns among users whether the assessment results are robust enough to use when making appropriate decisions. Both the LCI assessment and LCIA are subject to uncertainties, among them uncertainties due to temporal variability, spatial variability, input or modeling parameters, choice of functional unit, system boundary definition, LCIA method, allocation procedures, and cut-off rules [29]. Determining the uncertainties helps to understand how realistic the LCA results are. Through a sensitivity analysis, one can evaluate to what extent varying an input parameter affects the results. If a minor change in a parameter changes the overall results considerably, then the LCA model is sensitive to that parameter. It is critical to acknowledge possible sources of error and sensitive parameters, especially when using LCA to compare products.

11.3 LCA of Thermochemical Processing of Biomass

11.3.1 Overview of the Thermochemical Processing of Biomass

The thermochemical processing of biomass uses heat and/or catalysts to convert biomass to biofuels. It has several advantages over biochemical processing including a shorter residence reaction time, the ability to produce a diverse selection of fuels, greater feedstock flexibility, lower catalysts costs, the ability to recycle catalysts, and a high operating temperature (200–1300 °C) that precludes the sterilization usually required for biochemical processing [30].

Thermochemical conversion processes include direct combustion, torrefaction, carbonization, pyrolysis, gasification, HTL, and hydrothermal gasification [9, 11–14, 31]. Pyrolysis and gasification are the most widely studied, although hydrothermal processes have also been studied as an alternative for handling high-moisture biomass feedstocks (such as algae [32]), since the process eliminates the costly and energy-intensive drying step [12, 33]. The thermochemical processes produce various products, such as heat, power, liquid fuels, syngas, biochars, and chemicals. To optimize both the production and quality of one or more products, different reaction conditions (i.e. pressure, temperature, residence time, heating rate, purge gas flow rate, reactive or inert atmosphere, etc.) are used by each process.

11.3.2 The Use of LCA to Promote Low Carbon Technologies

LCAs have been extensively used to evaluate the environmental impacts of the thermochemical conversion of biomass to various products [34–65]. When LCA is used during the designing stages and throughout the entire research and development of a thermochemical processing technology, it can help to design a "greener" product [59]. LCA in theory provides a holistic approach to product assessment using all aspects of an environmental impact. A complete LCA study evaluates several impact categories simultaneously. Some of the more common impact indicators are acidification, ozone depletion, photochemical smog, resource depletion, eutrophication, land use, human health, water use, global warming, aquatic toxicity, and terrestrial toxicity. To facilitate the most appropriate sustainable outcome, all the environmental impact indicators as well as the social and economic implications must be evaluated during the development of the technology.

11.3.3 Review of LCA Studies on Thermochemical Processing of Biomass

Only peer-reviewed publications were considered in this review. We focused primarily on recent publications on the LCA of different thermochemical technologies. While there are some recent reviews of LCA studies relevant to thermochemical processing of biomass [66, 67], this review considers a wider range of thermochemical processes. Table 11.1 gives an overview of the LCAs of different thermochemical conversion technologies for a wide range of biomass. These studies considered many different impact categories, with global warming potential (GWP) and net energy ratio (NER) assessed most often. GHGs and GWP are usually reported as CO_2 equivalents, which include CO_2, CH_4, and N_2O.

For LCAs of biomass combustion, Gagnon et al. [43], for example, conducted an LCA of the most common options for electricity generation considering a number of environmental indicators. The best options identified are hydropower, nuclear power, and wind power, with biomass performing relatively poorly in comparison, especially in terms of NO_x emissions, land requirements, and energy payback ratios. This is due to the uncertainties involved in carbon neutrality and from the fact that biomass plantations require the most land per unit of energy delivered. Biomass showed emissions of 664 kt CO_2eq/TWh and 26 t SO_2/TWh, land requirements from 533 to 2200 km^2/TWh, and an energy output-to-input ratio of 5. Eriksson et al. [40] compared waste incineration, biomass combustion, and natural gas combustion for both combined heat and power (CHP) plants and exclusively heat production plants. The study comprises two options for energy recovery CHP or heat only), two alternatives for external, marginal electricity generation (fossil lean or intense), and two options for the alternative waste management (landfill disposal or material recovery). Their results indicated that the combustion of biofuel in a CHP is environmentally favorable and robust with respect to the avoided type of electricity and waste management. They concluded that waste incineration is often the preferable choice when incineration substitutes landfill disposal of waste. It is, however, never the best choice (and often the worst) when incineration substitutes recycling [40].

Sebastián et al. [57] compared GHG emissions from biopower plants for four feedstock scenarios: dedicated biomass firing and biomass co-firing for three different feedstocks. The main observation was that a high net electric efficiency (>28%) is required to make the biomass-only fired power plant more advantageous than biomass co-firing from the final GHG emissions reduction point of view. Zuwala [65] presented a case for the co-firing of biomass in a CHP plant. The study shows that a partial substitution of coal with biomass (0–20%) will decrease GHG emissions by 16% in the case of electricity production and 15% in the case of heat production compared with the same plant operating only with coal.

Royo et al. [56] explored the possibility of biomass co-firing to decrease GHG emissions on a large scale in Spain. The paper states that forest and agricultural residues collected near power plants could generate up to 7.7% of the energy demand for the country and reduce GHGs by 87 tCO_2eq for every 1 TJ$_{th}$ of biomass used instead of coal. The country could reduce emissions by 3.4 Mt CO_2eq/year. Mann and Spath [53] presented results for co-firing wood residue at 5% and 15% of the capacity of a coal plant. GHG emissions fell by 5.4% and 18.2% for each case, and SO_2, NO_x, non-methane hydrocarbons, particulates, and CO emissions also dropped. Moreover, energy consumption fell by 3.5% and 12.4% for the 5% and 15% co-firing capacities, respectively. Heller et al. [45] studied the environmental impact of co-firing biomass at a rate of 10% in a coal plant to produce electricity. It was estimated that for the co-firing scenario, the system NER increases by 8.9%, while the net global warming potential decreases by 7–10% and SO_2 emissions decrease by 9.5%.

Table 11.1 List of LCA studies on the thermochemical processing of biomass.

Process type	Feedstock	End product	Purpose	Modeling	Function unit	Scope	LCIA method/Data/Software	Impact categories	Uncertainty/sensitivity analysis	Ref
Biomass combustion	Wood biomass	Heat	Comparative	Consequential	42 PJ	Cradle-to-grave	SimaPro 5, Eco-indicator 99, Ecotax 02, EPS 2000	NER, GWP	Only sensitivity	[40]
	Forestry wastes	Electricity	Comparative	Attributional	1 kWh	Cradle-to-grave		Energy payback ratio, GHG, NO_x, SO_2, direct land requirements	None	[43]
Torrefaction	Scots pine	Wood pellets (WP), torrefied pellets (TP)	Comparative	Attributional	1 ton of WP, 1 MJ of TP	Cradle-to-gate	SimaPro 7.3, Ecoinvent 2.0	GHG, PMF, land use, net energy, fossil fuel depletion	Only sensitivity	[68]
	Olive husk	Torrefied pellets	Comparative	Attributional	1 ton of torrefied olive husk	Cradle-to-grave	GaBi, CML 2001	GWP, AP, EP, ODP, ADP	None	[69]
Bio-oil upgrading	Forest residue	Diesel, Gasoline	Comparative	Attributional	1 km, 1 MJ	Cradle-to-grave	SimaPro v 7.3, Ecoinvent v.2.2, Greet	GHG	Only uncertainty	[46]
	Corn stover	Biofuel (gasoline, diesel)	Comparative	Attributional	1 MJ	Well-to-pump	Greet	Energy demand, GWP	None	[39]
	Microalgae	Diesel	Comparative	Attributional	1 MJ	Well-to-pump	Greet 2013	NER, GWP	Only sensitivity	[34]
Co-firing	Herbaceous and woody agricultural residue, energy crops	Electricity	Comparative	Attributional	1 TJ	Cradle-to-gate	SimaPro 5.1	GHG	None	[57]
	Willow chips, wood chips	Electricity and heat	Comparative	Attributional	1 TJ heat, 1 MWh electricity	Cradle-to-gate	Eco-indicator 99, CML 2 baseline 2000, EPS 2000	GHG	Only Sensitivity	[65]

Feedstock	Product	Purpose	Approach	Functional unit	System boundary	Method/Software	Impact categories	Sensitivity/Uncertainty	Ref.
Whole trees, forest and agriculture residues	Electricity	Comparative	Attributional	1 MWh electricity	Cradle-to-grave	IPCC 100	NER, GHG, ARP, GOP	Only sensitivity	[70]
Herbaceous and woody agricultural residue, primary forestry residues, energy crops	Electricity	GHG reduction potential	Attributional	1 TJ	Cradle-to-gate	SimaPro 7.3.3	GHG	Only sensitivity	[56]
Wood residues	Electricity	Comparative	Attributional	1 kWh	Cradle-to-grave	TEAM	GHG, NO_x, SO_2 and other emissions, energy consumption	Sensitivity, not uncertainty	[53]
Willow biomass and wood residues	Electricity	Identify environmental impacts	Attributional	1 MWh	Cradle-to-grave	TEAM, DEAM	Air emissions, energy requirements, GWP	None	[45]
Biochar (from torrefaction of rice straw)	Electricity	Identify environmental impacts	Attributional	1 kWh	Cradle-to-grave	SimaPro 7.2, IMPACT 2002+	AA, AE, C, E_a, GW, IR, LO, ME, NC, NE, OL, RI, RO, TE, TA	Sensitivity analysis	[47]
Whole trees, forest and agriculture residues	HDRD	Comparative	Attributional	1 MJ	Well-to-wheel	Bottom-up analysis	Water consumption	Yes	[62]

Fast pyrolysis

Feedstock	Product	Purpose	Approach	Functional unit	System boundary	Method/Software	Impact categories	Sensitivity/Uncertainty	Ref.
Wood biomass	Bio-oil	Comparative	Attributional	1 MJ	Cradle-to-grave	GREET	GHG	Only sensitivity	[61]
Corn grain, corn stover	Ethanol, biochar, bio-oil	Identify environmental impacts	Attributional	1 ha	Cradle-to-grave	EPA, GREET	GHG	Only sensitivity	[49]
N/A	MoS_2/NiS on Al_2O_3, Ru/C catalyst	Identify environmental impacts	Attributional	1 kg fresh catalyst	Cradle-to-gate		GHG	None	[59]
Municipal solid waste	Bio-oil	Identify environmental impacts	Attributional	1 kg organic components in MSW		CML, GaBi (Software)	EP, HTP, GWP, ODP, POCP, TEP	Only sensitivity	[60]

(Continued)

Table 11.1 Continued

Process type	Feedstock	End product	Purpose	Modeling	Function unit	Scope	LCIA method/Data/Software	Impact categories	Uncertainty/sensitivity analysis	Ref
	Perennial grasses (switchgrass and miscanthus)	Renewable fuel	Comparative	Attributional	1 MJ renewable fuel	Well-to-wheel	IPCC, CED (cumulative energy demand)	GHG, EROI	Yes	[63]
	Corn stover and forest residue	Gasoline	Comparative	Attributional	1 MJ	Well-to-wheel	GREET	GHG	Only sensitivity	[44]
	Forest residue	Diesel, gasoline	Identify environmental impacts	Attributional	1 km, 1 MJ	Cradle-to-grave	SimaPro v 7.3, Ecoinvent v.2.2, Greet	GHG, NEV	Only uncertainty	[46]
	Poplar biomass	Diesel, gasoline, char	Identify environmental impacts	Attributional	1 ton	Cradle-to-gate	CML, SimaPro 7 (Software)	CED, GWP, OL, POF, LC, AP, EP	None	[48]
	Hybrid poplar	Bio-oil	Identify environmental impacts	Attributional	1 MJ	Cradle-to-gate	Aspen Plus	GHG	None	[55]
	Corn stover	Bio-oil	Identify environmental impacts	Attributional	1 km traveled by light-duty vehicle	Well-to-wheel	GREET, SimaPro 7.3	GWP	Only sensitivity	[64]
	Poplar, logging residue, willow and waste wood	Pyrolysis oil for electricity generation	Comparative	Attributional	1 kWh	Well-to-wheel	SimaPro 7.1	GHG	Only sensitivity	[41]
	Corn stover	Biofuel (gasoline, diesel)	Comparative	Attributional	1 MJ	Well-to-pump	Greet	Energy demand, GWP	None	[39]
	Miscanthus	Bio-hydrocarbons	Comparative	Attributional	1 ton of bio-hydrocarbons	Cradle-to-gate	Renewable Energy Directive (RED)	GHG	Only sensitivity	[58]
	Hybrid poplar	Bio-oil	Comparative	Attributional	2.7 kg hybrid poplar biomass (1 kg bio-oil)	Well-to-wheel	CML, SimaPro 7.3	ADP, AP, EP, GWP, ODP, CED	None	[54]
Fast pyrolysis/gasification	Whole forest, forest and agriculture residues	Biohydrogen	Comparative	Attributional	1 kg H_2	Cradle-to-gate	IPCC 100	NER, GHG	Only sensitivity	[71]

Process	Feedstock	Product	Study type	Approach	Functional unit	System boundary	Software/Method	Impact categories	Sensitivity analysis	Ref.
Fast pyrolysis/ hydrothermal liquefaction	Microalgae	Diesel	Comparative	Attributional	1 MJ	Well-to-pump	Greet 2013	NER, GWP	Only sensitivity	[34]
	Oil palm empty fruit bunch	Bio-oil	Comparative	Attributional	1 kg bio-oil	Cradle-to-gate	CML	GWP, AP, EP, HT, POF	Only sensitivity	[36]
	Logging residues/forest thinnings	Gasoline, diesel	Comparative	Attributional	1 g CO_2e/MJ	Well-to-wheel	Simparo 8.0, IPCC GWP 100	GHG	None	[72]
Gasification	Energy crop	Hydrogen	Comparative	Attributional	1 MJ	Cradle-to-grave	ECO-INDICATOR 95 (SimaPro database)	GHG, OL, AP, EP, energy, solid waste and others.	None	[35]
	Waste biomass from cotton, olive, rice, corn	Hydrogen	Comparative	Attributional	1 MJ	Cradle-to-gate	Eco-indicator 95	GHG, AP, EP	None	[50]
	Poplar	Syngas with high % of H_2	Identify environmental impacts	Attributional	1 MJ	Cradle-to-grave	EEA. EMEP/CORINAIR	GHG	Only sensitivity	[38]
Hydrothermal Liquefaction	Oil palm biomass	Bio-oil	Comparative	Attributional	1 kg	Cradle-to-gate	EIO-LCA	GWP, HT, AP, EP, POF	Only sensitivity	[37]
	Algae	Bio jet-fuel	Comparative	Attributional	1 GJ	Cradle-to-wake	SimaPro 7.3.3, Traci 2.0	GHG	Yes	[42]
	Algae	Biofuel (gasoline, diesel)	Identify environmental impacts	Attributional	1 MJ	Cradle-to-gate	Greet	EROI, GHG	Yes	[51]
	Forest residue	Biofuel	Comparative	Attributional	1 MJ	Well-to-wheel	GHGenius 4.03.	GHG	Only sensitivity	[73]
Hydrothermal Gasification	Waste biomass	SNG	Identify environmental impacts	Attributional	1 MJ	Cradle-to-gate	Eco-indicator 99, Eco-Scarcity	GWP	None	[52]
	Palm oil	Biodiesel	Comparative	Attributional and consequential	14 million l/d	Cradle-to-grave	ReCipe2008, SimaPro 8.0., Ecoinvent v2.2	EP, GW, HT, POF, TA	Only sensitivity	[74, 75]
	Corn, biomass	Gasoline, ethanol	Comparative	Attributional	1 MJ	Cradle-to-grave	Worldwide agricultural model	GW	Only sensitivity	[76]

AA, aquatic acidification; ADP, abiotic depletion potential; AE, aquatic ecotoxicity; AP, acidification potential; ARP, acid rain precursors; C, carcinogens; CED, cumulative energy demand; E_a, eutrophication (aquatic); EP, eutrophication potential; EROI, energy return on investment; GOP, ground level ozone precursors; GW, global warming; GWP, global warming potential; HTP, human toxicity potential; IR, ionizing radiations; LC, land competition; LO, land occupation; ME, mineral extraction; NC, non-carcinogens; NE, non-renewable energy; NER, net energy ratio. NEV, net energy value; ODP, ozone depletion potential; OL, ozone layer depletion; POCP, photochemical oxidant creation potential; POF, photochemical oxidant formation; RI, respiratory inorganics; RO, respiratory organics; TA, terrestrial acidification; TE, terrestrial ecotoxicity; TEP, terrestrial eutrophication potential.

A number of LCA studies have evaluated the environmental performance of energy production from biomass. The values in literature vary widely, depending on different considerations [77]. The types of feedstocks used, the selection of system boundary [78], transportation distance, energy conversion technologies [79], integration of land-use change [79, 80], land management [81], and other factors affect the overall results. These are important, especially when comparing the environmental impacts from bioenergy with other sources of energy that could be renewable or non-renewable sources.

Kabir and Kumar [70] compared the life cycle energy and environmental performance of nine different biomass/coal co-firing pathways for electricity generation. Three feedstocks (agricultural residue, whole tree, and forest residue) in various formed of chip, pellet, and bale were analyzed using two co-firing approach (direct and parallel co-firing). They reported that the NERs for all pathways were in the range of 0.37–0.42, with agricultural residue-based co-firing pathway having the lowest NER of 0.37–0.38. The GHG emissions were in the range of 957–1083 kg CO_2eq/MWh, with forest residue having the lowest GHG emissions of 957–1004 kg CO_2eq/MWh [70]. The acid rain precursor (ARP) emissions reported were in the range of 5.16–5.93 kg SO_2eq/MWh, while the ground-level ozone precursor (GOP) emissions were in the range of 1.79–1.91 kg (NO_x + VOC)/MWh for all the nine pathways [70].

The environmental impacts associated with production and delivery of torrefied wood pellets and wood pellets using cradle-to-gate LCA was conducted by Adams et al. [68]. Their results show that the relative benefits of torrefaction over wood pellets were higher on the basis of MJ delivered due to the higher calorific value of torrefied pellets. Particulate matter formation, fossil depletion, and global warming impacts for the torrefied pellets were also lower than that of the wood pellets; however, land requirements for torrefied pellets were higher than for wood pellets due to mass losses during production [68].

Christoforou and Fokaides [69] performed an LCA on torrefaction of olive husk using GaBi software and the CML 2001 impact assessment method. The environmental impact categories compared in their analysis are global warming potential, acidification potential, eutrophication potential, ozone layer depletion potential, and abiotic depletion potential. Feedstock drying was important to improving energy consumption in the overall process of torrefaction.

A study by Hsu [46] describes an LCA of transportation fuels produced through pyrolysis, including a hydroprocessing component. The CO_2 equivalent emissions were 117 g/km and the net energy value (NEV) was 1.09 MJ/km for gasoline, while for diesel the GHG emissions were 98 g/km and the NEV 0.92 MJ/km. Even with uncertainty results, all the values were lower than for conventional gasoline in 2005. Hsu claimed that the values could be lowered further if the electricity and hydrogen needed to operate the system were provided by from renewable energy. Dang et al. [39] considered three cases for biofuel production through fast pyrolysis and hydroprocessing; each case considers different sources for the hydrogen used during the hydroprocessing phase. Compared to conventional gasoline and diesel, case 1, in which hydrogen was provided by natural gas reforming, showed a reduction in net non-renewable energy demand by 67.5% and a reduction in net GWP by 69.1%. For case 2, where hydrogen was produced from steam reforming of about 35% of the bio-oil aqueous phase, the net non-renewable energy demand and net GWP were reduced by 76.6%

and 73.0%, respectively. Case 3, in which surplus hydrogen was obtained as a co-product from steam reforming the whole aqueous phase of bio-oil, showed reductions of 147.5% and 119.4%, respectively.

Wong et al. [62] analyzed the environmental impacts caused by the production of hydrogenation-derived renewable diesel (HDRD) through pyrolysis, with a focus on countries with cold climates, such as Canada. The GHG emissions estimates are from 35.4 to 42.3 g CO_2eq/MJ of HDRD, a reduction of 53.4–61.1% compared to fossil fuel-based diesel. The energy ratios are between 1.55 and 1.90, with the lowest case being the one where forest residue feedstock is used and the highest is where agricultural residue feedstock is used.

Bennion et al. [34] analyzed the HTL and pyrolysis pathways followed by hydroprocessing to produce biodiesel. The results for the energy consumed over energy produced (NER) and GHG emissions for both pathways were presented. The NER for HTL was 1.23 and the GHG emissions were −11.4 g CO_2eq/MJ. The NER for pyrolysis was 2.27 and the GHG emissions were 210 g CO_2eq/MJ. HTL has a clear advantage over pyrolysis, based on GHG emissions.

Comparative life cycle GHG emissions were estimated for transportation fuels produced from fast pyrolysis and HTL with upgrading and hydrotreating pathways, respectively, in a study conducted by Tews et al. [72] for the US Department of Energy. Their results show that the emissions for HTL are lower than for pyrolysis due to lower electricity consumption, which stems primarily from the fact that HTL oil is less oxygenated than pyrolysis oil and therefore requires less hydrogen for upgrading per unit of fuel energy produced. The conversion stage contributes mostly to the GHG emissions for both pyrolysis and HTL pathways.

Carpentieri et al. [35], on the other hand, focused on integrated biomass gasification combined cycle (IBGCC) with CO_2 removal. The study compared this method with a more classical integrated coal gasification-combined cycle (ICGCC). The biomass-based system shows an efficiency of 33.94% and GHG emissions of 178 kg/MWh, while the coal system shows 38.8% efficiency and GHG emissions of 130 kg/MWh. These results suggest that the IBGCC could be competitive in the market. Corti and Lombardi [38] compared IBGCC with CO_2 removal with the traditional, coal-based ICGCC. When the sensitivity of CO_2 removal is considered, the IBGCC pathway shows 35–36% efficiency, while ICGCC shows 38–39%. The IBGCC also has higher GHG emissions, 170 kg/MWh versus 130 kg/MWh from the ICGCC.

Because demand for hydrogen as a fuel source is increasing, Koroneos et al. [50] estimated the impact of hydrogen production via gasification. Of the two methods considered, the biomass–gasification–electricity–electrolysis route is less efficient than the biomass–gasification–steam reforming–PSA route. An energy input of 4.2 TJ per TJ of hydrogen generated was reported in the first case and 2.4 TJ per TJ of hydrogen in the second. However, in terms of environmental impact, the biomass–gasification–electricity–electrolysis route uses 92.9% of renewable resources, while biomass–gasification–steam reforming–PSA route uses only 54.2%.

Kabir and Kumar [71] investigated the life cycle NERs and the environmental impact of nine different pathways for the production of biohydrogen from three different feedstocks (agricultural residue, whole tree, and forest residue). The three technologies considered are the Battelle Columbus Laboratory (BCL) gasifier, the Gas Technology Institute (GTI) gasifier, and fast pyrolysis. Their results show that the GHG emissions lie in the range of 1.20–8.1 kg CO_2eq/kg H_2 and the NERs for the nine biohydrogen pathways lie in the range of 1.3–9.3, with the lowest limit for GHG emissions and maximum NER corresponding to the forest residue based biohydrogen production pathway using the GTI technology [71]. They concluded that the utilization of biohydrogen can improve overall energy efficiency and environmental performance since the NERs for biohydrogen are always higher than the fossil fuel-based hydrogen [71].

The innovative study of Chan et al. [37] analyzed the process parameters and LCA of the HTL of oil palm biomass for bio-oil production. The LCA showed 2.29 kg CO_2eq per litre of bio-oil, while other impacts such as acidification, toxicity, and eutrophication were relatively small. The sensitivity analysis concluded that the yield had more influence on the environmental impact than other variables.

An LCA of transportation biofuels from HTL of forest residues in British Columbia was investigated by Nie and Bi [73]. Three pathways were evaluated: conversion in a central integrated refinery (Fr-CIR), conversion to bio-oil in distributed biorefineries and subsequent upgrading in a central oil refinery (Bo-DBR), and densification of forest residues in distributed pellet plants and conversion in a central integrated refinery (Wp-CIR). Bo-DBR provides the lowest GHG emissions, 17.0 g CO_2eq/MJ, followed by Wp-CIR (19.5 g CO_2eq/MJ). The Fr-CIR pathway relatively has the highest impact (20.5 g CO_2eq/MJ). The conversion stage contributed more than 50% to total GHG emissions in all scenarios considered. When compared with petroleum fuels, they offer 78–82% GHG emissions reduction. They also considered a potential credit from biochar applied for soil amendment, which can cause the further achievement of a reduction of 6.8 g CO_2eq/MJ [73].

Fortier et al. [42] studied the production of bio-jet fuel through the HTL of algae. Their LCA shows a GWP of 35.2 kg CO_2eq/GJ for the HTL of algae compared to 86.5 kg CO_2eq/GJ for HTL produced at a refinery. The sensitivity analysis included the extent of heat integration, the heat source, and the solid contents of the feedstock, and showed that GWP could be reduced by 76%.

Liu et al. [51] used pilot-scale data to analyze gasoline and diesel production through the HTL of algae. Three scenarios were set up: laboratory, pilot, and prediction of a full-scale production plant. In all three scenarios, algae-based fuels showed much lower GHG emissions than fossil fuel-based fuels, while the energy output/input in terms of energy return on investment (EROI) for the algae-based fuels was lower (1 to 3 MJ/MJ) than for fossil fuels (currently around 5 MJ/MJ).

Luterbacher et al. [52] studied the possibility of using hydrothermal gasification to convert waste biomass to synthetic natural gas (SNG). Three scenarios based on production scale were assumed: large and small scale (both for manure feedstock), and medium scale (for wood feedstock). It was shown that whatever the scenario, 10% fossil fuel was necessary for the process. N_2O emissions can be reduced in the conversion of manure, and GHG emissions are lower when manure (−0.06 kg CO_2eq/MJ_{SNG}) is the feedstock instead of wood biomass (−0.02 kg CO_2eq/MJ_{SNG}).

11.4 Discussion on the Application of LCA for Thermochemical Processing of Biomass

11.4.1 Establishing Goal and Scope

Depending on the intended application, LCA is carried out for different purposes. One can apply LCA to define the environmental impact of an individual product system during its entire life cycle, with the aim of identifying the stages that contribute most to a particular environmental impact. The results could be used for product development or improvement. LCA is also widely used to compare the environmental performance of two or more goods or services that provide the same functions. The use of LCA in a bioenergy system is no exception. As the summary table shows (see Table 11.1), the majority of studies are comparative LCAs that focus on product systems (e.g. biofuel vs. fossil fuel), energy technologies (e.g. pyrolysis vs. HTL), and feedstock alternatives (e.g. wood vs. forest residue).

When performing comparative LCA, one carefully considers how to choose the reference system, the system boundary, the function unit, and other aspects. If the comparative LCA is performed to assess the potential environmental benefit of displacing fossil fuel by bioenergy, how would one select the fossil fuel to substitute? The potential environmental saving depends on the type of fossil fuel used in the comparison. It is crucial to identify first the type of fossil fuel that can realistically be displaced by the bioenergy product. When an exact substitute cannot be determined, the best available fossil-based energy system or an average fossil fuel energy system could be used. Depending on the assumptions and other considerations, the comparative environmental results of a bioenergy system vs. fossil fuel could widely differ. For example, if the study aims to estimate the GHG emissions savings from substituting fossil fuel-derived electricity by bioenergy, depending on the sources of electricity (natural gas, coal, hydropower, wind, or other), net benefits could differ. The net savings will be higher when it is assumed that coal-based electricity (high GHG emissions intensity) is substituted rather than natural gas (lower GHG emissions intensity than coal). Therefore, the definition of the reference fossil fuel source needs to be clear [82, 83].

In most reviewed papers, energy values (i.e. MJ or kWh) are used as a functional unit to provide a reference against which all inputs and outputs are normalized. The choice of functional unit is another important aspect, especially when the LCA is used to compare the environmental performance of different products. When the biomass is used as a biodiesel and compared with gasoline from fossil fuel, all the inputs should reflect the end-use function, which is to transport something a certain distance (person-km or vehicle-km). In that case, the reference system should be an equivalent amount of fossil-based gasoline that provides the same transportation service. The choice of functional unit and reference system could affect the final results and interpretation.

Depending on available data and intended application of the LCA, the boundary could be set to include the whole life cycle (cradle-to-grave/well-to-wheel) or exclude the use and end-of-life phases (cradle-to-gate/well-to-pump). However, according to the ISO, omitting life cycle stages or processes is only permitted if they are assumed to have little significance on the overall findings, and their omission must be justified and clearly stated [18].

Most of the studies reviewed here used as their system boundary cradle-to-grave/well-to-wheel and cradle-to-gate/well-to-pump. Even though all upstream processes affect the environmental performance of biofuels greatly, combustion emissions, which depend on fuel quality (chemical composition and physical properties), are also relevant. Comparisons based on well-to-pump or on energy content only could bring about misleading conclusions.

11.4.2 Life Cycle Inventory Analysis

11.4.2.1 Allocation

In LCA, allocation is used to attribute the contribution of environmental impacts between the main product and its co-products. Most bioenergy processes are multifunctional, and therefore the allocation procedure needs to be applied according to the ISO hierarchy. According to the ISO, allocation can be avoided whenever possible by expanding the system boundaries. Different allocation procedures could alter the results considerably. By-products from bioenergy processes have many alternative applications: animal feed, fertilizer, heat production, and electricity production, etc. The environmental trade-offs from substituting co-products can be modeled by including the life cycle impacts of the product and subtracting them from bioenergy production. When there is no alternative product and the system expansion/substitution method is not feasible, material and energy inputs and emissions and waste outputs between the main product and the co-products can be allocated in proportion to the mass or energy balance or the economic value.

11.4.2.2 Attributional vs Consequential

Almost all the studies presented here are attributional LCAs that identify environmental impacts, mainly GHG emissions and net energy consumption. But consequential LCA is more useful to decision makers as it addresses possible economic changes through a change in production. Bioenergy production from forest or energy crops can have an impact on land use and biodiversity. By-products from energy production pathways such as dry distiller grain could be used as animal feed, fertilizer, or energy source, however, and in most LCA studies the environmental benefits from the use of co-products are discounted by crediting the impacts from the displaced products. Yet the potential market consequences due to the substitution are not usually addressed. Consequential LCA allows the user to evaluate the environmental benefits of future energy policies in a given economic system. For example, one can perform a consequential LCA to estimate the environmental impact consequences of changing an agricultural system to increase energy crop production and the potential effects on the energy sector, such as increasing the share of renewable energy sources in the grid mix. Consequential LCA helps to understand the possible future environmental consequences when biofuel production increases following changes in environmental regulations toward cleaner diesel production. It addresses questions such as: is there a potential market to absorb the by-products of biofuel production? What are the environmental benefits of using by-products to substitute natural gas for space heating? How will by-products affect energy demand and supply in the long run?

Another important factor in modeling environmental impacts of bioenergy is land-use change [84]. Bioenergy production requires the use of land as input to the production system. The impact associated with land-use change is either direct or indirect. Direct land-use change is the conversion of a natural ecosystem for an economic activity. The resulting GHG emissions from this conversion can be directly attributed to this new land use. An example of direct land-use change is plowing pasture land to grow corn for biofuels production, which is accompanied by an increase in carbon dioxide emissions to the atmosphere as a result of soil carbon loss. Indirect land-use change is the conversion of a natural ecosystem to crop land to alleviate local food shortages arising from the conversion of crop land in another part of the world from food production to biofuels production. In this case, the resulting GHG emissions are attributed not to the local land conversion to food production but to the remote land conversion to biofuels production. The concept of ILUC, first proposed in 2008, was advanced to predict the effect of an expanding grain ethanol industry on worldwide GHG emissions. It was hypothesized that diverting grain to ethanol production would cause food prices to rise, encouraging the conversion of forest and grassland around the world into cropland, which analysis suggested would release more GHG emissions into the atmosphere than were avoided through biofuels production. Although the original study encouraged the perception that biofuels were worse than fossil fuels in terms of GHG emissions, subsequent studies demonstrated that GHG emissions from ILUC are extremely sensitive to the many assumptions inherent in the analysis and were probably overstated in the original analysis [85].

More recently, the concept was applied to an attributional and consequential LCA study to identify potential environmental consequence of biodiesel production from palm oil vs. conventional diesel in Thailand [74, 75]. This study demonstrated the influence of modeling choice, consideration of co-product allocation and land-use change (LUC), and conversion technology on the overall results. The authors concluded that emissions from ILUC are the most relevant contributors to terrestrial acidification and marine eutrophication, when use phase emissions are excluded. A study on the establishments of a national Low Carbon Fuel Standard suggests that the inclusion of ILUC would result in an additional 1.3–2.6% reduction in GHG emissions between 2009 and 2027 [86]. This review found few studies that explicitly addressed the environmental impact of direct land-use change [43, 47, 68] and indirect land-use change [74–76] on thermochemical production of fuels and chemicals despite intense interest in the subject by policymakers in recent years [87]. This omission may reflect the high uncertainty in ILUC analyses and the absence of consistent and harmonized LCA methods to deal with indirect land-use change [88–90]. GHG emissions due to indirect land-use change could vary from 10 to 340 g CO_2eq/MJ depending on the type and location of feedstock used, LCA models applied, assumptions, scenarios, and data use and variability in amortization periods [90].

11.4.3 Life Cycle Impact Assessment

11.4.3.1 Impact Categories

Most LCA studies on the thermochemical processing of biomass are limited to addressing GHG emissions and net energy assessment. Only a few studies consider other impact categories. GHG emissions and energy balances are prominent issues on the global political agenda. More importantly, climate change mitigation and reducing fossil fuel dependency are key factors in recent bioenergy development, but comparing products' environmental performance through a single impact indicator would result in a misleading conclusion. It is important to include other impact categories that are equally relevant so that a full picture can be seen. In this regard, the environmental assessment for energy systems should go beyond simply assessing GHG emissions and include several midpoint impact categories. This will help to understand the environmental trade-offs in the system.

Only a few studies include the impact of land-use change on GHG emissions. The land-use change impact category is of particular importance for forest resource- or energy crop-based bioenergy production. A land-use change impact analysis focuses on any process that potentially alters landscapes and consequently affects the amount of carbon stored in plants and soil (carbon stock) and the biodiversity. A comprehensive LCA study of energy production from biomass via thermochemical conversion needs to address the land-use change impact. The impacts of bioenergy on food and water availability are also important issues that are not well covered in the reviewed papers. Bioenergy products could affect water resource availability (because the conversion process is very water intensive) and could affect water quality (due to eutrophication) [91].

11.4.3.2 Carbon and Climate Neutrality

Bioenergy production from sustainably managed forest biomass is usually considered to be carbon neutral, based on the assumption that the amount of CO_2 released into the atmosphere during product use is offset by the amount sequestered during plant growth. However, the climate neutrality of bioenergy production is the subject of much debate. There is a time difference between release into and withdrawal from the atmosphere. Tree growth rate, biomass decomposition rate, rotation time, and land-use change are key factors that need to be taken into account in assessing climate neutrality. The biomass carbon stock and the temporal dynamic of the CO_2 sink and emissions are usually neglected in traditional LCA. Traditional LCA is based on a steady-state assumption (i.e. the assumption is that emitting a large amount of pollutant all at once has the same impact as slowly emitting the same amount of pollutants over a certain interval in a given time) and ignoring temporal information in its assessment. ISO 14040 also acknowledges that ignoring the time variability in the impact modeling could reduce the environmental relevance for some results [17]. A dynamic LCA approach is the way forward as it takes into consideration the temporal variability of GHG emissions by implementing the time-dependent characterization factor to determine the global warming potential for different time horizons [92, 93].

11.5 Conclusions

There is significant potential in the thermochemical processing of biomass to convert biomass into value-added products. In this study, we carried out an extensive review of several LCA studies on the thermochemical processing of biomass to understand the main gaps in performing an LCA. While a few of the papers identify environmental impacts associated with different thermochemical pathways, most are comparative studies. The comparative LCAs focus on energy technologies (e.g. pyrolysis vs. HTL), product systems (e.g. biofuel vs. fossil fuel), and feedstock alternatives (e.g. wood vs. forest residue). Most of the LCAs reviewed here are limited to addressing GHG emissions and net energy assessment. Other local and global impacts are not well covered. Moreover, while there are many LCAs for pyrolysis, biomass combustion, and gasification, LCAs for hydrothermal processes and bio-oil upgrading are limited.

LCA has shown huge development since its conception in the early 1900s. Consequential LCA, dynamic LCA, environmentally extended input–output-based LCA, and social LCA are all advanced LCA approaches that have been widely applied for decision making. But their applications in bioenergy technologies are limited.

Land-use change is not widely studied in the LCA of the thermochemical processing of biomass. The impact of land-use change is an important category that needs to be assessed for forest resource- or energy crop-based bioenergy production. Future LCA of the thermochemical processing of biomass should focus on this aspect.

A few of the papers reviewed here included water footprints or water consumption in their impact categories. The environmental assessment of the thermochemical processing of biomass should include water footprint as an impact category because the production of bioenergy is water intensive and could lead to water scarcity in the future.

Acknowledgements

The authors are grateful to the NSERC/Cenovus/Alberta Innovates Associate Industrial Research Chair Program in Energy and Environmental Systems Engineering and the Cenovus Energy Endowed Chair in Environmental Engineering at the University of Alberta for financial support for this research. Astrid Blodgett is thanked for editorial assistance.

References

1. IPCC 2014. Climate Change 2014: Synthesis Report. Contribution of Working Groups I, II and III to the Fifth Assessment Report of the Intergovernmental Panel on Climate Change.Geneva, Switzerland: IPCC.
2. United Nations 2015. Paris Agreement. http://unfccc.int/paris_agreement/items/9485.php (accessed 30 November 2017).
3. European Union 2009. Directive 2009/28/EC of the European Parliament and of the Council of 23 April 2009 on the promotion of the use of energy from renewable sources and amending and subsequently repealing Directives 2001/77/EC and 2003/30/EC (Text with EEA relevance). European Parliament, Council of the European Union.
4. Federal Register 2015. Executive Order 13693—Planning for Federal Sustainability in the Next Decade. United States Environmental Protection Agency. https://www.epa.gov/greeningepa/executive-order-13693-planning-federal-sustainability-next-decade (accessed 10 November 2017).
5. IRENA 2017. Untapped potential for climate action: Renewable energy in Nationally Determined Contributions. Abu Dhabi. http://irena.org/-/media/Files/IRENA/Agency/Publication/2017/Nov/IRENA_Untapped_potential_NDCs_2017.pdf (accessed 10 November 10, 2017).
6. UNEP 2017. Global Trends in Renewable Energy Investment 2017. http://fs-unep-centre.org/sites/default/files/publications/globaltrendsinrenewableenergyinvestment2017.pdf (accessed 1 November 1, 2017).
7. REN 21 2016. Renewables 2016 – Global Status Report. REN21 (Renewable Energy Policy Network for the 21st Century. http://www.ren21.net/status-of-renewables/global-status-report/renewables-2016-global-status-report (accessed 10 November 10, 2017).
8. REN 21 2015. Renewables 2015 –Global Status Report. REN21 Secretariat: Paris, France. http://www.ren21.net/status-of-renewables/global-status-report/renewables-2015-global-status-report (accessed 10 November 2017).

9. Oyedun, A.O., Gebreegziabher, T., and Hui, C.W. (2013). Mechanism and modelling of bamboo pyrolysis. *Fuel Processing Technology* **106**: 595–604.
10. Agbor, E., Oyedun, A.O., Zhang, X., and Kumar, A. (2016). Integrated techno-economic and environmental assessments of sixty scenarios for co-firing biomass with coal and natural gas. *Applied Energy* **169**: 433–449.
11. Bridgwater, A. and Peacocke, G. (2000). Fast pyrolysis processes for biomass. *Renewable and Sustainable Energy Reviews* **4** (1): 1–73.
12. Kumar, M., Oyedun, A.O., and Kumar, A. (2017). Hydrothermal liquefaction of biomass for the production of diluents for bitumen transport. *Biofuels, Bioproducts and Biorefining* **11**: 811–829.
13. Lam, K.L., Oyedun, A.O., and Hui, C.W. (2012). Experimental and modelling studies of biomass pyrolysis. *Chinese Journal of Chemical Engineering* **20** (3): 543–550.
14. Zhang, L., Xu, C.C., and Champagne, P. (2010). Overview of recent advances in thermo-chemical conversion of biomass. *Energy Conversion and Management* **51** (5): 969–982.
15. Oyedun, A.O. and Kumar, A. (2017). An assessment of the impacts of biomass processing methods on power generation. In: *Encyclopedia of Sustainable Technologies* (ed. M. Abraham), 153–170. Amsterdam, The Netherlands: Elsevier Inc.
16. Yi, Z., Oyedun, A.O., Maojian, W. et al. (2014). Modeling, integration and optimization of biomass and coal co-gasification. *Energy Procedia* **61**: 113–116.
17. ISO 2006. ISO 14040:2006: International Standard – Environmental management – Life cycle assessment – Principles and framework. Geneva, Switzerland: International Organization for Standardization. https://www.iso.org/standard/37456.html (accessed 5 November 2017).
18. ISO 2006. ISO 14044:2006: International Standard – Environmental management – Life cycle assessment – Requirements and guidelines. Geneva, Switzerland: International Organization for Standardization. https://www.iso.org/standard/38498.html (accessed 5 November 2017).
19. Thomassen, M.A., Dalgaard, R., Heijungs, R., and de Boer, I. (2008). Attributional and consequential LCA of milk production. *The International Journal of Life Cycle Assessment* **13** (4): 339–349.
20. Sonnemann, G. and Bruce, V. 2011. Global guidance principles for life cycle assessment databases. UNEP/SETAC Life Cycle Initiative. http://www.lifecycleinitiative.org/wp-content/uploads/2012/12/2011%20-%20Global%20Guidance%20Principles.pdf (accessed 10 October 2017).
21. UNEP 2011. Global guidance principles for life cycle assessment databases: a basis for greener processes and products. http://www.lifecycleinitiative.org/wp-content/uploads/2012/12/2011%20-%20Global%20Guidance%20Principles.pdf (accessed 30 October 2017).
22. Hauschild, M.Z. and Huijbregts, M.A.J. (2015). Introducing life cycle impact assessment. In: *Life Cycle Impact Assessment* (ed. W. Klöpffer and M.A. Curran), 1–16. Springer.
23. Guinée, J.B., Gorrée, M., Heijungs, R., et al. 2002. Handbook on life cycle assessment. Operational guide to the ISO standards. I: LCA in perspective. IIa: Guide. IIb: Operational annex. III: Scientific background. Dordrecht: Kluwer Academic Publishers.
24. Goedkoop, M. and Sprinsma, R. 2001. The Eco-Indicator'99. A Damage Oriented Method for Life Cycle Impact Assessment, Methodology Report. Amersfoort. PRé Consultants. https://www.pre-sustainability.com/download/EI99_annexe_v3.pdf (acessed 10 November 2017).
25. Jolliet, O., Margni, M., Charles, R. et al. (2003). IMPACT 2002+: a new life cycle impact assessment methodology. *The International Journal of Life Cycle Assessment* **8** (6): 324.
26. Goedkoop, M., Heijungs, R., Huijbregts, M. et al. (2013). *ReCiPe 2008 – A life cycle impact assessment method which comprises harmonised category indicators at the midpoint and the endpoint level*. The Hague, The Netherlands: Dutch Ministry of the Environment.
27. Bare, J. (2011). TRACI 2.0: the tool for the reduction and assessment of chemical and other environmental impacts 2.0. *Clean Technologies and Environmental Policy* **13** (5): 687–696.
28. Rosenbaum, R.K., Bachmann, T.M., Gold, L.S. et al. (2008). USEtox—the UNEP-SETAC toxicity model: recommended characterisation factors for human toxicity and freshwater ecotoxicity in life cycle impact assessment. *The International Journal of Life Cycle Assessment* **13** (7): 532.

29. Huijbregts, M.A.J. (1998). Application of uncertainty and variability in LCA. *The International Journal of Life Cycle Assessment* **3** (5): 273.
30. Stevens, C. (2011). *Thermochemical Processing of Biomass: Conversion into Fuels, Chemicals and Power*. Wiley.
31. Patel, M., Oyedun, A.O., Kumar, A., and Gupta, R. (2018). A techno-economic assessment of renewable diesel and gasoline production from Aspen hardwood. *Waste and Biomass Valorization* 1–16. https://doi.org/10.1007/s12649-018-0359-x.
32. Pankratz, S., Oyedun, A.O., Zhang, X., and Kumar, A. (2017). Algae production platforms for Canada's northern climate. *Renewable and Sustainable Energy Reviews* **80**: 109–120.
33. Kumar, M., Oyedun, A.O., and Kumar, A. (2018). A review on the current status of various hydrothermal technologies on biomass feedstock. *Renewable and Sustainable Energy Reviews* **81** (Part 2): 1742–1770.
34. Bennion, E.P., Ginosar, D.M., Moses, J. et al. (2015). Lifecycle assessment of microalgae to biofuel: comparison of thermochemical processing pathways. *Applied Energy* **154**: 1062–1071.
35. Carpentieri, M., Corti, A., and Lombardi, L. (2005). Life cycle assessment (LCA) of an integrated biomass gasification combined cycle (IBGCC) with CO 2 removal. *Energy Conversion and Management* **46** (11): 1790–1808.
36. Chan, Y.H., Tan, R.R., Yusup, S. et al. (2016). Comparative life cycle assessment (LCA) of bio-oil production from fast pyrolysis and hydrothermal liquefaction of oil palm empty fruit bunch (EFB). *Clean Technologies and Environmental Policy* **18** (6): 1759–1768.
37. Chan, Y.H., Yusup, S., Quitain, A.T. et al. (2015). Effect of process parameters on hydrothermal liquefaction of oil palm biomass for bio-oil production and its life cycle assessment. *Energy Conversion and Management* **104**: 180–188.
38. Corti, A. and Lombardi, L. (2004). Biomass integrated gasification combined cycle with reduced CO 2 emissions: performance analysis and life cycle assessment (LCA). *Energy* **29** (12): 2109–2124.
39. Dang, Q., Yu, C., and Luo, Z. (2014). Environmental life cycle assessment of bio-fuel production via fast pyrolysis of corn stover and hydroprocessing. *Fuel* **131**: 36–42.
40. Eriksson, O., Finnveden, G., Ekvall, T., and Björklund, A. (2007). Life cycle assessment of fuels for district heating: a comparison of waste incineration, biomass-and natural gas combustion. *Energy Policy* **35** (2): 1346–1362.
41. Fan, J., Kalnes, T.N., Alward, M. et al. (2011). Life cycle assessment of electricity generation using fast pyrolysis bio-oil. *Renewable Energy* **36** (2): 632–641.
42. Fortier, M.-O.P., Roberts, G.W., Stagg-Williams, S.M., and Sturm, B.S. (2014). Life cycle assessment of bio-jet fuel from hydrothermal liquefaction of microalgae. *Applied Energy* **122**: 73–82.
43. Gagnon, L., Belanger, C., and Uchiyama, Y. (2002). Life-cycle assessment of electricity generation options: the status of research in year 2001. *Energy Policy* **30** (14): 1267–1278.
44. Han, J., Elgowainy, A., Dunn, J.B., and Wang, M.Q. (2013). Life cycle analysis of fuel production from fast pyrolysis of biomass. *Bioresource Technology* **133**: 421–428.
45. Heller, M.C., Keoleian, G.A., Mann, M.K., and Volk, T.A. (2004). Life cycle energy and environmental benefits of generating electricity from willow biomass. *Renewable Energy* **29** (7): 1023–1042.
46. Hsu, D.D. (2012). Life cycle assessment of gasoline and diesel produced via fast pyrolysis and hydroprocessing. *Biomass and Bioenergy* **45**: 41–47.
47. Huang, Y.-F., Syu, F.-S., Chiueh, P.-T., and Lo, S.-L. (2013). Life cycle assessment of biochar cofiring with coal. *Bioresource Technology* **131**: 166–171.
48. Iribarren, D., Peters, J.F., and Dufour, J. (2012). Life cycle assessment of transportation fuels from biomass pyrolysis. *Fuel* **97**: 812–821.
49. Kauffman, N., Hayes, D., and Brown, R. (2011). A life cycle assessment of advanced biofuel production from a hectare of corn. *Fuel* **90** (11): 3306–3314.

50. Koroneos, C., Dompros, A., and Roumbas, G. (2008). Hydrogen production via biomass gasification—a life cycle assessment approach. *Chemical Engineering and Processing Process Intensification* **47** (8): 1261–1268.
51. Liu, X., Saydah, B., Eranki, P. et al. (2013). Pilot-scale data provide enhanced estimates of the life cycle energy and emissions profile of algae biofuels produced via hydrothermal liquefaction. *Bioresource Technology* **148**: 163–171.
52. Luterbacher, J.S., Fröling, M., Vogel, F. et al. (2009). Hydrothermal gasification of waste biomass: Process design and life cycle assessment. *Environmental Science & Technology* **43** (5): 1578–1583.
53. Mann, M. and Spath, P. (2001). A life cycle assessment of biomass cofiring in a coal-fired power plant. *Clean Products and Processes* **3** (2): 81–91.
54. Peters, J.F., Iribarren, D., and Dufour, J. (2015). Life cycle assessment of pyrolysis oil applications. *Biomass Conversion and Biorefinery* **5** (1): 1–19.
55. Peters, J.F., Iribarren, D., and Dufour, J. (2015). Simulation and life cycle assessment of biofuel production via fast pyrolysis and hydroupgrading. *Fuel* **139**: 441–456.
56. Royo, J., Sebastián, F., García-Galindo, D. et al. (2012). Large-scale analysis of GHG (greenhouse gas) reduction by means of biomass co-firing at country-scale: application to the Spanish case. *Energy* **48** (1): 255–267.
57. Sebastián, F., Royo, J., Serra, L., and Gómez, M. (eds.) 2007. *Life Cycle Assessment of Greenhouse Gas Emissions from Biomass Electricity Generation: Cofiring and Biomass Monocombustion.* Proceedings of the 4th Dubrovnik Conference on Sustainable Development of Energy Water and Environment Systems, Dubrovnik, Croatia.
58. Shemfe, M.B., Whittaker, C., Gu, S., and Fidalgo, B. (2016). Comparative evaluation of GHG emissions from the use of Miscanthus for bio-hydrocarbon production via fast pyrolysis and bio-oil upgrading. *Applied Energy* **176**: 22–33.
59. Snowden-Swan, L.J., Spies, K.A., Lee, G., and Zhu, Y. (2016). Life cycle greenhouse gas emissions analysis of catalysts for hydrotreating of fast pyrolysis bio-oil. *Biomass and Bioenergy* **86**: 136–145.
60. Wang, H., Wang, L., and Shahbazi, A. (2015). Life cycle assessment of fast pyrolysis of municipal solid waste in North Carolina of USA. *Journal of Cleaner Production* **87**: 511–519.
61. Winjobi, O., Shonnard, D.R., Bar-Ziv, E., and Zhou, W. (2016). Life cycle greenhouse gas emissions of bio-oil from two-step torrefaction and fast pyrolysis of pine. *Biofuels, Bioproducts and Biorefining* **10** (5): 576–588.
62. Wong, A., Zhang, H., and Kumar, A. (2016). Life cycle water footprint of hydrogenation-derived renewable diesel production from lignocellulosic biomass. *Water Research* **102**: 330–345.
63. Zaimes, G.G., Soratana, K., Harden, C.L. et al. (2015). Biofuels via fast pyrolysis of perennial grasses: a life cycle evaluation of energy consumption and greenhouse gas emissions. *Environmental Science & Technology* **49** (16): 10007–10018.
64. Zhang, Y., Hu, G., and Brown, R.C. (2013). Life cycle assessment of the production of hydrogen and transportation fuels from corn stover via fast pyrolysis. *Environmental Research Letters* **8** (2): 025001.
65. Zuwała, J. (2012). Life cycle approach for energy and environmental analysis of biomass and coal co-firing in CHP plant with backpressure turbine. *Journal of Cleaner Production* **35**: 164–175.
66. Patel, M. and Kumar, A. (2016). Production of renewable diesel through the hydroprocessing of lignocellulosic biomass-derived bio-oil: a review. *Renewable and Sustainable Energy Reviews* **58**: 1293–1307.
67. Patel, M., Zhang, X., and Kumar, A. (2016). Techno-economic and life cycle assessment on lignocellulosic biomass thermochemical conversion technologies: a review. *Renewable and Sustainable Energy Reviews* **53**: 1486–1499.
68. Adams, P., Shirley, J., and McManus, M. (2015). Comparative cradle-to-gate life cycle assessment of wood pellet production with torrefaction. *Applied Energy* **138**: 367–380.
69. Christoforou, E.A. and Fokaides, P.A. (2016). Life cycle assessment (LCA) of olive husk torrefaction. *Renewable Energy* **90**: 257–266.

70. Kabir, M.R. and Kumar, A. (2012). Comparison of the energy and environmental performances of nine biomass/coal co-firing pathways. *Bioresource Technology* **124**: 394–405.
71. Kabir, M.R. and Kumar, A. (2011). Development of net energy ratio and emission factor for biohydrogen production pathways. *Bioresource Technology* **102** (19): 8972–8985.
72. Tews, I., Zhu, Y., Drennan, C. et al. (2014). *Biomass Direct Liquefaction Options: Techno Economic and Life Cycle Assessment, PNNL 23579*. Richland, WA: Pacific Northwest National Laboratory.
73. Nie, Y. and Bi, X. (2018). Life-cycle assessment of transportation biofuels from hydrothermal liquefaction of forest residues in British Columbia. *Biotechnology for Biofuels* **11** (1): 23.
74. Prapaspongsa, T. and Gheewala, S.H. (2017). Consequential and attributional environmental assessment of biofuels: implications of modelling choices on climate change mitigation strategies. *International Journal of Life Cycle Assessment* **22** (11): 1644–1657.
75. Prapaspongsa, T., Musikavong, C., and Gheewala, S.H. (2017). Life cycle assessment of palm biodiesel production in Thailand: impacts from modelling choices, co-product utilisation, improvement technologies, and land use change. *Journal of Cleaner Production* **153**: 435–447.
76. Searchinger, T., Heimlich, R., Houghton, R.A. et al. (2008). Use of US croplands for biofuels increases greenhouse gases through emissions from land-use change. *Science* **319** (5867): 1238–1240.
77. Amponsah, N.Y., Troldborg, M., Kington, B. et al. (2014). Greenhouse gas emissions from renewable energy sources: a review of lifecycle considerations. *Renewable and Sustainable Energy Reviews* **39**: 461–475.
78. Martin, E.W., Chester, M.V., and Vergara, S.E. (2015). Attributional and consequential life-cycle assessment in biofuels: a review of recent literature in the context of system boundaries. *Current Sustainable/Renewable Energy Reports* **2** (3): 82–89.
79. Wang, M.Q., Han, J., Haq, Z. et al. (2011). Energy and greenhouse gas emission effects of corn and cellulosic ethanol with technology improvements and land use changes. *Biomass and Bioenergy* **35** (5): 1885–1896.
80. Meyer-Aurich, A., Schattauer, A., Hellebrand, H.J. et al. (2012). Impact of uncertainties on greenhouse gas mitigation potential of biogas production from agricultural resources. *Renewable Energy* **37** (1): 277–284.
81. Qin, Z., Canter, C.E., Dunn, J.B. et al. (2018). Land management change greatly impacts biofuels' greenhouse gas emissions. *GCB Bioenergy* **10** (6): 370–381.
82. Cherubini, F. (2010). GHG balances of bioenergy systems – Overview of key steps in the production chain and methodological concerns. *Renewable Energy* **35** (7): 1565–1573.
83. Cherubini, F. and Strømman, A.H. (2011). Life cycle assessment of bioenergy systems: state of the art and future challenges. *Bioresource Technology* **102** (2): 437–451.
84. Melillo, J.M., Reilly, J.M., Kicklighter, D.W. et al. (2009). Indirect emissions from biofuels: how important? *Science* **326** (5958): 1397–1399.
85. Dumortier, J., Hayes, D.J., Carriquiry, M. et al. (2011). Sensitivity of carbon emission estimates from indirect land-use change. *Applied Economic Perspectives and Policy* **33** (3): 428–448.
86. Khanna, M., Wang, W., Hudiburg, T.W., and DeLucia, E.H. (2017). The social inefficiency of regulating indirect land use change due to biofuels. *Nature Communications* **8**: 15513.
87. Ahlgren, S. and Di Lucia, L. (2014). Indirect land use changes of biofuel production – a review of modelling efforts and policy developments in the European Union. *Biotechnology for Biofuels* **7** (1): 35.
88. Gawel, E. and Ludwig, G. (2011). The iLUC dilemma: how to deal with indirect land use changes when governing energy crops? *Land Use Policy* **28** (4): 846–856.
89. Finkbeiner, M. (2014). Indirect land use change – Help beyond the hype? *Biomass and Bioenergy* **62**: 218–221.
90. Panichelli, L. and Gnansounou, E. (2017). Chapter 9 – Modeling Land-Use Change Effects of Biofuel Policies: Coupling Economic Models and LCA. *Life-Cycle Assessment of Biorefineries*, 233–258. Amsterdam: Elsevier.

91. Gerbens-Leenes, P.W., Hoekstra, A.Y., and van der Meer, T. (2009). The water footprint of energy from biomass: a quantitative assessment and consequences of an increasing share of bio-energy in energy supply. *Ecological Economics* **68** (4): 1052–1060.
92. Levasseur, A., Lesage, P., Margni, M. et al. (2010). Considering time in LCA: dynamic LCA and its application to global warming impact assessments. *Environmental Science & Technology* **44** (8): 3169–3174.
93. Helin, T., Sokka, L., Soimakallio, S. et al. (2013). Approaches for inclusion of forest carbon cycle in life cycle assessment – a review. *GCB Bioenergy* **5** (5): 475–486.

Index

acetic acid 309
acetogens 316–317
adiabatic reactors 150
air-blown gasifiers 95–97
air-to-fuel ratio (A/F) 87
alcoholysis 262
alkali and alkaline earth metals (AAEMs) 28
anhydro-oligosaccharides 30
ash effects
   demineralization 180
   environmental conditions 180
   inorganic phosphorus 183
   liquid yield and liquid's composition 179
   moisture content 181
   non-metal S 182
   organic yield 181
   phase separation 184
   sulfur and phosphorus 180
   total acid number (TAN) 185

Becke-three parameter-Lee-Yang-Parr (B3LYP) functional 30
bio-based chemicals 11
bioenergy 87
biofuels 10–11
biogas/landfill 53
biomass combustion
   baseload power generation 50
   combustion properties
      composition of 59–62
      density and particle size 65
      heating value 63–64
      moisture content 62

combustion stoichiometry
   air/fuel ratio 66
   equilibrium 68
   flame temperature 66–67
   rates of reaction 68–71
   simplified global reaction 65–66
combustors, types of
   alternative combustion and power generation concepts 57–58
   co-firing 56–57
   power and heat generation, large-scale systems for 54–56
   small-scale systems 53–54
electricity generation 51
firewood gathering 50
fuel types
   gaseous fuel 52–53
   solids 51–52
fundamentals of 59
magnetohydrodynamic energy conversion 51
plant photosynthesis and respiration 50
pollutant emissions and environmental impacts
   dioxin-like compounds 74–76
   greenhouse gas emissions 77
   heavy metals 76
   incomplete combustion 74
   oxides of nitrogen and sulfur 72–74
   particulate matter 74
   radioactive species 76–77
renewable and zero-carbon resources 51
biomass energy and carbon capture and sequestration (BECCS) 340

Thermochemical Processing of Biomass: Conversion into Fuels, Chemicals and Power, Second Edition.
Edited by Robert C. Brown.
© 2019 John Wiley & Sons Ltd. Published 2019 by John Wiley & Sons Ltd.

biomass integrated gasification combined cycles (BIGCCs)   56
biomass particles   10
bio-methanol   11
bio-oil
  aging   211–212
  catalytic FP (CFP)
    aqueous phase   218
    biomass feedstock   221–222
    catalyst   221
    catalyst deactivation   223
    chemistry   221
    of lignocellulosic biomass   223
    lower pyrolysis temperature and/or improve selectivity   218
    oil phase   218
    process configuration   222
    process parameters   222–223
    in situ *and* ex situ *configuration*   220
  catalytic hydrotreating
    condensation and polymerization reactions   214
    deep hydrotreating   217–220
    hydrogen consumption   215
    nitrogen physisorption analysis   216
    Ru metal catalyst   215
    stabilization reactions   216
    sulfur poisoning   216
  characteristics and quality
    biomass composition   210
    bio-oil viscosity   209
    boiler combustion applications   208
    hemicellulose degradation   208
    inorganic content and composition   209–210
    water content of biomass   210
  chemicals   232–235
  composition of
    operating temperature effects   210
    vapor residence time   211
  diesel
    hydrotreated bio-oil   232
    raw-bio oil   231–232
  fractionation   318
  gasoline
    FCC upgraded bio-oil   230
    hydrotreated higher oxygen content bio-oil   229–230
    hydrotreated low oxygen content bio-oil   229
    oxygenated model compounds   230
  hydrogen production   235
  jet fuel   230–231
  norms and standards   212–213
  physical pre-treatment
    fractionation   214
    physical filtration   213
    solvent addition   213–214
  solvent liquefaction   263–264
  upgrading strategies
    catalyzed reactions with alchol   227–228
    liquid bio-oil zeolite co-processing   223, 226
  viscosity   209
biopower   9–10
Biot number   30
bubbling fluidized-bed (BFB)   95, 101–102

Carbo-V process   105
catalytic fast pyrolysis (CFP)
  aqueous phase   218
  biomass feedstock   221–222
  catalyst   221
  catalyst deactivation   223
  chemistry   221
  of lignocellulosic biomass   223
  lower pyrolysis temperature and/or improve selectivity   218
  oil phase   218
  process configuration   222
  process parameters   222–223
  in situ *and* ex situ *configuration*   220
catalytic partial oxidation (CPO)   200
cellulose conversion reactions   36–38
cell wall structure, role of
  computer simulation   32
  conventional heat transfer calculations   30
  lignocellulosic biomass   30
  volatile product formation and vaporization   31
  volatile product mass transfer   32
char and tar   109–110
charcoal   52
char formation   41
circulating fluidized-bed (CFB)   95, 103–104
climate mitigation strategy   87
coal-fired boilers   340
co-firing   56–57
combined heat and power (CHP)   117, 361
condensed phase reactions, thermal deconstruction
  alkali and alkaline earth metals (AAEMs)   28
  cell structure and arrangement   19
  cellular microstructure   18
  cell wall structure, role of
    computer simulation   32
    conventional heat transfer calculations   30

lignocellulosic biomass 30
    volatile product formation and
        vaporization 31
    volatile product mass transfer 32
combustion 27
computational methodology 34–35
error propagation 19
fast pyrolysis 24
gasification 26
investigation of
    anhydro-oligosaccharides 30
    Becke-three parameter-Lee-Yang-Parr
        (B3LYP) functional 30
    computational chemistry 30
    devolatilization and liquid droplet
        ejection 28
    low volatility and short reaction times
        28
    surface tension-driven coalescence 29
    water-soluble anhydro-oligosaccharides
        30
liquid products, formation of
    alkali and alkaline earth metals 40
    cellulose conversion reactions 36–38
    char formation 41
    gaseous products, formation of 40
    hemicellulose conversion reactions 38
    large-scale polymers 41
    lignin conversion reactions 38–39
    water evolution 39
physical and chemical transformations 19
slow pyrolysis 20, 22
solvent liquefaction 23–24
torrefaction 20, 22–23
crude pyrolytic substrate 318–320

decarbonylation 262
decarboxylation 261
decomposition reactions 17
dehydration 262
dehydrogenation 262
demineralization 180
depolymerization 17
direct liquefaction 348
downdraft gasifier 100
dry scrubbing 58
Dynamotive Technologies Corp 190–192

electric power generation
    biomass gasification systems 339
    direct combustion 338–339
    pyrolysis power generation 339–340
Empyro pyrolysis plant 190
Ensyn plant 187
Ensyn's technology 186–188

entrained-flow gasifier 104–106
esterification 349
ethanol and mixed alcohols
    alcohol production rates 156
    catalyst 159
    chemistry 157–159
    direct synthesis 156
    Fischer–Tropsch and methanol synthesis
        156
    reactor designs 157
    rhodium-based catalysts 156

fast pyrolysis 12
    ash effects
        demineralization 180
        environmental conditions 180
        inorganic phosphorus 183
        liquid yield and liquid's composition
            179
        moisture content 181
        non-metal S 182
        organic yield 181
        phase separation 184
        sulfur and phosphorus 180
        total acid number (TAN) 185
    biomass and intermediate products
        carbohydrate and lignin fractions 178
        cellulose 177
        endothermic/exothermic reaction
            enthalpies 177
        guaiacyl-syringyl lignin 178
        hemicellulose 176
        kinetic expressions 177
        levoglucosan and cellobiosan 178
        lignin 177
        lignocellulosic biomass 176
        technical lignin 179
    and bio-oil fermentation 346–347
    BTG-BtL (rotating-cone) 188–190
    catalytic fast pyrolysis 345–346
    catalytic pyrolysis 202
    char and non-condensable gases 175
    co-FCC option
        hydrotreated pyrolysis liquids 198–201
        pyrolysis liquids and crude oil derivatives
            196–198
    combustion conditions 192–193
    co-refining options 194–196
    cost-effective method 175
    diesel engines 193–194
    Dynamotive Technologies Corp 190–192
    Ensyn's technology 186–188
    gasification to fuels 346
    and hydroprocessing 344–345
    of lignocellulosic biomass

Index

fast pyrolysis (*continued*)
    acetic acid   309
    lignin derivatives   309
    pyrolytic substrates   308
    pyrolytic sugars   309
    thermal decomposition   176
    Valmet/UPM (CFB)   188
Fischer-Tropsch (FT) synthesis   58, 87, 96, 137, 340–341
fixed-bed gasifiers   99–101
fluid catalytic cracking (FCC) units
    hydrotreated pyrolysis liquids
        chemicals   200–201
        gasification   199–200
    pyrolysis liquids and crude oil derivatives
        carbon yields   198
        conventional crude-derived oils   196
        fluid bed regenerator reactor   197
        small-scale laboratory reactors   197
fluid coking technology   96

gas chromatography (GC) analysis   184
gaseous fuel   52–53
gaseous products, formation of   40
gasification   126
    air-blown gasifiers   95–97
    applications of gasification plants   111–117
    bubbling fluidized bed   101–102
    char and tar   109–110
    circulating fluidized bed (CFB)   103–104
    combined heat and power (CHP)   117
    entrained flow   104–106
    feedstock properties
        alkali and phosphorus   94
        biomass and coals   92–93
        HCN   93
        nitrogen, chlorine, and sulfur   93–94
        partial-oxidation gasification   91
        partial-oxidation gasifiers   94
        potassium   94
        proximate analysis   91
        pyrolysis   95
        refuse-derived fuel (RDF) and sewage sludge   95
        solid fuels and biomass   92
        solid–liquid phase behavior   94
    fixed-bed gasifiers   99–101
    fuel and chemical synthesis   117–118
    fundamentals
        gas-phase reactions   91
        gas–solid reactions   90
        heating and drying   89
        pyrolysis   89–90
    gaseous products   106–109

    gasification terminology   87–88
    history of
        bioenergy   87
        climate mitigation strategy   87
        Fischer–Tropsch (FT) synthesis   87
        illation and heating applications   85
        natural gas production and distribution   86
        "pneumatic trough," invention of   86
        storage capability   86
        technological improvements   86
        war economics and petroleum supply restrictions   86
    indirectly heated gasifiers
        fluid coking technology   96
        partial-oxidation reactors   96
        Range Fuels gasifier   98
        separate gasification and combustion reactor compartments   98
        ThermoChem Recovery International (TRI) gasifier   98
    liquid fuels
        fast pyrolysis liquids   344–348
        Fischer-Tropsch liquids   340–341
        gasoline   342–343
        mixed alcohols   341–342
        PHA and co-product hydrogen   343–344
        SF process analyses   343
    oxygen/steam-blown gasifiers   96
    pressurized gasification   106
    process heat   111, 117
gasification–syngas fermentation process
    mass transfer limitation and bioreactor design   321–322
    syngas and gas cleanup   320–321
gasifiers   53
gasoline   342–343
    FCC upgraded bio-oil   230
    hydrotreated higher oxygen content bio-oil   229–230
    hydrotreated low oxygen content bio-oil   229
    oxygenated model compounds   230
gas-phase reactions   91
gas–solid reactions   90

Hansen solubility parameter   285–286
heating and drying   89
heating value   63–64
heat recovery steam generator (HRSG)   117
hemicellulose conversion reactions   38
herbaceous fuels   62
higher alcohol synthesis (HAS)
    catalyst composition   161–162

process and performance   160
reactor designs   157
thermodynamic constraints   158
Hildebrand solubility parameter   284–285
hot gas efficiency (HGE)   87
hybrid processing
  biochemical and thermochemical processes   307
  biomass gasification, syngas production   309–310
  commercialization   322–323
  fast pyrolysis, of lignocellulosic biomass
    acetic acid   309
    lignin derivatives   309
    pyrolytic substrates   308
    pyrolytic sugars   309
  gasification–syngas fermentation process
    mass transfer limitation and bioreactor design   321–322
    syngas and gas cleanup   320–321
  gasification–syngas fermentation route   308
  microbial utilization   308
  pyrolysis-based hybrid processing   308
  pyrolysis–fermentation process
    bio-oil fractionation   318
    crude pyrolytic substrate   318–320
    pyrolytic substrates detoxification   318–319
  pyrolytic substrates, fermentation of
    acetic acid   312
    lignin derivatives   312–313
    pyrolytic sugars   310–312
  sulfur and chloride contaminants   308
  syngas fermentation   308
    acetogens   316–317
    metabolic pathway in   313–316
hydrodeoxygenation   262
hydrogenation   262
hydrogenation-derived renewable diesel (HDRD)   367
hydrolysis   262
hydrothermal liquefaction (HTL)   261

indirectly heated gasifiers
  fluid coking technology   96
  partial-oxidation reactors   96
  Range Fuels gasifier   98
  separate gasification and combustion reactor compartments   98
  ThermoChem Recovery International (TRI) gasifier   98
inorganic solvents
  hydrothermal liquefaction
    aqueous   268–269
    of lignin   271
    of saccharides   270–271
    of whole biomass   270
  non-aqueous   268
integrated biomass gasification combined cycle (IBGCC)   367
ionic liquids   266
  for conversion of lignin   277
  for conversion of saccharides   276–277
  whole biomass   277
isothermal reactor   150–151

jet fuel   230–231

Kamlet-Taft parameter   286–287

LanzaTech   118
large-scale polymers   41
life cycle assessment (LCA)   12
  allocation   370
  attributional *vs.* consequential   370–371
  carbon and climate neutrality   372
  environmental sustainability and socioeconomic implications   356
  goal and scope definition   357
  impact categories   371–372
  life cycle impact assessment   358–359
  life cycle interpretation   359
  life cycle inventory (LCI) analysis   357–358
  and life cycle thinking   356–357
  product development/improvement   369
  sensitivity and uncertainty analyses   359–360
  thermochemical processing, of biomass
    biomass combustion   361
    CO_2 equivalent emissions   366
    combined heat and power (CHP) plants   361
    comparative life cycle GHG emission   367
    energy consumption   361
    Fr-CIR pathway   368
    gasoline and diesel production   368
    GHG emissions   361
    hydrogenation-derived renewable diesel (HDRD)   367
    integrated biomass gasification combined cycle (IBGCC)   367
    life cycle energy and environmental performance   366
    low carbon technologies   360
    review studies   362–365
    sensitivity analysis   368
    synthetic natural gas (SNG)   368

life cycle assessment (LCA) (*continued*)
lignin conversion reactions   38–39
lignin derivatives   309
lignocellulosic biomass   3, 4, 17
Lignocol   340
liquid fuels
   direct liquefaction   348
   esterification   349
   gasification
      fast pyrolysis liquids   344–348
      Fischer-Tropsch liquids   340–341
      gasoline   342–343
      mixed alcohols   341–342
      PHA and co-product hydrogen 343–344
      SF process analyses   343
low-carbon energy technologies   355

methanol to gasoline (MTG) process   153
mixed alcohols   341–342
moisture content   62
municipal solid waste (MSW)   51

non-disruptive technique   186
non-polar solvents   266–268
   coal liquefaction   277
   ethanol and formic acid   278
   low reactivity and high thermal stability   278

oxygen/steam-blown gasifiers   96

Paris Convention   355
partial-oxidation reactors   96
polar aprotic solvents   265–268
   for lignin   275–276
   for saccharides   274–275
   for whole biomass   276
polar protic solvents
   lignin in   273–274
   saccharides in   273
   whole biomass   271–272
pressurized gasification   106
process heat   111, 117
pyrolysis   89–90
pyrolysis–fermentation process
   bio-oil fractionation   318
   crude pyrolytic substrate   318–320
   pyrolytic substrates detoxification   318–319
pyrolytic substrates   308
pyrolytic substrates detoxification   318–319
pyrolytic sugars   309

Range Fuels gasifier   98
refuse-derived fuel (RDF)   95
renewable energy   355

second-generation biofuels   10, 12
sewage sludge   95
smokestacks belching pollutants   1
solids   51–52
solvent liquefaction   12
   chemical effects
      reaction energetics   288–289
         solvent–catalyst interactions   291–292
         solvent–product interactions   292
         solvent–reactant interactions   289–291
   definition   257
   feedstock types   258
   high pressure feed systems   292–293
   history of   257–258
   liquid phase processing
      acid and base solvents   261
      biomass degradation products   259
      hydrothermal liquefaction (HTL)   261
      non-equilibrium conditions   259
      non-volatile products   260
      polar aprotic solvents   261
      solvolysis reactions   260
      vapor-phase recovery   260
   physical effects   283
   process conditions, influence of
      catalysts   282–283
      feedstock-to-solvent ratio   281
      pressure   279
      residence time   279–281
      temperature   278–279
      water content   281–282
   processing conditions   262–263
   processing solvents
      alcohols and carboxylic acids   266
      inorganic solvents   268–271
      ionic liquids   266, 274–277
      non-polar solvents   266–268, 277–278
      polar aprotic solvents   265–268, 274–276
      polar protic solvents   271–274
      polar solvents   265
   reaction types   261–262
   solid residue, separation of   293
   solubility parameters
      Hansen solubility parameter   285–286
      Hildebrand solubility parameter   284–285
      Kamlet-Taft parameter   286–287
   solvent recovery and recycle   293–294
   structural effects   287
   target products

bio-oil 263–264
co-products 265
fuels and chemicals productions 264–265
steam reforming 200
Stirling engine 58
surface tension-driven coalescence 29
syngas cleanup and conditioning
  alkali metal content 127
  alkalis and heavy metals 132–133
  ammonia decomposition and HCN removal 132
  aromatic hydrocarbons 128
  chlorides 133–134
  commercial quench systems 128
  Fischer–Tropsch synthesis
    chemistry 146–148
    cobalt catalysts 148–149
    copper catalyst 149
    product distributions 145
    product selectivity 145
    reactor design and process development 145
    tubular steam-reforming reactors 146
  gas-phase impurities 126
  hydrogen
    bio-hydrogen processes 138
    chemistry 139–140
    conventional steam-reforming catalysts 140
    HTS catalyst 140
    membrane separation 141
    pressure swing adsorption 140–141
    renewable hydrogen technologies 138
    steam reforming of methane 138–139
    Sulfur-tolerant or "sour shift" catalysts 140
  liquid scrubbing/absorption systems 127
  methanol
    adiabatic reactors 150
    catalyst activity 150
    catalyst formulations 151–152
    chemistry 151
    dimethyl ether 155
    ethanol and mixed alcohols 155–159
    isothermal reactor 150–151
    methanol to gasoline (MTG) process 153
    methanol-to-olefins (MTO) synthesis 154–155
    reforming and autothermal reforming 149
    thermodynamic constraint 150
    topsøe integrated gasoline synthesis (TIGAS) 154
  operating temperature 127
  particulates
    barrier filters 129
    char and attrited bed material 128
    electrostatic precipitators (ESPs) 130
    fabric filters 130
    fluid catalytic cracking (FCC) units 130
    gas cyclone technology 129
    high-temperature gas streams 129
    operating temperature and product gas composition 129
    sulfur, chlorine, and alkali metal salts 129
    wet scrubbing systems 130
  Rectisol process 127
  substitute/synthetic natural gas 141–145
  sulfur
    RTI's transport reactor desulfurization system 131
    ZnO-based guard bed materials 130–131
  tar sampling protocols 128
  tars and soot
    hydrogenation 136
    steam reforming 136–137
    thermal cracking 135
    wet scrubbing 134–135
  thermal efficiency 127
  utilization 138–139
Syngas fermentation 118
syngas fermentation 308, 343
  acetogens 316–317
  metabolic pathway in 313–316
synthetic natural gas (SNG) 368

thermal decomposition 176
thermal deconstruction
  computer simulation 32
  conventional heat transfer calculations 30
  lignocellulosic biomass 30
  volatile product formation and vaporization 31
  volatile product mass transfer 32
thermochemical conversion technologies
  direct combustion 5–6
  gasification 6–7
  pyrolysis 7–8
  solvent liquefaction 8
thermochemical process
  and biochemical process 2
  biodiesel 2–3
  cellulosic biomass 2
  first-generation biofuels 2

thermochemical process (*continued*)
  glycosidic bonds 3
  greenhouse gas (GHG) emissions 2
  high-temperature combustion and gasification 5
  lignin 4
  lignocellulosic biomass 3
  thermal energy and catalysts 4
  triglycerides 2
ThermoChem Recovery International (TRI) gasifier 98
thermolysis 261
topsøe integrated gasoline synthesis (TIGAS) 154

torrefaction 52
total acid number (TAN) 185

updraft gasifier 100

Valmet/UPM (CFB) 188

waste-to-energy (WTE) systems 52
water evolution 39
water gas shift (WGS) reaction 126
water-soluble anhydro-oligosaccharides 30